Fluid Mechanics and Its Applications

Volume 108

Series editor

André Thess, Stuttgart, Germany

Founding Editor

René Moreau, Ecole Nationale Supérieure d'Hydraulique de Grenoble, Saint Martin d'Hères Cedex, France

For further volumes:
http://www.springer.com/series/5980

Aims and Scope of the Series

The purpose of this series is to focus on subjects in which fluid mechanics plays a fundamental role.

As well as the more traditional applications of aeronautics, hydraulics, heat and mass transfer etc., books will be published dealing with topics which are currently in a state of rapid development, such as turbulence, suspensions and multiphase fluids, super and hypersonic flows and numerical modeling techniques.

It is a widely held view that it is the interdisciplinary subjects that will receive intense scientific attention, bringing them to the forefront of technological advancement. Fluids have the ability to transport matter and its properties as well as to transmit force, therefore fluid mechanics is a subject that is particularly open to cross fertilization with other sciences and disciplines of engineering. The subject of fluid mechanics will be highly relevant in domains such as chemical, metallurgical, biological and ecological engineering. This series is particularly open to such new multidisciplinary domains.

The median level of presentation is the first year graduate student. Some texts are monographs defining the current state of a field; others are accessible to final year undergraduates; but essentially the emphasis is on readability and clarity.

Bernard Zappoli · Daniel Beysens
Yves Garrabos

Heat Transfers and Related Effects in Supercritical Fluids

Springer

Bernard Zappoli
Centre National d'Etudes Spatiales
Toulouse Cedex 9
France

Daniel Beysens
Ecole Supérieure de Physique et Chimie Industrielle de la Ville de Paris and Commissariat à l'Energie Atomique
Paris
France

Yves Garrabos
Institut de Chimie de la Matière Condensée de Bordeaux
Centre National de la Recherche Scientifique and Université de Bordeaux 1
Pessac Cedex
France

ISSN 0926-5112 ISSN 2215-0056 (electronic)
ISBN 978-94-017-9186-1 ISBN 978-94-017-9187-8 (eBook)
DOI 10.1007/978-94-017-9187-8
Springer Dordrecht Heidelberg New York London

Library of Congress Control Number: 2014942063

Printed on acid-free paper

Springer is part of Springer Science+Business Media (www.springer.com)

Preface

Over the past 50 years, the problem of supercritical or near-critical fluids has turned from a particular aspect of phase transitions described in books on thermodynamics to an active domain of research involving a variety of fundamental topics in statistical physics, fluid dynamics, chemistry, as well as many applications in different industrial processes. Supercritical fluids have provided a canonical example in the study of critical phenomena and anomalous exponents. They have displayed remarkable effects in out-of-equilibrium physics and hydrodynamics where they have made possible laboratory experiments in an extreme parameter range usually achieved only in geophysical or astrophysical flows. They also involve rapidly developing applications in engineering such as the ones taking benefit of the peculiar properties of chemical reactions in supercritical fluids.

Although there exist excellent books on each facet of the above subjects, a consistent account of the field is needed since it has become increasingly difficult to follow the literature on all these different topics. In addition, most books on supercritical fluids consider either the fundamental problems statistical of statistical physics or the aspects related to engineering processes in fluid dynamics or chemistry without making connection between these fields. This book will thus fill a gap in the existing literature. The authors have made a major effort to introduce the fundamental concepts, both in statistical physics and in hydrodynamics to readers who have no previous knowledge of these fields. They then present more specialized material on heat transfer, boiling, and hydrodynamic instabilities in supercritical fluids with emphasis on related microgravity experiments. The authors have been most actively engaged in various studies on supercritical fluids in a remarkable collaborative effort over the past 30 years. This book is a witness to this fruitful collaboration that provided many results. In particular, pioneering studies on the piston-effect that is a mechanism of heat transfer characteristic of near-critical fluids, fluids are widely described in the book.

I hope that this book will provide a profitable introductory text addressed to graduate students but will also be useful to researchers studying one of the many aspects of supercritical fluids.

Paris, February 2014 Stephan Fauve

Contents

Nomenclature

Latin Symbols

a or a_v	Oscillatory amplitude
a	Specific Helmhotz energy
a	van der Waals specific attractive parameter
$\bar{a}$	van der Waals molar attractive parameter
a_m	Modulus of the magnetic acceleration vector
A	Cross-sectional area
b	van der Waals specific covolume
$\bar{b}$	van der Waals molar covolume
B	Amplitude of the magnetic field vector
B_f	Dimensionless thermo-oscillatory number
c_S	Velocity of sound
$\overline{c_p}$	Molar heat capacity at constant pressure
c_p	Specific heat at constant pressure
$\overline{c_V}$	Molar heat capacity at constant volume
c_V	Specific heat at constant volume
CBL	Cooling boundary layer
C	Convective parameter
d	Inner diameter of an anisotropic thermal boundary layer
d	Dimension of the space
d_c	Upper critical dimension of space
d_0	Thermistor radius
D	Outer diameter of an anisotropic thermal boundary layer
D	Coefficient of diffusion
D_T	Thermal diffusivity
$\bar{e}$	Molar energy
e	Specific energy
e	Length, or thickness, of the system
E_{vdW}	van der Waals solid–liquid interaction energy (per unit area)
f	Frequency
HBL	Heating boundary layer
g	Specific Gibbs free energy

g	Modulus of the residual gravitational acceleration
g or $1g$	Residual gravitational acceleration expressed in unit of g_0
g_0	Modulus of the Earth's gravitational acceleration
g^*	Effective gravitational acceleration
h	Specific enthalpy
I	Current intensity
j	Specific heat flux (or heat flux per area unit)
K_m	Typical wave vector of phase-separating domains
K_m^*	Reduced typical wave vector of phase-separating domains
ℓ_c	Capillary length
l_v	Viscous boundary layer
L	Latent heat
L_m	Pseudo-wavelength between phase-separating domains
m	Mass
$\overline{m}$	Molar mass
$m_{\bar{p}}$	Molecular (or particle) mass
M	Order parameter
$M^{+,-}$	Order parameter in the inhomogeneous region
n	Dimension of the order parameter
n	Refractive index
n	Number density
$\bar{n}$	Molar number
p	Pressure
P_r	Prandtl number
q or $\dot{q}$	Heating power per unit volume or surface
q or Q	Heating power or heat flux
q or Q	Energy
Q	Heating power
r	Radius of the system
r	Spatial variable
r	Space variable
r	Specific gas constant
R	Universal constant of ideal gases
R_D	Universal ratio for thermal diffusivity
R_0	Typical cell radius
Ra_v	Vibrational Rayleigh number
Ra or Ra_0	Rayleigh number
Re or Re_0	Reynolds number
$\bar{s}$	Molal entropy
s	Specific entropy
S	Entropy
t	Time
t_D	Heat diffusion time
t_{eq}	Adiabatic temperature equilibrium timescale

t'_c or t_0 or t_{PE}	Piston effect time
t_ξ	Typical critical fluctuation time
t^*	Time reduced by the typical critical fluctuation time
t_{PE}	Piston-effect timescale
T	Temperature
T_c	Critical temperature
T_{cx}	Coexistence temperature
u	Specific internal energy
u	Molal internal energy
u, v, w	Velocity components on Cartesian axes
v	Specific volume
$\bar{v}$	Molar volume
V	Volume
x, y, z	Cartesian coordinates
x_η	Critical exponent for the shear dynamic viscosity
x_λ	Critical exponent for the thermal conductivity
$Y_\eta(=\nu x_\eta)$	Critical exponent of the shear dynamic viscosity

Greek Symbols

ϵ or ε	Small parameter of asymptotic expansions
α	Angle of vibration with respect to the direction of the temperature gradient
β_p	Isobaric thermal expansion coefficient
ν	Molar volume
ρ	Density
ρ_c	Critical density
$\bar{\rho}$	Molar density
κ_T	Isothermal compressibility
χ_B	Magnetic susceptibility
χ_T	Isothermal susceptibility
ϕ	Volume fraction
Λ	Thermal conductivity
ξ	Correlation length of fluctuations
γ	Ratio of specific heats at constant pressure and constant volume
$\dot{\gamma}$	Acceleration
τ	Reduced temperature
ν	Kinematic viscosity
Ξ	Amplitude of vibration
μ	Shear dynamic viscosity
μ_B	Bulk dynamic viscosity
Π	Disjoining pressure
η_b	Bulk viscosity
η	Shear viscosity

σ	Gas–liquid surface tension
Ω	$=0.5\left(\beta'_P a'\omega'\right)^2$ Vibration parameter
Ra	Rayleigh number
Θ	Time reduced by the piston effect time
θ	Time reduced by heat diffusion time
μ^*	Chemical potential
μ_0	Magnetic permeability of vacuum
δ	Thickness of the thermal boundary layer
ζ	Distance
α	Critical exponent of the critical isochore for the specific heat at constant volume
β	Critical exponent of the critical isochore of the coexistence curve
γ	Critical exponent of the critical isochore of the isothermal susceptibility
ν	Critical exponent of the critical isochore of the correlation length
$\tilde{\delta}$	Critical exponent of the critical isotherm of the susceptibility (χ)
η	Fisher's critical exponent of the correlation function for order parameter fluctuations at the CP
Δ	Corrections-to-scaling exponent
ψ	Generic critical exponent
Ψ	Generic critical amplitude
ω	Angular frequency
$\tilde{x}$	Generic exponent of the spatial evolution of the thermal boundary layer
ϑ	Period of vibration

Vectors and Tensors

$\mathbf{u}$	Fluid velocity vector
$\mathrm{X}(x, y, z)$	Spatial location vector
∇	Gradient operator
$\mathbf{n}$	Unit vector for the direction of vibration
$\vec{\tau}$	Viscosity stress vector
$\bar{\bar{\tau}}$	Viscosity stress tensor
Ξ	Vibration vector
$\mathbf{o}$	Rotational part of vector

General

$\Delta\ldots$, $\delta\ldots$	Difference (e.g., ΔT, δT are temperature differences)

Subscripts and Superscripts

a	Averaged quantity
b	Background quantity
b	Boundary layer quantity
b	Bulk quantity
c	Critical quantity
cx	Coexistence quantity
exp	Experimental quantity
g	Gas phase quantity
IG	Ideal gas quantity
l	Liquid phase quantity
MF	Mean field approximation
mod	Model quantity
p	Pulsating quantity
p	Quantity at constant pressure
V	Quantity at constant volume
v	Saturated vapor quantity
vdW	van der Waals quantity
w	Value at the sample wall
$-$	Cold part, below T_c
$+$	Hot part, above T_c
$\parallel$	Parallel
$\perp$	Perpendicular
0	Leading amplitude

Chapter 1
General Introduction to Near-Critical and Supercritical Fluids

Abstract A general introduction on near-critical and supercritical fluids is given including the van der Waals equation of state and the critical region. The notions of order parameter, critical fluctuations and fluctuation correlation length (a natural lengthscale for critical point phenomena) are exposed. Mean-field and renormalized critical exponents are defined and discussed. Concerning critical dynamics, the phenomenon of critical slowing-down is presented as the main aspects of phase separation dynamics, including wetting and adsorption properties. The very strong effects due to earth gravity are discussed, as the means to get rid of it (free fall, satellites, magnetic compensation, Plateau method, etc.). A way to fit experimental values within the approximated van der Waals approach is discussed in details. The critical parameters of many fluids of current use are also given (^3He, pH_2, N_2, O_2, Xe, CO_2, SF_6, H_2O). Specific aspects concerning scaling laws, universality and renormalization-group, the upper critical dimensionality and the conventional theories of nucleation and spinodal decomposition are given in the annex part.

The properties of fluids have been investigated for a long time. One of the most famous experiments (and the oldest mentioned in the literature) was that of Baron Charles Cagniard de Latour [15], a French gentleman and scientist—he was the inventor of the siren. He was studying the phase equilibrium between equal volumes of gas and liquid CO_2 in a sealed gun. While checking for the presence of both phases by listening for the splashing of the liquid phase when the gun was agitated, he discovered with surprise that the splashing disappeared above a certain temperature: he was crossing a "critical point" (CP), a surface phase point (i.e., a set of conditions such as temperature, pressure and density) above which a gas-liquid phase boundary ceases to exist. In this case, the phase boundary between the gas and liquid phases of CO_2 vanished at the critical temperature; the density of the fluid is the same as that of the liquid. This disappearance of the meniscus between the gas and the liquid can be now observed in an optical cell, such as that shown in Fig. 1.1.

More generally, a pure liquid and its vapor can only coexist at equilibrium for certain combinations of pressure and temperature that together constitute a *saturation curve* (see Fig. 1.2). This line is a projection on the pressure–temperature plane of the

B. Zappoli et al., *Heat Transfers and Related Effects in Supercritical Fluids*, Fluid Mechanics and Its Applications 108, DOI: 10.1007/978-94-017-9187-8_1,

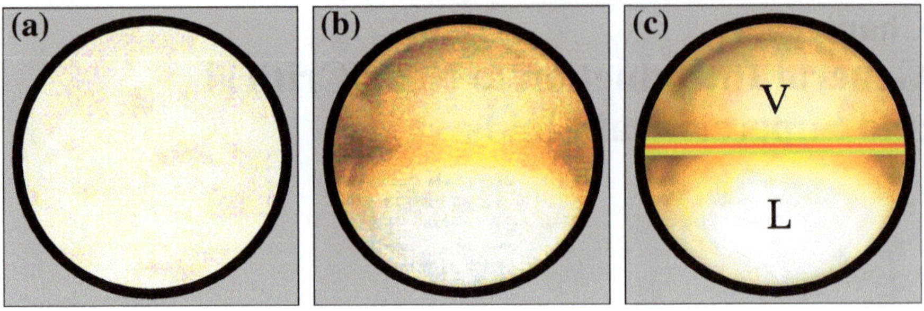

Fig. 1.1 A sample of SF_6 at critical density observed in transmission on Earth (i.e., with gravity directed downwards). **a** 10 K above the critical point (45.5 °C; 3.76 MPa; 742 kg m^{-3}). **b** 1 mK above the critical point; the sample is compressed under its own weight. Critical density can only be reached in the middle of the sample, where a thin opalescent layer is visible. This phenomenon of "critical opalescence" is the hallmark of enhanced, critical fluctuations at the scale of the wavelength of light; **c** 1 mK below its critical point: both the liquid (L) and the vapor (V) phases are compressed under their own weight (note that the horizontal liquid–vapor interface has been artificially lightened for better visualization) (*Photo* Johannes Straub)

three-dimensional pressure–volume–temperature (p, V, T) surface. Such a diagram can be constructed by measuring a series of isotherms in which the pressure is determined as a function of volume (or density $\rho = \frac{M}{V}$, where M is the mass of the fluid system). The isotherms exhibit horizontal regions at all temperatures at which the vapor and liquid can coexist, and the locus of the ends of these tie lines is the vapor–liquid coexistence curve, which is the T, V (or T, ρ) projection of the full diagram. At temperatures and densities that lie within the curve, a system will consist of two phases whose densities are given by the values on the curve.

The maximum of the curve is the vapor–liquid critical point: the point at which the distinction (i.e., the phase boundary) between the liquid and vapor phases is lost. Above this critical combination of temperature and pressure, fluids are said to be *supercritical*, and they exhibit a variety of properties (high density, low viscosity, large diffusivity) which indicate that they are gases with the densities of liquids. In addition, their isothermal compressibilities can become very large, especially when they approach the critical point. Those properties of common fluids such as CO_2, in the supercritical state make them rather useful as very efficient and nonpolluting solvents for organic materials [22]. Also, at a moderately high temperature, supercritical H_2O can be used as the host in a chemical oxidation process with a very high yield [60].

Table 1.1 lists various fluids that will be discussed in this book, along with their critical temperatures, pressures and densities.

In the following, we will mostly consider the fluids hydrogen (H_2), carbon dioxide (CO_2), and sulfur hexafluoride (SF_6). Let us briefly recall the molecular structures of these fluids. CO_2 and H_2 molecules have linear structures. CO_2 consists of a central carbon atom joined to the two oxygen atoms by polar covalent double bonds. The interatomic distances of these linear molecules are generally much smaller than

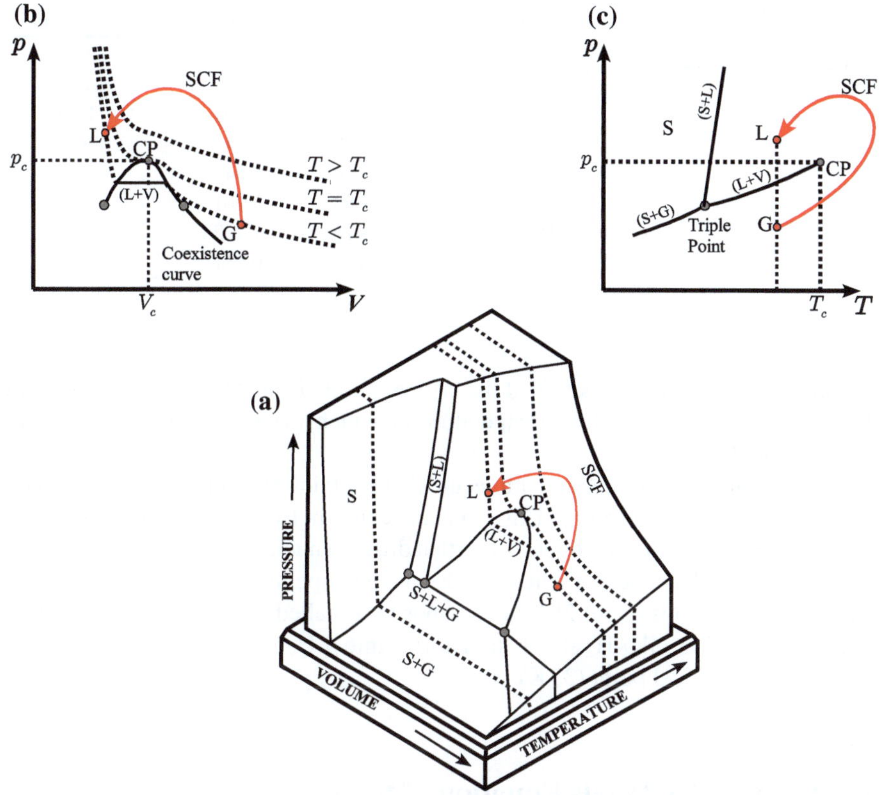

Fig. 1.2 **a** The pVT surface for the equilibrium states of CO_2. The solid line GL is a thermodynamic path where the continuous transformation of the gas into a liquid does not require the phenomenon of condensation to cross the liquid–vapor coexistence curve at constant temperature. This line crosses the supercritical fluid region, where the pressure and temperature are greater than the critical pressure and the critical temperature, respectively, of the critical point. *CP* stands for critical point, *SCF* for supercritical fluid, *S* for solid, *V* for the saturated vapor that coexists with its liquid phase, *L* for liquid, and *G* for gas. **b** Isothermal lines and liquid–vapor coexistence curve in the pV projection of (**a**). **c** Gas–solid, liquid–solid and liquid–vapor coexistence curves in the pT projections of (**a**). The point at which the gas, liquid and solid all coexist is known as the *triple point*

the mean distance between molecules, so they can be considered to be spherical molecules. The SF_6 molecule has a sulfur atom that is covalently bound to six fluorine atoms, leading to an approximately spherical structure. These three molecules are nonpolar, small, and inert; they repel each other at close range and attract each other at greater separation due to London forces. Although they are very weak, these forces have a strong influence on the physical properties of a fluid substance, particularly on its melting and boiling points, and cause nonpolar gases to condense into liquids and—at very low temperatures—freeze into solids. It is precisely this competition between thermal excitation and short-range intermolecular interactions that leads to

Table 1.1 Temperature, pressure and density at the critical point for some fluids

Substance	T_c (K)	p_c (MPa)	ρ_c (kg m^{-3})
H_2O	647.1	22.06	322.2
CO_2	304.13	7.378	467.8
N_2	126.24	3.398	313.9
O_2	154.58	5.043	436.2
Xe	289.74	5.840	1113
SF_6	318.63	3.761	742
H_2	32.976	1.2928	31.43
^3He	3.317	0.117	41.45

the liquid–vapor phase transition at the critical point. Thus, the critical temperature T_c corresponds to the highest temperature at which the vapor and the liquid are distinguishable.

H_2 is a homonuclear diatomic molecule that contains magnetically active centers (generated by the proton spin). This molecule exists in two isomeric forms (parahydrogen, p-H_2, and orthohydrogen, o-H_2) that differ in nuclear spin configuration. The proportion of each isomer present depends on the temperature. At room temperature, the para/ortho ratio is about 25 % (this vapor is so-called normal hydrogen, n-H_2). When H_2 is cooled, all o-H_2 slowly transforms into p-H_2, reaching a ratio para/ortho of 96 % at the critical point (33 K).

1.1 The van der Waals Equation of State

Let us consider N particles of a pure fluid of total mass M that fill a total volume V at temperature T. The equation of state relates the variables (p, T, V), (p, T, ρ), or (p, T, n), where $\rho = \frac{M}{V}$ is the mass density and $n = \frac{N}{V}$ is the number density. This equation is thus fundamental to investigations of the properties of a fluid, especially its external pressure p. Historically, the most simple gas model was that of *ideal gases*, where the ratio $\frac{pV}{T}$ is a constant (leading to the *Mariotte* or *Gay-Lussac equation*):

$$pV = \bar{n}RT \tag{1.1}$$

or

$$\begin{aligned} p\bar{v} &= RT \\ p &= \rho\left(\frac{R}{\bar{m}}\right)T = \rho r T\,. \end{aligned} \tag{1.2}$$

Here, $\bar{n} = \frac{N}{N_A}$ is the number of moles, $\bar{v} = \frac{V}{\bar{n}}$ is the molar volume, $\bar{m} = \frac{M}{\bar{n}} = \frac{\rho V}{\bar{n}}$ is the molar mass, $R = k_B N_A = 8.3144\,\mathrm{J\,K^{-1}\,mol^{-1}}$ is the molar universal constant for ideal gases, $r = \frac{R}{\bar{m}}$ is the specific constant of the selected fluid, $k_B = 1.3806 \times 10^{-23}\,\mathrm{J\,K^{-1}}$ is the Boltzmann constant, and $N_A = 6.02217 \times 10^{23}$ is the

Avogadro number. This simplest equation (where molecules are considered as point particles) gained interest due to the Maxwell's kinetic theory of gases (see [43]).

This model can be applied to gases at temperatures such that $k_B T$ is small compared with the electronic excitation energies of the molecules (i.e., $T \lesssim 10^4$ K), and at low or moderate pressures. However, when the gas is highly pressurized, or its temperature is decreased by a large amount, the distance between molecules cannot reach zero because the molecules repulse each other. The volume ultimately occupied by a vapor molecule is therefore finite, which leads to the more realistic relations

$$p(V - \bar{n}\overline{b}) = \bar{n}RT \tag{1.3}$$

or

$$p(\bar{v} - \overline{b}) = RT. \tag{1.4}$$

Here, $\overline{b}$ is the molar covolume (the nonzero volume of N_A molecules). In addition, for moderate intermolecular distances, other (attractive) intermolecular forces that arise due to the dipolar nature of the polarizability cannot be neglected. These are the so-called *van der Waals forces*, which vary with the intermolecular distance d as d^{-6}, and thus correspond to a decrease in the overall pressure as V^2. The related equation of states can be then written as

$$p = \frac{\bar{n}RT}{V - \bar{n}\overline{b}} - \frac{\bar{n}^2\overline{a}}{V^2} \tag{1.5}$$

or

$$p = \frac{RT}{\bar{v} - \overline{b}} - \frac{\overline{a}}{\bar{v}^2} \tag{1.6}$$

Here $\overline{a}$ is a measure of the cohesion in the fluid due to the van der Waals attractive forces between molecules. Note that the second term on the right-hand side of Eq. (1.5) represents the forces that tend to oppose the fluid expansion, and correspond to an "internal" pressure $\frac{\bar{n}^2\overline{a}}{V^2}$. The external pressure p in a real vapor is then less than that of an ideal gas because intermolecular attractions decrease the speeds at which the molecules approach the wall. Since binary interactions predominate, this effect is proportional to $\bar{n}^2$.

Equation (1.5) is called the *van der Waals (vdW) equation of state*. Historically speaking, it was the first equation that was able to describe the phase separation and the critical point. It was developed by van der Waals [68] (see also [69]). Other useful forms of this equation are obtained when considering, either the molar density

$$\bar{\rho} = \frac{1}{\bar{v}} = \frac{\rho}{\bar{m}}, \tag{1.7}$$

or the mass density

$$\rho = \bar{\rho}\bar{m}, \tag{1.8}$$

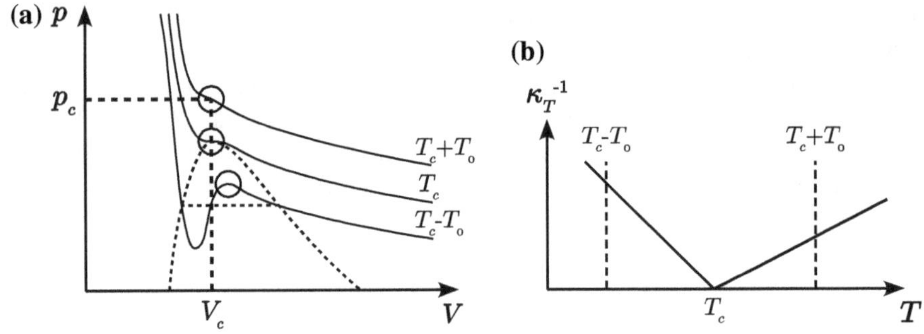

Fig. 1.3 **a** Isothermal lines and liquid–vapor coexistence curve in the pressure–volume diagram of a van der Waals fluid. **b** Temperature dependence of the inverse isothermal compressibility along the critical isochore above and below the critical temperature where the value reaches zero, which corresponds to an infinite value for the isothermal compressibility [(see Eq. (1.24)]

which lead, either to

$$p = \frac{RT\bar{\rho}}{1 - \bar{b}\bar{\rho}} - \bar{a}\bar{\rho}^2. \tag{1.9}$$

or

$$p = \frac{rT\rho}{1 - b\rho} - a\rho^2. \tag{1.10}$$

with $a = \frac{\bar{a}}{\bar{m}^2}$ and $b = \frac{\bar{b}}{\bar{m}}$.

Some isotherms calculated from Eq. (1.5) are plotted in the p–V diagram of Fig. 1.3a. At high temperatures, the equation has a single real root and p decreases monotonically with V. As the temperature is decreased, the isotherms become flatter and eventually develop loops (also called "van der Waals loops"). There are thus a range of pressures for which three real values of density can be found for a given T.

The existence of three real roots of the equation can be associated with the two-phase region. The temperature at which they first appear can be identified as the vapor–liquid critical temperature T_c. Below T_c, each isotherm has a minimum and a maximum. The minima move closer together with increasing temperature and merge at the critical point, where

$$\left(\frac{\partial p}{\partial \bar{v}}\right)_T^c = 0 \text{ and } \left(\frac{\partial^2 p}{\partial \bar{v}^2}\right)_T^c = 0. \tag{1.11}$$

The following critical coordinates can be found from Eq. (1.11), leading to the following relations

$$\bar{v}_c = \frac{1}{\bar{\rho}_c} = 3\bar{b}, \qquad T_c = \frac{8\bar{a}}{27\bar{b}R}, \qquad p_c = \frac{\bar{a}}{27\bar{b}^2}. \tag{1.12}$$

where are used the molar quantities, with

$$Z_{c,\mathrm{vdW}} = \left(\frac{p_c \bar{v}_c}{RT_c}\right)_{\mathrm{vdW}} = \left(\frac{p_c \bar{m}}{R\rho_c T_c}\right)_{\mathrm{vdW}} = \frac{3}{8} \tag{1.13}$$

Here we have defined $\bar{v}_c$, $\bar{n}_c$, and $\bar{\rho}_c$ as, respectively, the critical molar volume, the critical mole number and the critical molar density, with

$$\bar{v}_c = \frac{1}{\bar{\rho}_c} = \frac{\bar{m}}{\rho_c} = \frac{V}{\bar{n}_c}. \tag{1.14}$$

In [1, 63], the authors stress (as discussed below) that the van der Waals equation of state, although very useful, does not properly describe the behavior of a fluid near its critical point. In particular, it is impossible to fit the three critical coordinates of a real fluid using the two adjustable parameters $\overline{a}$ and $\overline{b}$. For instance, let us consider CO_2, with $T_c = 304.13$ K, $\rho_c = 467.8\,\mathrm{kg\,m^{-3}}$, and $\bar{m} = 0.044$ kg. From Eq. (1.12), the critical pressure should obey

$$p_{c,\mathrm{vdW}} = \frac{3}{8}\frac{RT_c}{\bar{v}_c} = \frac{3}{8}\frac{R\rho_c T_c}{\bar{m}}. \tag{1.15}$$

Using the experimental value $\bar{v}_c = \frac{\bar{m}}{\rho_c} = 9.406 \times 10^{-5}\,\mathrm{m^3\,mol^{-1}}$ in Eq. (1.15), it is found $p_{c,\mathrm{vdW}} \simeq 10$ MPa. This value is larger than the actual critical pressure ($p_c = 7.378$ MPa, see Table 1.1) by about 30 %. As a consequence, in the following, when we consider the van der Waals equation of state, we will calibrate the critical point coordinates using the experimental values for T_c and ρ_c and consider the critical pressure as deduced from Eq. (1.15).

1.2 The Critical Region in the van der Waals Approach

Following the practical approach given in [1], we reduce pressure, temperature and density by their respective values at the critical point:

$$\tilde{p} = \frac{p}{p_c}; \quad \tilde{T} = \frac{T}{T_c}; \quad \tilde{\rho} = \frac{\bar{\rho}}{\bar{\rho}_c} = \frac{\rho}{\rho_c}. \tag{1.16}$$

The van der Waals equation can be written in the reduced form:

$$\tilde{p} = \frac{8\tilde{T}\tilde{\rho}}{3 - \tilde{\rho}} - 3\tilde{\rho}^2. \tag{1.17}$$

Near the critical point as defined by Eq. (1.11), the Helmholtz free energy per unit volume $f = F/V$ (where F is the total Helmholtz free energy), expressed in its practical dimensionless form

$$\tilde{f} = \frac{F}{p_c V}, \tag{1.18}$$

can be expanded around its critical value $\tilde{f}_c$ as an analytical power series of the reduced variables

$$\begin{aligned} \tau &= \tilde{T} - 1 = \tfrac{T-T_c}{T_c}, \\ \Delta\tilde{\rho} &= \tilde{\rho} - 1 = \tfrac{\rho-\rho_c}{\rho_c} = \tfrac{\bar{\rho}-\bar{\rho}_c}{\bar{\rho}_c}, \\ \Delta\tilde{p} &= \tilde{p} - 1 = \tfrac{p-p_c}{p_c} \end{aligned} \tag{1.19}$$

such that

$$\Delta\tilde{f} = \tilde{f} - \tilde{f}_c = \frac{1}{2} a_0 \tau (\Delta\tilde{\rho})^2 + \frac{b_0}{4!} (\Delta\tilde{\rho})^4 + \cdots \tag{1.20}$$

The pressure behavior can be deduced from this equation and compared to the corresponding expansion of the van der Waals equation near the critical point:

$$\Delta\tilde{p} = \left(\frac{\partial \tilde{f}}{\partial \tau}\right)_{\tilde{\rho}} = p_1 \tau + a_0 \tau (\Delta\tilde{\rho}) + \frac{b_o}{3!} (\Delta\tilde{\rho})^3 + \cdots, \tag{1.21}$$

with, at the critical point,

$$p_1 = \left[\frac{T_c}{p_c}\left(\frac{\partial p}{\partial T}\right)_{\rho_c}\right]_{\text{vdW}} = 4,\ a_0 = 6,\ b_0 = 9. \tag{1.22}$$

For $T < T_c$, along the liquid (subscript ℓ)–vapor (subscript v) coexistence curve, the T, ρ coordinates are given by minimizing $\Delta\tilde{f}$. The solutions obtained are

$$\Delta\tilde{\rho}_{\ell,v,\text{vdW}} = \tfrac{\rho_\ell - \rho_v}{2\rho_c} = \pm B_{\text{vdW}} |\tau|^{\beta_{\text{MF}}} = \pm 2 |\tau|^{\beta_{\text{MF}}}, \tag{1.23}$$

with $B_{\text{vdW}} = 2$ and the classical (or mean field) exponent $\beta_{\text{MF}} = 1/2$ (the subscript "MF" stands for the "mean field", as discussed below in Sect. 1.3). From the expansion of the free energy [see Eq. (1.20)], it is clear that the vicinity of the critical point is characterized by a number of unusual properties, such as a decrease in the difference in density $\Delta\tilde{\rho}_{\ell,v}$ upon the coexistence of liquid and vapor [see Eq. (1.23)], or the following divergence of the isothermal compressibility $\kappa_T = \frac{1}{\rho}\left(\frac{\partial \rho}{\partial p}\right)_T$:

$$\kappa_{T,\text{vdW}} = \begin{cases} \Gamma^+_{0,\text{vdW}} \tau^{-\gamma_{\text{MF}}} = \frac{1}{6 p_{c,\text{vdW}}} \tau^{-\gamma_{\text{MF}}} & (T > T_c, \rho = \rho_c) \\ \Gamma^-_{0,\text{vdW}} \tau^{-\gamma_{\text{MF}}} = \frac{1}{12 p_{c,\text{vdW}}} |\tau|^{-\gamma_{\text{MF}}} & (T < T_c, \rho = \rho_{\ell,v}) \end{cases} \tag{1.24}$$

Here, $\Gamma^+_{0,\text{vdW}} = \frac{1}{6 p_{c,\text{vdW}}}$, $\Gamma^-_{0,\text{vdW}} = \frac{1}{12 p_{c,\text{vdW}}}$, and the mean field exponent $\gamma_{\text{MF}} = 1$.

The molar heat capacity at constant volume $\bar{c}_V$ is derived from the molar entropy $\bar{s}$ according to the general definition

$$\bar{c}_V = T\left(\frac{\partial \bar{s}}{\partial T}\right)_{\bar{v}}. \tag{1.25}$$

Introducing the practical dimensionless form $\tilde{s} = S/p_c V$ of the entropy per unit volume (S is the total entropy), Eq. (1.20) leads to

$$\tilde{s} = -\frac{1}{T_c}\left(\frac{\partial \bar{f}_c}{\partial T}\right)_{\bar{v}} - \frac{3}{T_c}(\Delta\tilde{\rho})^2. \tag{1.26}$$

On the critical isochore, $\Delta\tilde{\rho} = 0$ for $T - T_c > 0$ and $\Delta\tilde{\rho} = \Delta\tilde{\rho}_{\ell,v}$ for $T - T_c < 0$. Thus, for a van der Waals fluid, the molar heat capacity only exhibits a jump at T_c such that

$$\bar{c}_{V,\mathrm{vdW}}(T < T_c) - \bar{c}_{V,\mathrm{vdW}}(T > T_c) = \frac{9}{2}R. \tag{1.27}$$

This value can be compared with the values of the molar heat capacity at constant volume for ideal monoatomic gases, $\bar{c}_{V,\mathrm{IG}} = 3R/2$, and for ideal polyatomic gases, $\bar{c}_{V,\mathrm{IG}} = 5R/2$. Note that the regular behavior of $\bar{c}_{V,\mathrm{vdW}}(T > T_c)$ and $\bar{c}_{V,\mathrm{vdW}}(T < T_c)$ corresponds to the mean field value $\alpha_{\mathrm{MF}} = 0$ for the critical exponent α of the specific heat, as discussed below in Sect. 1.3 (see also [1] for example).

Another quantity of interest is the molar heat capacity at constant pressure $\bar{c}_p = T\left(\frac{\partial \bar{s}}{\partial T}\right)_p$, the behavior of which can be derived from the following thermodynamic relations:

$$\bar{c}_p - \bar{c}_V = \frac{T}{\bar{\rho}}\left(\frac{\partial p}{\partial T}\right)_\rho^2 \kappa_T = \frac{T}{\bar{\rho}}\beta_p\left(\frac{\partial p}{\partial T}\right)_\rho. \tag{1.28}$$

Here, $\beta_p = -\frac{1}{\rho}\left(\frac{\partial \rho}{\partial T}\right)_p$ is the isobaric thermal expansion coefficient. Near the critical point, on the critical isochore ($\Delta\tilde{\rho} = 0$), $\bar{c}_{p,\mathrm{vdW}}$ for a van der Waals fluid shows the same asymptotic behavior as $\kappa_{T,\mathrm{vdW}}$:

$$\bar{c}_{p,\mathrm{vdW}} = \bar{c}_{V,\mathrm{vdW}} + R\,(1+\tau)\,\tau^{-\gamma_{\mathrm{MF}}}, \tag{1.29}$$

with $\gamma_{\mathrm{MF}} = 1$. Thus, in the van der Waals fluid case, the ratio $\gamma_0 = \frac{\bar{c}_p}{\bar{c}_V}$ of the molar heat capacities is written as follows:

$$\gamma_{0,\mathrm{vdW}} = \frac{\bar{c}_{p,\mathrm{vdW}}}{\bar{c}_{V,\mathrm{vdW}}} = 1 + \frac{R}{\bar{c}_{V,\mathrm{vdW}}}(1+\tau)\,\tau^{-\gamma_{\mathrm{MF}}} \tag{1.30}$$

Therefore, the van der Waals equation of state has many advantages, such as its analytical nature and universal dimensionless behavior, and it captures the main aspects of critical point phenomena. However, its main disadvantage is that it fails to describe the close vicinity of the critical point, leading (near T_c) to deviations between the calculated and experimental values that cannot be neglected (see Sect. 1.11 and Fig. 1.8). In order to characterize these discrepancies, in the following we analyze the characteristics of the neighborhood of the critical point.

1.3 Neighborhood of the Critical Point

A fluid is termed *critical* or *near-critical* when its temperature and pressure are sufficiently close to its critical point to allow its distinctive critical properties to be clearly distinguished from its regular (or "background") contributions. This point is emphasized below in Sect. 1.10. In a wide domain of temperature and pressure around the critical point, important physical parameters (for example, the isothermal compressibility, the densities of the vapor and liquid phases, or the surface tension) obey universal power laws along a few particular thermodynamic paths (for example, the critical isochore $\rho = \rho_c$, the coexistence curve $\rho = \rho_{v,\ell}$, or the critical isotherm $T = T_c$). The parameters are easily varied or scaled by changing the temperature or density. Those highly variable properties of near-critical fluids give them unusual and often paradoxical thermocompressible effects, which are analyzed in this book.

1.3.1 Order Parameter and Critical Fluctuations

An important aspect of the critical region, and one that can already be seen in the van der Waals equation of state presented above, is that most of the anomalies of the thermodynamic and transport properties can be set in the form of scaled, universal functions (power laws) with respect to the critical point parameters [66]. Any results obtained with one single fluid can then be immediately rescaled to describe any particular member of a whole class of systems, the "class of fluids" (see Chap. 18 for details). This class of fluids includes, for example, liquid mixtures, polymer melts, polymer solutions, microemulsions, molten salts, and monotectic liquid metals. This universal scaling is fundamental in nature and stems from the universal behavior that the singular part $F - F_c$ of the free energy (where F_c is the regular part) must asymptotically obey at the critical point in order to fulfill the conditions of a second-order phase transition (see Fig. 1.4).

In this sense, universality and scaling are generic to all critical point phenomena. The order parameter (OP) of the transition is the parameter M that undergoes the largest fluctuations and then has a susceptibility that diverges as $T - T_c$ goes to zero. It also defines the critical point coordinates by being zero at temperatures above the critical point (where the system is homogeneous) and nonzero below (where the system consists of two different phases). In pure fluids, the order parameter is the reduced density difference

$$M \equiv \Delta\tilde{\rho} = \frac{\rho - \rho_c}{\rho_c}. \tag{1.31}$$

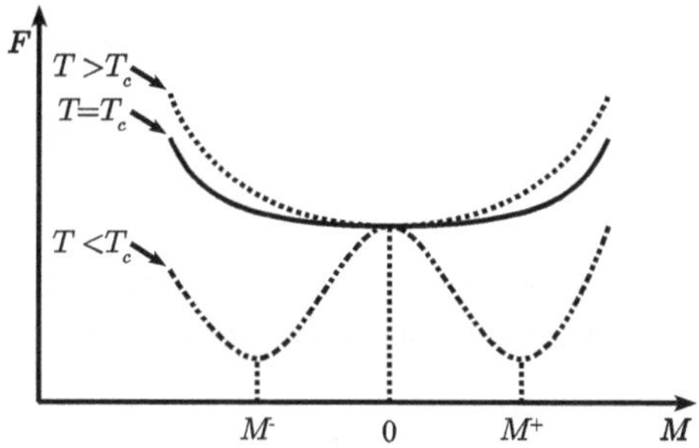

Fig. 1.4 Singular free energy $F - F_c$ at different temperatures around the critical point as a function of the order parameter M. $M = (\rho - \rho_c)/\rho_c$ is the order parameter of the liquid-vapor transition

Supercritical fluids on the critical isochore correspond to $M = 0$, and the liquid and vapor phases at coexistence below the critical point correspond to $M = M^+$ or M^-; that is, $M^+ = \Delta\tilde{\rho}_\ell = \frac{\rho_\ell - \rho_c}{\rho_c}$ and $M^- = \Delta\tilde{\rho}_v = \frac{\rho_v - \rho_c}{\rho_c}$. The simplest free-energy expansion that describes the transition is the Landau energy or mean-field energy ("mean field" because the true spatial extent of the fluctuations is neglected for the selected property), as already described using Eq. (1.20) for a van der Waals fluid.

1.3.2 Correlation Length of Fluctuations as a Natural Length Scale

As the critical point is approached, the order parameter fluctuates more and more strongly. This behavior is in agreement with the divergence of the isobaric thermal expansion β_p, which means that the minute spontaneous thermal fluctuations can give rise to very large density fluctuations near T_c. The vicinity of the critical point for a second-order phase transition is then characterized by the presence of large fluctuations of the order parameter M. These fluctuations give rise to refractive index fluctuations and unusually strong scattering of light: the "critical opalescence" of fluids (Fig. 1.1).

In order to account for the order parameter fluctuations in the vicinity of the critical point, it is necessary to include the corresponding contribution (~∇M^2), with amplitude d_0 [54], in the dimensionless Landau free energy of the fluid volume:

$$\tilde{f} = \tilde{f}_c + \int_V \left(\frac{a_0}{2} \tau M^2 + \frac{b_0}{4!} M^4 + \frac{d_0}{2} |\nabla M|^2 \right) d^3 r. \tag{1.32}$$

The gradient term naturally introduces a new length scale $\sim d_0^{-1/2}$. The space variable r is then explicitly incorporated into the free energy and the fluid properties (such as the isothermal compressibility, the specific heat, or the liquid and vapor densities as defined in Sect. 1.2) become space dependent.

The susceptibility is related to the isothermal compressibility by the relationship

$$\chi = \left(\frac{\partial^2 F}{\partial M^2} \right)^{-1} = \rho^2 \kappa_T, \tag{1.33}$$

whose Fourier transform can be easily deduced from Eq. (1.32):

$$\chi(q) = \left(\frac{\partial^2 F_q}{\partial M^2}\right)^{-1} = \frac{1}{a_0}\frac{\tau^{-1}}{\xi^{-2}+q^2}, \tag{1.34}$$

where q is the wavenumber associated with r. Using the inverse Fourier transform of $\chi(q)$, the following correlation function $\chi(r)$ can be obtained:

$$\chi(r) \equiv \langle \delta M(r)\delta M(0)\rangle \sim \frac{e^{-\frac{r}{\xi}}}{r}. \tag{1.35}$$

In Eq. (1.34), the quantity $\xi = (d_0\chi)^{1/2}$ is the correlation length of the order parameter fluctuations. It characterizes the spatial extent of the fluctuations. Therefore, using Eqs. (1.24) and (1.33), the temperature dependence in the mean field theory ξ_{MF} is given by the following power laws:

$$\xi_{\mathrm{MF}} = \left(\frac{d_0}{a_0}\right)^{1/2} \tau^{-\nu_{\mathrm{MF}}} \quad (M = 0;\ T > T_c) \tag{1.36}$$
$$\xi_{\mathrm{MF}} = \left(\frac{d_0}{2a_0}\right)^{1/2} \tau^{-\nu_{\mathrm{MF}}} \quad (M = M^{+,-};\ T < T_c),$$

with the mean field exponent $\nu_{\mathrm{MF}} = \frac{\gamma_{\mathrm{MF}}}{2} = \frac{1}{2}$. This shows that the correlation length diverges at the critical point and thus appears to be the natural length scale for the critical point phenomena. This important phenomenological result of a mean field approach is at the heart of the universality of the critical point, as it derives from the universal asymptotic form of the free energy and thus ignores the particular details of the intermolecular interactions at the molecular length scale.

1.4 Upper Critical Dimensionality, Renormalization, Exponents, and Two Scale Factor Universality

However, the Landau free energy neglects fluctuations where the size of the fluctuation diverges at the critical point. This would be self-consistent only within the condition (called the *Ginzburg criterion* [29]) that the space-averaged amplitude $\langle \delta M^2 \rangle$ remains finite at T_c. A simple derivation of the Ginzburg criterion is given in Chap. 19 At or very near to T_c, this condition naturally introduces another condition for the space dimensionality d:

$$d > d_c = 4, \tag{1.37}$$

which is the necessary condition for the mean-field theory to remain valid. Below this *upper critical dimensionality* d_c, and especially at $d = 3$, the role of fluctuations

Table 1.2 Universal power laws for thermodynamic and correlation properties used in this book, with the related Ising values of the critical exponents for the universality class $d = 3$; $n = 1$ (from [35])

Property	Power law	Thermodynamic path	Critical exponent Ising $d = 3$; $n = 1$	Mean field
Compressibility	$\kappa_T \sim \|\tau\|^{-\gamma}$	Critical isochore	$\gamma \simeq 1.239$	1
Specific heat at constant pressure	$c_p \sim \|\tau\|^{-\gamma}$	Critical isochore	$\gamma \simeq 1.239$	1
Specific heat at constant volume	$c_V \sim \|\tau\|^{-\alpha}$	Critical isochore	$\alpha \simeq 0.110$	0 (*jump*)
Density of two phases	$(\rho_\ell - \rho_v) \sim \|\tau\|^{\beta}$	Coexistence curve	$\beta \simeq 0.326$	1/2
Surface tension	$\sigma \sim \|\tau\|^{-\mu}$	Coexistence curve	$\mu = 2\nu \simeq 1.26$	1
Pressure	$\|\Delta p\| \sim \|\Delta\rho\|^{\delta}$	Critical isotherm	$\delta \simeq 4.80$	3
Correlation function	$H(r) \sim r^{2-d-\eta}$ $(\xi^{-1} = 0)$	Critical point, large r	$\eta \simeq 0.034$	0
Correlation length	$\xi \sim \|\tau\|^{-\nu}$	Critical isochore	$\nu \simeq 0.630$	1/2

d is the space dimension of the system (here, $d = 3$ for one-component fluids) and n is the dimension of the order parameter (here, $n = 1$ for the scalar density). $H(r)$ is the pair correlation function, where r is the distance between two fluid particles. The last column presents the mean field values of the critical exponents (see text)

must be taken into account. This is done by "renormalizing" the fluctuations [72]. For the coexistence curve, the behaviors of the density, the susceptibility, the isobaric and isochoric specific heats, the correlation length, etc. also correspond to power laws with modified values [also called Ising values; see below and Eq. (18.1)] of the critical exponents (see Tables 1.2 and 1.3). For instance, the Ising values of the critical exponents β, γ, and ν for the density difference $\Delta\tilde{\rho}_{\ell,v}$ on the coexistence curve, the susceptibility χ [or the isothermal compressibility κ_T, from Eq. (1.33)], and the correlation length ξ, respectively, are [35]

$$\beta \simeq 0.326\left(\frac{1}{2}\right), \qquad \gamma \simeq 1.239\ (1)\,, \qquad \nu \simeq 0.630\left(\frac{1}{2}\right). \tag{1.38}$$

In Eq. (1.38), the mean-field values are shown in parentheses. Moreover, near T_c at $\rho = \rho_c$, the slope of the isochore $[p\,(T)]_{V=cte}$ (i.e., $\gamma_V = \left(\frac{\partial p}{\partial T}\right)_{\rho=\rho_c}$) is a constant, so the divergences of the isothermal compressibility $\kappa_T = \frac{1}{\rho}\left(\frac{\partial\rho}{\partial p}\right)_T$ and the isobaric thermal expansion coefficient $\beta_p = -\frac{1}{\rho}\left(\frac{\partial\rho}{\partial T}\right)_p$ are similar (exponent $\gamma = 1.239$), due to the thermodynamic relation $\gamma_V \kappa_T \left(\beta_p\right)^{-1} = 1$. Now, considering the thermodynamic relations of Eq. (1.28) between $c_p - c_V$, κ_T and β_p, we can see that c_p diverges as κ_T, since the divergence exponent of c_V ($\alpha = 0.110$) is weaker than that of κ_T or β_p (exponent $\gamma = 1.239$). The amplitude of the c_p divergence can be

Table 1.3 Behavior of some static quantities of interest at $\rho = \rho_c$ ($\tau > 0$, amplitudes with a superscript +) or at phase coexistence ($\tau < 0$, amplitudes with a superscript −). The Ising values of the exponents are from [35]

Quantity	Leading amplitude	Leading ising divergence	Correction to scaling	Background	Exponent value
Order parameter M=	B	$B\,\|\tau\|^{\beta} \times$	$\left(1 + a_M \tau^{\Delta} + \cdots\right)$	$+M_{\mathrm{reg}}(\tau)$	$\beta \simeq 0.326$
Specific heat (const. V) $\left(\frac{\rho_c T_c}{p_c}\right) c_V^{+,-} =$	$\frac{A^{+,-}}{\alpha}$	$\frac{A^{+,-}}{\alpha}\,\|\tau\|^{-\alpha} \times$	$\left(1 + \alpha a_c^{+,-} \tau^{\Delta} + \cdots\right)$	$+B_{cr} + B_{reg}(\tau)$	$\alpha \simeq 0.110$
Susceptibility $p_c \kappa_T^{+,-} =$	$\Gamma^{+,-}$	$\Gamma^{+,-}\,\|\tau\|^{-\gamma} \times$	$\left(1 + a_{\Gamma}^{+,-} \tau^{\Delta} + \cdots\right)$		$\gamma \simeq 1.239$
Correlation length $\xi^{+,-} =$	$\xi_0^{+,-}$	$\xi_0^{+,-}\,\|\tau\|^{-\nu} \times$	$\left(1 + a_{\xi}^{+,-} \tau^{\Delta} + \cdots\right)$		$\nu \simeq 0.630$
Specific heat (const. p) $c_p^{+,-} =$	$c_{p0}^{+,-}(1+\tau) = \frac{T_c}{\rho_c}\left(\frac{\partial p}{\partial T}\right)_{\rho_c}^2 \frac{\Gamma^{+,-}}{p_c}(1+\tau)$	$c_{p0}^{+,-}(1+\tau)\,\|\tau\|^{-\gamma} \times$	$\left(1 + a_{\Gamma}^{+,-} \tau^{\Delta} + \cdots\right)$	$+c_V(\tau)$	$\gamma \simeq 1.239$
Ratio of specific heats $\gamma_0 = \frac{c_p^{+,-}}{c_V^{+,-}} =$	$\gamma_{00}^{+,-}(1+\tau) = \left[\frac{T_c}{p_c}\left(\frac{\partial p}{\partial T}\right)_{\rho_c}\right]^2 \frac{\alpha \Gamma^{+,-}}{A^{+,-}}(1+\tau)$	$\gamma_{00}^{+,-}(1+\tau)\,\|\tau\|^{-\gamma+\alpha} \times$	$\left(\frac{1 + a_{\Gamma}^{+,-} \tau^{\Delta} + \cdots}{1 + \alpha a_c^{+,-} \tau^{\Delta} + \cdots}\right)$	$+1$	$\gamma - \alpha \simeq 1.13$
Speed of sound $c_s = \sqrt{\frac{1}{\rho \kappa_S}} =$	$c_{s0}^{+}(1+\tau) = \left[\left(\frac{\partial p}{\partial T}\right)_{\rho_c}^2 \frac{\alpha T_c^2}{\rho_c p_c A^{+,-}}\right]_c^{1/2}(1+\tau)$	$c_{s0}^{+,-}(1+\tau)\,\|\tau\|^{\frac{\alpha}{2}} \times$		$+c_{s,reg}(\tau)$	$\frac{\alpha}{2} \simeq 0.055$
Interfacial tension $\sigma =$	σ_0	$\sigma_0\,\|\tau\|^{-2\nu} \times$		$+\sigma_{reg}(\tau)$	$2\nu \simeq 1.260$

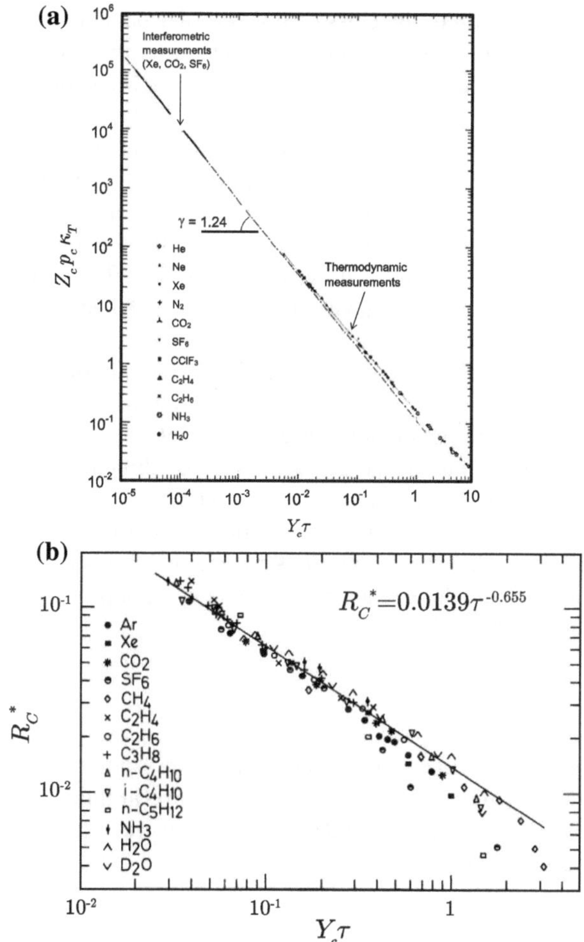

Fig. 1.5 Two scale factor universality **a** Isothermal susceptibility case (see Refs. [24–26, 28]). **b** Transport property case (see Ref. [41])

determined by Eq. (1.28), where the c_V term is considered a background term. More generally, because such relationships exist between the thermodynamics coefficients that undergo critical anomalies, a number of relations between the universal exponents and the related amplitudes of the power law divergences can be derived. It can be shown that only two exponents are independent. For the same reason, only two amplitudes are independent. As an important consequence, the behavior of any system can be expressed as scaled, universal functions with only two system-dependent amplitudes (see for example, Fig. 1.5). This is known as *two scale factor universality* [67]. All of the above results apply to systems that belong to the same universality class, as defined by the space dimensionality d and the dimensionality n of the order parameter that rules the free energy formulation. The class of fluids is defined by

$d = 3$, $n = 1$. In addition to the class of fluids (order parameter: density), there is also the class of partially miscible liquid mixtures (order parameter: concentration), including polymer–solvent and microemulsion systems. Many other systems belong to this class, such as the three-dimensional uniaxial Ising model (order parameter: magnetization), the current representative of the class that provides the Ising values of the critical exponents and the universal amplitude combinations.

The influence of fluctuations on the critical behavior can be observed in two different ways. Very near the critical point, the growth of fluctuations can be *suppressed* by external influences (e.g., a shear flow; see [8, 50–53]). The Ginzburg criterion is thus recovered, and mean-field behavior becomes relevant because the upper critical dimensionality is lowered.

Far from T_c, fluctuations become negligible, so the van der Waals equation of state becomes valid. "Far" corresponds to a distance that is sufficiently far from the critical point that the fluctuation correlation length becomes sufficiently small and the Ginzburg criterion can apply in a space of dimensionality $d = 3$ [55]. Crossover functions have been proposed for the region that is neither "close" nor "far" from the critical point [2–5, 27]. In most practical situations, the implementation of "nonanalytical" corrections or "corrections-to-scaling" terms is enough to describe the data. The temperature divergence, with exponent ψ, of property Y is modified to

$$\tau^{\psi} \rightarrow \tau^{\psi}(1 + a_Y^{(1)}\tau^{\Delta} + a_Y^{(2)}\tau^{2\Delta} + \cdots) \tag{1.39}$$

The exponent Δ is universal ($\Delta \simeq 0.51$), but the amplitudes $a_Y^{(i)}$ are not. Later in this chapter (Sect. 1.10), we will consider how to obtain the numerical values for a given fluid in more detail. We will also provide data on several fluids with "low" critical temperatures (n- and p-H_2), "moderate" critical temperatures (Xe, CO_2, SF_6) and "high" critical temperatures (H_2O).

1.5 Critical Slowing Down: Timescale

It is possible to define universal dynamics scaling laws for the transport coefficients in a similar way to the above universal scaling laws for thermodynamic properties. The renormalization theory is more delicate, and the universality classes are not as large as those for statics. The dynamics of the order parameter fluctuations appear to be the natural timescale, just as it is the correlation length for space, and they are dominated by the decay time of a fluctuation of size ξ. The fluctuation vanishes through a diffusion process with a diffusion coefficient D, which can be estimated from the Brownian diffusion of a cluster of size ξ [37]:

$$D = \frac{k_B T}{6\pi\mu\xi} \sim \tau^{\nu + Y_\eta}. \tag{1.40}$$

Here μ is the shear viscosity, which asymptotically diverges at the critical point with the weak exponent $Y_\eta \simeq 0.043$ (see Table 1.5):

$$\mu \sim \tau^{-Y_\eta}. \tag{1.41}$$

The typical time t_ξ is naturally defined as the diffusion time of a fluctuation of size ξ on the typical length scale ξ,

$$t_\xi = D^{-1}\xi^2 = \frac{6\pi\mu\xi^3}{k_B T}. \tag{1.42}$$

From the temperature dependence of D, it is easy to determine the temperature dependence of t_ξ:

$$t_\xi \sim \tau^{-(3\nu+Y_\eta)} = \tau^{-1.93}. \tag{1.43}$$

Equation (1.43) shows that the order parameter fluctuations relax more and more slowly as the system approaches its critical point. This is the well-known phenomenon of *critical slowing down*.

For a fluid, both density (the order parameter) and temperature fluctuations slow down. However, pressure equilibrates near-instantaneously (at the velocity of sound). Anticipating the discussion of Sect. 1.10, the thermal diffusion coefficient $D_T = \frac{\Lambda}{\rho c_p}$ can thus also be written as

$$D_T = \frac{\Lambda}{\rho c_p} = \frac{\Lambda_b}{\rho c_p} + \frac{\Lambda_c}{\rho c_p} = D_{T,b} + D_{T,c} \tag{1.44}$$

In Eq. (1.44), the thermal conductivity coefficient $\Lambda = \Lambda_b + \Lambda_c$ is written in an additive form of its background (Λ_b) and critical (Λ_c) terms. The latter is expected to have the critical divergence $\Lambda_c \sim \tau^{-Y_\Lambda}$ with the universal critical exponent Y_Λ. Then, the additive form $D_T = D_{T,b} + D_{T,c}$ of the isothermal diffusion coefficient contains the respective background term $D_{T,b} = \frac{\Lambda_b}{\rho c_p}$ and $D_{T,c} = \frac{\Lambda_c}{\rho c_p}$ critical term. $c_p = \frac{\bar{c}_p}{m}$ is the specific heat at constant pressure. As c_p diverges with the exponent γ along the critical isochore, by identifying the critical divergence $D \sim \tau^{\nu+Y_\eta}$ given by Eq. (1.40) to the one of

$$D_{T,c} = \frac{\Lambda_c}{\rho c_p} \sim \tau^{-Y_\Lambda+\gamma} \tag{1.45}$$

it is easy to infer the diverging behavior of the thermal conductivity from this relationship:

$$\Lambda \sim \Lambda_c \sim \tau^{-\gamma+\nu+Y_\eta} = \tau^{-Y_\Lambda} = \tau^{-0.57}. \tag{1.46}$$

In Eq. (1.46), the critical exponent $Y_\Lambda \simeq 0.57$ of the thermal conductivity is linked to the critical exponent Y_η of the viscosity through the scaling law $Y_\Lambda + Y_\eta = \nu\,(\varepsilon - \eta)$, with $\varepsilon = 4 - d$ (here $\varepsilon = 1$), and where $\eta \simeq 0.034$ is the critical exponent of the singular behavior of the correlation function at the critical point when $\xi^{-1} = 0$ (see Table 1.2). Such universal critical dynamics are also obtained from the mode coupling theory of critical dynamic phenomena (see [32] for example).

To give an idea of the importance of the critical slowing down of the heat diffusion process, let us consider the thermal equilibration of a CO_2 sample with an inner thickness $2e = 1$ cm at 1 mK from T_c. The typical time t_D needed for equilibration is

$$t_D \simeq \frac{e^2}{D_T} \sim 2 \times 10^6 \text{ s}, \tag{1.47}$$

where $D_T = 1.27 \times 10^{-11}$ m^2 s^{-1} (the numerical values can be calculated from the data given in Table 1.6). This result means that thermal equilibration by diffusion alone would require nearly three weeks!

At this point, we should emphasize again that the lifetime of the fluctuations appears to be the *natural timescale* of critical point phenomena, just as the fluctuation correlation length is the natural length scale. When properly scaled by ξ and t_ξ, most of the processes are found to be universal.

1.6 Phase Separation Dynamics

Phase separation in fluids and liquid mixtures is a common process that is important in many areas of science and technology [30] (see Chap. 20 for a brief presentation of the conventional theory of nucleation and spinodal decomposition). However, the connection between the domain morphology and the growth laws is not straightforward [7]. A typical phase separation experiment (see Fig. 1.6) involves *quenching the temperature*; in other words, changing the temperature at the highest possible rate, and in any case faster than the evolution of the system. Thus, the final state of the system can be considered to be the same as its initial homogeneous state with density ρ, but at a final temperature of T_f. The initial and final states are characterized by the order parameter M and temperature T_i (initial stage, homogeneous phase), and the order parameter $M^{+,-}$ and temperature T_f (final stage, two phases).

Two questions then arise. (i) How do $M^{+,-}$ regions appear in this system that is very far from equilibrium? This is known as the *nucleation problem*. (ii) Once such heterogeneous regions are formed, how rapidly and by which mechanisms do they grow? This is termed the *growth problem*.

The volume fraction of the final phase is a key parameter. In the critical region, it is easy to continuously vary the physical parameters. The relative volumes of the liquid and vapor phases are given by the lever rule hereafter illustrated using the *symmetrized* coexistence curve given in Fig. 1.6. The vapor volume fraction ϕ_v can be written as

$$\phi_v = \frac{M^+ - M}{M^+ - M^-} = \frac{\rho_\ell - \rho}{\rho_\ell - \rho_v}. \tag{1.48}$$

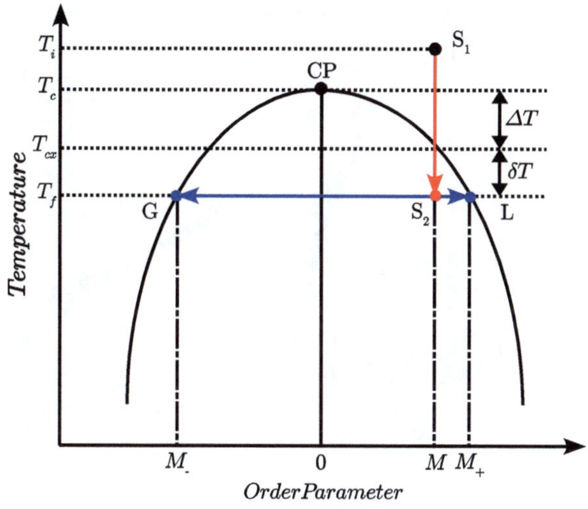

Fig. 1.6 Temperature quench procedure employed at a constant mean order parameter to induce a phase transition. S_1: Initial state; S_2: Final state after temperature quench

The equilibrium volume fraction ϕ of the minority phase varies with the temperature quench depth. ϕ is a function of T_c, the coexistence temperature T_{cx} corresponding to the initial density ρ, and the quench depth $\delta T = T_{cx} - T_f$:

$$\phi = \frac{1}{2}\left\{1 - \left(1 + \frac{\delta T}{\Delta T}\right)^{-\beta}\right\}. \tag{1.49}$$

Here, $\beta = 0.326$ is the universal exponent and $\Delta T = T_c - T_{cx}$ is calculated from the coexistence curve expression (Tables 1.2 and 1.3).

As will be discussed in subsequent Chaps. 4, 5 and 9, the "piston effect" speeds up thermalization (at the cost of forming a thin boundary layer), so that thermal quenches very close to T_c are limited only by the thermal response of the thermostat. Critical slowing down is nevertheless effective in droplet evolution. It allows a detailed investigation of the mechanisms involved in the separation process to be performed.

It is important to note that the typical scenario for the formation of a new phase does not correspond to the conventional processes of nucleation or spinodal decomposition (see Chap. 20). In this conventional nucleation approach, only sufficiently large fluctuations can overcome the energy barrier corresponding to the formation of the interface of a nucleus in the new phase. In spinodal decomposition, all fluctuations are unstable and grow with a rate that depends on their size, the fastest imposing a well defined wavelength to the phase separating pattern. Near the critical point, the large fluctuations of the order parameter induce another process, the so-called *generalized nucleation*, which is intermediate between spinodal decomposition and

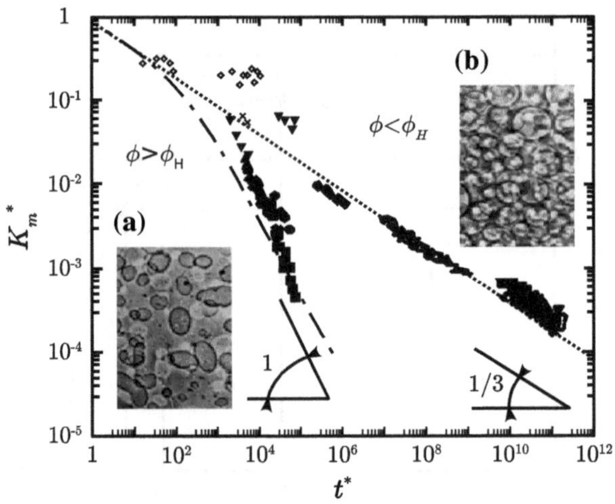

Fig. 1.7 Universality of phase transition growth laws and pattern morphologies [7]. The growth laws in SF_6 and CO_2 are expressed in the scaled units $K_m^* = \frac{2\pi\xi^-}{L_m}$ and $t^* = \frac{t}{t_{\xi^-}}$. L_m is the average distance between the domains, ξ^- is the correlation length of the order parameter fluctuations in the two-phase region, t_{ξ^-} is the corresponding typical fluctuation decay time, and ϕ is the volume fraction of the minority phase. ϕ_H is a threshold ($\approx$0.3). The *dotted and dashed lines* refer to liquid mixtures under reduced gravity. The data points are for SF_6 and CO_2 under microgravity. The *lower curve* ($\phi < \phi_H$) corresponds to a "fast" growth law and an interconnected morphology (pattern in **a**) and the reduced *upper curve* ($\phi > \phi_H$) refers to a "slow" growth law and a disconnected morphology (pattern in **b**)

nucleation [11, 12, 34]. In generalized nucleation, the large order parameter fluctuations of size ξ grow in amplitude, reach local equilibrium, and give rise to droplets of the minority phase. These droplets then grow at the expense of the majority phase and reach equilibrium. In addition, coalescence between drops, provoked by either Brownian motion or local flows induced by the coalescence process itself, play a key role. At very small volume fractions, these interactions should be very rare. Only a Lifshitz–Slyozov growth process, where the smallest domains evaporate and condense on the largest ones [42], takes place.

Experiments free of gravity effects show [7, 56] that when $M = 0$ (i.e., when the volume fraction of the minority phase is $\phi = \frac{1}{2}$), an interconnected pattern of domains that coalesce continuously is formed (see Fig. 1.7a). This interconnected pattern with well defined wavelength is often misleadingly associated with the conventional spinodal decomposition process. The characteristic length L_m of the domains can be defined as a pseudo-period between the phases. Late on in the process, the following growth law is observed:

$$L_m \sim \frac{\sigma}{\mu}t. \tag{1.50}$$

Here σ is the vapor–liquid surface tension, which scales with $\left(\xi^+\right)^2$ as

Table 1.4 Universal ratios between amplitudes and corrections to scaling at $\rho = \rho_c$ ($\tau > 0$, amplitudes with a superscript +) or at phase coexistence ($\tau < 0$, amplitudes with a superscript −). The numerical values are from [18, 57])

Universal ratios	Value
$\frac{A^+}{A^-}$	0.53
$\frac{\Gamma^+}{\Gamma^-}$	4.77
$\frac{\xi_0^+}{\xi_0^-}$	1.96
$R_C^+ = \frac{A^+\Gamma^+}{B^2}$	0.0565
$R_C^- = \frac{A^-\Gamma^-}{B^2}$	0.0223
$R_\xi^+ = \xi_0^+(\frac{\rho_c}{k_B T_c}A^+)^{1/3}$	0.266
$R_{\sigma\xi}^+ = \frac{\sigma_0}{k_B T_c}(\xi_0^+)^2$	0.36
$\frac{a_\xi^+}{a_\chi^+}$	0.65
$\frac{a_M}{a_\chi^+}$	0.9
$\frac{a_C^+}{a_\chi^+}$	8.6
$\frac{a_\chi^+}{a_{\chi^-}}$	0.32
$\frac{a_{\xi^+}}{a_\xi^-}$	≈1[a]

[a]Estimate (still under consideration)

$$\sigma = R_{\sigma\xi}^+ \frac{k_B T}{(\xi^+)^2}, \tag{1.51}$$

where $R_{\sigma\xi}^+ \approx 0.36$ is a universal amplitude ratio (see Table 1.4). When expressed in the scaled units K_m^* and t^* such that

$$K_m^* = \frac{2\pi}{L_m}\xi^-; \quad t^* = \frac{t}{t_\xi^-}, \tag{1.52}$$

all the data can be reasonably placed on the same scaled master curve (here the superscript-means that the correlation length is taken below T_c; the ratio of the correlation lengths below and above T_c is universal, $\frac{\xi^+}{\xi^-} = 1.96$, see Table 1.5). The curve can be described by the phenomenological equation given by Furukawa [23]:

$$(K_m^{*-1} - 1) - \left[\sqrt{\frac{A}{B}}\left\{\tan^{-1}\left(K_m^*\sqrt{\frac{B}{A}}\right) - \tan^{-1}\left(\sqrt{\frac{B}{A}}\right)\right\}\right] = Bt^*, \tag{1.53}$$

where A (≈0.14) and B (≈0.022) are adjustable parameters.

When the volume fraction of the domains is small and gravitational effects are negligible, the growth law is characterized by an exponent of 1/3 (see Fig. 1.7). All of the data can be placed on the same master curve:

Table 1.5 Behavior of some dynamic quantities at $\rho = \rho_c$ ($\tau > 0$, amplitudes with a superscript +) or at phase coexistence ($\tau < 0$, amplitudes with a superscript −). The values of the exponents are from [18], [57])

Property	Amplitude	Divergence	Correction to scaling	Additive background	Asymptotic exponent value
Shear viscosity $\mu =$	$\mu_b \times$	$\tau^{-Y_\eta} \times$	$(1 + a_\mu \tau^{\Delta}) \times$		$Y_\eta = 0.04$
Bulk viscosity[a] $\mu_B =$	$\mu_{B,c} \times$	$\tau^{\alpha - 3\nu - Y_\eta}$			$3\nu + Y_\eta - \alpha = 1.82$
Thermal conductivity $\Lambda^{+,-} =$	$\Lambda_0^{+,-} \times$	$\tau^{-Y_\Lambda} \times$	$(1 + a_\Lambda \tau^{\Delta}) +$	Λ_{b0}	$Y_\Lambda = \gamma - \nu - Y_\eta = 0.57$
Thermal diffusivity $D_T^{+,-} =$	$D_0^{+,-} \times$	$\tau^{\nu + Y_\eta} \times$	$(1 + a_D \tau^{\Delta}) +$	$D_{b0} \tau^{\gamma} \left(D_{b0} = \frac{\Lambda_b}{\rho_c c_{p0}} \right)$	$\gamma - Y_\Lambda = \nu + Y_\eta = 0.67$

[a] The divergence of the bulk viscosity is calculated in [49], but it has not been experimentally assessed yet, and is still subject to debate

$$K_m^* \approx 0.9\,(t^*)^{1/3}\,. \tag{1.54}$$

All the above data show that $K_m^* \approx 1$ when $t^* \approx 1$, which means that nucleation proceeds from fluctuations on the order of the correlation length, in agreement with the concept of generalized nucleation [11, 12].

Growth is mostly due to the coalescence of domains (bubbles or droplets). Depending on the volume fraction, coalescence can occur through collisions resulting from the Brownian diffusion of the domains ($t^{1/3}$ growth law, isolated domains, see Fig. 1.7b) or through hydrodynamical interactions resulting from the coalescence process itself (see Fig. 1.7a). In this latter process, a coalescence event induces a flow that can cause another coalescence event, and so on, in a kind of chain reaction [46]. This results in a viscous flow limited growth regime and a t^1 growth law with interconnected domains. The transition between the two regimes appears at a well-defined volume fraction of around 30 %.

It is not clear at the present time whether the hypercompressible nature of fluids near their critical point can lead to important effects other than those that are directly connected to the "piston effect", as analyzed in Chaps. 5 and 9. This thermal effect leads to the formation of a hot boundary layer, the density of which is markedly smaller than in the bulk, and localized nucleation phenomena can occur in this layer [7].

1.7 Wetting and Adsorption Properties

The behavior of pure fluids and fluid mixtures near a wall, which can be considered a third phase, changes in the proximity of a critical point. In the one-phase region, the adsorption of fluid molecules causes the density to vary in a manner that scales with the correlation length in a universal *critical adsorption* profile. If z is the distance measured perpendicularly to the wall, then if

$$\begin{array}{lll} T = T_c & & : M(z) \sim z^{-\beta/\nu} \\ T > T_c\,; & \begin{array}{l} z < \xi \\ z > \xi \end{array} & \begin{array}{l} : M(z) \sim z^{-\beta/\nu} \\ : M(z) \sim e^{-z/\xi}\,. \end{array} \end{array} \tag{1.55}$$

This adsorption profile may compete with and modify the thermal boundary layers, as discussed in Chap. 4.

In the two-phase region, a transition to complete wetting is expected. This transition can be altered by long-range forces, such as the van der Waals force. There is a great deal of literature on this subject (see, e.g., [61]). Here, we note only that wetting transitions have indeed been observed for liquid mixtures in contact with high-energy walls (e.g., silica). However, the transition is much more difficult to observe with simple fluids, which already completely wet (or dry) the wall at the triple point. Pure fluids like ^{4}He, Ne and H_2 on low-energy substrates like cesium (Cs) and rubidium (Rb) are the few systems where wetting transitions have been observed so far [21, 31].

A simple argument was provided by Cahn [16] for the necessity of a wetting transition near the critical point. In the simplest case where long-range and gravitational forces can be neglected, the balance of tension at the triple line liquid–vapor–wall leads to the well-known Young equation:

$$\sigma cos\theta = \sigma_{\ell 0} - \sigma_{v0}. \tag{1.56}$$

Here, θ is the contact angle and σ, $\sigma_{\ell 0}$, and σ_{v0} are the liquid–vapor, liquid–wall, and vapor–wall surface tensions, respectively. We note that $\sigma_{\ell 0} - \sigma_{v0}$ behaves as a two-dimensional surface density difference, which goes to zero near T_c as

$$(\rho_\ell - \rho_v)_s \sim |\tau|^{\beta_s}. \tag{1.57}$$

In this case, $\beta_s = 0.8$ is the critical exponent of the two-dimensional (surface) coexistence curve; this value differs from the that ($\beta = 0.326$) of the critical exponent of the three-dimensional (bulk) coexistence curve. From Eq. (1.51), the interfacial tension σ also goes to zero as $\left(\xi^+\right)^{-2}$ or $\tau^{2\nu}$. The following temperature behavior of the contact angle can then be deduced:

$$cos\theta \sim |\tau|^{\beta_s - 2\nu} \sim |\tau|^{-0.5}, \tag{1.58}$$

which diverges at T_c. Therefore, there should be a temperature at which $\theta = 0$ (the *wetting transition temperature*), and above which one of the two coexisting phases completely wets the wall.

The competition between the gravitational and wetting forces leads to the definition of the *capillary length* ℓ_c, which determines the shape of the vapor–liquid interface near a wall. With $g_0 = 9.81\ \mathrm{m\,s^{-2}}$ (the Earth's gravitational acceleration at its surface), the capillary length ℓ_c is defined as

$$\ell_c = \sqrt{\frac{\sigma}{g_0 \Delta\rho_{\ell v}}}, \tag{1.59}$$

and it vanishes at the critical point as

$$\ell_c \sim |\tau|^{\nu - \frac{\beta}{2}} \sim |\tau|^{0.467}. \tag{1.60}$$

When ℓ_c is much smaller than the horizontal extension of a vessel, the interface is flat. Close to the walls, however, the interface has a curvature that scales with ℓ_c. As a result, when the temperature is increased to T_c, the effect of gravity dominates, regardless of the values of g_0 or $\rho_\ell - \rho_v$. This effect looks paradoxical. The vicinity of the critical point means that the densities of the vapor and liquid are nearly equal, which should lower the effect of gravity. However, the surface tension also goes to zero with a strong temperature dependence: this results in a large effect of gravity.

1.8 Some Effects Due to the Earth's Gravity

Buoyancy does not affect the behavior of a homogeneous phase. However, the high compressibilities of fluids in the supercritical state mean that they are compressed under their own weights, and a density gradient is established with the least dense fluid at the top of the vessel and the densest fluid at the bottom (see Fig. 1.1 for example). The density at equilibrium is thus not uniform; it varies with height (z) with a local gradient of

$$\frac{d\rho}{dz} \simeq \rho^2 g\chi \sim \tau^{-\gamma}, \tag{1.61}$$

leading to a gravity-induced averaging of thermophysical properties in a cell of finite height at constant (mean) density and constant temperature. On Earth, the divergence of the correlation length will eventually be limited by hydrostatic pressure effects across the correlated volume. The relative magnitude of such gravity-induced effects can be estimated via the fluid-dependent critical height

$$h_0 = \frac{p_c}{\rho_c g_0},$$

which characterizes the gravitational scale height for each one-component fluid [40].

The density profile takes longer to establish as the critical point is approached. Quite naturally, the range of the effect scales with the correlation length ξ and the time taken to establish equilibrium, with the lifetime of the fluctuations t_ξ.

1.8.1 Reaching the Critical Point on Earth: An Impossible Task

As the temperature approaches T_c, a density profile develops where the density is critical in only an infinitely small layer—a layer on the order of the correlation length (the natural length scale, see [64]). The thickness of this small layer is comparable to the vapor–liquid interface thickness below the critical point, which is on the order of a few correlation lengths [9, 71]. It thus becomes impossible to experimentally determine whether the observed profile is a gravity-induced gradient or the meniscus between the vapor and liquid phases. In particular, this phenomenon limits the precision with which the critical temperature can be determined. Practically speaking, in a CO_2 sample with a height of 1 cm, T_c cannot be determined on Earth to a precision of better than 1 mK.

1.8.2 Buoyancy and Phase Separation

When a temperature quench is performed (see Fig. 1.6) from the one-phase region (temperature T_i) to the two-phase region (temperature T_f), both liquid-phase and vapor-phase domains appear. Because the phases have different densities, gravity will eventually induce sedimentation and flows. A natural length scale for gravitational effects is the capillary length ℓ_c, which represents the competition between capillary and gravitational forces [Eq. (1.59)]. Because $\ell_c \sim \tau^{0.5}$ goes to zero at T_c, the influence of gravity is more pronounced near T_c. When the size of a vapor bubble or a liquid droplet is on the order of ℓ_c, heavy liquid droplets start to exhibit sedimentation, and the buoyancy of the vapor bubbles causes them to rise. To give an example, after a quench at $T_c - 1\,\mathrm{mK}$ in CO_2, the limiting droplet or bubble size is on the order of $\ell_c \simeq 5\,\mu\mathrm{m}$; that is, a few correlation lengths. In practice, turbulent convection is observed, which does not allow domain growth to be studied.

1.8.3 Convective Flows and Thermal Instabilities

As noted above, near the critical point, the isobaric thermal expansion coefficient β_p diverges as

$$\beta_p = -\frac{1}{\rho}\left(\frac{\partial\rho}{\partial T}\right)_p \sim \tau^{-\gamma}. \tag{1.62}$$

This divergence means that a small temperature gradient can induce a noticeable density gradient, which can easily lead to convection under the influence of the Earth's gravitational field. Because these supercritical fluids are extremely unstable with respect to even minute temperature gradients, "isothermal" convection can occur (see [47] and Chap. 13 which treat the Rayleigh–Bénard configuration). Although this Rayleigh–Bénard configuration (a fluid layer of thickness e, heated from below, and enclosed between two isothermal plates with a temperature difference of ΔT) is rarely encountered in real situations, the value of the Rayleigh number Ra is nevertheless a good measure of a fluid's sensitivity to gravity-induced convection:

$$\mathrm{Ra} = \frac{g_0\left(-\frac{\partial\rho}{\partial T}\right)_p e^3\Delta T}{\mu D_T} = \frac{g_0\rho\beta_p e^3\Delta T}{\mu D_T} = \frac{g_0\rho c_p\beta_p e^3\Delta T}{\nu\Lambda} \tag{1.63}$$

However, this formulation ignores the gravity-induced density gradient. It can thus be split into a classical hydrodynamic expression Ra_0 and a scaling function that retains the temperature dependence:

$$\mathrm{Ra} = \mathrm{Ra}_0\tau^{-(\gamma+\nu)} \sim \tau^{-1.87}. \tag{1.64}$$

We see that Ra diverges at the critical point, so that the fluids are indeed very sensitive to temperature gradients. For example, if we consider a 1 cm high CO_2 fluid sample heated from below, where the top of the sample is at T_c, the critical value ($\text{Ra}_c \simeq 1{,}700$) of the Rayleigh number that defines the onset of convection [47] will be reached for a temperature gradient of only 10^{-14} K cm^{-1}! However, we will see in Chap. 13 that when the fluid is close to its critical point, the diverging isothermal compressibility means that the Rayleigh criterion should be replaced by the Schwarzschild criterion [62]. This criterion is also used for normally compressible gases, as encountered in atmospheric science [65]. This means, in particular, that the destabilizing temperature gradient does not go to zero according to Eq. (1.64), but to a constant value.

1.9 Suppressing the Effects of Gravity

It is impossible to escape the influence of gravity. One can, however, try to suppress the effect of gravity by (for example) stirring, density matching, studying the system while in free fall, or by counterbalancing the weight with a volume force, such as a magnetic or centrifugal force. All of these methods have advantages and drawbacks, as we shall see below.

1.9.1 Effects of Stirring

The most obvious way to homogenize a sample is to stir it [44]. However, the density becomes homogeneous at the cost of velocity gradients or local shear flows. Such gradients. The shear flow $S = \frac{dV}{dr}$, where V is the fluid velocity, affects the behavior through a convective effect associated with the velocity gradient [8]. This effect is important if there is enough time for the fluctuations, of mean spatial extension ξ, to be convected over a distance of around ξ during their lifetimes t_ξ; that is, $S\,\xi\,t_\xi \gtrsim \xi$, or

$$S\,t_\xi \gtrsim 1\,. \tag{1.65}$$

Since t_ξ is temperature dependent, this condition corresponds to a crossover temperature T_s below which the shear flow becomes efficient. Neglecting the weak critical divergence of the shear viscosity, μ, the inequality of Eq. (1.65) can be rewritten as

$$\frac{T_s - T_c}{T_c} < \left[\frac{6\pi\,\mu\xi_0^3}{k_B T_c}\right]^{\frac{1}{3\nu}} S^{\frac{1}{3\nu}}\,. \tag{1.66}$$

The effect of shear is to reduce the fluctuations and make them anisotropic. The reduced fluctuation amplitudes modify the free energy of the system, the mean field becomes relevant, and critical exponents are changed (this corresponds to a lowering of the upper critical dimensionality $d_c \simeq 2.4$ [50–53]). At the same time, the reduction affects the critical temperature, which in turn is reduced and becomes shear dependent.

In addition to these shear flow effects, stirring has another consequence. Because of the hydrostatic pressure, a fluid element convected by stirring feels different pressures depending on its height in the sample. Its temperature changes adiabatically as it travels to different heights (the temperature changes at the velocity of sound, which is much faster than the stirring velocity), resulting in a temperature gradient in the sample,

$$\frac{dT(z)}{dz} = -\rho g_0 \left(\frac{\partial T}{\partial p}\right)_\rho . \tag{1.67}$$

This gradient is on the order of $1\,\mathrm{mK\,cm^{-1}}$ for Xe, $0.4\,\mathrm{mK\,cm^{-1}}$ for CO_2 and $0.1\,\mathrm{mK\,cm^{-1}}$ for SF_6. As a consequence of the above gradient, a sample could be homogenized by producing a temperature gradient of the same amplitude that would induce a constant density in the sample without stirring. The temperature-induced density gradient counterbalances the gravity-induced density gradient. An example can be found in [6] where a temperature gradient of a few $0.1\,\mathrm{mK\,cm^{-1}}$ is applied in a CO_2 sample. When the temperature gradient is directed along the direction of Earth's gravitational acceleration, it enhances the gravity-induced gradients. When it is directed in the opposite direction, the gravity effects are nearly suppressed.

One could also homogenize a sample by using an electric field to produce electrostriction effects [44, 74, 75] to counterbalance the gravitational force. However, an electric field can also affect the fluid at the level of critical fluctuations and modify the critical transition. For instance, a small shift in the critical temperature is expected [48, 50].

1.9.2 Matching Densities of Binary Liquids by Isotopic Substitution

Some mixtures of two liquids (binary liquids) exhibit a partial miscibility gap as a function of temperature. The concentrations of both phases follow a coexistence curve that ends with a dissolution critical point that belongs to the same class as the fluids. The order parameter is the concentration and pressure corresponds to the osmotic compressibility. Choosing liquids with very similar densities can then allow gravitational effects to be compensated for. This is the case for isobutyric acid and water or cyclohexane and methanol, for example. Nevertheless, near the critical point, the capillary length always tends to zero, so closely matching the densities cannot remove the gravitational effects. A further density-matching step involves the use of a mixture of isotopes for one of the components, such as partially deuterated cyclohexane with methanol. Experiments that get as close as $1\,\mathrm{mK}$ from T_c have been performed in this way, free from gravitational effects [33].

1.9.3 Free Fall

In free fall towers with heights H that typically range from 1 to 100 m high, such as Zarm in Bremen (Germany) or CEA in Grenoble (France), the payload is left to accelerate under high vacuum for a time t_F such that

$$t_F = \left(\frac{2H}{g_0}\right)^{1/2}, \tag{1.68}$$

which corresponds to a few seconds. When a mine shaft in Japan (1,000 m deep) was used, the duration of weightlessness was increased to 14 s. The remaining disturbances to the acceleration, expressed in units of $g_0 = 9.81\,\mathrm{m^2\,s^{-1}}$ (the gravitational acceleration at the Earth's surface), are very weak (less than $10^{-5}\,g_0$). Aircraft like the CNES-ESA Zero-g Airbus A300 from Novespace in Bordeaux (France) can achieve parabolic trajectories that allow approximately 20 s of free fall with disturbances on the order of a few $10^{-2}\,g_0$. Longer free fall durations, from 120 to 900 s, can be provided by sounding rockets like the MiniTEXUS, TEXUS, MAXUS, and MASER rockets that are launched in Kiruna (Sweden). The level of weightlessness is excellent (on the order of $10^{-5}\,g_0$) over the total duration of free fall.

1.9.4 Satellites

For a satellite, the gravitational force is counterbalanced by the centrifugal force. The duration of weightlessness can range from days to months or years. A strict balance is realized only at the center of mass of the satellite. For a satellite payload located at a distance X from the center of mass, where the satellite is a distance h from the surface of the Earth, the residual acceleration δg, expressed in units of g_0, is

$$\frac{\delta g}{g_0} = \frac{X}{R+h}, \tag{1.69}$$

where R is the Earth's radius. For typical values ($X \simeq$ 1–10 m, $h \simeq 200$ km, $R \simeq 6{,}370$ km, an acceleration of 10^{-7} to $10^{-6}\,g_0$ is permanently imposed on the payload. Although this is very weak, its permanent nature can induce sedimentation. Mechanical vibrations in unmanned satellites lead to noise of around $10^{-4}\,g_0$. Time-dependent acceleration due to human activities in manned satellites (the former MIR orbital station, or the International Space Station) peaks at 10^{-3} to $10^{-2}\,g_0$, with an average of around $10^{-3}\,g_0$.

1.9.5 Magnetic Compensation

A volumic force can also be provided by a magnetic field gradient. In paramagnetic matter (like O_2) or diamagnetic matter (the most usual case), the volumic force as induced by a magnetic field gradient can counterbalance the gravitational force. The force per mass density (the "magnetic acceleration" $\mathbf{a}_m$) is

$$\mathbf{a}_m = \frac{\alpha}{\mu_0}\nabla\left(\frac{1}{2}\mathbf{B}^2\right). \tag{1.70}$$

Here, $\mathbf{B}$ is the magnetic field vector, $\mu_0 = 4\pi \times 10^{-7}\,\mathrm{H\,m^{-1}}$, and α is the magnetic susceptibility vector per mass density.

The magnetic force can also be used to vary the effective acceleration that acts on the sample. However, there are obvious experimental difficulties in this case, since a high (supraconductive) magnetic field is needed. Such a solution was evaluated in detail for two-phase p-hydrogen [73]; a deeper discussion of this is given in Chap. 23, Sect. 23.3. Exact compensation for the gravitational forces appears to be limited by the inhomogeneity of the magnetic force. It can be shown [59] that complete bulk homogenization violates the Maxwell equations of electromagnetism. Thus, the solution can only be approximate. For the simplest situation, where the magnetic gradient at the end of a magnetic coil is used, only one or two compensation points can exist. If z is the axis of symmetry of the magnetic coil, the compensation occurs at a height z_0, with a resulting acceleration that cancels as $(z - z_0)^2$. In addition, one finds that there is a radial acceleration field centered at z_0 that increases linearly with the distance from this point. For small samples a few millimeters in size, residual acceleration at the wall sample does not exceed $10^{-2}\,g_0$.

Such inhomogeneities can be quantified using an associated capillary length ℓ_c^*, corresponding to an effective remaining gravity field $\mathbf{g}^* = \mathbf{g} + \mathbf{a}_m$. The effects from this residual field remain negligible when ℓ_c^* is larger than the size of the cell. Near the critical point, ℓ_c^* tends to zero and the vapor–liquid interface can be deformed by the residual magnetic forces. For normal H_2, radial deformations are observed for temperatures lower than 1 mK ($\tau < 3 \times 10^{-5}$) at length scales on the order of 10 mm.

1.10 Background and Critical Behavior

1.10.1 Statics

The divergence (or convergence to zero) of the thermodynamic parameters and transport coefficients is due to the long-range part of the intermolecular correlation function, and is called the "critical" contribution. The molecular short-range part is nevertheless always present too, and acts as a regular background contribution. We will consider this part to be the "non-ideal gas" or "mean-field" contributions to the real fluid behavior. It is generally useful to separate out these two contributions. This separation is classically performed by assuming an additive form where the critical part is added to the regular part, as described below and already used for the case of the thermal conductivity coefficient (see Sect. 1.5). However, this is not always the case. For instance, the background contribution to the shear viscosity is a multiplicative prefactor of the critical contribution [17, 39] [(see Eq. (1.91)]; in this case, one can consider adding the logarithms of the selected parameters.

Anyway, the basic method is to fit the experimental data for a given quantity $Y(x)$ by using two additive regular (subscript: reg) and critical (subscript: c) contributions. Here, the reduced variable x stands for temperature ($x = \tau$) at $\rho = \rho_c$, or for density ($x = \frac{\rho}{\rho_c} - 1$) at $T = T_c$. For the sake of simplicity, we will only consider the case $x = \tau$ at $\rho = \rho_c$

$$Y = Y_c(\tau) + Y_{\text{reg}}(T) = Y_c(\tau) + Y_{\text{reg}}(\tau) \tag{1.71}$$

The regular contribution is an analytical function $Y_{\text{reg}}(T)$, or identically $Y_{\text{reg}}(\tau)$, which can be expanded around T_c:

$$\begin{aligned} Y_{\text{reg}}(T) &= A_0 + A_1 T + A_2 T^2 + \cdots \\ Y_{\text{reg}}(\tau) &= B_0 + B_1 \tau + B_2 \tau^2 + \cdots \end{aligned} \tag{1.72}$$

The critical contribution is nonanalytical. It takes the form of a power law with a generic universal exponent $\pm\psi$ and a system-dependent critical amplitude $\Psi_{Y0}^{+,-}$ along a few particular thermodynamic paths: the isochoric path $\rho = \rho_c$, above (indicated by the subscript +) or below T_c (indicated by the subscript −), which then includes the coexisting phase line, and the isothermal path $T = T_c$ (not considered here):

$$Y_c(|\tau|) = \Psi_{Y0}^{+,-} |\tau|^{\pm\psi} \tag{1.73}$$

Sometimes a critical contribution with a zero exponent appears, corresponding to a constant critical contribution (also call critical background). This is the case for the (negative) critical-fluctuation-induced constant of the specific heat at constant volume c_V [13, 14, 45], noted B_{cr}. It is of the same order as the leading specific heat amplitudes noted $A^{+,-}$ and of opposite sign (in suitable units), and that is thus the critical background to the specific heat at constant volume c_V [see below Eq. (1.79)].

As previously mentionned, the critical contribution of Eq. (1.71) is only valid asymptotically close to the critical point, where fluctuations at all scales renormalize the value of the critical exponent from mean field (where fluctuations are neglected) to non mean field. When the system is not asymptotically close to the critical point, some deviations must be accounted for. The critical behavior is complicated by a so-called crossover function (see for example [27]), which reproduces the Ising-like asymptotic behavior very close to the critical point, and exhibits mean field type critical behavior far from it [2, 63]. However, for most cases, it is enough to describe the crossover function through "non-analytical corrections" or "corrections-to-scaling" terms [70]:

$$Y_c\left(|\tau|\right) = \Psi_{Y0}^{+,-}\,|\tau|^{\pm\psi}\left(1 + a_Y^{(1)}\,|\tau|^{\Delta} + a_Y^{(2)}\,|\tau|^{2\Delta} + \cdots\right) \quad (1.74)$$

Unlike the $a_Y^{(i)}$, the exponent Δ is universal ($\Delta \simeq 0.502$). In most cases, it is enough to fit the data to the first term of the series. As noted above in Sect. 1.4, the functions $Y_c\left(|\tau|\right)$ for statics are not independent and knowledge of two of them enables the other functions to be readily determined. For the leading power law terms, such a situation introduces universal combinations between the amplitude terms $\Psi_{Y0}^{+,-}$. Only two of them (for example, the coexistence curve non-dimensional amplitude (noted B) and the isothermal compressibility non-dimensional amplitude noted Γ^+) are independent [2, 63]. For the non-analytical corrections, that induces universal ratios between the first order nonanalytical amplitudes. For instance, the ratio of the first order nonanalytical amplitudes of ξ and κ_T is $\frac{a_\xi^{(1)}}{a_\chi^{(1)}} \simeq 0.65$. Table 1.3 summarizes critical behavior (power law and corrections to scaling to within first order) and Table 1.4 shows the universal ratios between leading amplitudes and first order correction amplitudes to scaling. In these tables, we make use of the well-known relationship between $c_p - c_V$ and κ_T or β_p [Eq. (1.28)], as previously mentioned. The two next sections give additional details of interest for the following analyses in the particular cases of the ratio $\gamma_0 = \frac{c_p}{c_V}$ of the specific heats and the velocity of sound c_s.

1.10.2 Ratio of Specific Heats at Constant Pressure and Constant Volume

As already mentioned in Sect. 1.1 for the case of the van der Waals equation of state, the ratio $\gamma_0 = \frac{c_p}{c_V}$ of the specific heats is classically deduced from Eq. (1.28). Considering for example the behavior of γ_0 when $\tau \to 0$ along the critical isochore in the one-phase domain ($\tau > 0$), that needs to use the following expressions, which account for the respective behaviors of κ_T, c_V and c_p in their respective non-dimensional forms (see Table 1.3)

$$\kappa_T^* = p_c\kappa_T = \Gamma^+\tau^{-\gamma}\left(1 + a_\Gamma^{(1)}\tau^{\Delta} + a_\Gamma^{(2)}\tau^{2\Delta} + \cdots\right) \quad (1.75)$$

$$c_V^* = \left(\frac{\rho_c T_c}{p_c}\right)c_V = c_{V,b}^* + \frac{A^+}{\alpha}\tau^{-\alpha}\left[1 + \alpha\left(a_c^{(1)}\tau^{\Delta} + a_c^{(2)}\tau^{2\Delta} + \cdots\right)\right] \quad (1.76)$$

with

$$c_p^* = \left(\frac{\rho_c T_c}{p_c}\right)c_p = c_{p,b}^* + c_{p,c}^* = c_V^* + c_{p,c}^* \quad (1.77)$$

and

$$c_{p,c}^{*} = Z_c \left[\frac{T_c}{p_c} \left(\frac{\partial p}{\partial T} \right)_{\rho} \right]^2 (1+\tau)\, p_c \kappa_T = c_{p0}^{+} (1+\tau)\, \tau^{-\gamma} \left(1 + a_{\Gamma}^{(1)} \tau^{\Delta} + a_{\Gamma}^{(2)} \tau^{2\Delta} + \cdots \right) \quad (1.78)$$

where $Z_c = \frac{p_c \bar{m}}{R \rho_c T_c}$ is the critical compressibility factor and where $\frac{T_c}{p_c}\left(\frac{\partial p}{\partial T}\right)_{\rho}$ takes a constant critical value at the critical point. In Eq. (1.76), $c_{V,b}^{*}$ is a non-dimensional term including the regular contribution of an analytical function $c_{V,\text{reg}}(T)$ and the critical contribution of the (negative) constant critical background B_{cr}, generally written in the additive form $c_{V,b}^{*} = B_{cr} + \left(\frac{\rho_c T_c}{p_c}\right) c_{V,\text{reg}}(T)$. Therefore, when the regular behavior is expanded in an analytical power series of τ, $c_{V,b}^{*}$ can be written as follows

$$c_{V,b}^{*} = B_{cr} + c_{V,b,0}^{*} + c_{V,b,1}^{*}\tau + c_{V,b,2}^{*}\tau^2 + \cdots \quad (1.79)$$

In Eq. (1.77), c_V^{*} is considered as being the background contribution $c_{p,b}^{*}$ to c_p^{*}, while the critical contribution of Eq. (1.78) (with $c_{p0}^{+} = Z_c \left[\frac{T_c}{p_c} \left(\frac{\partial p}{\partial T} \right)_{\rho} \right]^2 \Gamma^{+}$) is similar to the singular behavior of κ_T^{*} [see Eq. (1.75)]. Introducing Eqs. (1.75), (1.76) and (1.79) in Eq. (1.28), it directly follows

$$\gamma_0 = 1 + Z_c \left[\frac{T_c}{p_c} \left(\frac{\partial p}{\partial T} \right)_{\rho} \right]^2 (1+\tau) \frac{\Gamma^{+}\tau^{-\gamma}\left(1+a_{\Gamma}^{(1)}\tau^{\Delta}+\cdots\right)}{\frac{A^{+}}{\alpha}\tau^{-\alpha}\left(1+\alpha a_C^{(1)}\tau^{\Delta}+\cdots\right)+B_{cr}+c_{V,b0}^{*}+c_{V,b1}^{*}\tau+c_{V,b2}^{*}\tau^2+\cdots} \quad (1.80)$$

The derivative $\left(\frac{\partial p}{\partial T}\right)_{\rho}$ along the selected isochoric thermodynamic path $\rho = \rho_c$ (with $T > T_c$) can be approximated by its contant critical value approaching the critical point. The form of Eq. (1.80) can be also rearranged to highlight the asymptotic divergence as follows

$$\gamma_0 = 1 + Z_c \left[\frac{T_c}{p_c} \left(\frac{\partial p}{\partial T} \right)_{\rho} \right]^2 \frac{\alpha \Gamma^{+}}{A^{+}} (1+\tau)\, \tau^{-\gamma+\alpha} \left[1 - \frac{A^{+}}{\alpha} \left(B_{cr} + c_{V,b0}^{*} \right) \tau^{\alpha} + \left(a_{\Gamma}^{(1)} - \alpha a_C^{(1)} \right) \tau^{\Delta} + \cdots \right] \quad (1.81)$$

Equation (1.81) written such as the additive form $\gamma_0 = \gamma_{0b} + \gamma_{0c}$, shows that the background contribution of the specific heat at constant volume is mainly given by $\gamma_{0b} = 1$, while the leading critical contribution can be expressed when $\tau \to 0$ as follows

$$\gamma_{0c} = \gamma_{00}^{+} (1 + \tau) \tau^{-\gamma+\alpha} \tag{1.82}$$

where the prefactor γ_{0c}^{+} of the linear temperature dependence of the leading amplitude is defined as follows

$$\gamma_{00}^{+} = Z_c \left[\frac{T_c}{p_c} \left(\frac{\partial p}{\partial T} \right)_{\rho} \right]^2 \frac{\alpha \Gamma^{+}}{A^{+}} \tag{1.83}$$

1.10.3 Velocity of Sound

The quadratic value of the velocity of sound, c_s, is classically written as

$$c_s^2 = \left(\frac{\partial p}{\partial \rho} \right)_S = \frac{1}{\rho \kappa_S} \tag{1.84}$$

For ideal gases,

$$c_{s,\mathrm{IG}}^2 = \gamma_{0,\mathrm{IG}} \left(\frac{R}{\bar{m}} \right) T_c \tag{1.85}$$

where $\gamma_{0,\mathrm{IG}} = \frac{c_{p,IG}}{c_{V,IG}} = 1.4$ is the ratio of the specific heats at constant pressure and at constant volume for ideal gases.

For real fluids, making use of the relation

$$\gamma_0 = \frac{c_p}{c_V} = \frac{\kappa_T}{\kappa_S} \tag{1.86}$$

and Eq. (1.28), we obtain

$$c_s^2 = \frac{1}{\rho \kappa_T} + \frac{T}{\rho^2 c_V} \left(\frac{\partial p}{\partial T} \right)_{\rho}^2 \tag{1.87}$$

We now highlight the leading divergences of κ_T and c_V along the critical isochore (see Table 1.3):

$$c_s^2 = \left[\frac{p_c}{\rho_c \Gamma^{+}} \right] \tau^{\gamma} + \left[\left(\frac{\partial p}{\partial T} \right)_{\rho}^2 \frac{\alpha T_c^2}{\rho_c p_c A^{+}} \right] (1 + \tau) \tau^{\alpha} \tag{1.88}$$

The first term tends to zero near T_c with a stronger exponent ($\gamma = 1.239$) than the second term ($\alpha = 0.110$) (Table 1.3). As a result, the asymptotic behavior of the velocity of sound along the critical isochore in the one-phase domain ($\tau > 0$) tends to zero as

$$c_s \simeq \left[\left(\frac{\partial p}{\partial T} \right)_\rho^2 \frac{\alpha T_c^2}{\rho_c p_c A^+} \right]_c^{1/2} (1+\tau)\,\tau^{\alpha/2} = c_{s0}^+ (1+\tau)\,\tau^{\alpha/2} \tag{1.89}$$

The exponent $\frac{\alpha}{2} = 0.055$ is very small. The leading amplitude shows a linear temperature dependence $c_{s0}^+ (1+\tau)$, with $c_{s0}^+ = \left[\left(\frac{\partial p}{\partial T} \right)_\rho^2 \frac{\alpha T_c^2}{\rho_c p_c A^+} \right]_c^{1/2}$. Thus, within the typical experimental range, the velocity of sound decreases only slightly. For the critical isochore of CO_2, c_s starts from about 100 m s^{-1} a few mK above the critical temperature and reaches 146 m s^{-1} at 1 K from T_c.

1.10.4 Dynamics

The transport coefficients also obey power law divergence. In Table 1.5, only the shear viscosity μ, the thermal conductivity Λ and the thermal diffusivity D_T are considered. In Table 1.5, we also indicate the expected divergence of the bulk viscosity μ_B as given in [36, 38, 58]. However, this has not been experimentally assessed yet, and is still subject to debate [19, 20, 49, 50]. In order to infer the critical and background terms of μ, Λ and D_T, we make use of Eqs. (1.40) and (1.44). In the mode coupling theory of transport properties, the shear viscosity and the thermal diffusivity are linked by the two definitions *of* D_T given by Eqs. (1.40) and (1.44), leading to identify the following singular behaviors

$$D_T = \frac{\Lambda}{\rho c_p} \equiv \frac{k_B T}{6\pi \mu \xi} = D \sim \tau^{\nu + Y_\eta} \tag{1.90}$$

close to the critical point [see Eq. (1.40)]. As previously mentionned introducing Eq. (1.44), the thermal diffusion critical term $D_{T,c} = \frac{\Lambda_c(\tau)}{c_P(\tau)}$ is the ratio of the critical thermal conductivity behavior $\Lambda_c \sim \tau^{-Y_\Lambda} \sim \tau^{-0.57}$, and the whole specific heat behavior $c_p \sim \tau^{-\gamma} \sim \tau^{-1.24}$, leading to the power law annulation $D_{T,c} = D_0^{+,-} \tau^{\gamma - Y_\Lambda} \sim \tau^{0.67}$ with a value of the critical exponent smaller than the one of the the power law annulation $D_{T,b} = D_{b0}\tau^{\gamma} \sim \tau^{1.24}$ of the background term $D_{T,b} = \frac{\Lambda_b(\tau)}{c_P(\tau)}$, which exhibits the power law behavior of $c_p^{-1} \sim \tau^{1.24}$. Here it was made use of the scaling law $Y_\Lambda + Y_\eta = \nu(1-\eta)$, with $Y_\eta = 0.04$, $\nu = 0.630$ and $\eta \simeq 0.034$.

The behavior of the shear viscosity is generally expressed as the product of a regular viscosity μ_b and a critical divergence with a very small exponent $Y_\eta = 0.04$. In the mode-coupling theory, a cut-off wavevector Q_0 is introduced. This corresponds to the upper limit of the mode-coupling integral for viscosity enhancement due to critical fluctuations and the relative influences of the background and critical thermal conductivities on the fluctuation decay rate [10]. The resulting critical behavior of μ is

$$\mu = \mu_b(Q_0\xi)^{x_\eta} = \mu_b(Q_0\xi_0)^{x_\eta}\tau^{-Y_\eta}. \tag{1.91}$$

The critical exponent $x_\eta = \frac{Y_\eta}{\nu} = 0.067$. As x_η is so small, the value of $(Q_0\xi_0)^{x_\eta}$ varies only weakly for different fluids and remains practically equal to $(Q_0\xi_0)^{x_\eta} = 0.77$. For the sake of simplicity, we will therefore consider only the constant quantity

$$\mu_{b0} = \mu_b(Q_0\xi_0)^{x_\eta}, \tag{1.92}$$

where we will adjust μ_b to the experimental critical value of the viscosity, as it contains both the critical and background parameters.

Finally, for the thermodynamic parameters and transport coefficients, fitting experimental data to Eqs. (1.71) and (1.74) (see also Table 1.2) enables both the regular value at the critical point $[Y_{reg}(x = x_c)]$ and the critical contribution $Y(x - x_c)$ to be deduced. Of course, the accuracy of the determination is linked to the precision of the data and the temperature region in which they were measured. In particular, it is important that the data were taken sufficiently close to the critical point to be able to determine the critical behavior, and far enough from it to account for the regular behavior. As the critical exponents are now well known, their Ising values will generally be imposed on the fit, which greatly increases the accuracy of the critical and regular amplitudes.

We now describe in more detail how such key quantities can be deduced from experiments. We consider five model fluids covering a broad temperature range: *n*- and *p*-H_2 ($T_c \approx 33$ K), Xe ($T_c \approx 289$ K), CO_2 ($T_c \approx 304$ K), SF_6 ($T_c \approx 319$ K), and H_2O ($T_c \approx 647$ K). We show the partitioning between the regular and critical behavior, including some first-order corrections to scaling as shown in Table 1.3, for the following quantities:

- *Statics (thermodynamic properties)*
  - Isothermal compressibility $\kappa_T = \frac{\chi_T}{\rho^2}$
  - Specific heat at constant pressure c_P
  - Specific heat at constant volume c_V
  - Ratio $\gamma_0 = \frac{c_P}{c_V}$
  - Correlation length ξ
- *Dynamics (transport coefficients)*
  - Shear viscosity μ
  - Thermal diffusivity D_T
  - Thermal conductivity Λ

An example is shown in Fig. 1.8. The results for the fluids cited above are provided in Table 1.6.

From these data, it is possible to infer the background and critical terms of the most important parameters. In the following chapters, we will scale these quantities by the corresponding values extrapolated from the ideal gas state. This means that we will normalize any value $Y(\tau)$ by its background value $Y_b(0)$ at the critical point:

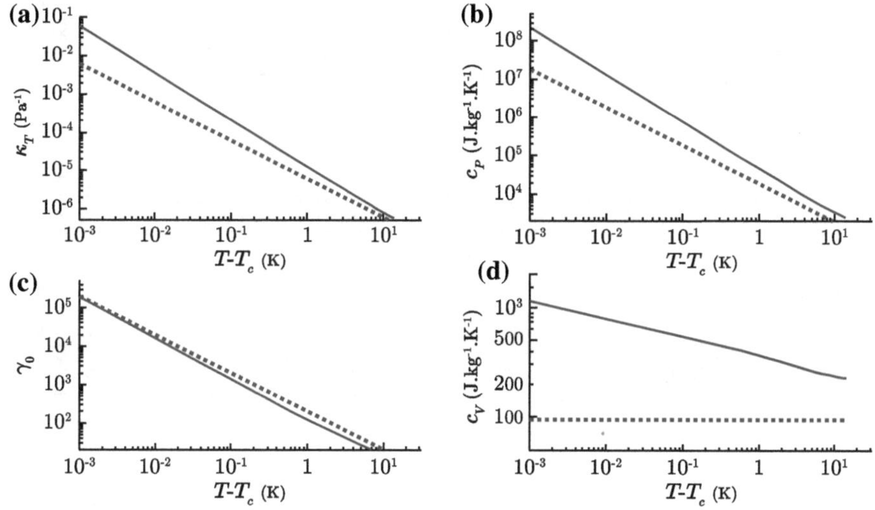

Fig. 1.8 Temperature behavior at $\rho = \rho_c$ of the isothermal compressibility κ_T, the specific heats at constant pressure c_p and at constant volume c_V, and the ratio of specific heats $\gamma_0 = \frac{c_p}{c_V}$ (where xenon is selected as an example here). The *solid line* refers to the experimental data fitted by theoretical Ising laws [27], while the *dotted line* corresponds to the mean-field behavior (see Sect. 1.11)

$$Y(\tau) \Longrightarrow \frac{Y(\tau)}{Y_b(0)}. \tag{1.93}$$

1.11 Fitting Experimental Values Within a van der Waals Approach

In the work developed in the following chapters, the analytical van der Waals equation of state [see Eqs. (1.5), (1.9)] will often be used. This equation captures the main qualitative aspects of static fluid behavior and enables much simpler calculations to be performed. However, it induces unavoidable discrepancies from the real fluid parameters, as already noted above in Sect. 1.2. We will quantify these discrepancies below and describe how to fit the main fluid parameters using a van der Waals approach along the critical isochore (ignoring the superscripts +, − that distinguish the $T > T_c$ and $T < T_c$ ranges, respectively).

1.11.1 Critical Coordinates T_c, ρ_c, $p_{c,vdW}$

As explained in Sect. 1.1, due to the relationship between p, T, and ρ, one of the three coordinates for the critical point will differ from the real value. We choose to

Table 1.6 Useful numerical values for the parameters of various fluids near their critical points

Parameter	Units	p-H_2	n-H_2	Xe	CO_2	SF_6	H_2O
Molar mass $\bar{m}$	$kg\,mol^{-1}$	0.020159	0.020159	0.131313	0.044009	0.14605	0.018048
Ideal gas constant $\frac{R}{\bar{m}}$	$J\,kg^{-1}\,K^{-1}$	4124.42	4124.42	63.3176	188.925	56.9284	460.683
Critical temperature T_c	K	32.976	33.19	289.74	304.14	318.735	647.067
Critical pressure p_c	10^6 Pa	1.2928	1.3152	5.84	7.3753	3.754	22.046
Critical density ρ_c	$kg\,m^{-3}$	31.43	30.11	1110	467.8	742	322.788
$\left(\frac{\partial p}{\partial T}\right)_{\rho_c}$ at $T = T_c$	$10^5\,Pa\,K^{-1}$	1.8887	1.8397	1.1900	1.7000	0.8350	2.67632
B	–	1.61	1.61		1.60	1.60	
σ_0	$10^{-2}\,N\,m^{-1}$	0.691	0.542	4.38	6.72	4.42	21.1
B_0	$J\,kg^{-1}\,K^{-1}$	−16477	−11825	−341	−1400	144	−2425
B_1	$J\,kg^{-1}\,K^{-1}$	0	54	2	110	−13	6
$c_{V,IG}$	R	$\frac{5}{2}$	$\frac{5}{2}$	$\frac{3}{2}$	$\frac{5}{2}$	$\frac{5}{2}$	$\frac{5}{2}$
A_0^+	$J\,kg^{-1}\,K^{-1}$	21474	15434	373.549	1566	491.66	4089.55
$\frac{\rho_c T_c}{p_c} A_0^+ = \mathcal{A}$	–	1.8937	1.29002	2.26287	3.32395	3.407	4.2619
a_C^+	–	1.34	1.34	1.55	1.71	1.73	1.83
ξ_0^+	10^{-10} m (Å)	1.54	1.745	1.813	1.500	1.892	1.233
a_ξ^+	–	1.20	1.20	1.395	1.54	1.55	1.65
Γ_0^+	$10^{-9}\,Pa^{-1}$		57.5602	9.9754	6.415	12.203	9.9754
$p_c\Gamma_0^+ = \Gamma^+$	–		0.075703	0.0582564	0.04731	0.04581	0.04813
a_Γ^+	–		1.36	1.55	1.72	1.73	1.83
μ_{b0}	10^{-6} Pa s	2.7	4.6	51.3	32.7	39.9	40.7
c_{p0}^+	$J\,kg^{-1}\,K^{-1}$		2147	36.873	120.53	36.547	313.493
D_0^+	$10^{-8}\,m^2\,s^{-1}$	4.4	5.11	2.92	5.997	4.020	2.9208
D_{b0}^+	$10^{-8}\,m^2\,s^{-1}$	47.4	10.67	4.447	7.981	13.20	4.4467
Λ_0	$10^{-3}\,W\,m^{-1}\,K^{-1}$		3.3034	1.19547	3.38138	1.09025	12.4305
Λ_{b0}	$10^{-2}\,W\,m^{-1}\,K^{-1}$		6.8975	1.82000	4.4998	3.5799	21.4000

consider the real T_c and ρ_c, and calculate a van der Waals critical pressure $p_{c,\text{vdW}}$ according to Eq. (1.12); i.e.,

$$p_{c,\text{vdW}} = \frac{3}{8}\rho_c T_c \frac{R}{\bar{m}}. \tag{1.94}$$

This critical pressure will be larger than the real value p_c as shown in Tables 1.6 and 1.7. In addition to T_c, ρ_c, and $p_{c,\text{vdW}}$, the van der Waals equation allows the molar (specific) coefficients $\bar{a}$ (a) and $\bar{b}$ (b) to be evaluated, with the covolumes ($\bar{b}$, b) determined from the critical densities [see for example Eq. (1.14)], and the attractive interaction parameters ($\bar{a}, a$) from the critical temperature [see for example Eq. (1.12)].

1.11.2 Derivative $\left(\frac{\partial p_{vdW}}{\partial T}\right)_\rho$

The derivative $\left(\frac{\partial p_{\text{vdW}}}{\partial T}\right)_\rho$ comes from the derivation of the van der Waals equation of state [Eq. (1.22)], and its value along the critical isochore is lower than the experimental value (see Tables 1.6 and 1.7):

$$\left(\frac{\partial p_{\text{vdW}}}{\partial T}\right)_{\rho_c} = 4\,\frac{p_{c,\text{vdW}}}{T_c}. \tag{1.95}$$

1.11.3 Coexistence Curve

The amplitude of the coexistence curve B_{vdW} (=2) can be readily obtained from Eq. (1.23), and is reported in Table 1.7.

1.11.4 Isothermal Compressibility $\kappa_{T,vdW}$

The isothermal compressibility $\kappa_{T,\text{vdW}}$ exhibits the following τ^{-1} temperature variation:

$$\kappa_{T,\text{vdW}} = \Gamma_{0,\text{vdW}}\tau^{-1} = \left(\frac{1}{6p_{c,\text{vdW}}}\right)\tau^{-1}, \tag{1.96}$$

which differs from the experimental behavior. Moreover, the amplitude $\Gamma_{0,\text{vdW}} = \frac{1}{6p_{c,\text{vdW}}}$ is smaller than the real leading amplitude Γ_0. In Fig. 1.8, it is apparent that,

Table 1.7 Useful numerical values for fluids near their critical points (van der Waals/mean-field approximation)

Parameter		Unity	H_2	Xe	CO_2	CO_2-bis	SF_6
Critical temperature	T_c	K	33.19	289.74	304.14	304.14	318.735
Critical density	ρ_c	$kg\,m^{-3}$	30.11	1110	467.8	467.8	742
Critical pressure	$p_{c,\mathrm{vdW}}$	10^6 Pa	1.54565	7.63637	10.0799	10.0799	5.04886
$\left(\frac{\partial p_{\mathrm{vdW}}}{\partial T}\right)_{\rho_c}$ at $T = T_c$	$4\frac{p_{c,\mathrm{vdW}}}{T_c}$	$10^5\,Pa\,K^{-1}$	1.86279	1.05424	1.32569	1.32569	0.633613
$\Delta\tilde{\rho}_{\ell,v,\mathrm{vdW}}$ critical amplitude	B_{vdW}		2	2	2	2	2
$\kappa_{T,\mathrm{vdW}}$ critical amplitude	$\Gamma_{0,\mathrm{vdW}} = \frac{1}{6p_{c,\mathrm{vdW}}}$	$10^{-8}\,Pa^{-1}$	10.7830	2.18254	1.65346	1.65346	3.3011
$c_{V,\mathrm{vdW}}$ critical amplitude	$\frac{3\mathrm{or}5}{2}\frac{R}{\overline{m}}$	$J\,kg^{-1}\,K^{-1}$	10311	94.9764	472.313	472.313	142.321
ξ_{MF} critical amplitude	$\xi_{0,\mathrm{MF}}$	10^{-10} m	0.9624	1.16353	1.07798	2.59	1.37874
$D_{T,\mathrm{MF}}$ critical amplitude	$D_{0,\mathrm{MF}}$	$10^{-8}\,m^2\,s^{-1}$	7.1260	3.5538	6.317	2.629	4.2418
Background	$D_{b0,\mathrm{MF}}$	$10^{-7}\,m^2\,s^{-1}$	4.7277	1.9804	3.7255	0.257	6.3016
$c_{p,c,\mathrm{vdW}}$ critical amplitude	$c_{p0,\mathrm{vdW}} = \frac{R}{\overline{m}}$	$J\,kg^{-1}\,K^{-1}$	4124.42	63.3176	188.925	188.925	56.9284
Background	$c_{V,\mathrm{vdW}} = \frac{3\text{ or }5}{2}\frac{R}{\overline{m}}$	$J\,kg^{-1}\,K^{-1}$	10311	94.9764	472.313	472.313	142.321
Λ_{MF} critical amplitude	$\Lambda_{0,\mathrm{MF}}$	$10^{-3}\,W\,m^{-1}\,K^{-1}$	7.5300	1.9100	4.0849	1.7	1.3322
Background	$\Lambda_{b0,\mathrm{MF}}$	$10^{-2}\,W\,m^{-1}\,K^{-1}$	5.87123	1.39187	3.29259	0.227	2.66185
$\mu_{b,\mathrm{MF}}$ background	μ_{b0}	10^{-6} Pa s	2.7	51.3	32.7	32.7	39.9
$\bar{a}$		$m^3\,mol^{-1}\,K$	0.020785	0.32061	0.267632	0.267632	0.58683
$\bar{b}$		$10^{-5}\,m^3\,mol^{-1}$	2.23171	3.94333	3.13588	3.13588	6.56111
$a = \frac{\bar{a}}{\overline{m}^2}$		$m^3\,kg^{-1}\,K$	51.14605	18.56352	138.1831	138.1831	27.51118
$b = \frac{\bar{b}}{\overline{m}}$		$10^{-5}\,m^3\,kg^{-1}$	110.7054	30.0300	71.25543	71.25543	44.9237

The alternative values listed under CO_2-bis correspond to the values adopted in the simulations throughout this book (see text) for xenon fluid, the value of $\kappa_{T,\mathrm{vdW}}$ is always below the experimental value, with a deviation from T_c that increases from a factor of 2 at 1 K to a factor of 10 at 1 mK.

1.11.5 Specific Heat at Constant Volume $c_{V,vdW}$

Let us now consider the specific heat at constant volume. This does not exhibit a critical divergence [it shows a jump of $\frac{9}{2}R$ when crossing the critical point; see Eq. (1.27)]. We take the ideal gas value $c_{V,\mathrm{IG}}$ above T_c; that is,

$$c_{V,\mathrm{vdW}} = c_{V,\mathrm{IG}} = \frac{3}{2}\frac{R}{\bar{m}} \text{ or } \frac{5}{2}\frac{R}{\bar{m}}, \tag{1.97}$$

where the numerator is 3 for monoatomic fluids (such as Xe) and 5 for polyatomic fluids (such as CO_2). Figure 1.8 shows the deviation between the experimental value and the van der Waals value for xenon. The discrepancy is similar to that of the isothermal compressibility: the van der Waals value is lower than T_c by a factor of 4 at 1 K and by a factor of 20 at 1 mK.

1.11.6 Specific Heat at Constant Pressure $c_{p,vdW}$

The specific heat at constant pressure is classically deduced from Eq. (1.28), leading to the additive form $c_{p,\mathrm{vdW}} = c_{p,b,\mathrm{vdW}} + c_{p,c,\mathrm{vdW}}$. It directly follows that the background part $c_{p,b,\mathrm{vdW}}$ of the specific heat is simply

$$c_{p,b,\mathrm{vdW}} = c_{V,\mathrm{vdW}} = c_{V,\mathrm{IG}} = \left(\frac{3 \text{ or } 5}{2}\right)\frac{R}{\overline{m}}. \tag{1.98}$$

where $c_{V,\mathrm{IG}} = \left(\frac{1}{\gamma_{0,\mathrm{IG}}-1}\right)\frac{R}{\overline{m}}$ does not exhibit any critical divergence. The critical part is proportional to $\kappa_{T,\mathrm{vdW}}$. Following Eqs. (1.94) and (1.96), the critical behavior of $c_{p,c,\mathrm{vdW}}$ can be written as

$$c_{p,c,\mathrm{vdW}} = c_{p0,\mathrm{vdW}}\,(1+\tau)\,\tau^{-1}, \tag{1.99}$$

where

$$c_{p0,\mathrm{vdW}} = \left[Z_c\left(\frac{p_{c,\mathrm{vdw}}}{T_c}\frac{\partial p_{\mathrm{vdW}}}{\partial T}\right)^2_{\rho_c}\right]\Gamma^+_{\mathrm{vdW}}\frac{R}{\overline{m}} = \frac{R}{\overline{m}} = r.$$

We note that the amplitude $c_{p0,\mathrm{vdW}}$ differs from $c_{p,\mathrm{IG}} = \left(\frac{\gamma_{0,\mathrm{IG}}}{\gamma_{0,\mathrm{IG}}-1}\right)\frac{R}{\overline{m}}$, which corresponds to the specific heat at constant pressure of an ideal gas. In the case of xenon, the plot in Fig. 1.8 of the experimental and van der Waals behaviors of the specific heat at constant pressure shows a deviation from T_c similar to the one for the isothermal compressibility, ranging from a factor of 3 at 1 K to a factor of 10 at 1 mK.

1.11.7 The Ratio $\gamma_{0,vdW}$ of Specific Heat Coefficients

The ratio $\gamma_{0,\mathrm{vdW}} = \frac{c_{p,c,\mathrm{vdW}}}{c_{V,\mathrm{vdW}}}$ can be expressed as

$$\gamma_{0,\mathrm{vdW}} = 1 + \gamma_{0,c,\mathrm{vdW}} \tag{1.100}$$

where

$$\gamma_{0,c,\mathrm{vdW}} = \left(\gamma_{0,\mathrm{IG}} - 1\right)(1+\tau)\,\tau^{-1} \tag{1.101}$$

is the mean field singular part equivalent to the Ising one of Eq. (1.82), with $\gamma^{+}_{00,\mathrm{vdW}} = \gamma_{0,\mathrm{IG}} - 1$, $\gamma_{\mathrm{MF}} = 1$ and $\alpha_{\mathrm{MF}} = 0$. In the case of xenon, the plots in Fig. 1.8 for the experimental and the van der Waals specific heat ratios show only small differences between the curves, with the van der Waals data being higher than the experimental data by a factor 1.5 at 1 K but the two curves having the same values at 1 mK.

1.11.8 Velocity of Sound $c_{s,vdW}$

For a unit mass of a van der Waals fluid, Eq. (1.87) takes the following general form:

$$c^2_{s,\mathrm{vdW}} = \left(\frac{r}{c_V} + 1\right)\frac{rT}{(1-\rho b)^2} - 2a\rho. \tag{1.102}$$

However, ignoring in Eq. (1.87) the contribution of the term $\propto \frac{1}{\kappa_T} \sim \tau$, which goes to zero for $\tau \to 0$ along the critical isochore, and using $c_{V,\mathrm{vdW}} = c_{V,\mathrm{IG}} = \left(\frac{1}{\gamma_{0,\mathrm{IG}}-1}\right) r$, we can express the constant value $c_{\mathrm{sc,vdW}}$ of the velocity of sound at the critical point of the van der Waals fluid as being proportional to the value $c_{s,\mathrm{IG}} = \sqrt{\gamma_{0,\mathrm{IG}} r T_c}$ of the velocity of sound $c_{s,\mathrm{IG}}$ in an ideal (polyatomic) gas at the critical temperature [see Eq. (1.85)]:

$$c_{\mathrm{sc,vdW}} = \sqrt{\frac{T_c}{\rho_c^2 c_{V,\mathrm{IG}}}\left[\left(\frac{\partial p}{\partial T}\right)^2_{\rho=\rho_c}\right]_{c,\mathrm{vdW}}} = \frac{3}{2}\sqrt{\frac{r}{c_{p,\mathrm{IG}}}}\,c_{s,\mathrm{IG}} \simeq 0.8\,c_{s,\mathrm{IG}}, \tag{1.103}$$

In Eq. (1.103), $c_{p,\mathrm{IG}} = \left(\frac{\gamma_0}{\gamma_{0,\mathrm{IG}}-1}\right) r$ and $\gamma_{0,\mathrm{IG}} = \frac{c_{p,\mathrm{IG}}}{c_{V,\mathrm{IG}}} = 1.4$ is the ratio of specific heats at constant pressure and constant volume of an ideal (polyatomic) gas. To obtain the numerical value of the proportionality factor $\frac{3}{2}\sqrt{\frac{r}{c_{p,\mathrm{IG}}}} \simeq 0.8$ of Eq. (1.103), we have used Eqs. (1.94) and (1.95). As the value of this factor is close to unity, we can approximate the critical value of the velocity of sound of a van der Waals fluid by the value of the velocity of sound of an ideal (polyatomic) gas, i.e.,

$$c_{\mathrm{sc,vdW}} \simeq c_{s,\mathrm{IG}} \tag{1.104}$$

1.11.9 Correlation Length ξ_{MF}, Viscosity μ_{MF}, Thermal Conductivity Λ_{MF}, and Thermal Diffusivity D_{MF}

To estimate the mean field divergence of the correlation length ξ_{MF}, there is no way to infer the power law

$$\xi_{\mathrm{MF}} = \xi_{0,\mathrm{MF}} \tau^{-\frac{1}{2}} \tag{1.105}$$

from the van der Waals equation of state due to the unknown value of the amplitude d_0 in Eqs. (1.32) and (1.36). However, there is an universal dynamic relation that correlates the Einstein–Stokes-like diffusion coefficient of a fluctuation of size ξ [see Eq. (1.40)] with the usual thermal diffusivity D_T [see Eq. (1.90)]. One difficulty using such a relation is the mixing of experimental data, thermodynamic van der Waals values, and transport coefficients, with the latter being absent in the static van der Waals approach. We choose to be consistent with the following mean-field behaviors for the transport coefficients (viscosity and thermal conductivity):

$$\begin{aligned} \Lambda_{\mathrm{MF}} &= \Lambda_{c,\mathrm{MF}} + \Lambda_{b,\mathrm{MF}} = \Lambda_{0,\mathrm{MF}} \tau^{-1/2} + \Lambda_{b0,\mathrm{MF}} \\ \mu_{\mathrm{MF}} &= \mu_{b,\mathrm{MF}} = \text{constant} = \mu_{b0} \end{aligned} \tag{1.106}$$

Here, the subscripts c and b stand for the critical and background contributions, respectively. The viscosity, whose experimental divergence is very weak, is deduced from the experimental background contribution μ_{b0} [(see Eq. (1.92), Table 1.6] when the experimental data are fitted to Eq. (1.91). Writing then Eq. (1.90) as the form

$$\Lambda \mu = \frac{\rho c_p k_B T}{6\pi \xi} \tag{1.107}$$

it is easy to see that the mean field power law behavior of the correlation length [see Eq. (1.105)] and the specific heat at constant pressure [see Eq. (1.99)] must lead to the following mean-field behaviors of the thermal diffusivity written as the following two-term power law:

$$D_{T,\mathrm{MF}} = D_{c,\mathrm{MF}} + D_{b,\mathrm{MF}} = D_{0,\mathrm{MF}}\tau^{1/2} + D_{b0,\mathrm{MF}}\tau \tag{1.108}$$

In Eq. (1.108), the background amplitude $D_{b0,\mathrm{MF}}$ can be written from Eq. (1.99) as

$$D_{b0,\mathrm{MF}} = \frac{\Lambda_{b0,\mathrm{MF}}}{\rho_c c_{p0,\mathrm{vdW}}} = \frac{\overline{m}}{\rho_c R}\Lambda_{b0,\mathrm{MF}}. \tag{1.109}$$

The evaluation of the critical term $D_{0,\mathrm{MF}}\tau^{1/2}$ is more subtle (due to the fact that hyperscaling is not accounted for in a mean field theory, see Chap. 18). Considering only the asymptotic behavior very near the critical point, Eq. (1.107) becomes

$$\Lambda_{0,\mathrm{MF}}\mu_{b0} = \rho_c c_{p0,\mathrm{vdW}}\frac{k_B T_c}{6\pi\xi_{0,\mathrm{MF}}} = \rho_c c_{p0,\mathrm{vdW}} D_{0,\mathrm{MF}}\mu_{b0}. \tag{1.110}$$

We can see different combinations between the constant parameters (μ_{b0}, $\frac{k_B}{6\pi}$, ρ_c, and T_c), the van der Waals amplitude $c_{p0,\mathrm{vdW}}$ and the mean-field amplitudes ($\Lambda_{0,\mathrm{MF}}$, $D_{0,\mathrm{MF}}$, and $\xi_{0,\mathrm{MF}}$). Deducing $\xi_{0,\mathrm{MF}}$ from this relation is necessarily a compromise, as the experimental data do not fit the mean-field behavior and the van der Waals equation of state well. However, the right-hand side of Eq. (1.107) corresponds to statics (thermodynamics) and the left-hand side to dynamics (transport coefficients), leading to distinguish the errors due to the thermodynamic data fitting and the transport property data fitting. For example, in the van der Waals equation of state, the critical pressure is overestimated. As $c_{p0,\mathrm{vdW}}$ is proportional to $p_{c,\mathrm{vdW}}$ [see Eq. (1.99)], a systematic error is made calculating $\xi_{0,\mathrm{MF}}$. This error, on the order of $\frac{p_{c,\mathrm{vdW}}}{p_{c,\exp}}$, can be corrected using the following expression

$$\xi_{0,\mathrm{MF}} = \frac{k_B T_c}{6\pi\mu_{b0}}\frac{p_{c,\exp}}{p_{c,\mathrm{vdW}}}\frac{\rho_c c_{p0,\mathrm{vdW}}}{\Lambda_{0,\mathrm{MF}}} = \frac{k_B T_c}{6\pi\mu_{b0}}\frac{p_{c,\exp}}{p_{c,\mathrm{vdW}}}\frac{\rho_c R}{\overline{m}}\frac{1}{\Lambda_{0,\mathrm{MF}}}. \tag{1.111}$$

Because the thermal conductivity is the key parameter that determines the heat flux at the border between the fluid and the heating wall during the short time transient period of the fluid thermal heating (see Chap. 5), we can first prioritize this transport property to remain the closest to the experimental data using a fitting procedure to obtain the best values of $\Lambda_{b0,\mathrm{MF}}$ and $\Lambda_{0,\mathrm{MF}}$.

Then the procedure starts from this best fit of the experimental thermal conductivity obtained from Eq. (1.106) in the temperature range $30\,\mathrm{mK} \lesssim T - T_c \lesssim 3\,\mathrm{K}$. The data in Fig. 1.9 show that imposing this mean-field critical behavior on the thermal conductivity data is a rather satisfactory approach, with the mean-field value of 1/2 for the exponent giving relatively small discrepancies from its exact Ising value $\gamma - \nu - Y_\eta = 0.57$. As a last step, the correlation length is deduced from Eq. (1.111). The corresponding values of some useful parameters for this van der Waals/mean-field description of the selected fluids are reported in Table 1.7.

Note that we chose in the above approach to emphasize the fit to the experimental thermal conductivity. We could also have chosen to emphasize the fit to the thermal

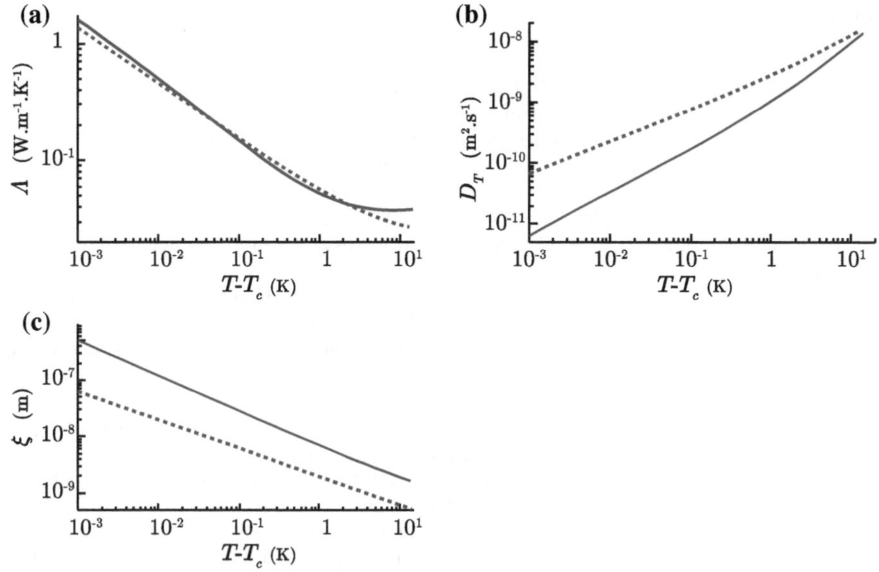

Fig. 1.9 Temperature behavior at $\rho = \rho_c$ of the thermal conductivity Λ, the thermal diffusivity D_T, and the correlation length ξ. The *solid lines* refer to the best fits to the experimental data in the case of xenon, and the *dotted lines* correspond to their calculated van der Waals or mean-field behaviors (see Table 1.7)

diffusivity for the CO_2 case (for example when we are mainly interested analyzing the different time scales of the thermal equilibration, see Chaps. 3–5). In that approach, the fitting procedure leads to the best values of $D_{b0,\text{MF}}$ and $D_{0,\text{MF}}$, while the $\xi_{0,\text{MF}}$ and $\Lambda_{0,\text{MF}}$ values can be estimated using the following expressions

$$\xi_{0,\text{MF}} = \frac{k_B T_c}{6\pi \mu_{b0}} D_{0,\text{MF}} \text{ and } \Lambda_{0,\text{MF}} = \rho_c c_{p0,\text{vdW}} D_{0,\text{MF}}$$

The corresponding data are included in the CO_2-bis column of Table 1.7 and will be used in the Chaps. 3, 4 and 5 dealing with the Navier–Stokes modeling, although they are slighly different from the above one of CO_2 that corresponds to the best fitting of the experimental thermal conductivity in the temperature range $T - T_c \leq 1$ K. These different fits show the difficulty involved in determining the proper fluid-dependent coefficients when performing numerical hydrodynamic simulations in a mean-field approximation.

Therefore, it should be noted that the above approaches to fitting the experimental data with the van der Waals equation of state and mean-field behavior are only approximate and are not unique. However, although there are still discrepancies from the experimental data in all cases, the mean-field approach has the advantage of capturing the essential features of the physical phenomena and greatly simplifying the numerical calculations.

References

1. Anisimov MA, Sengers JV (2000) Critical region. In: Equations of state for fluids and fluid mixtures, vol. I. Elsevier, Amsterdam
2. Bagnuls C, Bervillier C (1984) Nonasymptotic critical behaviour from field theory for ising-like systems in the homogeneous phase: theoretical framework. J Phys Lett 45(3):L95–L100
3. Bagnuls C, Bervillier C (1985) Nonasymptotic critical behavior from field theory at $d = 3$: the disordered-phase case. Phys Rev B 32(11):7209–7231
4. Bagnuls C, Bervillier C (2002) Classical-to-critical crossovers from field theory. Phys Rev E 65(6):066132
5. Bagnuls C, Bervillier C, Meiron DI, Nickel BG (1987) Nonasymptotic critical behavior from field theory at d=3. ii. The ordered-phase case. Phys Rev B 35(7):3585–3607
6. Beysens D (1995) Scaling and gravity effects in critical point phenomena. Microgravity Q 5:35–43
7. Beysens D, Garrabos Y, Chabot C (1998) Hydrodynamics and phase separation in simple fluids. In: Tokuyama M, Oppenheim I (eds) The 8th Tohwa University international symposium on slow dynamics in complex systems. American Institute of Physics, Woodbury, pp 222–235
8. Beysens D, Gbadamassi M, Moncef-Bouanz B (1983) New developments in the study of binary fluids under shear flow. Phys Rev A 28(4):2491–2509
9. Beysens D, Robert M (1987) Thickness of fluid interfaces near the critical point from optical reflectivity measurements. J Chem Phys 87:3056–3061; ibid (1990) Erratum. J Chem Phys 93:6911
10. Bhattacharjee JK, Ferrell RA, Basu RS, Sengers JV (1981) Crossover function for the critical viscosity of a classical fluid. Phys Rev A 24(3):1469–1475
11. Binder K (1984) Nucleation barriers, spinodals, and the Ginzburg criterion. Phys Rev A 29(1):341–349
12. Binder K (1991) Spinodal decomposition. In: Material science and technology: phase transitions in materials, Chapter 5. VCH Verlagsgesellschaft, Weinheim, pp 405–471 (and references therein)
13. Brezin E (1982) In: LÃ©vy M, Le Guillou JC, Zinn-Justin J (eds) Phase transitions, CargÃšse 1980. Plenum, New York, p 339
14. Brezin E, Le Guillou JC, Zinn-Justin J (1974) Universal ratios of critical amplitudes near four dimensions. Phys Lett A 47:285–287
15. Cagniard de la Tour C (1822) Experimente unter hohem drucke. Ann Chim Phys 21:127–131, 178–181
16. Cahn JW (1997) Critical point wetting. J Chem Phys 66:3667–3672
17. Calmettes P (1977) Critical transport properties of fluids. Phys Rev Lett 39:1151–1154
18. Campostrini M, Hasenbusch M, Pelissetto A, Rossi P, Vicari E (2001) Critical behavior of the three-dimensional xy universality class. Phys Rev B 63(21):214503
19. Carles P (1998) The effect of bulk viscosity on temperature relaxation near the critical point. Phys Fluids 10:2164–2176
20. Carles P, Dadzie K (2005) Two typical time scales of the piston effect. Phys Rev E 71:066310
21. Cheng E, Cole MW, Dupont-Roc J, Shaam WF, Treiner J (1993) Novel wetting behaviour in quantum films. Rev Mod Phys 65:557–567
22. Eckert CA, Knutson BL, Debenedetti PG (1996) Supercritical fluids as solvents for chemical and material processing. Nature 383:313–318
23. Furukawa H (1985) A dynamical scaling assumption for phase separation. Adv Phys 34:703–750
24. Garrabos Y (1982) Contribution à l'étude des propriétés d'état des fluides purs dans leur région critique. Thèse de Doctorat d'Etat, Université de Paris 6
25. Garrabos Y (1985) Phenomenological scale factors for the liquid-vapor critical transition of pure fluids. J Phys (Paris), 46:281–206 (see also http://fr.arxiv.org/abs/condmat/0512347)
26. Garrabos Y (1986) Scaling behavior of the fluid subclass near the liquid-gas critical point. J Phys (Paris) 47:197–206

27. Garrabos Y, Bervillier C (2006) Mean crossover functions for uniaxial 3d ising-like systems. Phys Rev E 74(2):021113 (16 p)
28. Garrabos Y, Lecoutre C, Palencia F, LeNeindre B, Erkey C (2008) Master crossover functions for one-component fluids. Phys Rev E 77:021116
29. Ginzburg VL (1961) Some remarks on phase transitions of the 2nd kind and the microscopic theory of ferroelectric materials. Sov Phys Solid State 2:1824–1834
30. Gunton JD, San Miguel M, Sahni PS (1983) The dynamics of first-order phase transitions. In: Phase transitions and critical phenomena, Chap 3, vol 8. Academic, New York, p 269
31. Hess GB, Sabatini MJ, Chan MHW (1997) Nonwetting of cesium by neon near its critical point. Phys Rev Lett 78(9):1739–1742
32. Hohenberg PC, Halperin BI (1977) Theory of dynamic critical phenomena. Rev Mod Phys 49:435–479
33. Houessou C, Guenoun P, Gastaud R, Perrot F, Beysens D (1985) Critical behavior of the binary fluids cyclohexane-methanol, deuterated cyclohexane-methanol and of their isodensity mixture: application to micrograwity simulations and wetting phenomena. Phys Rev A 32(3):1818
34. Jayalakshmi Y, Khalil B, Beysens D (1991) Phase separation under a weak concentration gradient. Phys Rev Lett 69:3088–3091
35. Justin JZ (2002) Quantum field theory and critical phenomena, 4th edn. Oxford University Press, Oxford
36. Kadanoff LP, Swift J (1968) Transport coefficients near the liquid-gas critical point. Phys Rev 166:89
37. Kawasaki K (1970) Kinetic equations and time correlation functions of critical fluctuations. Ann Phys 61(1):1–56
38. Kawasaki K (1970) Sound attenuation and dispersion near the liquid-gas critical point. Phys Rev A 1:1750
39. Lee SP (1978) Need research. Chem Phys Lett 57:611
40. Levelt-Sengers JMH (1975) Experimental thermodynamics, Chap 14, vol 2. Butterworths, London
41. Le Neindre B, Garrabos Y, Tufeu R (1991) The critical thermal-conductivity enhancement along the critical isochore. Int J Thermophys 12:307–321
42. Lifshitz IM, Slyosov VV (1961) J Phys Chem Solids 50:19–35
43. Maxwell JC (1890) The scientific papers of James Clark Maxwell. Cambridge University Press, Cambridge
44. Moldover MR, Sengers JV, Gammon RW, Hocken RJ (1979) Gravity effects in fluids near the gas-liquid critical point. Rev Mod Phys 51(1):79–99
45. Nicoll JF, Albright PC (1986) Background fluctuations and Wegner corrections. Phys Rev B 34:1991
46. Nikolayev VS, Beysens D, Guenoun P (1996) New hydrodynamic mechanism for drop coarsening. Phys Rev Lett 76:3144–3147
47. Normand C, Pomeau Y, Vellarde MG (1977) Convective instabilities: a physicist's approach. Rev Mod Phys 49(3):581–624
48. Onuki A (1984) Electric field effects in fluids near the critical point. Europhys Lett 29:611–616
49. Onuki A (1997) Bulk viscosity near the critical point. J Phys Soc Jpn 66:511–513
50. Onuki A (2002) Phase transition dynamics. Cambridge University Press, Cambridge
51. Onuki A, Kawasaki K (1978) Fluctuations in nonequilibrium steady states with laminar shear flow: classical fluids near the critical point. Prog Theor Phys Suppl 64:436–441
52. Onuki A, Kawasaki K (1979) Light scattering by critical fluids in the presence of a uniform shear flow. Phys Lett A 72:233–235
53. Onuki A, Kawasaki K (1979) Nonequilibrium steady state of critical fluids under shear flow: a renormalization group approach. Ann Phys 121:456–528
54. Ornstein LS, Zernike F (1918) Contributions to the kinetic theory of solids. The thermal pressure of isotropic solids. Proc Acad Sci Amsterdam 17:793–803 (see Phys Z 19:134)
55. Patashinskii AZ, Pokroviskii VL (1979) Fluctuation theory of phase transitions. Pergamon, Oxford

56. Perrot F, Beysens D, Garrabos Y, Fröhlich T, Guenoun P, Bonetti M, Bravais P (1999) Morphology transition observed in a phase separating fluid. Phys Rev E 59(3):3079–3083
57. Privman V, Honenberg PC, Aharony A (1991) Phase transitions and critical phenomena, Universal critical point amplitude relations, Chap, vol 14. Academic, New York, pp 1–134
58. Quentrec B (1979) A new analysis of sound propagation near the critical point of xenon. J Phys Lett Paris 40(13):257–261
59. Quettier L, Felice H, Mailfert A, Chatain D, Beysens D (2005) Magnetic compensation of gravity forces in liquid-gas mixtures: surpassing intrinsic limitations of a superconducting magnet by using ferromagnetic inserts. Eur Phys J Appl Phys 32:167–175
60. Savage PE, Gopalan S, Mizan TI, Martino CJ, Brock EE (1995) Reactions at supercritical conditions: applications and fundamentals. AIChE J 41(7):1723–1778
61. Schick M (1990) Introduction to wetting phenomena. In: Liquids at interfaces. North Holland, Amsterdam
62. Schwarzchild K (1992) Gesammelete Werke (collected works), vol 1. Springer, Berlin
63. Sengers JV, Levelt Sengers JMH (1978) In: Progress in liquid physics. Wiley, New York
64. Sikkenk JH, van Leeuwen JMJ, Sengers JV (1986) Gravity effects on the fluctuations in a vapor-liquid interface close to the critical temperature. Phys A 139(1):1–27 (and references therein)
65. Spiegel EA (1965) Convective instabilities in a compressible atmosphere. Astrophys J 141:1068
66. Stanley HE (1971) Introduction to phase transitions and critical phenomena. Clarendon, Oxford
67. Stauffer D, Ferer M, Wortis M (1972) Universality of second-order phase transitions: the scale factor for the correlation length. Phys Rev Lett 29(6):345–349
68. van der Waals JD (1887) In: Rowlinson JS (ed) On the continuity of the gaseous and liquid states. PhD thesis, Leiden University, Leiden
69. van der Waals JD (1988) On the continuity of the gaseous and liquid state. Dover Publications Inc, New York
70. Wegner FJ (1972) Corrections to scaling laws. Phys Rev B 5(11):4529–4536
71. Widom B (1965) Surface tension and molecular correlations near the critical point. J Chem Phys 43:3892–3897
72. Wilson KG, Kogut J (1974) The renormalization group and the ϵ expansion. Phys Rep 12(2):75–199
73. Wunenburger R, Chatain D, Garrabos Y, Beysens D (2000) Magnetic compensation of gravity forces in (p-)hydrogen near its critical point: application to weightless conditions. Phys Rev E 62(1):469–476
74. Zimmerli GA, Wilkinson RA, Ferrell RA, Moldover MR (1999) Electrostriction of a near-critical fluid in microgravity. Phys Rev E 59(5):5862–5869
75. Zimmerli GA, Wilkinson RA, Ferrell RA, Moldover MR (1999) Electrostriction of near-critical sf_6 in microgravity. Phys Rev Lett 82(26):5253–5256

Part I
Thermomechanical Effects

Chapter 2
Introduction to Thermomechanical Effects

This part of the book builds upon the classical conservation laws of fluid dynamics. These conservation laws (the conservation of mass, the conservation of linear momentum, and the conservation of energy, also known as the Navier–Stokes equations) are summarized in Chap. 21 after a brief presentation of the main notations and rules for the scalar, vector, and tensor quantities needed to define the fluid motion. In the present part of the book, as we are mainly interested in the transient and stationary nonequilibrium states of the near-critical fluids described in the previous introductory chapter, the most important assumption we make is local thermodynamic equilibrium, since local equilibrium must be present for the hydrodynamic equations presented in Chap. 21 to be valid. Readers interested in a more detailed discussion of the foundational axioms of near-critical fluid dynamics should refer to [1, 2].

The three chapters comprising Part I therefore focus on explicitly deriving the analytical methods needed to solve the Navier–Stokes equations that can be used to calculate the exact fluid motion of a one-dimensional van der Waals fluid in the absence of a gravitational field. The equations that describe the 1-D motion of the near-critical fluid are derived in Chap. 3 using the van der Waals equation of state and the mean field behaviors of the transport coefficients (see Sect. 1.11), in order to completely specify the problem as the critical point is approached. Therefore, this chapter considers the case of a compressible fluid described by the perfect gas equation of state. After that, in Chap. 4, we make use of the asymptotic expansions technique to explore the response of a near-critical one-component fluid in a 1-D, slab-like container to an increase in the wall temperature (parietal temperature step). We show that the bulk temperature evolves more quickly than heat diffusion due to acoustic heating. We obtain the analytical solutions of the equations on the three characteristic timescales of the problem: the acoustic time, the piston-effect time, and the heat diffusion time. Then, in the final chapter of this part (Chap. 5), we make use of the asymptotic expansions technique to explore the response of a near-critical one-

B. Zappoli et al., *Heat Transfers and Related Effects in Supercritical Fluids*,
Fluid Mechanics and Its Applications 108, DOI: 10.1007/978-94-017-9187-8_2,

component fluid in a 1-D, slab-like container to a heat flux at the boundary (parietal heat flux). We show that the bulk phase behaves as a thermal short circuit, and we obtain working formulae that can be used to analyze the corresponding experimental results; the latter are discussed in more detail in Part II.

References

1. Onuki A (2002) Phase transition dynamics. Cambridge University Press, Cambridge, England
2. Ortiz de Zarate JM, Sengers JV (2006) Hydrodynamics fluctuations in fluids and fluid mixtures. Elsevier, Amsterdam

Chapter 3
Bridging Gas and Near-Critical Fluid Dynamics

Abstract The mechanical model is first presented, considering the fluid as a Newtonian, viscous, heat conducting and compressible medium. The physical model is then detailed. Emphasis is placed on the expression of the transport coefficients as a function of temperature and density, written in the mean field framework, refinement being given when necessary. The equations that describe the one-dimensional motion of a near-critical fluid are then derived from the classical conservation laws of mechanics. When written on the acoustic time scale, the equations show a small parameter that is the ratio of the acoustic time to the diffusion time of the cavity. Some basic equations of fluid dynamics are recalled in the annex part. Chaps. 18–23.

3.1 The One-Dimensional Model

Near the critical point, the amplitude of the density fluctuations increase with the associated correlation length ξ. We may therefore question the assumption of local equilibrium at scales that are smaller than ξ. This indicates that the hydrodynamic equations apply fully at the scale

$$L \gtrsim \xi. \tag{3.1}$$

However, for smaller length scales, great care must to be taken. In this book, we restrict most of our analysis to the length scales that fulfill Eq. (3.1). The fluid state is then described by the hydrodynamic conservation equations written for a viscous, low heat diffusing and highly expandable fluid (also refered as belonging to the class of low Mach number fluids). The corresponding system of equations needs to be closed with the expressions for the stress tensor and for the heat flux as functions of the thermodynamic and dynamic dependent variables density, velocity and temperature. The description of the mechanical and thermal behavior of the fluid model is thus provided by the Newtonian behavior of the stress tensor [see Eq. (21.21)], also known as Newton Second Law of Motion) and by the Fourier law [see Eq. (21.26)] for the

B. Zappoli et al., *Heat Transfers and Related Effects in Supercritical Fluids*,
Fluid Mechanics and Its Applications 108, DOI: 10.1007/978-94-017-9187-8_3,

heat flux. However, even when the fluid is Newtonian, the equation of state has to be defined, as well as the viscosity and heat conductivity coefficients as a function of the local thermodynamical state. Such a description of the thermodynamic and transport properties of a near-critical fluid was introduced in Chap. 1, and this was notably different from the one for a dilute compressible (ideal) gas described by the perfect gas equation of state, where, in addition, the transport coefficients are given by the kinetic theory of gases. A complementary discussion of the static and dynamic situations of near-critical fluids can be also found in Ref. [2].

The following description of the one-dimensional fluid model presents the main physical parameters needed to better understand the pathway between ideal gas and near-critical fluid dynamics, especially when the real near-critical fluid can be modeled by a van der Waals fluid, the equation of state for which is given by Eq. (1.9). We recall here that the van der Waals equation of state per unit of mass of fluid can be expressed as follows:

$$p_{\mathrm{vdW}} = \frac{rT\rho}{1 - b\rho} - a\rho^2, \tag{3.2}$$

where the specific values of the fluid parameters a and b are given in Table 1.7 for the one-component fluids considered in this book. In Eq. (3.2), T and ρ are the two selected independent variables, whose values are equal to their respective experimental data, while p_{vdW} is the calculated pressure, which is different from the experimental pressure (as shown in Sect. 1.1).

3.1.1 1-D Mechanical van der Waals Model

As the fluid is assumed to be contained in a one-dimensional slab-like container, the Navier-Stokes equations presented in Chap. 21 take their 1-D forms. The fluid is initially at rest and at thermodynamic equilibrium on the critical isochore $\rho = \rho_c$, and slightly above the critical temperature such that

$$\tau = \left(\frac{T - T_c}{T_c}\right) \ll 1. \tag{3.3}$$

Therefore,

$$\frac{T}{T_c} = 1 + \tau,$$

where τ can be treated as a small parameter.

In its 1-D approximation, the fluid motion is described by four dependent variables: velocity u, pressure p, density ρ, and temperature T. These all depend on two independent variables: the spatial coordinate x and time t. Moreover, we recall that we assume the condition $g = 0$ in this Part I to suppress the effects due to Earth's gravity or any external acceleration field.

If we chose, *a priori*, to examine what happens at the slab length scale, even if other length scales will be found later on, the spatial variable is made nondimensional with respect to the width of the slab L:

$$x^* = \frac{x}{L}. \tag{3.4}$$

Now, thermodynamic and hydrodynamic equations need to be made nondimensional in order to be able to specify the relevant scale for each of them and to rewrite the equations in a simpler form. Using the experimental values of the critical temperature and critical density of the selected fluid, the nondimensional thermodynamic variables can be defined as follows:

$$T^* = \frac{T}{T_c} = 1 + \tau; \qquad \rho^* = \frac{\rho}{\rho_c}; \qquad p^* = \frac{p}{\rho_c r T_c}, \tag{3.5}$$

where $r = \frac{R}{\overline{m}}$ is the specific universal ideal gas constant, $\overline{m}$ is the molar mass of the fluid, and $\rho_c r T_c = p_{\text{IG}}$ is the pressure of a hypothetical ideal gas, which would be placed under the same critical temperature and critical density conditions. We must emphasize here that the above reference pressure p_{IG} differs from the experimental critical pressure p_c. However, such a procedure presents the advantage of identifying the numerical value of the prefactor resulting from the use of an equation of state such as the van der Waals equation of state, where the critical pressure $p_{c,\text{vdW}} = \frac{3}{8}\rho_c r T_c$ [see Eq. (1.94)] is also distinct from the experimental critical pressure. Therefore, the prefactor value of $\frac{3}{8}$ accounts for the order of magnitude of the critical compressibility factor $Z_c = \frac{p_c}{\rho_c r T_c}$ of a van der Waals one-component fluid. As a consequence, the nondimensional form of Eq. (3.2) is now written as

$$p^* = \frac{\rho^* T^*}{1 - \frac{1}{3}\rho^*} - \frac{9}{8}\left(\rho^*\right)^2, \tag{3.6}$$

where the explicit contribution of the above reference pressure p_{IG} is given by the amplitude change ($\frac{9}{8}$ in place of $\frac{3}{8}$) of the $(\rho^*)^2$ prefactor term when compared to the usual nondimensional form of Eq. (1.17). Equation (3.6) is the first non-dimensional equation of state of the van der Waals fluid.

As the smallest hydrodynamic characteristic time in our system is the (dimensional) acoustic time t_a (the time it takes for an acoustic disturbance to cross the slab of width L), time is referred to this value in order to describe the shortest hydrodynamic timescale. The velocity of sound c_s of a critical fluid slowly decreases as $c_s \propto \tau^{0.055}$ when approaching the critical temperature (see Sect. 1.10.3 and Table 1.3). We have also shown in Sect. 1.10.3 that the experimental behavior of c_s in the near-critical region can be approximated by a constant value which appears to be of the same order of magnitude as the critical velocity of sound $c_{s,\text{vdW}} \simeq \sqrt{\frac{T_c}{\rho_c^2 c_{V,\text{IG}}}\left[\left(\frac{\partial p}{\partial T}\right)^2_{\rho=\rho_c}\right]_{c,\text{vdW}}}$ of a van der Waals fluid [see Eq. (1.103)], or as the critical velocity of sound $c_{s,\mathrm{IG}} = \sqrt{\gamma_{0,\mathrm{IG}} r T_c}$ of an ideal gas at the critical temperature [see also Eq. (1.103)]. We recall that $\gamma_{0,\mathrm{IG}} = \frac{c_{p,\mathrm{IG}}}{c_{V,\mathrm{IG}}}$ is the ratio of the specific heats at constant pressure (with $c_{p,\mathrm{IG}} = \frac{\gamma_{0,\mathrm{IG}}}{\gamma_{0,\mathrm{IG}}-1} r$) and constant volume (with $c_{V,\mathrm{IG}} = \frac{1}{\gamma_{0,\mathrm{IG}}-1} r$) for an ideal (polyatomic) gas (anticipating the comparison with the results of experiments using carbon dioxide for example). Therefore, we first choose the following approximation

$$c_{V,\mathrm{vdW}} \simeq c_{V,\mathrm{IG}} = \frac{r}{\gamma_{0,\mathrm{IG}} - 1} = \frac{5}{2} r \tag{3.7}$$

to be the second dimensional equation of state of a near-critical van der Waals fluid. Since $\gamma_{0,\mathrm{IG}} = 1.4$ is a fixed constant, the value of the specific heat at constant volume is the only fluid-dependent quantity required to introduce a second approximation [see Eq. (1.104)]:

$$c_{s,\mathrm{vdW}} \simeq c_{s,\mathrm{IG}} = \sqrt{\gamma_{0,\mathrm{IG}} r T_c} \tag{3.8}$$

as the velocity of sound of a near-critical van der Waals fluid at the critical temperature. Thus, the time is made nondimensional with respect to the acoustic time in the ideal gas:

$$t^* = \frac{t}{t_a}; \qquad \text{with } t_a = \frac{L}{\sqrt{\gamma_{0,\mathrm{IG}} r T_c}}. \tag{3.9}$$

The natural reference velocity that now appears is $c_{s,\mathrm{IG}}$ of Eq. (3.8), and the nondimensional fluid velocity is then defined by

$$u^* = \frac{u}{c_{s,\mathrm{IG}}}$$

By replacing the dimensional variables in the 1-D Navier–Stokes equations (see Sect. 21.5) by their nondimensional counterparts, we obtain the following system of nonlinear partial differential equations (we will discuss how below making explicit reference to the needed equations given in Chap. 21):
Mass balance:

$$\rho^*_{t^*} + \left(\rho^* u^*\right)_{x^*} = 0$$

Momentum balance:

$$\rho^* \left(u^*_{t^*} + u^* u^*_{x^*}\right) = -\frac{1}{\gamma_{0,\mathrm{IG}}} p^*_{x^*} + \epsilon \frac{4}{3} \left(\mu^* u^*_{x^*}\right)_{x^*} \tag{3.10}$$

Energy balance:

$$\rho^* \left(e^*_{t^*} + u^* e^*_{x^*}\right) = -p^* u^*_{x^*} + \epsilon \left[\frac{\gamma_{0,\mathrm{IG}}}{\left(\gamma_{0,\mathrm{IG}} - 1\right) \mathrm{Pr}_{0,\mathrm{IG}}} \left(\Lambda^* T^*_{x^*}\right)_{x^*} + \frac{4}{3} \gamma_{0,\mathrm{IG}} \mu^* \left(u^*_{x^*}\right)^2 \right],$$

(subscripts t^* and x^* denote partial derivatives), which contains the nondimensional internal energy e^*, viscosity μ^*, thermal conductivity Λ^*, Prandtl number $\mathrm{Pr}_{0,\mathrm{IG}}$, and a new small parameter ϵ to be defined below. However, before that, we note that the right-hand side of the energy balance equation contains the term $-p^* u^*_{x^*}$, which is the contribution from the power of the internal pressure stresses imparted to the fluid element during its deformation and, in the parentheses multiplied by amplitude ϵ, the term $\frac{\gamma_{0,\mathrm{IG}}}{(\gamma_{0,\mathrm{IG}}-1)\mathrm{Pr}_{0,\mathrm{IG}}} \left(\Lambda^* T^*_{x^*}\right)_{x^*}$, which represents the contribution of the heat diffusion, as well as the term $\frac{4}{3} \gamma_{0,\mathrm{IG}} \mu^* \left(u^*_{x^*}\right)^2$, which is the power of the internal viscous stresses imparted to the fluid domain during its deformation. Obviously, the corresponding values of the dimensionless amplitudes of these terms are obtained through the reduction scheme discussed in the following.

The specific internal energy e is expressed in units of rT_c, in order to maintain T_c as a single fluid-characteristic parameter for the energy scale. Moreover, it is important to note that the complete thermodynamic description of the van der Waals fluid is closed in a consistent manner by the introduction of the two dimensional equations of state (3.2) and (3.7), in spite of their unsuitability to quantitatively fit the experimental behavior of the selected one-component fluid close to its liquid-gas critical point (see Chap. 1). Then, referencing to Sect. 21.3.2, the internal energy per unit mass as a function of T and ρ can be found from the general equation of state given by Eq. (21.38). The derivative $\left(\frac{\partial p}{\partial T}\right)_\rho = \frac{\rho r}{1-b\rho}$ is thus calculated from the van der Waals Eq. (3.2), leading to $\left(\frac{\partial e}{\partial \rho}\right)_T = -a$; note that this latter derivative value can also be obtained using the van der Waals form of the Helmholtz equation given by Eq. (21.46) (see Sect.21.3.2). By using Eq. (3.7) (the second dimensional equation of state) and $c_V = \left(\frac{\partial e}{\partial T}\right)_\rho$ [see Eq. (21.31)] to estimate the constant of the integrating process, the specific energy e of the related van der Waals fluid can be obtained:

$$e = c_V T - a\rho.$$

Therefore, the nondimensional form of the specific internal energy in rT_c units is

$$e^* = \frac{1}{\gamma_{0,\mathrm{IG}} - 1} T^* - \frac{9}{8} \rho^*, \tag{3.11}$$

which can be considered as being the second nondimensional equation of state in the thermodynamic description of a van der Waals fluid of unit mass, where T^* and ρ^* are the two independent variables and where $Z_{c,\mathrm{vdW}} = \frac{p^*_c}{\rho^*_c T^*_c} = \frac{3}{8}$.

As discussed in Sect. 1.11, the small divergence in the viscosity $\mu \propto \tau^{-0.04}$ given by Eq. (1.91) can be approximated by the constant mean-field viscosity $\mu_{b,\mathrm{MF}} = \mu_{b0}$ of a van der Waals fluid [see Eq. (1.106)], which is chosen here to be equal to the experimental value μ_b of the background term at the critical point defined by Eq. (1.92). Therefore, the nondimensional viscosity can be expressed in units of μ_{b0}, i.e.,

$$\mu^* = \frac{\mu}{\mu_{b0}}. \tag{3.12}$$

Similarly, considering the mean field behavior of the heat conductivity coefficient given by Eq. (1.106); i.e.,

$$\Lambda = \Lambda_{0,\mathrm{MF}}\tau^{-\frac{1}{2}} + \Lambda_{b0,\mathrm{MF}}, \tag{3.13}$$

the thermal conductivity coefficient Λ can be made nondimensional with respect to the mean field background term $\Lambda_{b0,\mathrm{MF}}$ (the determination of which from experimental data was discussed in Sect. 1.11.9). Using

$$\Lambda^* = \frac{\Lambda}{\Lambda_{b0,\mathrm{MF}}},$$

the nondimensional form of Eq. (3.13) becomes

$$\Lambda^* = 1 + \frac{\Lambda_{0,\mathrm{MF}}}{\Lambda_{b0,\mathrm{MF}}}\tau^{-\frac{1}{2}} = 1 + \Lambda_c^*\tau^{-\frac{1}{2}}. \tag{3.14}$$

Obviously, with such practical nondimensional forms, we obtain $\mu^* = 1$ as the constant dimensionless viscosity of any van der Waals fluid, while the background value of unity must be added to the mean field divergence $\sim \tau^{-\frac{1}{2}}$ (with characteristic amplitude $\Lambda_c^* = \frac{\Lambda_{0,\mathrm{MF}}}{\Lambda_{b0,\mathrm{MF}}}$) of its dimensionless thermal conductivity. In a first step, where the mean-field divergences of the specific heat at constant pressure and the thermal conductivity are not accounted for, we can also introduce a constant Prandtl number $\mathrm{Pr}_{0,\mathrm{IG}}$ of a ideal gas of constant heat capacity $c_{p,\mathrm{IG}}$, constant viscosity μ_{b0}, and constant thermal conductivity $\Lambda_{b0,\mathrm{MF}}$ at the critical conditions, defined as follows:

$$\mathrm{Pr}_{0,\mathrm{IG}} = \frac{\mu_{b0}c_{p,\mathrm{IG}}}{\Lambda_{b0,\mathrm{MF}}}, \tag{3.15}$$

$\mathrm{Pr}_{0,\mathrm{IG}}$ is thus considered equivalent to a background-like Prandtl number $\mathrm{Pr}_{0,b,\mathrm{MF}}$ in a mean field description of the van der Waals fluid, where the background-like term $(c_{V,\mathrm{IG}})$ for the specific heat at constant pressure is now written as follows

$$c_{V,\mathrm{IG}} = \frac{c_{p,\mathrm{IG}}}{\gamma_{0,\mathrm{IG}}} = \frac{1}{\gamma_{0,\mathrm{IG}} - 1}r$$

As shown in Table 3.1, this Prandtl background value at the critical conditions is around $\mathrm{Pr}_{0,\mathrm{IG}}$ ~2–3 for such a van der Waals model of any real one-component fluid. Recall that the Prandtl number value is $\mathrm{Pr}_{0,\mathrm{IG}} = \frac{\mu_{\mathrm{IG}}c_{p,\mathrm{IG}}}{\Lambda_{\mathrm{IG}}} \sim 1$ for an ordinary ideal gas at room temperature and atmospheric pressure (μ_{IG} is the viscosity and Λ_{IG} is the thermal conductivity of this ordinary ideal gas in normal conditions).

As a direct consequence of this practical nondimensional reduction scheme, Eq. (3.10) include the small parameter ϵ, which is defined as follows:

$$\varepsilon = \frac{\mu_{b0}}{L\rho_c c_{s,IG}}. \tag{3.16}$$

Comprehending the physical meaning of this parameter will be crucial to gaining a better understanding of the analytical developments of the next two chapters of this book.

As a matter of fact, any dimensional transport properties can be rescaled when the dimensionless spatial [see Eq. (3.4)] and temporal [see Eq. (3.9)] variables refer to the total length L of the model and the related characteristic time $t_a = \frac{L}{c_{s,\mathrm{IG}}}$ of a traveling acoustic wave. Indeed, as already discussed in Sect. 21.3.1, the dimensions of the (momentum and thermal) kinetic coefficients ν [see Eq. (21.25)] and D_T [see Eq. (21.27)], which govern the respective (momentum and thermal) diffusion processes, are $\left[\mathrm{L}^2\mathrm{T}^{-1}\right]$. Therefore, these kinetic coefficients can be made nondimensional using the ratio $\frac{L^2}{t_a} = L\sqrt{\gamma_{0,\mathrm{IG}} r T_c}$, which correspondingly means that the shear viscosity μ and the thermal conductivity Λ can be made nondimensional by the reference quantities $\mu_{\mathrm{ref}} = L\rho_c\sqrt{\gamma_{0,\mathrm{IG}} r T_c} = \rho_c c_{s,\mathrm{IG}} L$ and $\Lambda_{\mathrm{ref}} = \rho_c r L\sqrt{\gamma_{0,\mathrm{IG}} r T_c} = \rho_c r c_{s,\mathrm{IG}} L$, respectively. However, note that $\mu_{\mathrm{ref}} \neq \mu_{b0}$ and $\Lambda_{\mathrm{ref}} \neq \Lambda_{b0,\mathrm{MF}}$. As a consequence, $\frac{\mu_{b0}}{\mu_{\mathrm{ref}}} = \frac{1}{\mathrm{Re}_a}$, where

$$\mathrm{Re}_a = \frac{t_\mu}{t_a} = \frac{\rho_c c_{s,\mathrm{IG}} L}{\mu_{b0}}$$

is the acoustic Reynolds number, which compares the characteristic time for vorticity diffusion (i.e., the diffusive viscous relaxation of the momentum) as defined by

$$t_\mu = \frac{L^2 \rho_c}{\mu_{b0}} \tag{3.17}$$

to the characteristic time $t_a = \frac{L}{\sqrt{\gamma_{0,\mathrm{IG}} r T_c}}$ for acoustic wave propagation at the macroscopic length scale L of the model. On the other hand, the background-like Prandtl number Pr_0, previously defined by Eq. (3.15), can now be rewritten as the ratio $\mathrm{Pr}_{0,\mathrm{IG}} = \frac{t_\mu}{t_{D,b}}$, which compares the characteristic time for vorticity diffusion (t_μ) to that for regular thermal diffusion (t_{D_b}), as defined by

$$t_{D_b} = \frac{L^2}{D_{T0}}. \tag{3.18}$$

Here, regular diffusion means that the critical slowing down [$\sim \tau^{\frac{1}{2}}$, see Eq. (1.108)] of a mean-field behavior is not accounted for in the characteristic background-like time t_{D_b} of Eq. (3.18). The regular thermal diffusivity coefficient of the actual van der Waals fluid model is thus defined as follows:

$$D_{T0} = \frac{\Lambda_{b0,\mathrm{MF}}}{\rho_c c_{p,\mathrm{IG}}} = \frac{\Lambda_{b0,\mathrm{MF}}}{\rho_c r}\frac{\gamma_{0,\mathrm{IG}} - 1}{\gamma_{0,\mathrm{IG}}}, \tag{3.19}$$

and differs from the usual definition of the background thermal diffusivity term $D_{b,\mathrm{MF}} = \frac{\Lambda_{b,\mathrm{MF}}}{\rho_c c_{p,\mathrm{vdW}}} = D_{b0,\mathrm{MF}}\tau = \frac{\Lambda_{b0\mathrm{MF}}}{\rho_c c_{p,0,\mathrm{vdW}}}\tau = \frac{\Lambda_{b0,\mathrm{MF}}}{\rho_c r}\tau$, which decreases with τ in a mean-field approach [see Eqs. (1.108) and (1.109)]. However, D_{T0} differs from the amplitude $D_{b0,\mathrm{MF}} = \frac{\Lambda_{b0,\mathrm{MF}}}{\rho_c r}$ only by the constant ratio value of $\frac{\gamma_{0,\mathrm{IG}}-1}{\gamma_{0,\mathrm{IG}}} = \frac{2}{7}$. Now, D_{T0} of Eq. (3.19), just like t_{D_b} of Eq. (3.18), mainly represents the characteristic parameters of a thermal diffusion process occurring in an ideal gas described by the kinetic theory of compressible gases. A typical value of D_{T0} is given in Table 3.1.

As a final result, ϵ is the ratio of the acoustic timescale and the vorticity diffusion timescale, and is linked to $\mathrm{Pr}_{0,\mathrm{IG}}$ by the relation

$$\epsilon \equiv \frac{1}{\mathrm{Re}_a} = \frac{t_a}{t_\mu} = \mathrm{Pr}_{0,\mathrm{IG}} \frac{t_a}{t_{D_b}}. \tag{3.20}$$

Using Eq. (3.20), it is easy to compare the relative roles of the order of magnitude of the essential timescales in real fluids, namely the timescales t_μ and t_{D_b} of the regular vorticity and thermal diffusion phenomena involved in the mass transfer and energy transfer of van der Waals fluids, and the shortest hydrodynamic timescale t_a of the acoustic phenomena. For the one-component fluids considered in this book, ε is typically $\sim \{2\text{–}3\} \times 10^{-8}$ in a 1-D container 10 mm long, while $t_a \sim \{20\text{–}50\}$ μs and $t_\mu \sim \{2\text{–}3\} t_{D_b} \sim 1000\text{–}2000$ s, with $\mu_{b0} \sim 20\text{–}50$ μPa s.

To summarize, the characteristic parameters used in the nondimensional formulation of the fluid model are the 1-D cavity length L, the experimental values of T_c, ρ_c, μ_{b0} and $\Lambda_{b0,\mathrm{MF}}$, and the calculated ideal gas values of $c_{V,\mathrm{IG}} = \frac{r}{\gamma_{0,\mathrm{IG}}-1}$, $c_{p,\mathrm{IG}} = \frac{\gamma_{0,\mathrm{IG}} r}{(\gamma_{0,\mathrm{IG}}-1)}$ (or $\gamma_{0,\mathrm{IG}} = \frac{c_{p,\mathrm{IG}}}{c_{V,\mathrm{IG}}}$), and $c^2_{s,\mathrm{IG}} = \gamma_{0,\mathrm{IG}} r T_c$. This fluid-dependent parameter set introduces a single small parameter ϵ that is able to characterize the timescale of the viscous effects, and a constant Prandtl number $\mathrm{Pr}_{0,\mathrm{IG}}$ that is able to characterize the regular timescale of the thermal effects, by reference to the characteristic acoustic timescale. All the other numerical values of the dimensionless prefactors in Eq. (3.10) are well defined.

For the above system of Eq. (3.10), we now need to elucidate the singular mean-field behaviors of the thermodynamic and transport properties as functions of temperature [see Sec. (1.11)]. This is done in the next section.

3.1.2 Mean Field Behavior of the 1-D Physical Model

The choice of the equation of state is the result of a compromise between the need to use a reliable physical model and the need to obtain tractable analytical calculations. Of course, it is always possible to solve the above conservation equations numerically, but as we shall see in the following, analytical and asymptotic methods define subsets of spatial and temporal scales in which the equations have simplified and thus solvable forms that only retain the prominent physical phenomena. In the two next chapters, we will focus on the phenomenology of temperature equilibration (Chap. 4) and heat transport (Chap. 5), using the above 1-D physical model. Our goal is to build a hydrodynamic model of a near-supercritical fluid that departs slightly from equilibrium conditions, and from which we are able to extract this basic phenomenology. This means that we must consider initial conditions that are not too close to the critical point in order to fulfill Eq. (3.1), and we must consider the hydrodynamic equations defined in the previous section to be valid. This also means that we must deal with linear descriptions of the thermal and mechanical perturbations. However, when we need to explore more complex 2-D nonlinear problems, such as those encountered when the temperature equilibration and heat transport phenomena are subjected to the Earth's gravitational field [see Part (III)], we will need to run robust numerical methods on supercomputers.

The model selected should thus incorporate the main features of the criticality of pure fluids. It is also possible to use linear approximations of real equations of state whenever required so that it can be compared with experimental results. The model chosen, in which the equation of state is the van der Waals equation and the thermal conductivity coefficient is given by the mean-field theory, was shown to fulfill the above requirements and to be consistent in Sect. 1.10. Therefore, we will first concern ourselves with the mean-field behavior of the specific heat at constant pressure [see Eqs. (1.98) and (1.99)]; i.e.,

$$c_{p,\mathrm{vdW}} = c_{V,\mathrm{IG}} + r\,(1+\tau)\,\tau^{-1}. \tag{3.21}$$

By using the specific heats expressed in units of r, we obtain

$$c_p^* = c_{V,\mathrm{IG}}^* + (1+\tau)\,\tau^{-1} = \frac{1}{\gamma_{0,\mathrm{IG}} - 1} + (1+\tau)\,\tau^{-1}. \tag{3.22}$$

Correspondingly, the ratio $\gamma_{0,\mathrm{vdW}} = \frac{c_p^*}{c_V^*} = \frac{c_{p,\mathrm{vdW}}}{c_{V,\mathrm{vdW}}}$ [see Eq. (1.100)] becomes

$$\gamma_{0,\mathrm{vdW}} = 1 + (1+\tau)\,\tau^{-1}. \tag{3.23}$$

leading to the critical contribution [see Eq. (1.101)]

$$\gamma_{0c,\mathrm{vdW}} = (1+\tau)\,\tau^{-1}. \tag{3.24}$$

Using Eqs. (3.14) and (3.22) to account for the mean-field behavior of the thermal conductivity and the specific heat at constant pressure along the critical isochore, the related mean-field behavior of the heat diffusivity can be written $D_{T,\mathrm{MF}} = D_{0,\mathrm{MF}}\tau^{\frac{1}{2}} + D_{b0,\mathrm{MF}}\tau$ [see Eq. (1.106)]. The resulting (critical and background) amplitudes are then defined by

$$D_{0,\mathrm{MF}} = \frac{\Lambda_{0,\mathrm{MF}}}{\rho_c\left(c_{V,\mathrm{IG}}\tau + r\right)} \simeq \frac{\Lambda_{0,\mathrm{MF}}}{\rho_c r} \tag{3.25}$$

$$D_{b0,\mathrm{MF}} = \frac{\Lambda_{b0,\mathrm{MF}}}{\rho_c\left(c_{V,\mathrm{IG}}\tau + r\right)} \simeq \frac{\Lambda_{b0,\mathrm{MF}}}{\rho_c r}, \tag{3.26}$$

where the approximations on the right-hand sides of the equations are obtained by neglecting the contribution of the $\rho_c c_{V,\mathrm{IG}}\tau$ term when $\tau \ll 1$. We note that rT_c, r and $\Lambda_{b0,\mathrm{MF}}$ have already been selected as the units of energy, specific heat, and thermal conductivity, respectively. In addition, $\frac{\mu_{b0}}{\rho_c}$ and D_{T0} are the two kinetic coefficients that have already been used to characterize the regular diffusion processes of momentum and heat. These reference parameters lead to the appearance of $\gamma_{0,\mathrm{IG}}$ and ϵ in the nondimensional momentum balance equation, and of $\gamma_{0,\mathrm{IG}}$, ϵ, $\mathrm{Pr}_{0,\mathrm{IG}}$ and Λ^* in the nondimensional energy equation. It now remains to give the closing relations between the interrelated dimensional forms of the two above amplitudes of Eqs. (3.25) and (3.26) and the kinetic coefficients D_{T0} or $\frac{\mu_{b0}}{\rho_c}$ (using the Prandtl number $\mathrm{Pr}_{0,\mathrm{IG}}$), which can be written as follows:

$$D_{0,\mathrm{MF}} \simeq \frac{\Lambda_{0,\mathrm{MF}}}{\rho_c r} = D_{T0}\frac{\gamma_{0,\mathrm{IG}}}{\gamma_{0,\mathrm{IG}}-1}\Lambda_c^* = \frac{\mu_{b0}}{\rho_c}\frac{1}{\mathrm{Pr}_{0,\mathrm{IG}}}\frac{\gamma_{0,\mathrm{IG}}}{\gamma_{0,\mathrm{IG}}-1}\Lambda_c^*$$
$$D_{b0,\mathrm{MF}} \simeq \frac{\Lambda_{b0,\mathrm{MF}}}{\rho_c r} = D_{T0}\frac{\gamma_{0,\mathrm{IG}}}{\gamma_{0,\mathrm{IG}}-1} = \frac{\mu_{b0}}{\rho_c}\frac{1}{\mathrm{Pr}_{0,\mathrm{IG}}}\frac{\gamma_{0,\mathrm{IG}}}{\gamma_{0,\mathrm{IG}}-1}.$$

When $\tau \to 0$, the mean field behavior of the thermal diffusivity of the van der Waals fluid can be then expressed in the units of any regular diffusive amplitude (D_{T0} or $\frac{\mu_{b0}}{\rho_c}$), via the following relations:

$$D_T = D_{T0}\left[\frac{\gamma_{0,\mathrm{IG}}}{\gamma_{0,\mathrm{IG}}-1}\Lambda_c^*\tau^{\frac{1}{2}} + \frac{\gamma_{0,\mathrm{IG}}}{\gamma_{0,\mathrm{IG}}-1}\tau\right] \sim D_{T0}\frac{\gamma_{0,\mathrm{IG}}}{\gamma_{0,\mathrm{IG}}-1}\Lambda_c^*\tau^{\frac{1}{2}} \tag{3.27}$$

$$D_T = \frac{\mu_{b0}}{\rho_c}\frac{1}{\mathrm{Pr}_{0,\mathrm{IG}}}\left[\frac{\gamma_{0,\mathrm{IG}}}{\gamma_{0,\mathrm{IG}}-1}\Lambda_c^*\tau^{\frac{1}{2}} + \frac{\gamma_{0,\mathrm{IG}}}{\gamma_{0,\mathrm{IG}}-1}\tau\right] \sim \frac{\mu_{b0}}{\rho_c}\frac{1}{\mathrm{Pr}_{0,\mathrm{IG}}}\frac{\gamma_{0,\mathrm{IG}}}{\gamma_{0,\mathrm{IG}}-1}\Lambda_c^*\tau^{\frac{1}{2}} \tag{3.28}$$

For example, by using the following nondimensional form of the thermal diffusivity coefficient:

$$D_T^* = \frac{D_T}{D_{T0}} = \frac{\Lambda_b^{\mathrm{MF}}}{\rho_c c_{p,\mathrm{IG}}}\Lambda_c^*\frac{c_{p,\mathrm{IG}}}{r}\tau^{\frac{1}{2}} = \frac{\gamma_{0,\mathrm{IG}}}{\gamma_{0,\mathrm{IG}}-1}\Lambda_c^*\tau^{\frac{1}{2}}, \tag{3.29}$$

and ignoring the background term that decreases with τ in the respective brackets of Eqs. (3.27) and (3.28), we can define a supplementary dimensionless critical amplitude D_c, which is a fluid-dependent quantity that is proportional to Λ_c^* already introduced in Eq. (3.14) [see Sect. (3.1.1)]. Therefore D_c is written as follows:

$$D_c = \frac{\gamma_{0,IG}}{\gamma_{0,IG}-1}\frac{1}{\mathrm{Pr}_{0,\mathrm{IG}}}\Lambda_c^*. \tag{3.30}$$

that closes the thermodynamic and transport properties description of a van der Waals one-component fluid in a mean field approach.

In the following section and the two subsequent chapters, we will drop the asterisks in the notation for the nondimensional variables. Moreover, as the previous discussion clearly identified the origin of the ideal gas Prandtl number $\mathrm{Pr}_{0,\mathrm{IG}}$, we will also drop the IG extension in the related subscript. However, the IG extension maintains the distinction between the ideal gas constant ratio $\gamma_{0,\mathrm{IG}} = \frac{c_{p,\mathrm{IG}}}{c_{V,\mathrm{IG}}}$ and the diverging (real or van der Waals) ratio $\gamma_0 = \frac{c_p}{c_V}$ [see Eqs. (1.81) and (3.23)]. All properties will be assumed to be nondimensional unless otherwise specified.

Note that a more elaborate (Ising-like) description is needed to allow comparisons to be made with experimental results, or closer to the critical point, as will be illustrated in Part II of this book.

3.2 Basic One-Dimensional Hydrodynamic Equations

3.2.1 Governing Equations and Boundary and Initial Conditions

By introducing the mean-field behavior considered above into the set of Eq. (3.10), the basic nondimensional equations of motion for a super-near-critical fluid can be obtained as follows. Mass conservation or continuity equation:

$$\rho_t + (\rho u)_x = 0 \tag{3.31}$$

Momentum balance equation:

$$\rho\,(u_t + uu_x) = -\frac{1}{\gamma_{0,\mathrm{IG}}} p_x + \epsilon \frac{4}{3} u_{xx} \tag{3.32}$$

Energy balance equation:

$$\begin{aligned} \frac{\rho}{\gamma_{0,\mathrm{IG}} - 1}\,(T_t + uT_x) = &- \left(p + \frac{9}{8}\rho^2\right) u_x \\ &+ \epsilon \left\{ \frac{\gamma_{0,\mathrm{IG}}}{\left(\gamma_{0,\mathrm{IG}} - 1\right) \mathrm{Pr}_0} \left(\left[1 + \Lambda_c \tau^{-\frac{1}{2}}\right]\right) T_{xx} \right. \\ &\left. - \frac{1}{2} \Lambda_c T_x^2 \tau^{-\frac{3}{2}} \right\} + \frac{4}{3} \epsilon \gamma_{0,\mathrm{IG}} u_x^2 \end{aligned} \tag{3.33}$$

Equation of state:

$$p = \frac{\rho\,(1+\tau)}{1 - \frac{1}{3}\rho} - \frac{9}{8}\rho^2, \tag{3.34}$$

where the small parameter ε was defined in Eq. (3.16), and where the small parameter τ is the distance to the critical point, which satisfies the condition of Eq. (3.3) because the initial conditions are close to the critical point in the situations considered here. This small parameter will appear later in association with the partial derivative $\left.\frac{\partial p}{\partial \rho}\right|_T$ after the equations have been linearized.

This means that the above complete system of equations actually involves two small parameters: ε and τ. It is noticeable that the governing equations of Ref. [1] can be found back by replacing the heat conductivity and the equation of state by those of the ideal gas in the above system of equations.

The initial and boundary conditions, which are specific to each problem under study, must be combined with the above system of equations.

- Initial conditions:

$$\begin{array}{ll} \text{Initial temperature just above } T > T_c & T(x,0) = 1+\tau \\ \text{Initial critical density } \rho = \rho_c & \rho(x,0) = 1 \\ \text{Initial pressure given by Eq. (2.34)} & p(x,0) = \frac{3}{2}(1+\tau) - \frac{9}{8} \\ \text{Fluid initially at rest} & u(x,0) = 0 \end{array} \quad (3.35)$$

- Boundary conditions:

In the experiments, the fluid cell is in ideal contact with a thermostat, which imposes a temperature or a heat flux on the cell wall (which in turn has infinite conductivity or zero thickness). We present here the basic features of the piston effect when the temperature is imposed at $x = 0$. It is assumed to vary over time as follows:

$$T(0,t) = \begin{bmatrix} 1+\tau+\alpha f(t) & \text{for } t < t_0 \\ 1+\tau+\alpha f(t_0) = C^{ste} & \text{for } t \geqslant t_0 \end{bmatrix}, \quad (3.36)$$

where $f(t)$ is a function which remains on the order of 1. α is thus the order of magnitude of the increase in the wall temperature on the timescale defined by the acoustic scale. t_0 is a cutoff time on the acoustic timescale. Another formula for the boundary condition of Eq. (3.36) is

$$T(0,t) = 1+\tau+\alpha f(t) + \alpha H_{[t_0]}\left[f(t_0) - f(t)\right],$$

where $H_{[t_0]}$ is the Heaviside step function, defined as

$$H_{[t_0]} = \begin{cases} 0 & \text{for } t < t_0 \\ 1 & \text{for } t \geqslant t_0 \end{cases}.$$

For a thermally insulated plate at $x = 1$ (adiabatic wall),

$$T_x(1,t) = 0. \quad (3.37)$$

For rigid and impermeable plates at $x = 0$ and $x = 1$ (closed system),

$$u(0,t) = u(1,t) = 0. \quad (3.38)$$

It must also be noted that a boundary condition where the temperature is known and imposed necessitates a perfectly conducting wall. Actually, as the heat conduction of a near-critical fluid goes to infinity when the critical point is approached, this condition may not fully apply. The issue of the thermal nonideality of real walls will be addressed in Sect. 5.4

3.2.2 Typical Set of Parameters

To end this chapter, which presents the basic phenomenology, we will now consider the set of parameters summarized in Table 3.1, which correspond to the van der Waals model of a

Table 3.1 Summary of the values chosen for the van der Waals model of a critical fluid

Critical temperature	Critical density
$T_c = 304.14\,\mathrm{K}$	$\rho_c = 467.8\,\mathrm{kg\,m^{-3}}$
Kinematic viscosity	Thermal diffusivity
$\nu_0 = 7.3 \times 10^{-8}\,\mathrm{m\,s^{-1}}$	$D_{T0} = 3.25 \times 10^{-8}\,\mathrm{m^2\,s^{-1}}$
Critical pressure ($p_{c,\mathrm{vdW}}$)	Specific gas constant $r = \frac{R}{m}$
$p_{c,\mathrm{vdW}} = 10.08 \times 10^6\,\mathrm{Pa}$	$r = 188.9\,\mathrm{J\,Kg^{-1}\,K^{-1}}$
Prandtl number (IG)	Velocity of sound (IG)
$\mathrm{Pr}_0 = 2.24$	$c_{s,\mathrm{IG}} = 284\,\mathrm{m\,s^{-1}}$

critical fluid whose thermodynamics and transport properties take values close to those of carbon dioxide.

For an experiment performed 1.5 K above the critical temperature in a 10 mm container filled to critical density, a typical heating function will provide a temperature increase of 20 mK over a time period of 0.1 s. Considering that the typical acoustic time for the ideal gas $t_a = \frac{L}{c_{s0}} = 3.52 \times 10^{-5}$ s, the value of t_0 as defined in the boundary conditions is $t_0 = 2.84 \times 10^3$, and $\alpha = 2.31 \times 10^{-8}$. The function $f(t)$, which is always on the order of 1, may take any form that correctly represents the experiment under consideration. It can also take a typical form that is used to keep the equation simple without diminishing the generality and the utility of the theory. From Table 3.1, we get: $\varepsilon = 2.56 \times 10^{-8}$, $D_c = 1.15$ and $\Lambda_c = 0.75$.

References

1. Kassoy DR (1979) The response of a confined gas to a thermal disturbance. i. Slow transients. SIAM J Appl Math 36:3
2. Onuki A (2002) Phase transition dynamics. Cambridge University Press, Cambridge

Chapter 4
Temperature and Density Equilibration

Abstract The basic features of thermoacoustic effects in a near-critical fluid layer is presented for the simplest (one dimensional) configuration. A temperature step is imposed at one wall side of a slab-like container filled with a near-supercritical pure fluid. The other wall side of the container is thermally isolated. Gravity is assumed to be zero, so buoyant convection is absent from the system. The transient behavior of a given fluid mass is thus studied between two thermodynamic equilibrium states corresponding to the temperature step. It will be shown in particular that the respective relaxations of temperature and density towards equilibrium are uncoupled to first order. The piston effect homogenizes temperature on a timescale that is much shorter than the diffusion one, while density relaxes diffusively in a quasi-isothermal medium. A thermodynamic description of heat exchanges in a near critical fluid is given first. This shows the source of characteristic timescale of the adiabatic temperature equilibration. This characteristic time scale compares well with the characteristic time of the piston effect obtained from the detailed hydrodynamic approach presented hereafter.

4.1 Thermodynamic Approach of the Temperature Equilibration Timescale

We consider a one-dimensional (1-D) container of arbitrary size L at $T = 1 + \tau$, filled with an expandable and compressible near-supercritical fluid at critical density. The wall on the left-hand side is heated while that on the right-hand side is thermally insulated. Heat diffuses in a thin thermal boundary layer while the rest of the container is adiabatic. Between the initial time t_0 and time t, the boundary layer receives an amount Q_{BL} of heat. The first law of thermodynamics gives the variation in the internal energy, ΔE_{BL}, of the boundary layer between t_0 and t:

$$\Delta E_{BL} = Q_{BL} + W = Q_{BL} - p\Delta V_{BL}, \tag{4.1}$$

B. Zappoli et al., *Heat Transfers and Related Effects in Supercritical Fluids*, Fluid Mechanics and Its Applications 108, DOI: 10.1007/978-94-017-9187-8_4,

where $W = -p\Delta V_{BL}$ is the work done by the pressure p , and ΔV_{BL} is the change in volume of the boundary layer. The thermal boundary layer undergoes an isobaric process (the pressure is homogeneous to first order as it is smoothed out by pressure waves), and thus

$$(Q_{BL})_p = \Delta E_{BL} + \Delta(pV_{BL}) = \Delta H_{BL}. \tag{4.2}$$

The energy received is then

$$(Q_{BL})_p = \rho c_p \Sigma \delta \Delta T_{BL}, \tag{4.3}$$

where c_p is the specific heat at constant pressure, which exhibits a critical divergence ($c_p \propto \tau^{-\gamma}$, see Chap. 1) as the critical temperature is approached, δ is the thickness of the boundary layer at time t, and Σ is an arbitrary surface perpendicular to the direction of the 1-D x axis. If ΔE_B denotes the variation in the internal energy of the bulk fluid during the same time period, the variation in the internal energy for the container is

$$\Delta E_{tot} = \Delta E_{BL} + \Delta E_B. \tag{4.4}$$

Considering that the container does not exchange work due to pressure forces (the walls are nondeformable), it only receives energy through the boundary layer. We can write the total energy balance

$$\Delta E_{tot} = Q_{BL}. \tag{4.5}$$

We now assume that the fluid is described by a van der Waals equation of state that gives the following expression for the specific internal energy [see Eq. (3.11)]:

$$e = c_V T - a\rho. \tag{4.6}$$

We can express the detailed internal energy balances for the bulk (subscript B) and the boundary layer (subscript BL) as a function of T and ρ:

$$\begin{aligned} \Delta E_{BL} &= \rho(c_V \Delta T_{BL} - a\Delta\rho_{BL})\Sigma\delta \\ \Delta E_B &= \rho(c_V \Delta T_B - a\Delta\rho_B)\Sigma\,(L-\delta)\,, \end{aligned}$$

which lead to the following expression for the internal energy balance of the container:

$$\Delta E_{tot} = \rho c_V \Delta T_{BL}\Sigma\delta + \rho c_V \Delta T_B \Sigma(L-\delta) - a\rho\,[\Delta\rho_{BL}\Sigma\delta + \Delta\rho_B\Sigma(L-\delta)]\,.$$

If we account for the mass conservation

$$\Delta\rho_{BL}\Sigma\delta + \Delta\rho_B\Sigma(L-\delta) = 0,$$

the internal energy balance can be written as

$$\Delta E_{tot} = \rho c_V \Delta T_{BL}\Sigma\delta + \rho c_V \Delta T_B \Sigma(L-\delta), \tag{4.7}$$

which, taking into account Eqs. (4.3) and (4.5) , gives

$$c_V(\Delta T_{BL} - \Delta T_B)\delta + c_V L \Delta T_B = c_{pc}\Delta T_B \delta. \tag{4.8}$$

The fluid is thermalized when the bulk and the boundary layer temperature are equal:

$$\Delta T_{BL} = \Delta T_B,$$

so that Eq. (4.8) gives the relation

$$\frac{\delta}{L} = \frac{c_V}{c_p} = \frac{1}{\gamma_0}, \tag{4.9}$$

where the specific heat ratio $\gamma_0 \simeq \gamma_{0c} \propto \tau^{-\gamma-\alpha}$ diverges to infinity when $\tau \to 0$ (with $c_V \propto \tau^{-\alpha}$, $c_p \propto \tau^{-\gamma}$, $\gamma \simeq 1.24$ and $\alpha \simeq 0.11$; [see Eq. (1.82) and Table 1.2]). As the boundary layer grows by thermal diffusion, the temperature equilibration time, t_{eq}, is such that the thermal penetration length is

$$\delta = \sqrt{D_{Tc} t_{eq}}, \tag{4.10}$$

where the thermal diffusivity coefficient $D_{Tc} \simeq \frac{\Lambda}{\rho_c c_{pc}} \propto \tau^{\gamma - Y_\Lambda}$ goes to zero for $\tau \to 0$ (with the thermal conductivity $\Lambda \propto \tau^{-Y_\Lambda}$ and $Y_\Lambda \simeq 0.57$; [see Eq. (1.46) and Table 1.4]).

Equations (4.10) and (4.31) give

$$t_{eq} = \frac{L^2}{D_{Tc}\gamma_0^2}, \tag{4.11}$$

where the characteristic time t_{D_T} of the heat diffusion at the sample length scale L is

$$t_{D_T} = \frac{L^2}{D_{Tc}}. \tag{4.12}$$

Therefore, Eq. (4.11) can be written as

$$t_{eq} = \frac{t_{D_T}}{\gamma_0^2}, \tag{4.13}$$

Because γ_{0c} goes to infinity as $\gamma_{0c} \propto \tau^{-\gamma}$ when the critical point is approached, the heat equilibration time in a near-supercritical fluid becomes much shorter than the heat diffusion time. Indeed, instead of the expected critical slowing down

$$t_{D_T} \propto \tau^{Y_\Lambda - \gamma} \quad (\text{with } Y_\Lambda - \gamma \simeq -0.67)$$

of the heat diffusion time, it is a critical speeding up

$$t_{eq} \propto \tau^{Y_\Lambda + \gamma - 2\alpha} \quad (\text{with } Y_\Lambda + \gamma - 2\alpha \simeq -1.59)$$

of the adiabatic equilibration time that is observed [4]. The equilibration time as defined by Eq. (4.13) is different from $\frac{t_{D_T}}{(\gamma_0 - 1)^2}$ as obtained from the Onuki et al. approach [20]. Here different densities in the bulk fluid and in the boundary layer are accounted for. These two times are however asymptotically equivalent near the critical point. For a mean-field approach to a van der Waals fluid, we note that the value $Y_{\Lambda,\text{MF}} + \gamma_{\text{MF}} - 2\alpha_{\text{MF}} = -\frac{3}{2}$ of the related mean-field exponent of t_{eq} is very similar, with $Y_{\Lambda,\text{MF}} = \frac{1}{2}$, $\gamma_{\text{MF}} = 1$, and $\alpha_{\text{MF}} = 0$ (see Table 1.7). For ideal gases in normal conditions where $\gamma_{0,\text{IG}} = \frac{c_P}{c_V} \sim 1.4$, note, from Eq. (4.13), that the characteristic time for adiabatic temperature equilibration is of the same order of magnitude as the characteristic time for heat diffusion.

4.2 Fast Thermal Equilibration by the Piston Effect

The problem defined by the system of partial differential Eqs. (3.31)–(3.34) has no exact solution. However, the presence of the two small parameters $\varepsilon = \text{Pr}_0 \frac{t_a}{t_{D_0}} \ll 1$ and $\tau = \frac{T - T_c}{T_c} \ll 1$ makes it possible to obtain approximate solutions by asymptotic methods.

A full description of asymptotic techniques is beyond the scope of this monograph, but readers who would like more information on this topic can refer to [15]. In the following we only explain the main concepts involved in the use of asymptotic techniques.

4.2.1 About Asymptotic Techniques

4.2.1.1 Asymptotic Expansions to Solve Regular or Singular Problems in Fluid Mechanics

When the solution of a differential equation or a system of differential equations is not known, the asymptotic expansion of a given function $f(\varepsilon)$ with respect to a given asymptotic sequence $\nu_n(\varepsilon)$ is given by

$$f(\varepsilon) = \sum_{n=1}^{N} a_n \nu_n(\varepsilon) + o(\nu_N(\varepsilon)). \tag{4.14}$$

$o(\nu_n(\varepsilon))$ means that $\lim_{\varepsilon\to 0}\left(\frac{\nu_{n+1}(\varepsilon)}{\nu_n(\varepsilon)}\right) = 0$. The asymptotic sequence $\nu_n(\varepsilon)$ is such that $\nu_{n+1}(\varepsilon) = o(\nu_n(\varepsilon))$. In the above equation, f at the order $\nu_N(\varepsilon)$ is defined by $\sum_{n=1}^{N} a_n \nu_n(\varepsilon)$.

The coefficients a_n of the asymptotic expansion can be obtained step by step from the first to the kth by constructing the series

$$a_1 = \lim_{\varepsilon\to 0}\left(\frac{f(\varepsilon)}{\nu_1(\varepsilon)}\right)$$
$$a_k = \lim_{\varepsilon\to 0}\left(\frac{f(\varepsilon) - \sum_{n=1}^{ka_i} a_n \nu_n(\varepsilon)}{\nu_k(\varepsilon)}\right).$$

When it is possible to construct all of the terms of an asymptotic development for $N \to \infty$, we obtain an infinite series called an *asymptotic series*. Asymptotic series differ the asymptotic development limited to N terms that was previously defined. To give an overview of the technique, we will start by stating that the sum of the N first terms $F_N(\varepsilon)$ of a convergent series is such that

$$\lim_{\substack{N\to\infty \\ \varepsilon \text{ fixed}}} F_N(\varepsilon) = f(\varepsilon),$$

while the sum of the first N terms in the case of an asymptotic expansion is such that

$$\lim_{\substack{\varepsilon\to 0 \\ N \text{ fixed}}} F_N(\varepsilon) = f(\varepsilon). \tag{4.15}$$

The larger the value of N, the better the approximation given by a (convergent) asymptotic series for a given ε; the smaller the value of ε, the better the approximation given by a (possibly divergent) asymptotic expansion for a given N. Beyond their application to analytically known functions with short series, asymptotic expansions are widely used to obtain approximate solutions for systems of differential equations.

Practical problems often involve a variable x and a small parameter ε; the solution to such a problem, $f(x, \varepsilon)$, can be considered a function of x and ε. The parameter ε can be very small, but its value is fixed. This is why one way to obtain a better approximation of f is to calculate more terms in its convergent series expansion. However, this implies that the successive terms of the series must be calculated and its convergence has to be assessed. In real problems, it is often impossible to calculate many terms and to show their convergence. The interest in applying asymptotic expansions to practical problems is due to the potential of this approach to obtain good precision with only a few terms.

This asymptotic expansion technique is particularly well adapted to nonlinear differential problems derived from the Navier–Stokes system of equations when there is a small parameter due to the nondimensionalization. Let $f(x, \varepsilon)$ be a solution of a differential problem $\mathcal{D}(x, \varepsilon)$. The basic assumption of asymptotic methods is the following. For small values of the parameter ε, the asymptotic approximation of the exact solution $f(x, \varepsilon)$, denoted $f_\varepsilon(x)$, is the same as the exact solution of the approximate differential problem $\mathcal{D}_\varepsilon$ obtained by expanding the function as an asymptotic sequence like that in Eq. (4.14) (which needs ε to be small in $\mathcal{D}(x, \varepsilon)$). The approximate equations $\mathcal{D}_\varepsilon$ are generally much simpler, and it is possible to solve them analytically, unlike the exact problem $\mathcal{D}(x, \varepsilon)$.

4.2.1.2 The Particular Case of a Singular Perturbation Problem

Let us give an example of a singular perturbation problem where the exact solution is known. Consider the equation

$$\varepsilon \frac{d^2 f}{dx^2} + \frac{df}{dx} = a \qquad f(0) = 0, \ f(1) = 1. \tag{4.16}$$

This equation is singular at $x = 0$ for $\varepsilon \ll 1$ (except for $a = 1$). Therefore, its asymptotic expansion if x is on the order of 1 with $\varepsilon \ll 1$ is not valid when x goes to zero. More generally, a singular perturbation problem may be suspected if the highest order derivative vanishes when ε is set to zero. The singularity is easily demonstrated, as the exact solution of Eq. (4.16) is known explicitly:

$$f(x, \epsilon) = (1 - a)\frac{1 - e^{-\frac{x}{\varepsilon}}}{1 - e^{-\frac{1}{\varepsilon}}} + ax. \tag{4.17}$$

Recall that there is no exact solution for the Navier–Stokes equations.

Setting $\varepsilon = 0$ reduces Eq. (4.16) to a first-order equation, so both boundary conditions cannot be satisfied (unless $a = 1$). The exact solution shows that the condition at $x = 0$ is useless. The approximate solution for small ε is the following (see, for example, Fig. 4.1):

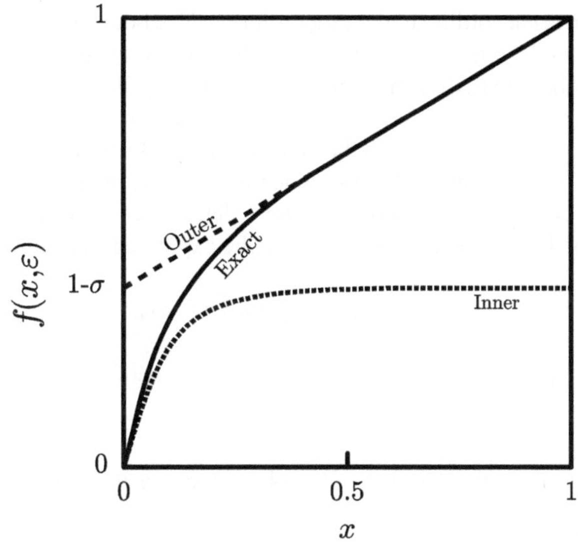

Fig. 4.1 Exact solution and approximate solution when $x \to 0$ for $\epsilon \ll 1$

$$\bar{f}(x,\varepsilon) \sim (1-a) + ax.$$

We observe that the approximation is good, except in the "boundary layer," where $x = O(\varepsilon)$. This solution, valid for $x = O(1)$, is called the *outer solution*, and x is the outer variable. The rescaling procedure consists of defining a magnified coordinate, called the inner variable:

$$\tilde{x} = \frac{x}{\varepsilon}, \tag{4.18}$$

which is appropriate for describing the boundary layer region, so that

$$f(x,\varepsilon) = f(\varepsilon\tilde{x},\varepsilon) = \tilde{f}(\tilde{x},\varepsilon). \tag{4.19}$$

Using Eqs. (4.18) and (4.19) in (4.16) transforms the latter into

$$\frac{d^2\tilde{f}}{d\tilde{x}^2} + \frac{d\tilde{f}}{d\tilde{x}} = a\varepsilon \qquad \tilde{f}(0) = 0, \quad \tilde{f}\left(\frac{1}{\varepsilon}\right) = 1. \tag{4.20}$$

If $\varepsilon = 0$ in Eq. (4.20), the solution that satisfies the boundary condition at $\tilde{x} = 0$ is

$$\tilde{f}(\tilde{x},0) = \alpha\left(1 - e^{-\tilde{x}}\right), \tag{4.21}$$

where α is any multiplicative factor. Equation (4.21) is the inner solution for which the factor α must be determined. The boundary condition at $\tilde{x} = \frac{1}{\varepsilon}$ gives $\alpha = 1$, which is not correct according to the exact solution. The boundary condition at $x = 1$ must be abandoned for the inner solution. In the same way, the boundary condition at $x = 0$ cannot be satisfied by the outer solution. These difficulties are overcome by the *matching principle*, which basically replaces the missing boundary conditions by matching the inner solution with the outer solution. The limit-matching principle, in its simplest formulation, is that the inner limit of the outer solution is equal to the outer limit of the inner solution. For more details, see for example [23]. In our present case, the limit matching principle can be expressed as follows:

$$\lim_{x \to 0} \bar{f}(x, \varepsilon) = \lim_{\tilde{x} \to \infty} \tilde{f}(\tilde{x}, \varepsilon), \tag{4.22}$$

from which we obtain

$$\alpha = 1 - a.$$

The approximations to the solution of Eq. (4.17) are thus the following:

- For the outer solution:

$$(1 - a) + ax \qquad \text{when } \varepsilon \to 0 \text{ with } x \text{ fixed.}$$

- For the inner solution:

$$(1 - a)\left(1 - e^{-\tilde{x}}\right) \qquad \text{when } \varepsilon \to 0 \text{ with } \tilde{x} = \frac{x}{\varepsilon} \text{ fixed.}$$

The uniformly valid approximation for any value of x, $\hat{f}(x, \varepsilon)$, is obtained using the additive composition procedure:

$$\hat{f}(x, \varepsilon) = \bar{f}(x, \varepsilon) + \tilde{f}\left(\frac{x}{\varepsilon}, \varepsilon\right) - \lim_{\tilde{x} \to \infty} \tilde{f}(\tilde{x}, \varepsilon).$$

In the example above, the additive composition procedure leads to

$$\hat{f}(x, \varepsilon) = (1 - a)\left(1 - e^{-\frac{x}{\varepsilon}}\right) + ax.$$

In the following, singular perturbation problems will be encountered when we solve the heat transfer problem $\mathcal{D}$ as defined by Eqs. (3.31)–(3.34), which contain two parameters ε and τ. This problem has no analytical solution if ε and τ are on the order of 1. It can, however, be changed into a simpler but approximate problem noted $\mathcal{D}_{\varepsilon\tau}$ as ε and τ are small parameters. The procedure that will be used initially involves finding the relevant asymptotic sequences through a scaling analysis. Once this is done, the expansion of the dependent variable is fed into $\mathcal{D}(x, \varepsilon, \tau)$ to obtain the approximate problem $\mathcal{D}_{\varepsilon\tau}$ and look for its exact solution. The behavior of this solution at different length and time scales gives information on the existence of singularities and thus of specific scales at which different physical phenomena are prominent. The general mathematical techniques that are employed to obtain asymptotic approximations (see, for example, [23]) are beyond the scope of this monograph. Only absolutely necessary information on this topic will be supplied when needed in this text.

4.2.2 Acoustic Period

4.2.2.1 Thermal Boundary Layer Expansion: The Piston

The analysis we will provide in this chapter derives from [2, 25, 29]. When heat is supplied to the fluid through a perfectly conducting wall at $x = 0$, it diffuses into a very thin thermal boundary layer, since the diffusion depth in the timeframes of some acoustic times is very small. It is thus necessary to rescale the space variable x to focus on what happens in this layer of characteristic thickness $\delta(\varepsilon)$, where the rescaled variable z, defined as

$$z = \frac{x}{\delta(\varepsilon)}, \tag{4.23}$$

is on the order of 1.

In order to define the above approximate differential problem $\mathcal{D}_{\varepsilon\tau}$, we must asymptotically expand the different dependent variables as a function of an asymptotic sequence that is unknown *a priori*. The dependent variables are decorated by a tilde and expanded as follows:

$$\begin{aligned}
&\widetilde{T}^a(z,t) = 1 + \tau + \eta_T T^a_{bl}(z,t) + o(\eta_T) && \\
&\widetilde{\rho}^a(z,t) = 1 + \eta_\rho \rho^a_{bl}(z,t) + o(\eta_\rho) && \eta_\rho \equiv \eta_\rho(\epsilon,\tau) \ll 1 \\
&\widetilde{u}^a(z,t) = 0 + \eta_u u^a_{bl}(z,t) + o(\eta_u) && \eta_u \equiv \eta_u(\epsilon,\tau) \ll 1 \\
&\widetilde{p}^a(z,t) = \tfrac{3}{2}(1+\tau) - \tfrac{9}{8} + \eta_p p^a_{bl}(z,t) + o(\eta_p) && \eta_P \equiv \eta_p(\epsilon,\tau) \ll 1,
\end{aligned} \tag{4.24}$$

where the superscript a stands for acoustic period, the subscript bl stands for boundary layer, and the functions η_T, η_ρ, η_u and η_p are the order of magnitudes of the different variables in the thermal boundary layer (to be determined by an order-of-magnitude analysis). The scaling functions $\eta_\rho(\epsilon,\tau)$, $\eta_u(\epsilon,\tau)$ and $\eta_p(\epsilon,\tau)$ form the asymptotic sequences. The scaling functions depend on the two small parameters of the problem ε and τ. The functions $T^a_{bl}(z,t)$, $\rho^a_{bl}(z,t)$, $u^a_{bl}(z,t)$ and $p^a_{bl}(z,t)$ are on the order of 1.

Since the boundary condition [see Eq. (3.36)] at $x = 0$ imposes a temperature perturbation with an order of magnitude of α, evaluating the order of magnitude η_T is straightforward: $\eta_T = \alpha$. Replacing the expansions in Eqs. (3.31)–(3.34) then gives the approximate equations (corresponding to the problem $\mathcal{D}_{\varepsilon\tau}(x,t)$). To first order, Eq. (3.31) can be written as

$$\eta_\rho \rho^a_{bl\,t} + \frac{\eta_u}{\delta} u^a_{bl\,z} = 0. \tag{4.25}$$

This equation must be satisfied by the perturbations for mass conservation to be satisfied, so the following relation must hold:

$$\eta_\rho = \frac{\eta_u}{\delta}. \tag{4.26}$$

To first order, Eq. (3.34) can be written as

$$\eta_P p^a_{bl} = \frac{3}{2}\alpha T^a_{bl} + \frac{9}{4}\tau \eta_\rho \rho^a_{bl}. \tag{4.27}$$

As heating is the driving force for the evolution, density can be coupled to temperature only if

$$\tau \eta_\rho = \alpha \Rightarrow \eta_\rho = \frac{\alpha}{\tau}. \tag{4.28}$$

It must be noted here that the first density perturbation in the thermal boundary layer is τ times larger in near-critical fluids than in ideal gases. This also imposes a first restriction on the linearized hydrodynamic theory:

$$\eta_\rho \ll 1 \Rightarrow \tau \gg \alpha. \tag{4.29}$$

In terms of the initial distance to the critical point, the condition is

$$\Delta T_W \ll T_0 - T_c,$$

where ΔT_W is the maximum increase in temperature at the boundary.

To obtain the orders of magnitude from the energy equation [see Eq. (3.33)], we must match the orders of magnitude of the diffusion terms $\tilde{T}^a_{zz}$ with the largest terms in the equation to effectively describe the diffusive process. As the other terms are much smaller, we then obtain

$$\begin{array}{ccccc} \frac{\alpha}{\gamma_{0,\mathrm{IG}}-1} T^a_{bl\,t} & = & -\frac{3}{2}\frac{\alpha}{\tau} u^a_{bl\,z} & + & D_c \frac{\varepsilon\alpha}{\sqrt{\tau}\delta^2} T^a_{bl\,zz} \\ \mathbf{I} & = & \mathbf{II} & + & \mathbf{III} \end{array}. \tag{4.30}$$

Term **I** corresponds to the rate of change in temperature; term **II** represents the power of the inner pressure forces due to the deformation of the fluid domain; term **III** represents heat diffusion. In ideal gases [13], the above equation can be derived by setting $\tau = 1$ in Eq. (4.30): term **II** is of the same order of magnitude as term **I**. In near-critical fluids, as $\tau \ll 1$, term **I** is much smaller than term **II**. We must thus impose matching between diffusion term **III** and term **II**. This gives a boundary layer thickness that tends to zero as $\tau^{\frac{1}{4}}$ when the critical point is approached:

$$\frac{\alpha}{\tau} = \frac{\varepsilon\alpha}{\sqrt{\tau}\delta^2} \Rightarrow \delta = \varepsilon^{\frac{1}{2}}\tau^{\frac{1}{4}}. \tag{4.31}$$

For an ideal gas, **III** is matched with both **II** and **I**, which leads to $\delta = \epsilon^{\frac{1}{2}}$. In this case, compressibility effects are small corrections to diffusion effects. The diffusion boundary layer in near-critical fluids is thus $\tau^{\frac{1}{4}}$ times thinner than it is in an ideal gas on the same timescale. This is consistent with the fact that the heat diffusivity tends to zero as $\tau^{\frac{1}{2}}$. The specific heat at constant pressure is indirectly present in the energy equation through the term for pressure forces. It would appear directly if the enthalpy balance was used instead of the internal energy balance.

From Eqs. (4.26), (4.28), and (4.31), we obtain

$$\delta(\varepsilon, \tau) = \epsilon^{\frac{1}{2}}\tau^{\frac{1}{4}}; \qquad \eta_\rho(\tau) = \frac{\alpha}{\tau}; \qquad \eta_u(\varepsilon, \tau) = \frac{\alpha\epsilon^{\frac{1}{2}}}{\tau^{\frac{3}{4}}}. \tag{4.32}$$

The order of magnitude $\eta_p(\varepsilon, \tau)$ of the pressure perturbation will be determined by the matching conditions with the bulk. This expresses the need for the solutions to be matched so as to represent a continuous approximation across the whole range of space and time. For any dependent variable $X(x, t)$, the matching condition to first order can be written as

$$\eta_{bl}(\varepsilon, \tau) \lim_{\frac{x}{\delta}\to\infty,\, x\ll 1} X^a_{bl}(z, t) = \eta_b(\varepsilon, \tau) \lim_{x\to 0} X^a_b(x, t), \tag{4.33}$$

where the functions $\eta_{bl}(\varepsilon, \tau)$ and $\eta_b(\varepsilon, \tau)$ are, respectively, the orders of magnitude of the approximate solutions for the perturbations in the boundary layer $X^a_{bl}(z, t)$ and in the bulk $X^a_b(x, t)$. Let us anticipate that the order of magnitude in the boundary layer will be found to conform to $\eta_p = \eta_u$. In other words, the approximation for the equation of state in the boundary layer is (note that any other choice for this order of magnitude would lead to awkward results):

$$T^a_{bl} = -\frac{3}{2}\rho^a_{bl}.$$

When included in Eqs. (4.25), (4.27), and (4.30), the following equations are obtained:

Energy equation (heat diffusion):

$$-\frac{3}{2}u^a_{bl\,z} + D_c T^a_{bl\,zz} = 0 \tag{4.34}$$

Equation of continuity:

$$\rho^a_{bl\,t} + u^a_{bl\,z} = 0 \tag{4.35}$$

Equation of state:

$$T^a_{bl} + \frac{3}{2}\rho^a_{bl} = 0. \tag{4.36}$$

The boundary condition at $x = 0$ for the bulk equations is replaced by the matching condition with the boundary layer solutions at $z \to \infty$. When $z \to \infty$, the temperature perturbation in the bulk must go to zero. At the acoustic scale, diffusion has not yet reached the bulk, and other perturbations—if any—should be of a higher order. The boundary conditions are thus

$$z = 0: \quad T^a_{bl}(0,t) = \begin{bmatrix} \frac{t}{t_0} & for \quad t \le t_0 \\ 1 & for \quad t > t_0 \end{bmatrix} = \frac{t}{t_0} - H_{[t_0]}\left(\frac{t}{t_0} - 1\right) \tag{4.37}$$

$$u^a_{bl}(0,t) = 0 \tag{4.38}$$

$$\lim_{z\to\infty} T^a_{bl}(z,t) = 0. \tag{4.39}$$

The initial conditions are deduced from Eq. (3.35):

$$T^a_{bl}(z,0) = u^a_{bl}(z,0) = \rho^a_{bl}(z,0) = 0. \tag{4.40}$$

We have chosen a linear increase in temperature to simplify the description, since we are only aiming in this chapter to describe the phenomenology. In Part II we will consider more realistic boundary conditions. By combining Eqs. (4.34), (4.35), and (4.36), an equation for diffusion in a semi-infinite medium is obtained:

$$T^a_{bl\,t} = D_c T^a_{bl\,zz}, \tag{4.41}$$

with the initial and boundary conditions of Eqs. (4.37), (4.39), and (4.40). The solution is obtained by the Laplace transform technique (see [7]):

$$T^a_{bl}(z,t) = 4\left[\frac{t}{t_0}\, i^2 \operatorname{erf} c(\tilde{\eta}) - H_{[t_0]}\left(\frac{t}{t_0} - 1\right) i^2 \operatorname{erf} c\left(\frac{\tilde{\eta}}{\sqrt{\left(1 - \frac{t}{t_0}\right)}}\right)\right]. \tag{4.42}$$

Here, $\tilde{\eta}$ is the usual diffusion variable:

$$\tilde{\eta} = \frac{z}{2\sqrt{D_c\, t}}.$$

In Eq. (4.42), $i^n \operatorname{erf} c(\tilde{\eta})$ [1] represents the repeated integrals of the complementary error function erfc.

Density is obtained from Eq. (4.36) and velocity from the integration of Eq. (4.35):

$$\rho^a_{bl}(z,t) = -\frac{8}{3}\left[\frac{t}{t_0}\, i^2 \operatorname{erf} c(\tilde{\eta}) - H_{[t_0]}\left(\frac{t}{t_0} - 1\right) i^2 \operatorname{erf} c\left(\frac{\tilde{\eta}}{\sqrt{\left(1 - \frac{t}{t_0}\right)}}\right)\right] \tag{4.43}$$

$$u^a_{bl}(z,t) = \frac{\sqrt{D_c}}{t_0}\left[\sqrt{t}\left(\frac{2}{\sqrt{\pi}} - i^0\,\mathrm{erf}\,c(\tilde{\eta})\right) - H_{[t_0]}\left(\sqrt{t-t_0}\right)\left(\frac{2}{\sqrt{\pi}} - i^0\,\mathrm{erf}\,c\left(\frac{\tilde{\eta}}{\sqrt{\left(1-\frac{t}{t_0}\right)}}\right)\right)\right]. \quad (4.44)$$

It must be emphasized that, just as for ideal gases [13], the velocity at the edge of the thermal boundary layer, denoted $u^a_{\mathrm{Piston}}(t)$, is nonzero:

$$u^a_{\mathrm{Piston}}(t) = \lim_{z\to\infty} u^a_{bl}(z,t) = \frac{2}{\sqrt{\pi}}\frac{\sqrt{D_c}}{t_0}\left[\sqrt{t} - H_{[t_0]}\sqrt{(t-t_0)}\right]. \quad (4.45)$$

This means that the boundary layer injects mass into the bulk. Given the scaling of the velocity in the boundary layer, as given by Eq. (4.32), this mass injection is $\tau^{-\frac{3}{4}}$ stronger than for an ideal gas. According to Ref. [13], at the sample length scale, the inflow appears to emerge from the boundary as if a piston had injected it through the wall. Moreover, after integrating the approximate energy equation in the boundary layer written in the form of Eq. (4.34), taking into account the boundary conditions of Eq. (4.38) and Eqs. (4.37), and (4.39), we obtain

$$u^a_{\mathrm{Piston}}(t) = -\frac{2}{3} D_c T^a_{bl\,z}(0,t). \quad (4.46)$$

This also means that *the velocity at the edge of the boundary layer varies in exactly the same way over time as the heat flux at the heated wall.* This is an important phenomenological signature of the mechanism of the piston effect, which will be illustrated throughout several analysis of the experimental results presented in this book. The reduced velocity $\frac{\tilde{u}^a}{\alpha}$ and the reduced temperature perturbation $\frac{\tilde{T}^a-(1+\tau)}{\alpha}$ [see Eq. (4.24)] at the end of the heating period are plotted in Fig. 4.2 with respect to z for different τ (using $t_0 = 2.8 \times 10^3$ and $\alpha = 6.6 \times 10^{-5}$, see Table 3.1 in Sect. 3.2.2).

4.2.2.2 The Acoustic Field Generated in the Bulk by the Piston

In this section, we describe the bulk phase on the acoustic timescale. The bar or the subscript b refers to the bulk properties. The general expression for the matching condition [see Eq. (4.33)] between the boundary layer and the bulk phase can be written as follows for the velocity:

$$\bar{u}^a(x=0,t) = \frac{\alpha\epsilon^{\frac{1}{2}}}{\tau^{\frac{3}{4}}}\lim_{z\to\infty} u^a_{bl}(z,t) = \frac{\alpha\epsilon^{\frac{1}{2}}}{\tau^{\frac{3}{4}}} u^a_{\mathrm{Piston}}(t). \quad (4.47)$$

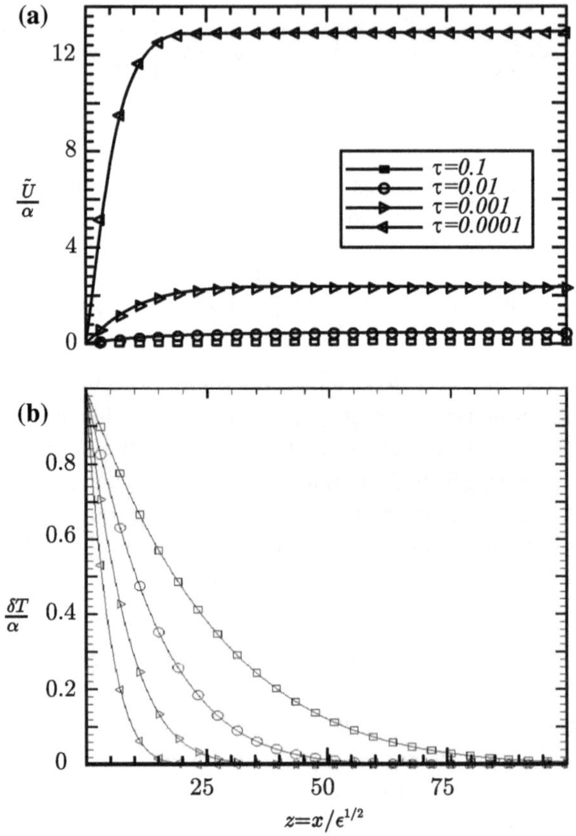

Fig. 4.2 Reduced velocity (**a**) and temperature perturbations (given by Eq. 4.42) (**b**) in the boundary layer on the acoustic timescale at the end of the heating period

Equation (4.47) means that the mass emerging from the boundary layer region provokes a mechanical disturbance, the order of magnitude of which is $\frac{\alpha\epsilon^{\frac{1}{2}}}{\tau^{\frac{3}{4}}}$. Accordingly, the solution for $\bar{u}^a$ must be sought in the form of the following asymptotic expansion:

$$\bar{u}^a(x,t) = \frac{\alpha\epsilon^{\frac{1}{2}}}{\tau^{\frac{3}{4}}} u_b^a(x,t) + o\left(\frac{\alpha\epsilon^{\frac{1}{2}}}{\tau^{\frac{3}{4}}}\right). \tag{4.48}$$

The mechanical disturbance provokes perturbations of the coupled thermodynamic variables, which are thus of the same order of magnitude. Accordingly, the appropriate expansions are

$$\begin{aligned}\bar{T}^a &= 1+\tau+\frac{\alpha\epsilon^{\frac{1}{2}}}{\tau^{\frac{3}{4}}}T(x,t)+o\left(\frac{\alpha\epsilon^{\frac{1}{2}}}{\tau^{\frac{3}{4}}}\right)\\ \bar{\rho}^a &= 1+\frac{\alpha\epsilon^{\frac{1}{2}}}{\tau^{\frac{3}{4}}}\rho_b^a(x,t)+o\left(\frac{\alpha\epsilon^{\frac{1}{2}}}{\tau^{\frac{3}{4}}}\right)\\ \bar{p}^a &= \tfrac{3}{2}(1+\tau)-\tfrac{9}{8}+\frac{\alpha\epsilon^{\frac{1}{2}}}{\tau^{\frac{3}{4}}}p_b^a(x,t)+o\left(\frac{\alpha\epsilon^{\frac{1}{2}}}{\tau^{\frac{3}{4}}}\right).\end{aligned} \tag{4.49}$$

As the order of magnitude of the pressure disturbance in the bulk is known, the matching condition (Eq. 4.33) ensures that the order of magnitude of the pressure perturbation in the boundary layer is also $\frac{\alpha\epsilon^{\frac{1}{2}}}{\tau^{\frac{3}{4}}}$. When included in the momentum equation for the boundary layer [see Eq. (3.3)], the latter becomes $p^a_{bl\,z}=0$, leading to a homogeneous pressure within the boundary layer.

Using the expansions of Eqs. (4.48) and (4.49) in the system of Eqs. (3.31)–(3.34), we obtain the following equations:

$$\rho^a_{b\,t}+u^a_{b\,x}=0 \tag{4.50}$$

$$u^a_{b\,t}=-\frac{1}{\gamma_{0,\mathrm{IG}}}p^a_{b\,x} \tag{4.51}$$

$$\frac{3}{2}T^a_b=p^a_b \tag{4.52}$$

$$\frac{3}{2}\left(\gamma_{0,\mathrm{IG}}-1\right)u^a_{b\,x}+T^a_{b\,t}=0. \tag{4.53}$$

The initial conditions are deduced from Eq. (3.35)

$$T^a_b(x,0)=u^a_b(x,0)=\rho^a_b(x,0)=0, \tag{4.54}$$

and the boundary conditions for temperature are given by Eq. (3.37). Elementary combination of Eqs. (4.50)–(4.53) gives a propagation equation for u^a_b:

$$u^a_{b\,tt}-c^2_{s0}u^a_{b\,xx}=0, \tag{4.55}$$

where

$$c^2_{s0}=\frac{9}{4}\frac{\gamma_{0,\mathrm{IG}}-1}{\gamma_{0,\mathrm{IG}}}. \tag{4.56}$$

The boundary condition at $x=0$ is given by matching conditions with the boundary layer:

$$u^a_b(x=0,t)=u^a_{\mathrm{Piston}}(t)\quad \text{at } x=0.$$

The absence of a thermal boundary layer (insulated boundary) implies that $u^a_b(1,t)=0$ at $x=1$. The solution of Eq. (4.55), found by the Laplace transform, can be written as

$$u_b^a(x,t) = \frac{2\sqrt{D_c}}{\sqrt{\pi}t_0}\left\{(1-x)\left(\sqrt{t} - H_{[t_0]}(t-t_0)^{\frac{1}{2}}\right) + \frac{1}{\pi}\sum_{n=1}^{\infty}\left[C_n(t,x) - H_{[t_0]}C_n(t-t_0,x)\right]\right\}, \tag{4.57}$$

where

$$C_n(t,x) = \frac{(-1)^n}{n}\sin\left[n\pi(1-x)\right]\int_0^t \frac{\cos\left[n\pi c_{s0}(t-u)\right]}{\sqrt{u}}du.$$

When incorporated into Eqs. (4.50)–(4.53), the latter gives

$$T_b^a(x,t) = \frac{2\left(\gamma_{0,\mathrm{IG}}-1\right)\sqrt{D_c}}{c_{s0}t_0\sqrt{\pi}} \times\left\{\frac{2}{3}c_{s0}\left(t^{\frac{3}{2}} - H_{[t_0]}(t-t_0)^{\frac{3}{2}}\right) + \frac{1}{\pi}\sum_{n=1}^{\infty}\left[S_n'(t,x) - H_{[t_0]}S_n'(t-t_0,x,0)\right]\right\}, \tag{4.58}$$

where

$$\begin{aligned} S_n'(t,x,0) &= S_n(t,x) \\ &= \frac{(-1)^n}{n}\cos\left[n\pi(1-x)\right]\int_0^t \frac{\sin\left[n\pi c_{s0}(t-u)\right]}{\sqrt{u}}du \end{aligned}$$

$$\rho_b^a(x,t) = \frac{2}{3}\frac{(\gamma_{0,\mathrm{IG}}-1)\sqrt{D_c}}{c_{s0}t_0\sqrt{\pi}} \times\left\{\frac{2}{3}c_{s0}\left(t^{\frac{3}{2}} - H_{[t_0]}(t-t_0)^{\frac{3}{2}}\right) + \frac{1}{\pi}\sum_{n=1}^{\infty}\left(S_n''(t,x,0) - H_{[t_0]}S_n''(t-t_0,x,0)\right)\right\} \tag{4.59}$$

$$p_b^a(x,t) = 3\frac{(\gamma_{0,\mathrm{IG}}-1)\sqrt{D_c}}{c_{s0}t_0\sqrt{\pi}} \times\left\{\frac{2}{3}c_{s0}\left(t^{\frac{3}{2}} - H_{[t_0]}(t-t_0)^{\frac{3}{2}}\right) + \frac{1}{\pi}\sum_{n=1}^{\infty}\left(S_n'(t,x,0) - H_{[t_0]}S_n'(t-t_0,x,0)\right)\right\}. \tag{4.60}$$

The boundary layer thus continuously provokes a linear acoustic perturbation in the bulk.

When the distance to the critical point τ is such that

$$\tau \gg \varepsilon^{\frac{2}{3}}, \tag{4.61}$$

Table 4.1 Basic mechanism of thermomechanical coupling

Boundary layer	Matching conditions	Bulk phase
Heat diffusion		Adiabatic compression
Thermal energy		
⇓		
Kinetic energy	$\lim_{z\to\infty} u^a_{bl}(z,t) = u^a_{\text{Piston}}(t)$	Kinetic energy
$D_c T^a_{bl\,zz} = \frac{3}{2} u^a_{bl\,z}$	$\lim_{x\to 0} u^a_b(z,t) = u^a_{\text{Piston}}(t)$	⇓
$\lim_{z\to\infty} u^a_{bl}(z,t) = u^a_{\text{Piston}}(t)$		
$= D_c T^a_{bl\,z}(0,t)$		Thermal energy
		$u^a_{b\,tt} = c_0^2 u^a_{b\,xx}$
		$u^a_b(0,t) = D_c T^a_{bl\,z}(0,t)$

we have

$$\frac{\alpha\epsilon^{\frac{1}{2}}}{\tau^{\frac{3}{4}}} \ll \alpha.$$

Accordingly, the matching condition [see Eq. (4.33)] at order α between the solutions for the temperatures in the bulk and in the boundary layer is fulfilled on the acoustic timescale, since

$$\alpha \times \lim_{z\to\infty} T^a_{bl}(z,t) + o(\alpha) = \alpha \times 0 + o(\alpha) = \frac{\alpha\epsilon^{\frac{1}{2}}}{\tau^{\frac{3}{4}}} \lim_{x\to 0} T^a_b(x,t). \qquad (4.62)$$

When the initial conditions are closer to the critical point than $\varepsilon^{\frac{2}{3}}$, the above matching condition is no longer satisfied. A new theoretical description will be provided in Sect. 4.4, in which heat propagates adiabatically at the speed of sound. The continuously generated compression wave that originates at the edge of the boundary layer provokes heating in the bulk phase. The boundary layer transforms thermal energy into kinetic energy. Fluid flows out of a very thin wall region that therefore experiences strong mass depletion.

When this mass flow enters the bulk phase, it induces heating by adiabatic compression. The energy exchange between the boundary layer and the bulk is a thermomechanical coupling. This thermomechanical coupling is basically the same as it is for ideal gases, except that the acoustic wave is more intense in near-critical fluids because of the diverging compressibility. This is sketched in Table 4.1, which simultaneously shows the energy equations in the bulk and in the boundary layer.

The reduced velocity perturbations $\frac{\overline{u}^a}{\alpha}$ [see Eq. (4.48)], and temperature perturbations $\frac{\overline{T}^a-(1+\tau)}{\alpha}$ [see Eq. (4.49)], are plotted in Fig. 4.3a, b as a function of x for different half acoustic times (previously reported in Fig. 4.2 for the boundary layer case). The boundary region is enlarged to show the velocity in this region. Figure 4.3b clearly shows the expansion of the fluid, which is accelerated rightward

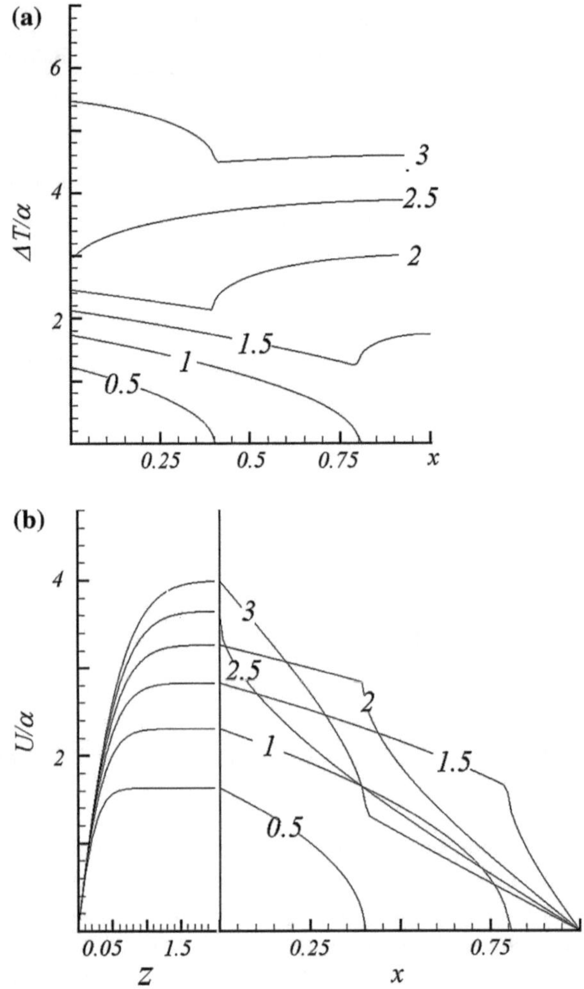

Fig. 4.3 Evolution in the bulk for each half acoustic time from 0.5 to 3: **a** reduced temperature perturbation; **b** reduced velocity (see Fig. 4.2 for notation)

in the boundary layer. The velocity reaches its maximum value at its edge, and this is the driving force for the acoustic field. Velocity is positive throughout the whole sample, since a compression wave is emitted that propagates and reflects on the walls. The compressive nature of the wave is confirmed by the fact that the temperature increases behind it (see Fig. 4.3a). Note in Fig. 4.3b that the functions are continuous, but the derivatives are not (first-order solution and matching).

4.2.3 A Fourth Temperature Equilibration Mechanism

4.2.3.1 Temperature Equilibration Timescale in Near-Critical Fluids

Equation (4.58) for the bulk temperature shows that the accumulated effect of the compression waves is a monotonic increase in the temperature that is proportional to $t^{\frac{3}{2}}$. When $t \to \infty$ in the acoustic timescale, the temperature in the bulk behaves as follows:

$$\lim_{t\to\infty} \bar{T}^a(x,t) = 1+\tau + \frac{4}{3}\frac{\alpha\epsilon^{\frac{1}{2}}}{\tau^{\frac{3}{4}}}\left[\frac{\sqrt{D_c}(\gamma_{0,\mathrm{IG}}-1)}{\sqrt{\pi}\,t_0}\sqrt{t} + \text{bounded acoustic terms}\right] + o\left(\frac{\alpha\epsilon^{\frac{1}{2}}}{\tau^{\frac{3}{4}}}\right). \quad (4.63)$$

The presence of the secular term that is proportional to the square root of time (which gives an infinite increase in the bulk temperature as $t \to \infty$) suggests that time must be rescaled to avoid nonphysical behavior. On the other hand, the temperature in the boundary layer remains of the same order of magnitude, because it is fixed by its value (order of magnitude: α) at the wall. As time increases, the first temperature perturbation in the bulk becomes on the same order as the temperature in the boundary layer. The matching condition [see Eq. (4.62)] is then no longer fulfilled, and time must be rescaled to satisfy this matching condition. The new timescale, Θ, needed to achieve this matching can be defined as

$$\Theta = \zeta t. \quad (4.64)$$

The value of ζ is such that the bulk and boundary layer temperatures are of the same order of magnitude when

$$\frac{\alpha\epsilon^{\frac{1}{2}}}{\tau^{\frac{3}{4}}}\zeta^{-\frac{1}{2}} = \alpha,$$

leading to $\zeta = \varepsilon\tau^{-\frac{3}{2}}$. Thus,

$$\Theta = \varepsilon\tau^{-\frac{3}{2}}t. \quad (4.65)$$

As long as $\zeta \ll 1$, the Θ timescale is longer than the acoustic timescale. We note that when $\tau = \epsilon^{\frac{2}{3}}$ (i.e., $\zeta = 1$), Θ exactly matches the acoustic timescale (see below, Sect. 4.4).

Some important features of the bulk behavior can be deduced from these relationships by comparing the acoustic (t) and the Θ timescales with the heat diffusion time. In this case, we denote the latter the θ timescale:

- The increase in bulk temperature occurs on a much shorter timescale than the heat diffusion timescale. Using Eqs. (3.29) and (3.17), the typical time needed for heat to diffuse across the sample is given by

$$t_D = \frac{L^2}{D_{T0}} \frac{r}{\Lambda_c c_{p0}} \tau^{-\frac{1}{2}} = t_{\mu 0} \frac{\mathrm{Pr}_0 r}{\Lambda_c c_{p0}} \tau^{-\frac{1}{2}},$$

so the time expressed in units of t_D becomes θ, with

$$\theta = \varepsilon\sqrt{\tau} t \qquad \tau \ll 1, \tag{4.66}$$

where t is the time expressed in units of t_a. This relation indicates that the heat diffusion timescale θ is $\tau^{\frac{1}{2}}$ longer than the corresponding timescale in ideal gases, εt, where $\epsilon = \frac{t_a}{t_{\nu 0}} \frac{\Lambda_c c_{p0}}{\mathrm{Pr}_0 r}$. This is consistent with the fact that the heat diffusion time tends to infinity as $\tau^{\frac{1}{2}}$.

- From Eqs. (4.65) and (4.66), the following relation holds between θ and Θ:

$$\theta = \tau^2 \Theta. \tag{4.67}$$

The Θ timescale is thus τ^2 shorter than the θ timescale:the bulk heating through mass addition (and the related adiabatic compression) causes a critical speeding up of temperature homogenization. As mass is injected into the bulk fluid by expanding the boundary layer, which acts as a piston [13], we have termed this thermomechanical effect the *piston effect*. Accordingly, the Θ timescale is called the piston-effect timescale. In the case of an ideal gas, the effect exists, but it manifests itself only on the diffusion timescale, and becomes a small correction to diffusion without giving any specific independent timescale to temperature equilibration [13]. Recall that in ideal gases, fluid expansion under boundary or bulk heating has permitted the origin of "transport anomalies" [26] observed during crystal growth from the vapor phase in zero gravity to be identified. We note that Eq. (4.67) is consistent with Eq. (4.11), since γ_{0c} diverges as τ^{-1} in the mean-field approach [see Eq. (1.101)].

The next section gives the corresponding detailed description of the hydrodynamics.

4.2.3.2 Piston-Effect Timescale

We must first find the orders of magnitude of the dependent variables. These orders of magnitude are given by the behavior of the solution obtained on the t scale when $t \to \infty$ with $\Theta = \varepsilon\tau^{-\frac{3}{2}} t \ll 1$; that is to say, close to the initial conditions on the Θ scale. In other words, we look for the matching conditions between two timescales just as we looked for the matching conditions between the bulk phase and the boundary layer region by applying the matching rule to the first order in Eq. (4.33). This condition is similar to Eq. (4.33), but the limits are taken in time instead of space. We thus expand Eqs. (4.57), (4.58), (4.59), and (4.60), for the bulk, density, velocity, temperature and pressure expressed on the acoustic timescale for $t \to \infty$ with $\Theta = \varepsilon\tau^{-\frac{3}{2}}t \ll 1$. We obtain the following expansions:

$$\lim_{t\to\infty,\, \varepsilon\tau^{-\frac{3}{2}}t\ll 1} u_b^a(x,t) \sim \frac{\sqrt{D_c}}{\sqrt{\pi}t_0}\frac{\alpha\,\varepsilon}{\tau^{\frac{3}{2}}}(1-x)\Theta^{-\frac{1}{2}} + \frac{\alpha\epsilon^{\frac{1}{2}}}{\tau^{\frac{3}{4}}}\text{ [bounded acoustic terms]}$$

$$\lim_{t\to\infty,\, \varepsilon\tau^{-\frac{3}{2}}t\ll 1} \overline{T}^a(x,t) \sim 1+\tau+2\left(\gamma_{0,\mathrm{IG}}-1\right)\frac{\sqrt{D_c}}{c_{s0}t_0\sqrt{\pi}}\alpha\sqrt{\Theta} + \frac{\alpha\epsilon^{\frac{1}{2}}}{\tau^{\frac{3}{4}}}\text{ [bounded acoustic terms]},$$

with similar developments for p_b^a and ρ_b^a.

The above equations represent the initial conditions for the piston-effect timescale solution. They involve a dominant term that only depends on Θ, which is added to bounded perturbative acoustic terms representing the legacy of the acoustic disturbances as described in the previous section. Ideally, one should look for the solution in the form of a multiple timescale analysis. This will be done in Chap. 7 to explicitly show the thermoacoustic nature of the piston effect.

It should be also emphasized that the condition

$$\varepsilon \ll \tau^{\frac{3}{2}}$$

must hold if the asymptotic expansion is to be valid. We shall come back to this aspect in Sect. 4.4 when we look at the lower limit of the critical speeding up upon approaching the critical point. Heat can ultimately propagate at the speed of sound in this limit.

Accordingly, we will only look for the Θ part of the solution, and we thus consider the following expansions:

$$\bar{T}^{pe}(x,\Theta) = 1+\tau+\alpha T_b^{pe}(x,\Theta)+o(\alpha)$$
$$\bar{\rho}^{pe}(x,\Theta) = 1+\alpha\rho_b^{pe}(x,\Theta)+o(\alpha)$$
$$\bar{p}^{pe}(x,\Theta) = \frac{3}{2}(1+\tau)-\frac{9}{8}+\alpha p_b^{pe}(x,\Theta)+o(\alpha)$$
$$\bar{u}^{pe}(x,\Theta) = \frac{\alpha\varepsilon}{\tau^{\frac{3}{2}}}u_b^{pe}(x,\Theta)+o\left(\frac{\alpha\varepsilon}{\tau^{\frac{3}{2}}}\right),$$

where the bulk variables are decorated by a bar, the superscript *pe* stands for "piston effect period," and the subscript *b* stands for "bulk."

Inserting these expansions into Eqs. (3.31)–(3.34), the following system is obtained:

$$\rho^{pe}_{b\,\Theta} + u^{pe}_{b\,x} = 0 \tag{4.68}$$

$$\frac{1}{\gamma_{0,\mathrm{IG}}} P^{pe}_{b\,x} = 0 \tag{4.69}$$

$$\frac{3}{2} T^{pe}_{b} = p^{pe}_{b} \tag{4.70}$$

$$\frac{3}{2}\left(\gamma_{0,\mathrm{IG}} - 1\right) u^{pe}_{b\,x} + T^{pe}_{b\,\Theta} = 0, \tag{4.71}$$

with the initial conditions

$$T^{pe}_{b}(x,0) = u^{pe}_{b}(x,0) = \rho^{pe}_{b}(x,0) = 0 \tag{4.72}$$

and the boundary and matching conditions

$$T^{pe}_{b}(0,\Theta) = \lim_{z\to\infty} T^{pe}_{bl}(z,\Theta)$$

$$T^{pe}_{b\,x}(1,\Theta) = 0 \tag{4.73}$$

$$u^{pe}_{b}(0,\Theta) = \lim_{z\to\infty} u^{pe}_{bl}(z,\Theta)$$

$$u^{pe}_{b}(1,\Theta) = 0 \tag{4.74}$$

$$p^{pe}_{b}(\Theta) = \lim_{z\to\infty} p^{pe}_{bl}(z,\Theta). \tag{4.75}$$

This system permits an obvious solution:

$$T^{pe}_{b}(x,\Theta) \equiv T^{pe}_{b}(\Theta) = \frac{2}{3} p^{pe}_{b}(\Theta) \tag{4.76}$$

$$u^{pe}_{b}(x,\Theta) \equiv u^{pe}_{b}(\Theta, x) = \frac{4}{9}\frac{1}{\gamma_{0,\mathrm{IG}} - 1} p^{pe}_{b\,\Theta}(\Theta).\,(1-x) \tag{4.77}$$

$$\rho^{pe}_{b}(x,\Theta) \equiv \rho^{pe}_{b}(\Theta) = \frac{4}{9}\frac{1}{\gamma_{0,\mathrm{IG}} - 1} p^{pe}_{b}(\Theta), \tag{4.78}$$

where $p^{pe}_{b}(\Theta)$ is still to be determined. However, even if the whole solution is yet to be obtained, Eqs. (4.76), (4.77), and (4.78) already indicate the specific properties of the piston effect:

1. The temperature as well as the coupled thermodynamic variables are homogeneous throughout the bulk.
2. Equation (4.65) indicates that the piston-effect timescale becomes shorter as the critical point is approached, although the heat diffusivity goes to zero.
3. The piston effect is thus a fourth temperature equilibration mechanism in hypercompressible media, in addition to diffusion, convection and radiation. We shall

see in Chap. 9 that, depending on the thermal boundary conditions, it can also be a new heat transfer mechanism.

4. The piston effect is an adiabatic effect, and we will comment on this question in Chap. 9. As a matter of fact, the heating process does not involve any dissipative phenomena to first order, since it is a bulk nonviscous process. The process is thus adiabatic, since the bulk fluid is not in thermal contact with any heat source. Fast temperature equilibration can also be obtained on purely thermodynamic grounds [4, 19, 20], and is called the "adiabatic heating effect."

The homogeneous pressure evolution in the bulk $p_b^{pe}(\Theta)$ in Eqs. (4.76)–(4.78) is determined by performing a boundary layer analysis. The above solution for the temperature cannot fulfill the boundary condition at $x = 0$. The spatial variable x must be rescaled in the thermal boundary layer, which has a characteristic dimension that is equal to the heat diffusion depth on the piston-effect timescale, such that

$$z = \frac{x}{\delta(\tau)} = \frac{x}{\tau}.$$

The wall at $x = 1$ is insulated and the temperature gradient is zero at this boundary: the homogeneous bulk temperature of Eq. (4.76) thus satisfies the boundary condition at $x = 1$, and there is no thermal boundary layer near this adiabatic wall.

Arguments similar to those used to find the orders of magnitude of variables in the thermal boundary layer (subscript bl) at $x = 0$ within the acoustic period lead us to expand the variables (decorated with a tilde) as follows:

$$\begin{aligned}
\tilde{T}^{pe} &= 1 + \tau + \alpha T_{bl}^{pe}(z, \Theta) + o(\alpha) \\
\tilde{\rho}^{pe} &= 1 + \frac{\alpha}{\tau}\rho_{bl}^{pe}(z, \Theta) + o\left(\frac{\alpha}{\tau}\right) \\
\tilde{p}^{pe} &= \frac{3}{2}(1 + \tau) - \frac{9}{8} + \alpha p_{bl}^{pe}(z, \Theta) + o(\alpha) \\
\tilde{u}^{pe} &= \frac{\alpha\varepsilon}{\tau^{\frac{3}{2}}} u_{bl}^{pe}(z, \Theta) + o\left(\frac{\alpha\varepsilon}{\tau^{\frac{3}{2}}}\right).
\end{aligned}$$

Feeding these expansions into the Navier–Stokes Equations (3.31)–(3.34) yields the following system:

$$\rho_{bl\,\Theta}^{pe} + u_{bl\,z}^{pe} = 0 \tag{4.79}$$

$$\frac{1}{\gamma_{0,\mathrm{IG}}} p_{bl\,z}^{pe} = 0 \tag{4.80}$$

$$\frac{3}{2}T_{bl}^{pe} + \frac{9}{4}\rho_{bl}^{pe} = p_{bl}^{e} \tag{4.81}$$

$$\frac{3}{2}u_{bl\,z}^{pe} - D_c T_{bl\,zz}^{pe} = 0, \tag{4.82}$$

with the initial conditions

$$T_{bl}^{pe}(x,0) = u_{bl}^{pe}(x,0) = \rho_{bl}^{pe}(x,0) = 0 \tag{4.83}$$

and the boundary and matching conditions

$$\begin{aligned} T_{bl}^{pe}(0,\Theta) &= H_{[0]}(\Theta) \\ \lim_{z\to\infty} T_{bl\,x}^{pe}(z,\Theta) &= T_b^{pe}(0,\Theta) \\ \lim_{z\to\infty} u_{bl}^{pe}(z,\Theta) &= u_b^{pe}(0,\Theta) \\ \lim_{z\to\infty} p_{bl}^{pe}(z,\Theta) &= p_b^{pe}(0,\Theta) = p_0(\Theta). \end{aligned} \tag{4.84}$$

Equations (4.80) and (4.84) indicate that the pressure is homogeneous throughout the boundary layer. Since it is also equal to the bulk value $p_0(\Theta)$, it is uniform throughout the whole sample. From Eqs. (4.79), (4.80), and (4.82), we obtain the following diffusion equation for the density in the boundary layer:

$$\rho_{bl\,\Theta}^{pe} = D_c \rho_{bl\,zz}^{pe}, \tag{4.85}$$

with boundary conditions at $z = 0$ of

$$\rho_{bl}^{pe}(0,\Theta) = -\frac{2}{3}\left[H_{[0]}(\Theta) - \frac{2}{3}p_0(\Theta)\right] \tag{4.86}$$

and the matching condition

$$\lim_{z\to\infty}\rho_{bl}^{pe}(z,\Theta) = -\lim_{z\to\infty}\frac{2}{3}\left[T_{bl}^{pe}(z,\Theta) - \frac{2}{3}p_0(\Theta)\right] = 0. \tag{4.87}$$

Using Duhamel's theorem [7], the solution of Eq. (4.85) with the boundary conditions shown in Eqs. (4.86) and (4.87) can be written as

$$\rho_{bl}^{pe}(z,\Theta) = \frac{2}{3}\,\mathrm{erfc}\left(\frac{z}{2\sqrt{\Theta}\sqrt{D_c}}\right) - \frac{4}{3\sqrt{\pi}}\int_{\frac{z}{2\sqrt{\Theta}\sqrt{D_c}}}^{\infty} e^{-v^2} p_0\left(\Theta - \frac{z^2}{4v^2\Theta}\right) dv. \tag{4.88}$$

By considering the continuity Eq. (4.79) in the boundary layer and integrating it over the boundary layer thickness, and by taking into account the solution of Eq. (4.77) for the velocity in the bulk, we obtain

$$\begin{aligned} \lim_{z\to\infty} u_{bl}^{pe}(z,\Theta) &= u_b^{pe}(0,\Theta) \\ &= \tfrac{4}{9}\tfrac{1}{\gamma_{0,\mathrm{IG}}-1}p_0(\Theta) \\ &= -\tfrac{2}{3}\textstyle\int_0^\infty \rho_{bl\,\Theta}^{pe}(z,\Theta)dz. \end{aligned}$$

By carrying Eq. (4.88) into the latter and rearranging, we obtain a convolution Volterra equation of the second kind for the homogeneous bulk pressure perturbation $p_0(\Theta)$:

$$p_0(\Theta) = \frac{(\gamma_{0,\mathrm{IG}} - 1)}{\sqrt{\pi}} \sqrt{D_c} \left(\sqrt{\Theta} - \frac{1}{2} \int_0^{\Theta} \frac{p_0(\Theta - u)}{\sqrt{u}} du \right), \tag{4.89}$$

which is solved using the Laplace transform:

$$p_0(\Theta) = \frac{3}{2} \left[1 - \exp\left((\gamma_{0,\mathrm{IG}} - 1)^2 D_c \Theta \right) \mathrm{erfc}\left((\gamma_{0,\mathrm{IG}} - 1) \sqrt{D_c} \sqrt{\Theta} \right) \right]. \tag{4.90}$$

Once the homogeneous pressure is known, the whole solution can be obtained from the above approximate boundary layer and bulk equations. They are written as follows:

$$\begin{aligned} T_{bl}^{pe}(z, \Theta) = & \left[1 - \exp\left((\gamma_{0,\mathrm{IG}} - 1)^2 D_c \Theta \right) \mathrm{erfc}\left((\gamma_{0,\mathrm{IG}} - 1) \sqrt{D_c} \sqrt{\Theta} \right) \right] \\ & + \exp\left((\gamma_{0,\mathrm{IG}} - 1) z^2 + (\gamma_{0,\mathrm{IG}} - 1)^2 D_c \Theta \right) \\ & \times i^0 \,\mathrm{erfc}\left((\gamma_{0,\mathrm{IG}} - 1) \sqrt{D_c} \sqrt{\Theta} + \frac{z}{2\sqrt{D_c}\sqrt{\Theta}} \right) \end{aligned} \tag{4.91}$$

$$\rho_{bl}^{pe}(z, \Theta) = \frac{4}{9} p_0(\Theta) - \frac{2}{3} T_{bl}^{pe}(z, \Theta) \tag{4.92}$$

$$\begin{aligned} u_{bl}^{pe}(x, \Theta) = & \frac{2}{3} D_c (\gamma_{0,\mathrm{IG}} - 1) \exp\left((\gamma_{0,\mathrm{IG}} - 1)^2 D_c \Theta + (\gamma_{0,\mathrm{IG}} - 1) z \right) \\ & \times \mathrm{erfc}\left((\gamma_{0,\mathrm{IG}} - 1) \sqrt{D_c} \sqrt{\Theta} + \frac{z}{2\sqrt{D_c}\sqrt{\Theta}} \right) \\ & - \frac{2}{3} D_c (\gamma_{0,\mathrm{IG}} - 1) \exp\left((\gamma_{0,\mathrm{IG}} - 1)^2 D_c \Theta \right) \\ & \times \mathrm{erfc}\left((\gamma_{0,\mathrm{IG}} - 1) \sqrt{D_c} \sqrt{\Theta} \right) + \frac{2}{3} \sqrt{D_c} \frac{1}{\sqrt{\pi}\sqrt{\Theta}} \left[1 - \exp\left(\frac{z^2}{4 D_c \Theta} \right) \right]. \end{aligned} \tag{4.93}$$

4.2.3.3 Uniformly Valid Description in Space on the Piston-Effect Timescale

The approximate solution for the differential problem describing the flow field on the piston-effect timescale has two parts, which are valid in the boundary layer and in the bulk, respectively, and are denoted $\widetilde{X}^{pe}(z, \Theta)$ and $\overline{X}^{pe}(x, \Theta)$. These are given by Eqs. (4.76), (4.77), (4.78) and Eqs. (4.91), (4.92), and (4.93). There is a classical procedure that uses Eq. (4.94) to construct a uniformly valid approximate solution across a whole domain of space. It is the additive composite of the approximate bulk and boundary layer solutions. This procedure involves constructing, for each dependent variable, the function

$$X^{pe}(x,\Theta) = \tilde{X}^{pe}\left(\frac{x}{\tau},\Theta\right) + \bar{X}^{pe}(x,\Theta) - \lim_{\frac{x}{\tau}\to\infty, x\ll 1} \tilde{X}^{pe}\left(\frac{x}{\tau},\Theta\right). \tag{4.94}$$

According to the matching condition

$$\lim_{\frac{x}{\tau}\to\infty,\ x\ll 1} \tilde{X}^{pe}\left(\frac{x}{\tau},\Theta\right) = \lim_{x\to 0} \bar{X}^{pe}(x,\Theta),$$

the function constructed with the additive composition procedure is an approximate solution of the Navier–Stokes equations throughout the whole sample.

The uniformly valid approximate solutions on the piston-effect timescale are

$$p^{pe}(x,\Theta) \equiv p^{pe}(\Theta) = \frac{3}{2}(1+\tau) - \frac{9}{8} + p_0(\Theta) \tag{4.95}$$

$$\begin{aligned}
T^{pe}(x,\Theta) &= \widetilde{T}^{pe}\left(\frac{x}{\tau},\Theta\right)\\
\rho^{pe}(x,\Theta) &= \widetilde{\rho}^{pe}\left(\frac{x}{\tau},\Theta\right) + \frac{4}{9}\frac{1}{\gamma_{0,\mathrm{IG}}-1}p^{pe}(\Theta) \qquad (4.96)\\
u^{pe}(x,\Theta) &= \frac{\alpha\varepsilon}{\tau^{\frac{3}{2}}}\left\{\frac{2}{3}\sqrt{D_c}(1-x)\left[\frac{1}{\sqrt{\pi}\sqrt{\Theta}} - (\gamma_{0,\mathrm{IG}}-1)\sqrt{D_c}\, e^{(\gamma_{0,\mathrm{IG}}-1)^2 D_c\Theta}\right.\right.\\
&\quad \left.\left.\times\, i\ \mathrm{erf}\,c\left(\left(\gamma_{0,\mathrm{IG}}-1\right)\sqrt{D_c}\sqrt{\Theta}\right)\right]\right\} - \frac{\alpha\varepsilon}{\tau^{\frac{3}{2}}}\frac{2}{3}\sqrt{D_c}\left[\frac{e^{-\frac{z^2}{4D_c\Theta}}}{\sqrt{\pi}\sqrt{\Theta}}\right.\\
&\quad - (\gamma_{0,\mathrm{IG}}-1)\, e^{(\gamma_{0,\mathrm{IG}}-1)^2 D_c\Theta + (\gamma_{0,\mathrm{IG}}-1)z}\\
&\quad \left.\times\,\mathrm{erf}\,c\left((\gamma_{0,\mathrm{IG}}-1)\sqrt{D_c}\sqrt{\Theta} + \frac{z}{2\sqrt{D_c}\sqrt{\Theta}}\right)\right].
\end{aligned}$$

4.2.3.4 Typical Evolution of the Hydrodynamic Variables

Let us consider a van der Waals fluid that is at critical density and is 1.5 K from the critical temperature in a 1 cm cell. The parameters that appear in the various expressions take the following values (see Table 3.1):

$$\begin{aligned}
D_c &= 1.15 & \Pr_0 &= 2.24\\
\varepsilon &= 2.45\times 10^{-8} & \tau &= 5\times 10^{-3}.
\end{aligned} \tag{4.97}$$

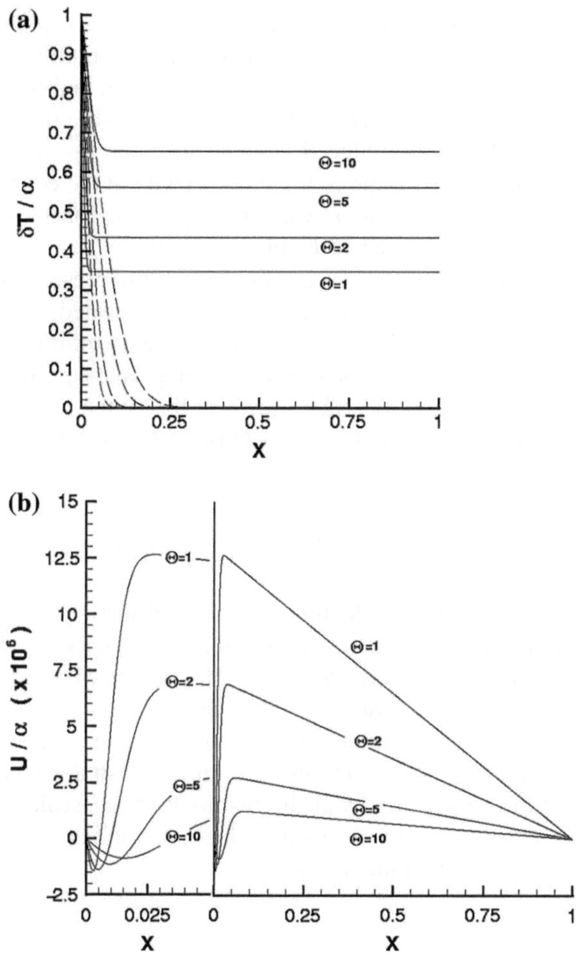

Fig. 4.4 Evolution in the bulk on the piston-effect timescale: **a** reduced temperature $\delta T^{*pe}(x)$ (*dashed lines* are for pure diffusion); **b** reduced velocity $u^{*pe}(x)$. The boundary layer region is enlarged

Under these conditions, the typical heat diffusion time as previously defined is $t_{D_0} = \frac{L^2 \rho_c c_{p0}}{\Lambda_b^{\mathrm{MF}}} = \frac{L^2}{D_{T0}} = 3076\,\mathrm{s}$ [see Eq. (3.18)]. From Eq. (4.67), the characteristic heat diffusion time $t_D = \frac{t_a}{\varepsilon\sqrt{\tau}}$ is equal to 17320 s (~5 h) in the near-critical fluid at 1.5 K above T_c. From Eq. (4.65), the characteristic time for the piston effect is

$$t_{PE} = \varepsilon \tau^{\frac{3}{2}} t_a \tag{4.98}$$

It is as short as 0.433 s.

The reduced temperature perturbation $\delta T^{*pe} = \frac{T^{pe}-(1+\tau)}{\alpha}$ is plotted in Fig. 4.4a as a function of x for different values of time on the piston-effect timescale. The purely diffusive process is also plotted on the same figure, and this clearly shows that the piston effect speeds up heat equilibration compared to the purely diffusive process. Figure 4.4b shows the reduced velocity $u^{*pe} = \frac{u^{pe}}{\alpha}$ in the bulk phase, and demonstrates the nonzero value of the fluid velocity at the edge of the thermal boundary layer (the maximum value), which is responsible for the bulk heating by adiabatic compression (the essence of the piston effect).

This mass flow comes from the expansion of the fluid present in the boundary layer region. The negative velocity that appears near the heated wall for increasing values of Θ is due to the diffusive backflow that begins to refill the mass-depleted thermal boundary layer area when the temperature begins to equilibrate.

The expansion velocity exactly reflects the heat flux at the wall. Indeed, the following relation holds between the rate of increase of the bulk temperature and the heat flux at the wall:

$$-(\gamma_{0,\mathrm{IG}}-1)D_c T_z^{pe}(0,\Theta) = T_b \Theta^{pe}(\Theta) = \frac{3}{2}(\gamma_{0,\mathrm{IG}}-1)\overline{u}(0,\Theta). \qquad (4.99)$$

This relation clearly illustrates the transformation of thermal energy into kinetic energy and then into heat by adiabatic compression. *This means that, if we cut off the heat flux at the boundary, the piston effect stops.* This property is used in experiments to tune the piston effect and to turn it on and off at will.

However, in the present situation, where a constant temperature is imposed at the wall, the heat flux at $x = 0$ decreases with time, and so does the piston effect. Ultimately, when time goes to infinity on the piston-effect timescale Θ, the boundary layer assumes macroscopic dimensions and the piston effect vanishes. The next section is devoted to this long time period.

4.3 Density Relaxation

The long time period, which is studied in [2], is addressed here through the validity of Eq. (4.96) when time goes to infinity on the piston-effect timescale.

4.3.1 The Behavior of the Piston-Effect Solution Over a Long Time Period

When $\Theta \to \infty$ in Eq. (4.96), the temperature, pressure and velocity perturbations take values that are of the same order of magnitude throughout the sample:

$$T^{pe} \approx 1 + \tau + \alpha - \alpha \frac{1}{\sqrt{D_c}\sqrt{\pi}} \frac{1}{\gamma_{0,\mathrm{IG}} - 1} \frac{1}{\sqrt{\Theta}} \left[1 - \exp\left(-\frac{x^2}{4D_c\Theta} \right) \right] \tag{4.100}$$

$$p^{pe} \approx \frac{3}{2}(1+\tau) - \frac{9}{8} + \alpha \frac{3}{2} \left(1 - \frac{1}{\gamma_{0,\mathrm{IG}} - 1} \frac{1}{\sqrt{\pi}\sqrt{D_c}} \frac{1}{\sqrt{\Theta}} \right) \tag{4.101}$$

$$\begin{aligned} u^{pe} \approx \frac{2}{3} \frac{\alpha\varepsilon}{\tau^{\frac{3}{2}}} \frac{1}{2\sqrt{\pi}} \frac{1}{\Theta^{\frac{3}{2}}\sqrt{D_c}} & \left[\frac{1}{(\gamma_{0,\mathrm{IG}} - 1)^2} (1 - x) \right. \\ & \left. - \exp\left(-\frac{z^2}{4D_c\Theta} \right) \left(\frac{1}{(\gamma_{0,\mathrm{IG}} - 1)^2} + \frac{1}{\gamma_{0,\mathrm{IG}} - 1} z \right) \right]. \end{aligned} \tag{4.102}$$

In contrast, density obeys different asymptotic behaviors depending on whether we consider the boundary layer or the bulk:

(1) In the boundary layer:

$$\widetilde{\rho}^{pe} \approx 1 - \frac{\alpha}{\tau} \frac{2}{3} \frac{1}{(\gamma_{0,\mathrm{IG}} - 1)\sqrt{D_c}\sqrt{\pi}} \frac{1}{\sqrt{\Theta}} \exp\left(-\frac{z^2}{4D_c\Theta} \right). \tag{4.103}$$

(2) In the bulk:

$$\overline{\rho}^{pe} \approx 1 + \alpha \frac{2}{3} \frac{1}{\gamma_{0,\mathrm{IG}} - 1} \left(1 - \frac{1}{(\gamma_{0,\mathrm{IG}} - 1)\sqrt{\pi}\sqrt{D_c}} \frac{1}{\sqrt{\Theta}} \right). \tag{4.104}$$

The matching condition between the density in the bulk and the density in the boundary layer is satisfied for finite values of Θ:

$$\lim_{z\to\infty} \widetilde{\rho}^{pe}(z,\Theta) = 1 + \frac{\alpha}{\tau} \lim_{z\to\infty} \rho_{bl}^{pe}(z,\Theta) + o\left(\frac{\alpha}{\tau}\right) = 1 \tag{4.105}$$

$$= 1 + 0 \times \frac{\alpha}{\tau} + \alpha\rho^{pe}(0,\Theta) + o(\alpha). \tag{4.106}$$

However, it is clear from the above equations that, when $\Theta \to \infty$, the value of the density in the boundary layer decreases continuously. From Eq. (4.103), for large values of z, it becomes on the order of α when Θ is on the order of τ^{-2}. The density in the boundary layer thus goes to zero at the order α (see the following equation), meaning that the above matching condition is no longer valid. More precisely,

$$\begin{aligned} \lim_{\Theta\to\infty} \lim_{z\to\infty} \widetilde{\rho}^{pe}(z,\Theta) &= 1 + \frac{\alpha}{\tau} \lim_{\Theta\to\infty} \lim_{z\to\infty} \rho_{bl}^{pe}(z,\Theta) + O(\alpha) \\ &= 1 - \alpha \lim_{\Theta\to\infty} \frac{1}{\sqrt{\tau^2\Theta}} = 1, \end{aligned}$$

and since

$$\bar{\rho}^{pe} = 1 + \alpha \frac{2}{3} \frac{1}{\gamma_{0,\mathrm{IG}} - 1} = C^{ste},$$

it turns out that

$$\lim_{\Theta \to \infty} \lim_{z \to \infty} \overset{\sim pe}{\rho_{bl}}(z, \Theta) \neq \lim_{\Theta \to \infty} \lim_{x \to 0} \bar{\rho}_{bl}^{pe}(z, \Theta). \tag{4.107}$$

Therefore, time must be rescaled again, just as it was rescaled at the end of the acoustic period, in order for temperature to remain continuous on the piston-effect timescale.

4.3.2 The Heat Diffusion Timescale

4.3.2.1 The Heat Diffusion Timescale as the Ultimate Timescale

It is clear from Eq. (4.107) that, on the diffusion timescale θ introduced by Eq. (4.67), i.e.,

$$\theta = \tau^2 \Theta, \tag{4.108}$$

the density becomes on the same order of magnitude throughout the whole sample. The necessary matching between the solution in the bulk and that in the boundary layer is thus fulfilled, and density relaxation occurs on this timescale. Taking into account the definition of the piston-effect timescale in Eq. (4.65) as a function of the acoustic scale t, we can retrieve Eq. (4.66).

4.3.2.2 First-Order Uncoupled Temperature and Density Relaxations

The matching conditions that were imposed between the acoustic and piston-effect time periods must also be imposed to obtain the initial conditions for evolution on the θ scale, as well as the relevant asymptotic sequences.

When Eqs. (4.103) and (4.104) are rewritten in terms of θ, we obtain the following matching forms for the density in the boundary layer:

$$\rho_{bl}^{pe} \approx \lim_{\Theta \to \infty, \tau^2 \Theta \ll 1} \rho_{bl}^{pe}(x, \Theta) = 1 - \alpha \frac{2}{3} \frac{1}{\sqrt{\pi}\left(\gamma_{0,\mathrm{IG}} - 1\right)\sqrt{D_c}} \frac{1}{\sqrt{\theta}} \times \exp\left(-\frac{x^2}{4 D_c \theta}\right) \tag{4.109}$$

and in the bulk phase

$$\rho_b^{pe} \approx \lim_{\Theta\to\infty,\tau^2\Theta\ll 1} \rho_b^{pe}(x,\Theta) = 1 + \alpha\frac{2}{3}\frac{1}{\gamma_{0,\mathrm{IG}}-1}\left(1 - \frac{1}{(\gamma_{0,\mathrm{IG}}-1)\sqrt{\pi}\sqrt{D_c}}\frac{\tau}{\sqrt{\theta}}\right). \tag{4.110}$$

These equations, which only display the bulk variable x, mean that diffusion now matters in the bulk. The thermal boundary layer has disappeared.

The matching conditions [see Eqs. (4.109) and (4.110)] are the initial conditions for the density $\rho^{\rho}(x,\theta)$ on the θ scale. From now on, the superscript ρ refers to the diffusion timescale period for any variable. At the order α, we obtain

$$\lim_{\theta\to 0} \rho^{\rho}(x,\theta) = -\alpha\frac{2}{3}\frac{1}{\gamma_{0,\mathrm{IG}}-1}\left[\frac{1}{\sqrt{\pi}\sqrt{D_c}}\frac{\exp\left(-\frac{x^2}{4D_c\theta}\right)}{\sqrt{\theta}} - 1\right]. \tag{4.111}$$

This initial condition is singular for $\theta \to 0$, since $\frac{e^{-\frac{x^2}{4D_c\theta}}}{\sqrt{\pi}\sqrt{D_c}\sqrt{\theta}}$ tends to the Dirac function $\delta(x)$ when θ goes to zero. The matching conditions have transformed a function of order $\frac{\alpha}{\tau}$ in the boundary layer for small θ into a Dirac function for $\theta = 0$.

When rewritten as functions of the variable θ, Eqs. (4.100), (4.101), and (4.102) give the matching forms for the temperature, the pressure and the velocity:

$$\lim_{\Theta\to\infty,\tau^2\Theta\ll 1} T^{pe}(x,\Theta) = \lim_{\theta\to 0} T^{\rho}(x,\theta) = 1 + \tau + \alpha - \alpha\tau\frac{1}{\sqrt{\pi}\sqrt{D_c}}\frac{1}{\gamma_{0,\mathrm{IG}}-1} \times \frac{1}{\sqrt{\theta}}\left(1 - \exp\left(-\frac{x^2}{4D_c\theta}\right)\right)$$

$$\lim_{\Theta\to\infty,\tau^2\Theta\ll 1} p^{pe}(x,\Theta) = \lim_{\theta\to 0} p^{\rho}(x,\theta) = \frac{3}{2}(1+\tau) - \frac{9}{8} + \alpha\frac{3}{2} - \alpha\tau\frac{3}{2}\frac{1}{\gamma_{0,\mathrm{IG}}-1}\frac{1}{\sqrt{\pi}\sqrt{D_c}}\frac{1}{\sqrt{\theta}} \tag{4.112}$$

$$\lim_{\Theta\to\infty,\tau^2\Theta\ll 1} u^{pe}(x,\Theta) = \lim_{\theta\to 0} u^{\rho}(x,\theta) = \frac{2}{3}\alpha\varepsilon^{\frac{1}{2}}\tau^{\frac{1}{2}}\frac{1}{\gamma_{0,\mathrm{IG}}-1}\frac{1}{2\sqrt{\pi}}\frac{1}{D_c}\frac{1}{\theta^{\frac{3}{2}}} \times x\exp\left(-\frac{x^2}{4D_c\theta}\right).$$

The initial conditions, although singular for $\theta = 0$, show that temperature has already relaxed to the order of magnitude $\alpha\tau$ at the end of the piston-effect period. Similarly, the asymptotic behavior indicates that the density perturbation decreases on the Θ scale from the order of magnitude α/τ to α. α is the order of magnitude of the density perturbation at the end of the heating period if the fluid is an ideal gas or a supercritical fluid far from its critical point.

The conditions $\Theta \gg 1$ and $\tau^2\Theta \ll 1$, or equivalently $\frac{\theta}{\tau^2} \gg 1$ and $\theta \ll 1$, define an intermediate time period during which both solutions (on the piston-effect timescale and the diffusion timescale) are simultaneously valid and are commonly represented by Eqs. (4.111) and (4.112).

4.3.2.3 Density Evolution on the θ Scale (Diffusion Timescale)

When rewritten as functions of the variable θ, the expansions for $\Theta \to \infty$ of the fluid properties as given by Eqs. (4.100), (4.101), (4.103), and (4.104) are

$$T^{pe} \approx 1 + \tau + \alpha - \alpha\tau \frac{1}{\sqrt{D_c}\sqrt{\pi}} \frac{1}{\gamma_{0,\mathrm{IG}} - 1} \frac{1}{\sqrt{\theta}} \left(1 - \exp\left(-\frac{x^2}{4D_c\theta}\right)\right) \tag{4.113}$$

$$p^{pe} \approx \frac{3}{2}(1+\tau) - \frac{9}{8} + \alpha\frac{3}{2} - \alpha\tau\frac{3}{2}\frac{1}{\gamma_{0,\mathrm{IG}} - 1}\frac{1}{\sqrt{\pi}\sqrt{D_c}}\frac{1}{\sqrt{\theta}} \tag{4.114}$$

$$u^{pe} \approx \frac{1}{3}\alpha\varepsilon\tau^{\frac{1}{2}}\frac{1}{\gamma_{0,\mathrm{IG}} - 1}\frac{1}{\sqrt{\pi}}\frac{1}{\sqrt{D_c}}\frac{1}{\theta^{\frac{3}{2}}}x\exp - \frac{x^2}{4D_c\theta} \tag{4.115}$$

$$\rho^{pe} \approx 1 + \alpha\frac{2}{3}\frac{1}{(\gamma_{0,\mathrm{IG}} - 1)}\left(1 - \frac{1}{\sqrt{D_c}\sqrt{\pi}}\frac{1}{\sqrt{\theta}}\exp - \frac{x^2}{4D_c\theta}\right). \tag{4.116}$$

As the matching principle imposes the following relation for T, p, u and ρ:

$$\lim_{\theta\to 0} X(x,\theta) = \lim_{\Theta\to\infty} X(x,\Theta),$$

then Eqs. (4.113)–(4.116) give the initial conditions for the evolution on the θ scale. They are the legacy of the piston effect on the diffusion timescale: strong mass depletion in the heated wall region and a slightly homogeneous overdensity in the bulk region. These equations are also singular; i.e., they can be written in the form of generalized functions. For example, the initial condition for ρ^{pe} is

$$\lim_{\theta\to 0} \rho^{pe}(x,\theta) = -\frac{2}{3}\frac{1}{\gamma_{0,\mathrm{IG}} - 1}\left[\delta(x) - 1\right].$$

Equations (4.113)–(4.116) also indicate that the solutions describing the evolution of the legacy of the piston effect on the θ scale must be sought in the form of the following asymptotic expansions, where the superscript ρ now indicates that density relaxes on the diffusion timescale:

$$
\begin{aligned}
T^{\rho} &= 1 + \tau + \alpha H_{[0]}(\theta) + \alpha\tau \widetilde{T}(x,\theta) \\
p^{\rho} &= \frac{3}{2}(1+\tau) - \frac{9}{8} + \frac{3}{2}\alpha H_{[0]}(\theta) + \alpha\tau \widetilde{p}(x,\theta) \\
\rho^{\rho} &= 1 + \alpha\widetilde{\rho}(x,\theta) \\
u^{\rho} &= \alpha\varepsilon\sqrt{\tau}\widetilde{u}(x,\theta).
\end{aligned}
\tag{4.117}
$$

Incorporating these expansions into the Navier–Stokes equations leads to the following system of equations:

$$\widetilde{\rho}_{\theta} + \widetilde{u}_x = 0$$

$$\widetilde{p}_x = 0 \tag{4.118}$$

$$\frac{3}{2}\widetilde{T} + \frac{9}{4}\widetilde{\rho} = \widetilde{p} \tag{4.119}$$

$$\frac{3}{2}\widetilde{u}_x - D_c\widetilde{T}_{xx} = \frac{1}{\gamma_{0,\mathrm{IG}} - 1}\delta(\theta). \tag{4.120}$$

These equations can be complemented by the singular initial conditions given by the limits for $\theta \to 0$ of Eqs. (4.113)–(4.116) and the boundary conditions

$$\widetilde{T} = \widetilde{u} = 0 \qquad \text{at } x = 0 \tag{4.121}$$

$$\widetilde{T}_x = 0, \quad \widetilde{u} = 0 \quad \text{at } x = 1 \tag{4.122}$$

It must be emphasized here that the energy equation [see Eq. (4.120)] involves the work of the inner (conservative) pressure forces due to the deformation of the fluid domain as well as diffusion. The large expansion coefficient still influences the bulk equations, even though, on earlier timescales, it only acts on the bulk through the boundary matching. Compared to the ideal gas behavior, there is strong mass addition in the bulk, but the acoustic equations remain unchanged. Indeed, at earlier times, the bulk process is governed by acoustic isentropic modes only, and thus by the isentropic compressibility, which shows weak critical divergence. On the diffusion timescale, the bulk phase no longer undergoes an isentropic process, and the large isothermal compressibility acts on the bulk phase as it did in the boundary layer, only on shorter timescales.

In order to solve the above system of equations, we proceed to make the following change of variables:

$$
\begin{aligned}
p_1 &= \tilde{p} + \frac{3}{2}\frac{1}{\sqrt{D_c}}\frac{1}{\sqrt{\pi}}\frac{1}{\gamma_{0,\mathrm{IG}}-1}\frac{1}{\sqrt{\theta}} \\
T_1 &= \tilde{T} + \frac{1}{\sqrt{D_c}}\frac{1}{\sqrt{\pi}}\frac{1}{\gamma_{0,\mathrm{IG}}-1}\frac{1}{\sqrt{\theta}}\left(1-e^{-\frac{x^2}{4D_c\theta}}\right) \\
\rho_1 &= \tilde{\rho} - \frac{2}{3}\frac{1}{\left(\gamma_{0,\mathrm{IG}}-1\right)}\left(1-\frac{1}{\sqrt{D_c}\sqrt{\pi}}\frac{1}{\sqrt{\theta}}e^{-\frac{x^2}{4D_c\theta}}\right) \\
u_1 &= \tilde{u} - \frac{2}{3}\frac{1}{\gamma_{0,\mathrm{IG}}-1}\frac{1}{2\sqrt{\pi}}\frac{1}{D_c}\frac{1}{\theta^{\frac{3}{2}}}xe^{-\frac{x^2}{4D_c\theta}}.
\end{aligned}
\tag{4.123}
$$

The boundary and initial conditions become:

$$
\begin{aligned}
T_1 &= 0 = u_1 = 0 \quad \text{at } x = 0 \\
T_1 &= \frac{1}{\sqrt{D_c}}\frac{1}{\sqrt{\pi}}\frac{1}{\gamma_{0,\mathrm{IG}}-1}\frac{1}{\sqrt{\theta}}\left(1-e^{-\frac{1}{4D_c\theta}}\right) \quad \text{at } x = 1 \\
u_1 &= -\frac{2}{3}\frac{1}{\gamma_{0,\mathrm{IG}}-1}\frac{1}{2\sqrt{\pi}}\frac{1}{D_c}\frac{1}{\theta^{\frac{3}{2}}}e^{-\frac{1}{4D_c\theta}} \quad \text{at } x = 1.
\end{aligned}
$$

$T_1 = p_1 = u_1 = \rho_1 = 0$ at $\theta = 0$. A straightforward combination of this system of equations shows that density obeys the diffusion equation

$$
\frac{\partial \rho_1}{\partial \theta} - D_c\frac{\partial^2 \rho_1}{\partial x^2} = \frac{2}{3}\frac{1}{\gamma_{0,\mathrm{IG}}-1}\delta(\theta).
$$

This equation is solved by the Laplace transform using classical transform tables. p, T, and u are deduced from the solution for density. When inserted into expansions of Eq. (4.117), they lead to the solution that is valid on the heat diffusion timescale θ. This solution can be written as follows:

$$
\begin{aligned}
T^{\rho} &= 1 + \tau + \alpha H_{[0]}(\theta) \\
&\quad + \alpha\tau\frac{2}{\gamma_{0,\mathrm{IG}}-1}\sum_{n=1}^{\infty} e^{-n^2\pi^2 D_c\theta}\left[-1 + (-1)^n\cos(n\pi(1-x))\right] \\
p^{\rho} &= \frac{3}{2}(1+\tau) - \frac{9}{8} + \frac{3}{2}H_{[0]}(\theta) \\
&\quad + \alpha\tau\frac{3}{2}\frac{1}{\gamma_{0,\mathrm{IG}}-1}\left[1 - \sum_{n=1}^{\infty}(-1)^n e^{-n^2\pi^2 D_c\theta}\right] \\
\rho^{\rho} &= 1 + \alpha\frac{3}{2}\frac{1}{\gamma_{0,\mathrm{IG}}-1} \\
&\quad \times\left[1 - 2\sum_{n=1}^{\infty}(-1)^n e^{-n^2\pi^2 D_c\theta}\cos(n\pi(1-x))\right]
\end{aligned}
\tag{4.124}
$$

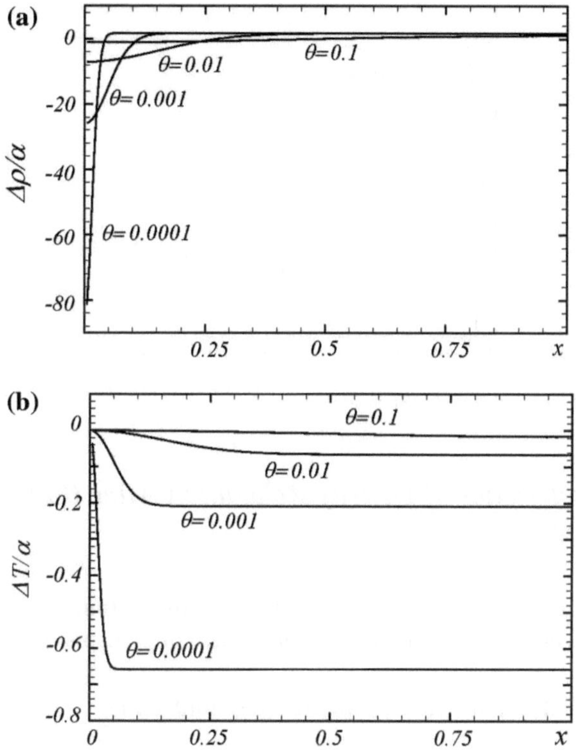

Fig. 4.5 Evolution on the diffusion timescale: **a** density; **b** temperature

$$u^\rho = -\alpha\varepsilon\sqrt{\tau}\frac{2}{3}\frac{1}{\gamma_{0,\mathrm{IG}}-1}$$
$$\times\left[(1-x)\,\delta(\theta) - 2\pi D_c \sum_{n=1}^{\infty}(-1)^n e^{-n^2\pi^2 D_c\theta}\sin(n\pi(1-x))\right].$$

Perturbations in the bulk are plotted in Figs. 4.5b, a and 4.6, respectively, for the reduced temperature $\delta T^{*\rho} = \frac{T^\rho-(1+\tau)}{\alpha}$, the reduced density $\delta\rho^{*\rho} = \frac{\rho^\rho-1}{\alpha}$, and the reduced velocity $u^{*\rho} = \frac{u^\rho}{\alpha}$. The singular behavior of the solutions on the θ scale is clearly seen. In particular, it is apparent that the density tends to $\frac{3}{2}\frac{1}{\gamma_{0,\mathrm{IG}}-1}\left[1-\delta(x)\right]$ as $\theta \to 0$. This singular behavior can be overcome by applying the additive composition procedure to the solutions on the Θ and θ timescales, as described in the next section. The density relaxation on the long timescale is that of a Dirac-like strong mass depletion at $x = 0$ in a medium that is initially not in equilibrium. This evolution is driven by the diffusion of the temperature inhomogeneities left by the piston effect.

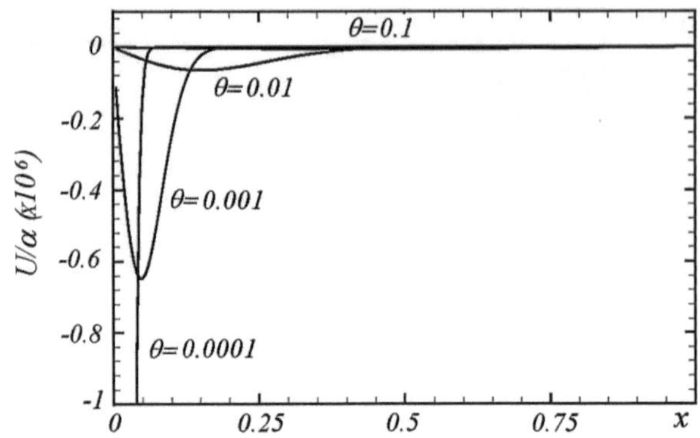

Fig. 4.6 Velocity on the diffusion timescale

4.3.3 Characteristics of Density Relaxation After Boundary Heating

As the above equations are invalid for small heat diffusion times, we must obtain the uniformly valid description for all timescales, including the piston-effect timescale.

4.3.3.1 Uniformly Valid Description in Space and Time

In Sect. 4.2.3.3, the additive composition procedure was used to construct a uniformly valid description in space on the piston-effect timescale. We now extend this procedure to the uniformly valid description in time. For any variable $X^{pe}(x, \Theta)$ or $X^{\rho}(x, \theta)$, we construct the following function:

$$X(x,\theta) = X^{pe}\left(x, \Theta = \frac{\theta}{\tau^2}\right) + X^{\rho}(x,\theta) - \lim_{\Theta \to \infty, \tau^2\Theta \ll 1} X^{pe}(x,\Theta).$$

This leads to the following expressions for the hydrodynamic variables:

$$\begin{aligned} T(x,\theta) = {} & 1+\tau \\ & +\alpha\left[1-\exp\left(\left(\gamma_{0,\mathrm{IG}}-1\right)^2 D_c \frac{\theta}{\tau^2}\right) i^0 \operatorname{erf} c\left(\left(\gamma_{0,\mathrm{IG}}-1\right)\sqrt{D_c}\times\frac{\sqrt{\theta}}{\tau}\right)\right] \\ & +\alpha\exp\left(\left(\gamma_{0,\mathrm{IG}}-1\right)\frac{x}{\tau}+\left(\gamma_{0,\mathrm{IG}}-1\right)^2\sqrt{D_c}\frac{\sqrt{\theta}}{\tau}\right) \\ & \times i^0 \operatorname{erfc}\left(\left(\gamma_{0,\mathrm{IG}}-1\right)\sqrt{D_c}\frac{\sqrt{\theta}}{\tau}+\frac{x}{2\sqrt{D_c}\sqrt{\theta}}\right) \\ & +\alpha\tau\frac{1}{\left(\gamma_{0,\mathrm{IG}}-1\right)\sqrt{D_c}} \end{aligned}$$

$$\times \left\{ 2\sqrt{D_c} \sum_{n=1}^{\infty} \exp\left(-n^2\pi^2 D_c\theta\right)\left[-1+(-1)^n \cos(n\pi(1-x))\right] \right.$$
$$\left. + \frac{1}{\sqrt{\pi}}\frac{1}{\sqrt{\theta}}\left(1-\exp\left(-\frac{x^2}{4D_c\theta}\right)\right)\right\} \tag{4.125}$$

$$\rho(x,\theta) = 1 - \frac{2}{3}\frac{\alpha}{\tau}\exp\left[\left(\gamma_{0,\mathrm{IG}}-1\right)\frac{x}{\tau} + \left(\gamma_{0,\mathrm{IG}}-1\right)^2 D_c \frac{\theta}{\tau^2}\right]$$
$$\times i^0 \operatorname{erf} c\left(\left(\gamma_{0,\mathrm{IG}}-1\right) \times \sqrt{D_c}\sqrt{\Theta} + \frac{x}{2\sqrt{D_c}\sqrt{\theta}}\right)$$
$$+ \alpha\,\frac{2}{3}\frac{1}{\gamma_{0,\mathrm{IG}}-1}\left\{1-\exp\left[\left(\gamma_{0,\mathrm{IG}}-1\right)^2 D_c \times \frac{\theta}{\tau^2}\right]\right\}$$
$$\times \operatorname{erf} c\left(\left(\gamma_{0,\mathrm{IG}}-1\right)\sqrt{D_c}\frac{\sqrt{\theta}}{\tau}\right)$$
$$+ \alpha\,\frac{2}{3}\frac{1}{\left(\gamma_{0,\mathrm{IG}}-1\right)\sqrt{D_c}}$$
$$\times \left\{\frac{1}{\sqrt{\pi}}\frac{1}{\sqrt{\theta}} \times \exp\left(-\frac{x^2}{4D_c\theta}\right)\right.$$
$$\left. - 2\sqrt{D_c}\sum_{n=1}^{\infty}(-1)^n \exp\left(-n^2\pi^2 D_c\theta\right) \times \cos(n\pi(1-x))\right\} \tag{4.126}$$

$$u(x,\theta) = \frac{\alpha\varepsilon}{\tau^{\frac{3}{2}}}\frac{2}{3}\sqrt{D_c}\left\{(1-x)\left[\frac{1}{\sqrt{\pi}\frac{\sqrt{\theta}}{\tau}} - \left(\gamma_{0,\mathrm{IG}}-1\right)\sqrt{D_c}\right.\right. \tag{4.127}$$
$$\left.\left. \times \exp\left(\left(\gamma_{0,\mathrm{IG}}-1\right)^2 D_c\frac{\theta}{\tau^2}\right) \times \operatorname{erf} c\left(\left(\gamma_{0,\mathrm{IG}}-1\right)\sqrt{D_c}\frac{\sqrt{\theta}}{\tau}\right)\right]\right\}$$
$$- \frac{\alpha\varepsilon}{\tau^{\frac{3}{2}}}\frac{2}{3}\sqrt{D_c} \times \left\{\frac{\exp\left(-\frac{\frac{x^2}{\tau^2}}{4D_c\theta}\right)}{\sqrt{\pi}\frac{\sqrt{\Theta}}{\tau}} - \left(\gamma_{0,\mathrm{IG}}-1\right)\sqrt{D_c}\right.$$
$$\times \exp\left(\left(\gamma_{0,\mathrm{IG}}-1\right)^2 D_c\frac{\theta}{\tau^2} + \left(\gamma_{0,\mathrm{IG}}-1\right)\frac{x}{\tau}\right)$$
$$\left. \times \operatorname{erf} c\left(\left(\gamma_{0,\mathrm{IG}}-1\right)\sqrt{D_c}\frac{\sqrt{\theta}}{\tau} + \frac{x}{2\sqrt{D_c}\sqrt{\theta}}\right)\right\}$$

$$+\alpha\varepsilon\sqrt{\tau}\frac{2}{3}\frac{1}{\gamma_{0,\mathrm{IG}}-1}\left\{\frac{1}{2\sqrt{\pi}}\frac{1}{\sqrt{D_c}}\frac{x\exp\left(-\frac{x^2}{4D_c\theta}\right)}{\theta^{\frac{3}{2}}}\right.$$
$$\left.-2\pi D_c\sum_{n=1}^{\infty}(-1)^n\exp\left(-n^2\pi^2D_c\theta\right)\sin(n\pi(1-x))\right\}.$$

4.3.3.2 Orders of Magnitude of the Time Periods

The piston-effect period appears as a *boundary layer in time* (e.g., a short time period near time zero), and so we define "the piston-effect period" as the values of θ such that

$$0 \ll \frac{\theta}{\tau^2} \ll 1.$$

It is always possible to perform matching with the acoustic period on the basis of the available solutions on this timescale. However, this approach will not provide any new insight into density relaxation.

The piston-effect period is followed by an intermediate period of time that corresponds to the matching zone between the solutions that describe the evolution on the Θ scale (piston effect) and the θ scale (diffusion). This overlapping period is defined by values of time θ such that

$$\frac{\theta}{\tau^2} \gg 1 \text{ and } \theta \ll 1.$$

It is followed by the long-time diffusion period, defined as

$$\theta = O(1).$$

4.3.3.3 Orders of Magnitude of Temperature and Density

We now discuss the orders of magnitude of the different properties, as they vary on different timescales. We will focus in particular on density and temperature. First, recall that, due to the diverging isothermal compressibility, the order of magnitude of the density perturbation is greater than that of the temperature perturbation by a factor τ. During the piston-effect time period, after the wall heating has stopped, the temperature inhomogeneity is on the order of α, while the density inhomogeneity is still on the order of α/τ. At the end of the piston-effect period (about 0.47 s for the conditions given in Sect. 3.2.2), as shown by Eq. (4.113), the temperature is homogenized at around $\alpha\tau$, while the density inhomogeneity has relaxed to around α: the order of magnitude that the density would have reached if the fluid was an ideal gas.

Then, during the heat diffusion period (a time of 4.6 h under the same conditions), temperature and density inhomogeneities relax diffusively to complete equilibrium. This means that the heat diffusion period involves only minute temperature differences on the order of $\alpha\tau$: the temperature relaxes first on the piston-effect timescale, while the density relaxes on the heat diffusion timescale.

Indeed, density and temperature perturbations are uncoupled in the bulk on the piston-effect timescale, as shown in the linearized equation of state:

$$\nu\delta p = \alpha\delta T + \alpha\tau\delta\rho \sim \alpha\delta T.$$

As the driving force for the evolution in the bulk phase during the piston-effect time period is an acoustic field, the density and temperature perturbations are of the same order of magnitude (that of the velocity at the edge of the boundary layer; i.e., α). In contrast, in the thermal boundary layer, or in the bulk on the heat diffusion timescale, the driving force is heat diffusion, so the density perturbations are τ times larger than the temperature ones, and they are coupled in the equation of state.

When gravity is present, these near-isothermal, high-density gradients lead to the striking concept of near-isothermal gravitational convection. This question will be addressed in Sect. 12.2, when we study the interaction between convection and thermoacoustic phenomena.

4.3.3.4 Temperature and Density Inhomogeneities

Both temperature and density relaxation effects are present simultaneously, but they are involved at different orders of magnitude in a given time period. The evolutions of the normalized density and temperature inhomogeneities, as defined by

$$\delta\rho^* = \frac{\rho(x=1,\theta) - \rho(x=0,\theta)}{\alpha} \text{ and } \delta T^* = \frac{T(x=0) - T(x=1,\theta)}{\alpha},$$

are plotted in Fig. 4.7 on a log-log scale as a function of heat diffusion time periods.

It is clear from this figure that there is always a ratio τ ($\simeq 5 \times 10^{-3}$ see Eq. (4.97) between reduced density and temperature perturbations due to the diverging compressibility. In the time period when solutions on both the piston-effect and the diffusion timescales are valid simultaneously, the time variable satisfies the condition

$$\frac{\theta}{\tau^2} = \Theta \gg 1 \text{ and } \theta \ll 1,$$

which extends from $\theta \simeq 2.5 \times 10^{-4}$ to $\theta \simeq 0.25$. For the latter value of θ, the density perturbation becomes on the order of 1, the value it would take if the fluid was an ideal gas. At the same time, the temperature reaches the order of magnitude $\alpha\tau$, which can be considered the new equilibrium to first order. In this time period, density and temperature follow the matching function in time (Eq. 4.113) written at

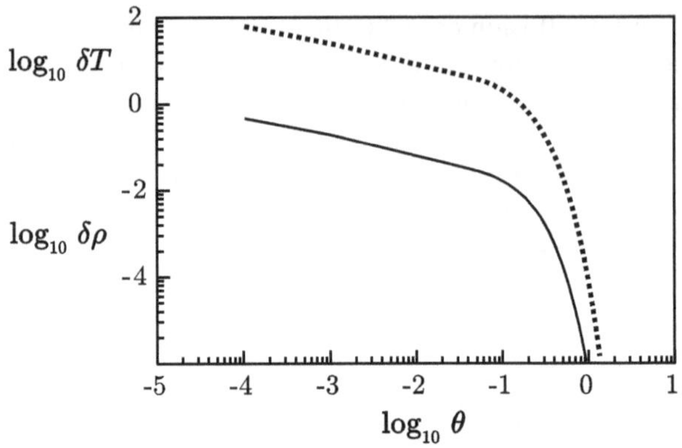

Fig. 4.7 Evolutions of reduced density ($\delta\rho^*(\theta)$, *dotted line*) and reduced temperature ($\delta T^*(\theta)$, *solid line*) inhomogeneities on the diffusion timescale

$x = 1$ that decreases as $\theta^{-\frac{1}{2}}$. This rate was said to be the bulk behavior by Hao and Ferrell [9]. For larger values of θ, the density inhomogeneity decays exponentially, which was also predicted by Onuki and Ferrel and detected as a "long diffusive tail" in experiments [11, 24]. This diffusive decay occurs in a quasi-isothermal medium.

4.3.3.5 Density Profiles in a 1-D Sample

The reduced density perturbation $\delta\rho^* = \frac{\rho(x,\theta)-1}{\alpha}$ is plotted in Fig. 4.8a for different times from the late piston-effect period ($\theta = 2.5 \times 10^{-4}$) to the early diffusion timescale ($\theta = 10^{-2}$). These values cover the intermediate time period. Significant mass depletion is present near the heated wall for small values of θ. This corresponds to the thermal boundary layer that lost mass during the piston-effect period. As time increases, the diffusion depth widens and backdiffusion begins to refill the depleted area. The increase in density over time in the bulk indicates that the latter is still compressed. This specific behavior is shown by the uniformly valid description (see Sect. 4.3.3.1) that accounts for both the piston effect and bulk diffusion in the intermediate period of time. While diffusion begins to drive the system back to equilibrium in the boundary layer area, the piston effect keeps driving the system out of equilibrium by compressing the bulk phase. The related reduced velocity $\delta u^* = \frac{u(x,\theta)}{\alpha}$ plot (see Fig. 4.8b) shows that a backflow (negative velocity) appears in the boundary layer while the velocity is still positive in the bulk, thus confirming both mass addition and the piston effect. Accordingly, there is an inversion of the velocity gradient in the sample that will be shown, in the next paragraph, to play an important role in the relaxation process. In contrast to the behaviors caused by this complete description, Fig. 4.5a, b, which correspond to the heat diffusion timescale solution only, show monotonically decreasing bulk density and a negative velocity.

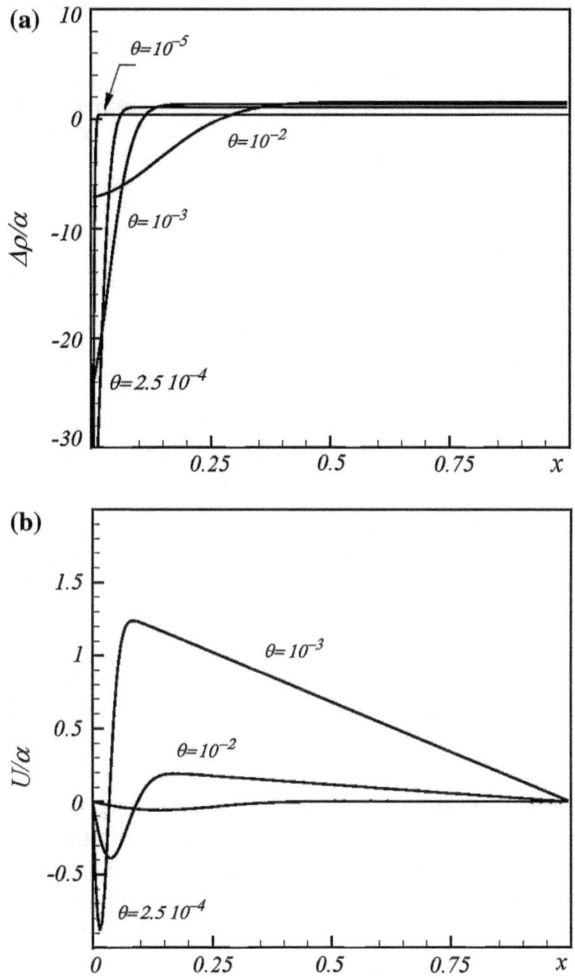

Fig. 4.8 Evolution on the diffusion timescale, as given by the uniformly valid description: **a** reduced density $\delta\rho^* = \frac{\rho(x,\theta)-1}{\alpha}$; **b** reduced velocity $\delta u^* = \frac{u(x,\theta)}{\alpha}$

For longer times, diffusion becomes the only driving force. The velocity is negative everywhere and the piston effect vanishes.

4.3.3.6 Density Evolution at Different Locations

In this section, we explore the evolution of density at different locations in the sample by studying the normalized density perturbation, as previously defined by $\delta\rho^* = \frac{\rho(x,\theta)-\rho(x,\theta=0)}{\alpha}$, where $\rho(x, \theta = 0) = 1$. Even though some details can fall

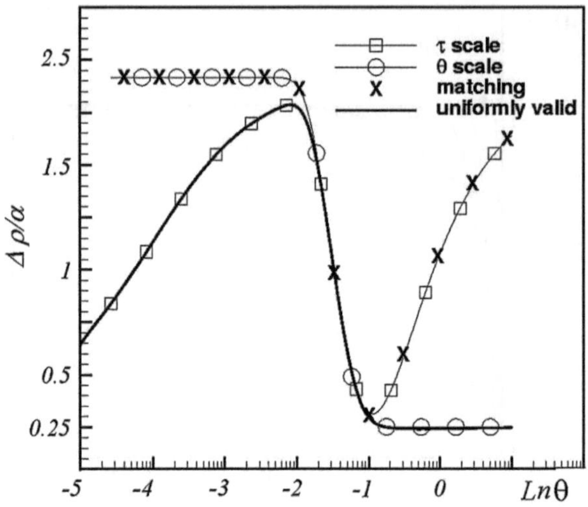

Fig. 4.9 Evolution of the reduced density $\delta\rho^*$ $(0.5, \theta)$ at $x = 0.5$. The solutions that are valid on the piston-effect and diffusion timescales are shown, as well as the matching function that corresponds to their common limit (see text)

below the experimental spatial resolution, we will still present them in order to assess the coherence of the whole theory. The different features relating to the density evolution at a given location will be correlated in the next paragraph with the velocity field evolution.

The evolution at $x = 0.5$ (middle of the sample), as given in Fig. 4.9, shows that density increases during the piston-effect period, according to the mass transfer from the boundary layer region to the bulk. Then, in the intermediate time period, density drops rapidly because the piston effect has stopped and the backflow refills the depleted zone near the origin. The density drop follows the matching function given by Eq. (4.116) for ρ. This function is also plotted in Fig. 4.9 along with the solutions that are valid on the Θ scale (piston effect) and the θ scale (diffusion) in order to illustrate the concept of asymptotic matching. In particular, it is absolutely clear that the part that the three functions have in common is the increasing branch of the function

$$\frac{1}{\sqrt{D_c}\sqrt{\pi}}\frac{1}{\sqrt{\theta}}e^{-\frac{x^2}{4D_c\theta}} - 1 \tag{4.128}$$

plotted at $x = 0.5$.

When the thermal boundary layer crosses a given location within the bulk, there is a steep drop in density towards the initial equilibrium [see Eq. (4.128)]. For larger values of θ, the density continues to relax monotonically. This part of the curve that extends approximately from $\theta = 0.1$ to infinity represents the long exponential diffusive tail that drives the system back to complete equilibrium. Recall that one characteristic diffusion time period (i.e., $\theta = 1$) corresponds to approximately 4.6 h.

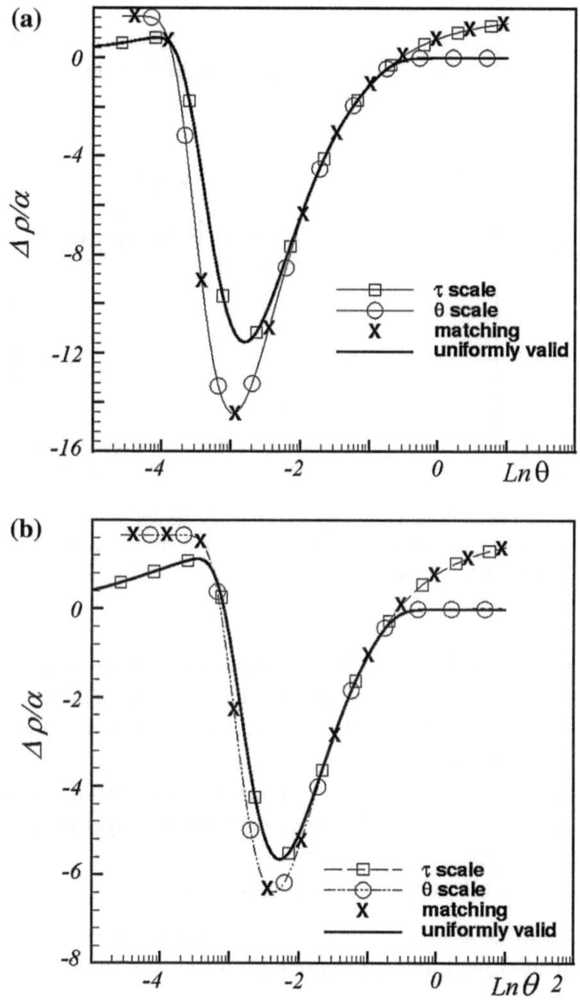

Fig. 4.10 Evolution of $\delta\rho^*(x, \theta)$ at two locations near the heated wall: **a** at $x = 0.05$; **b** at $x = 0.1$. The solutions that are valid on the piston-effect and the diffusion timescales are shown, as well as the matching function that corresponds to their common limit (see text)

The intermediate period of time thus plays an important role in the density relaxation process.

The evolutions of reduced density at $x = 0.05$ and $x = 0.1$ (thermal boundary layer), as given by Eq. (4.128), are plotted in Fig. 4.10a, b, respectively. In contrast with the bulk evolution, it is not monotonic. Density first increases due to the piston effect, and this increase is enhanced further from the boundary. When the diffusion layer extends to the location under consideration, the density decreases, which even leads to negative values of the perturbation. This location belongs to the initial,

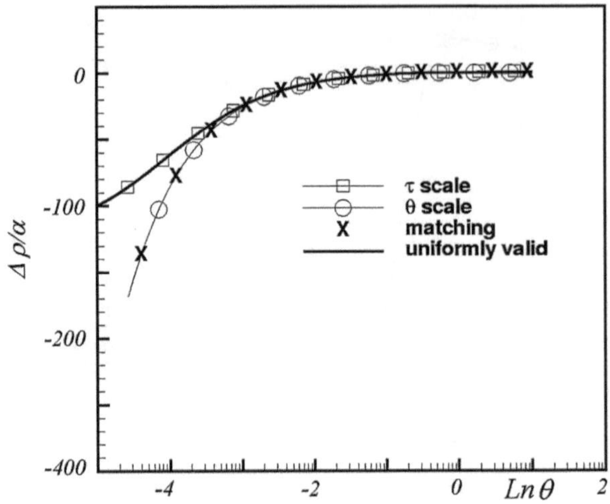

Fig. 4.11 Evolution of the reduced density $\delta\rho^*(0, \theta)$ at the heated wall $x = 0$. The solutions that are valid on the piston-effect and diffusion timescales are shown, as well as the matching function that corresponds to the common limit (see text)

low-density thermal boundary layer. The evolution for longer times displays another difference from the bulk behavior: when the backflow appears, the density increases again, following the long diffusive exponential tail to the final equilibrium value.

At $x = 0$ (heated wall), as shown in Fig. 4.11, the reduced density $\delta\rho^*(0, \theta)$—which initially decreases—relaxes (increases) monotonically towards equilibrium. In fact, it belongs to the thermal boundary layer at all times, and thus never experiences the piston effect.

This section can therefore be summarized as follows:

- At the heated wall, density only increases monotonically over time
- Close to the heated wall, density increases, decreases, and increases again over time
- In the bulk, density only increases and decreases.

In order to obtain a better understanding of the processes involved, we must correlate the density field with the hydrodynamic velocity field in the sample.

4.3.3.7 A Slowly Propagating Expansion–Compression Zone

The density perturbations $\delta\rho^*(x, \theta)$ at $x = 0$, $x = 0.05$, $x = 0.1$ and $x = 0.5$ are plotted in Fig. 4.12a (from Eq. (4.126), and they show a minimum. Further from the heated wall, the minimum is weaker and vanishes exponentially. In Fig. 4.12b, we plot the reduced velocity $\delta u^* = \frac{u(x,\theta)}{\alpha}$ profile at the times $\theta = 0.0016$ and $\theta = 0.0054$ when the density is minimum at $x = 0.05$ and $x = 0.1$, respectively. The velocity profile obviously exhibits a minimum at the measurement location under

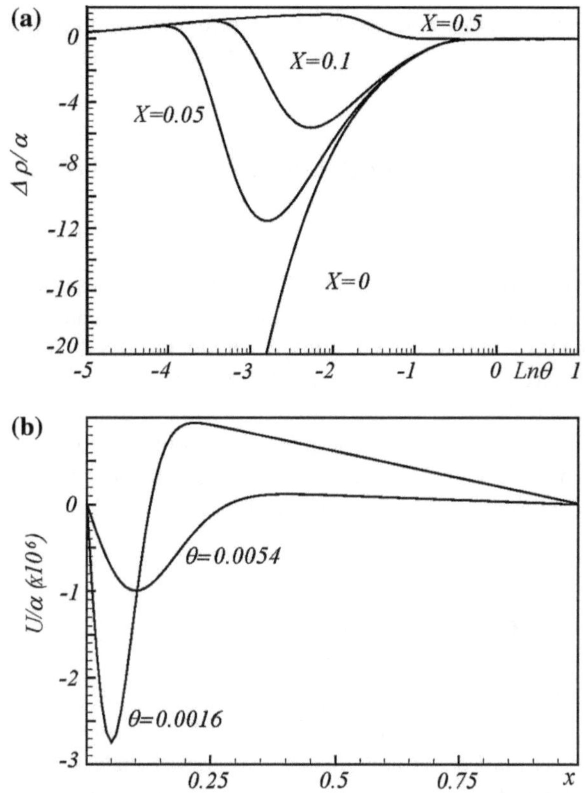

Fig. 4.12 **a** Reduced density $\delta\rho^*(x, \theta)$ and **b** reduced velocity $\delta u^* = \frac{u(x,\theta)}{\alpha}$ profiles at the times $\theta = 0.0016$ and $\theta = 0.0054$ when the density is minimum at $x = 0.05$ and $x = 0.1$, respectively

consideration. For values of x that are greater than the given location, the velocity gradient is positive (corresponding to an expansion zone), while for smaller values of x the velocity gradient is negative (corresponding to a compression zone). The density evolution at a given location is thus the result of an expansion–compression zone that follows a diffusion process and travels to the right. Density initially drops, as the fluid is accelerated to the left; then, as the expansion–compression zone propagates rightward, a compression occurs, as the velocity must be zero at the left boundary, and density increases again. While traveling rightward, the depth of the velocity minimum decreases before disappearing completely. Indeed, the low-density zone created at the heated wall is filled while moving rightward by the bulk density excess; it ultimately disappears before it reaches the other end of the sample cell. At $x = 0.5$, there is no longer a minimum, since the compression branch of the velocity profile has disappeared. The point beyond which the density evolution become monotonic is given by the location at which the minimum of the velocity profile disappears. Beyond this particular location, the bulk never sees the initial diffusion boundary layer, only the leftward diffusion that drives the bulk back to thermodynamic equilibrium.

4.4 Can Temperature Equilibrates at the Speed of Sound? Acoustic Saturation

Temperature equilibration by the piston effect speeds up as the critical point is approached. This leads to an interesting question: is there an upper limit to this critical speeding up when the critical point is neared, or does the temperature equilibration time decrease monotonically to zero? The answer to this question can be found theoretically: when the fluid is closer to the critical point than a certain threshold, the speeding up saturates and the heat propagates at the speed of sound. The material discussed in this section has been published in [27] but not yet been verified experimentally, although there have been some attempts to do so (see [16]).

4.4.1 Close to the Critical Point

The complete solution $F(x, t, \varepsilon, \tau)$ (where F represents p, ρ, u, T) of the governing Eqs. (3.31)–(3.34), as given in Sect. 3.2.1 where the variables and parameters can take any values, is not known analytically. An approximate system of differential equations was derived by performing a double asymptotic expansion, based on the fact that the parameters ε and τ take small values in the domain of physical interest. The analytical solution of this approximate system is the piston-effect solution given in Sect. 4.2. This solution indicates that the heat equilibration time decreases as $\tau^{\frac{3}{2}}$ as the critical point is neared. The relevant timescale for heat equilibration is [see Eq. (4.65)]

$$\Theta = \varepsilon\tau^{\frac{3}{2}}t, \tag{4.129}$$

where t is the acoustic time variable and Θ the piston-effect time variable. Can the speeding up of the heat transport be extrapolated to any distance from the critical point? This question addresses the singularity of the solution $F(x, t, \varepsilon, \tau)$ of the differential system of Eqs. (3.31)–(3.34). The system is singular if the limiting processes in the double expansion cannot be reversed; i.e.,

$$\lim_{\tau\to 0}\lim_{\varepsilon\to 0} F(x, t, \varepsilon, \tau) = F^{00}(x, t) \neq \lim_{\varepsilon\to 0}\lim_{\tau\to 0} F(x, t, \varepsilon, \tau) = F_{00}(x, t).$$

$F^{00}(x, t)$ is called the *first outer description* (f.o.d.) and $F_{00}(x, t)$ is called the *second outer description* (s.o.d.). We note that the definition of the piston-effect timescale as given by Eq. (4.129) requires the condition $\tau \gg \varepsilon^{\frac{2}{3}}$ [see Eq. (4.61)] to be fulfilled. We can thus conclude that the piston-effect double expansion $F_{PE}(x,t)$, as given in the present chapter, corresponds to the case when ε tends to zero faster than τ ($\varepsilon^{\frac{2}{3}} \ll \tau$ implies $\varepsilon \ll \tau$):

$$F_{PE}(x,t) = \lim_{\tau \to 0,\, \tau \gg \varepsilon^{\frac{2}{3}}} \lim_{\varepsilon \to 0} F(x,t,\varepsilon,\tau) = F^{00}(x,t).$$

The piston-effect solution, as previously obtained, is thus the f.o.d. of the differential problem, the asymptotic branch on which ε and τ both tend to zero with $\tau \gg \varepsilon$.

A sufficient condition for a double expansion to be singular is that the expansion does not remain well ordered when the largest parameter (τ here) becomes smaller than some power of the initial smaller parameter (ε here). To find this power, let us consider the f.o.d. as previously obtained. The expansion of Eq. (4.49) for the temperature in the bulk on the acoustic scale indicates that the perturbation is on the order of $\alpha \frac{\sqrt{\varepsilon}}{\tau^{\frac{3}{4}}}$, while the solution of Eq. (4.58) for this perturbation does not satisfy the thermal boundary condition at the wall. A thermal adaptation layer must be introduced to obtain the second-order perturbation in the boundary layer, which will therefore be on the order of $\alpha \frac{\sqrt{\varepsilon}}{\tau^{\frac{3}{4}}}$ (see for example [28]). Accordingly, the proper expansion (up to second order) for the temperature in the thermal boundary layer on the acoustic scale is

$$\widetilde{T}^a(z,t) = 1 + \tau + \alpha T^a_{bl}(z,t) + \alpha \frac{\epsilon^{\frac{1}{2}}}{\tau^{\frac{3}{4}}} T^{a(1)}_{bl}(z,t) + o\left(\alpha \frac{\epsilon^{\frac{1}{2}}}{\tau^{\frac{3}{4}}}\right). \tag{4.130}$$

This expansion is well ordered as long as

$$\tau \gg \varepsilon^{\frac{2}{3}}. \tag{4.131}$$

Under this condition, the second-order term $\alpha \frac{\epsilon^{\frac{1}{2}}}{\tau^{\frac{3}{4}}} T^{a(1)}_{bl}(z,t)$ is smaller than the first-order one $\alpha T^a_{bl}(z,t)$.

When $\tau \leq \varepsilon^{\frac{2}{3}}$, the expansion is not well ordered. We can thus conclude that the speeding up of the temperature equilibration cannot be extrapolated to any distance from the critical point. What happens closer to the critical point cannot be simply extrapolated from the straightforward limit $\tau \to 0$ because the f.o.d. is singular.

We now call the second outer description (s.o.d.) the asymptotic solution corresponding to the asymptotic branch on which ε and τ both tend to zero with $\tau \ll \varepsilon$; i.e., when τ tends to zero first and then ε tends to zero. We can say that the f.o.d. will never again give the s.o.d. when $\tau \to 0$ such that $\tau \ll \varepsilon$. The f.o.d. will thus provide an incorrect description when $\tau \ll \varepsilon$.

When ε is kept constant, the expansion of Eq. (4.130) is no longer well ordered, since its second-order term becomes of the same order of magnitude or larger than its first-order term. This demonstrates that the parameter ε is a singular perturbation parameter when the value of the parameter τ considered as a particular variable is zero. The asymptotic behavior of the complete solution of the problem $F(x, t, \varepsilon, \tau)$ is singular for $(\varepsilon, \tau) \to (0, 0)$; i.e., the order of the limit $(\varepsilon \to 0,\ \tau \to 0)$ cannot be reversed.

This situation is similar to that described at the beginning of this chapter, where the limit $(\varepsilon, x) \to (0, 0)$ was shown to be singular in the thermal boundary layer; i.e., for small values of x. A thermal boundary layer was introduced to study the close vicinity of the wall; its thickness in terms of the value of x was a function of ε. In the present situation, the exploration of the close vicinity of the critical point also needs the introduction of a boundary layer: its thickness in terms of the value of τ is a function of ε.

The important consequence of this singular behavior is that we cannot move continuously to the second outer description (s.o.d.) through the limit $\tau \to 0$ of the f.o.d. description of the critical speeding up, as it is invalid in the close neighborhood of the critical point. The speeding up of the temperature equilibration does not continue monotonically when nearing the critical point.

Using matched asymptotic description techniques [8], we can overcome this singularity provided that we look for an asymptotic solution of the problem for $\varepsilon \to 0$ with a fixed value of the ratio

$$\tilde{\tau} = \frac{\tau}{f(\varepsilon)},$$

where $f(\varepsilon)$ is a scaling factor. This description is called the *inner description*. In the present situation, choosing

$$f(\varepsilon) = \varepsilon^{\frac{2}{3}}$$

allows the construction of a well-ordered expansion. The two small parameters tend to zero simultaneously with a fixed constant ratio which ensures that the first- and second-order terms in the expansions [see Eq. (4.130)] merge throughout the limiting processes. Physically, we look for a solution where the acoustic timescale crosses the piston-effect timescale.

Once this inner description has been obtained, the matched asymptotic description technique ensures that both outer descriptions (f.o.d. and s.o.d.) can be found through the asymptotic behavior of the inner description when $\tilde{\tau}$ goes to zero (s.o.d.) or to infinity (f.o.d.). In particular, it becomes possible to explore the close vicinity of the critical point by expanding the inner solution for $\tilde{\tau} \to 0$ (so that $\tau \ll \varepsilon^{\frac{2}{3}}$). Symmetrically, the f.o.d. must again be found by expanding the inner solution for $\tilde{\tau} \to \infty (\tau \gg \varepsilon^{\frac{2}{3}})$. These features are summarized in Fig. 4.13.

Accordingly, if we wish to investigate closer to the critical point than $\varepsilon^{\frac{2}{3}}$, we need to:

- Asymptotically expand the functions for $\varepsilon \to 0$ with a fixed value of the ratio $\frac{\tau}{\varepsilon^{\frac{2}{3}}} = \tilde{\tau}$
- Expand the inner solution obtained for $\tilde{\tau} \to 0$.

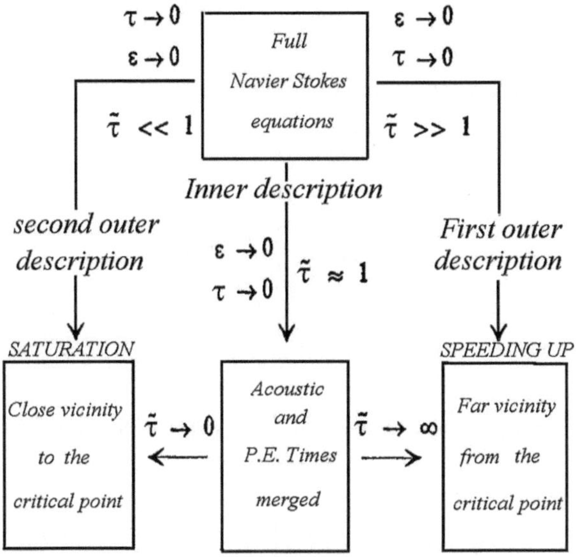

Fig. 4.13 Principle of matched asymptotic expansions. "Far from the critical point" and "close to the critical point" are meant relative to the value $\varepsilon^{\frac{2}{3}}$ of τ

4.4.2 Inner Description

We note first that removing the singularity from the parameter space (ε, τ) does not remove the singularity from real space. Therefore, it is necessary to proceed to boundary layer and bulk analyses.

4.4.2.1 The Boundary Layer

A similar reasoning to that used in Sect. 4.2.2, when looking for the speeding up solution, leads to the following expansions:

$$\begin{aligned}
\widetilde{T}^{in}(z,t) &= 1 + \tau + \alpha T_{bl}^{in}(z,t) + o(\alpha) \\
\widetilde{p}^{in}(z,t) &= \frac{3}{2}(1+\tau) - \frac{9}{8} + \alpha p_{bl}^{in}(z,t) + o(\alpha) \\
\widetilde{\rho}^{in}(z,t) &= 1 + \frac{\alpha}{\varepsilon^{\frac{2}{3}}} \rho_{bl}^{in}(z,t) + o\left(\frac{\alpha}{\varepsilon^{\frac{2}{3}}}\right) \\
\widetilde{u}^{in}(z,t) &= \alpha u_{bl}^{in}(z,t) + o(\alpha),
\end{aligned}$$

where the superscript in stands for inner description. The boundary layer thickness is given by

$$\delta = \varepsilon^{\frac{2}{3}},$$

which leads to the following system of equations:

$$u^{in}_{bl\,z}(z,t) + \rho^{in}_{bl\,t}(z,t) = 0 \tag{4.132}$$

$$p^{in}_{z}(z,t) = 0$$

$$u^{in}_{bl\,z}(z,t) = \frac{2}{3}\frac{1}{\sqrt{\tilde{\tau}}} D_c T^{in}_{bl\,zz}(z,t) \tag{4.133}$$

$$p^{in}_{bl}(z,t) = \frac{3}{2} T^{in}_{bl}(z,t) + \frac{9}{4}\tilde{\tau}\rho^{in}_{bl}(z,t),$$

with $D_c = \frac{\gamma_{0,\mathrm{IG}}}{\gamma_{0,\mathrm{IG}-1}} \frac{1}{\mathrm{Pr}_0 \Lambda_c}$ [see Eq. (3.30)]. This system is equivalent to a diffusion equation,

$$\left[T^{in}_{bl}(z,t) - \frac{2}{3} p^{in}_{bl}(z,t)\right]_t = \sqrt{\tilde{\tau}} D_c \left[T^{in}_{bl}(z,t) - \frac{2}{3} p^{in}_{bl}(z,t)\right]_{zz}, \tag{4.134}$$

associated with the condition

$$\lim_{z\to\infty} \left[T^{in}_{bl}(z,t) - \frac{2}{3} p^{in}_{bl}(z,t)\right] = 0 \tag{4.135}$$

imposed by the matching conditions and the approximate equations in the bulk [see the equation of state in Eq. (4.133)].

The momentum equation indicates that the pressure perturbation is homogeneous in the boundary layer. As the piston-effect timescale is merged with the acoustic timescale, only the heating part of the boundary conditions [see Eqs. (3.36) and (4.37)] is considered:

$$T^{in}_{bl}(z=0,t) = \alpha f(t). \tag{4.136}$$

In order to solve the governing equations, we use the Laplace transform, which involves exchanging, for any function $f(t)$, the original time variable for a new one (s) by performing the following integration:

$$Lf(s) = \int_0^{\infty} e^{-st} f(t)dt,$$

such that the function $Lf(s)$ is called the Laplace transform of $f(t)$. The derivation rule is a remarkable property of the Laplace transform:

$$L\frac{df(t)}{dt} = sF(s) - f(0).$$

This is very useful for solving linear partial differential equations.

Solving Eq. (4.134) with Eqs. (4.135) and (4.136) leads to the following in Laplace space:

$$LT_{bl}^{in}(z,s) = \left[Lf(s) - \frac{2}{3}Lp_{bl}^{in}(s)\right] \exp\left(-\tilde{\tau}^{-\frac{1}{4}}\sqrt{s}\frac{z}{\sqrt{D_c}}\right) + \frac{2}{3}Lp_{bl}^{in}(s). \quad (4.137)$$

At this stage of the calculation, the homogeneous pressure term $Lp_{bl}^{in}(s)$ is not yet known. The continuity and energy equations give the velocity after integrating over the boundary layer:

$$Lu_{bl}^{in}(z,s) = \frac{2}{3}\frac{\sqrt{D_c}}{\tilde{\tau}^{\frac{3}{4}}}\sqrt{s}\left[Lf(s) - \frac{2}{3}Lp_{bl}^{in}(s)\right]\left\{1 - \exp\left(-\tilde{\tau}^{-\frac{1}{4}} \times \sqrt{s}\frac{z}{\sqrt{D_c}}\right)\right\}.$$

Again, we find that the velocity at the edge of the boundary layer is nonzero. This "piston velocity" drives the acoustic field in the bulk.

4.4.2.2 The Bulk

The relevant expansions in the bulk region are obtained as in Sect. 4.2.2 (regarding the acoustic saturation):

$$\begin{aligned}
\overline{T}^{in}(x,t) &= 1 + \tau + \alpha T_b^{in}(x,t) + o(\alpha) \\
\overline{p}^{in}(x,t) &= \frac{3}{2}(1+\tau) - \frac{9}{8} + \alpha p_b^{in}(x,t) + o(\alpha) \\
\overline{\rho}^{in}(x,t) &= 1 + \alpha \rho_b^{in}(x,t) + o(\alpha) \\
\overline{u}^{in}(x,t) &= \alpha u_b^{in}(x,t) + o(\alpha).
\end{aligned}$$

Employing these expansions in the basic one-dimensional equations given in Sect. 3.2 gives the following system of approximate equations:

$$\begin{aligned}
\text{continuity:}\quad & u_{b\,x}^{in}(x,t) + \rho_{b\,t}^{in}(x,t) = 0 \\
\text{momentum:}\quad & p_{b\,x}^{in}(x,t) = -\gamma_{0,\mathrm{IG}} u_{b\,t}^{in} \\
\text{energy:}\quad & \frac{3}{2}(\gamma_{0,\mathrm{IG}} - 1)u_{b\,x}^{in} + T_{b\,t}^{in} = 0 \\
\text{state:}\quad & p_b^{in}(x,t) = \frac{3}{2}T_b^{in}.
\end{aligned}$$

These equations lead to the propagation of the velocity field, as described by

$$u_{b\,tt}^{in} - c_{s0}u_{b\,xx}^{in} = 0,$$

where $c_{s0}^2 = \frac{9}{4}\frac{\gamma_{0,\mathrm{IG}}-1}{\gamma_{0,\mathrm{IG}}}$ [see Eq. (4.56)], associated with the boundary conditions

$$u_b^{in}(x=0,t) = \lim_{z\to\infty} u_{bl}^{in}(z,t)$$
$$u_b^{in}(x=0,1) = 0.$$

The solution of the propagation equation in Laplace space is

$$Lu_b^{in}(z,s) = \frac{2}{3}\frac{\sqrt{D_c}}{\tilde{\tau}^{\frac{3}{4}}}\sqrt{s}\left[Lf(s) - \frac{2}{3}Lp_b^{in}(s)\right]\frac{\sinh\left[\frac{s}{c_{s0}}(1-x)\right]}{\sinh\left(\frac{s}{c_{s0}}\right)},$$

which gives, upon applying the momentum equation and the matching condition for pressure,

$$Lp_b^{in}(0,s) = \frac{3}{2}\gamma_{0,\mathrm{IG}}\sqrt{D_c}\sqrt{\frac{\gamma_{0,\mathrm{IG}}-1}{\gamma_{0,\mathrm{IG}}}}\tilde{\tau}^{-\frac{3}{4}}\sqrt{s}\left[Lf(s) - \frac{2}{3}Lp_b^{in}(s)\right]\frac{\cosh\left(\frac{s}{c_{s0}}\right)}{\sinh\left(\frac{s}{c_{s0}}c\right)}. \tag{4.138}$$

The pressure is deduced from the previous equation:

$$Lp_b^{in}(0,s) \equiv Lp_b^{in}(s) = \frac{\gamma_{0,\mathrm{IG}}\sqrt{D_c}\sqrt{\frac{\gamma_{0,\mathrm{IG}}-1}{\gamma_{0,\mathrm{IG}}}}\tilde{\tau}^{-\frac{3}{4}}\sqrt{s}\cosh\left(\frac{s}{c_{s0}}\right)}{\sinh\left(\frac{s}{c_{s0}}\right) + \frac{2}{3}\gamma_{0,\mathrm{IG}}\sqrt{D_c}\sqrt{\frac{\gamma_{0,\mathrm{IG}}-1}{\gamma_{0,\mathrm{IG}}}}\tilde{\tau}^{-\frac{3}{4}}\sqrt{s}\cosh\left(\frac{s}{c_{s0}}\right)} \times Lf(s). \tag{4.139}$$

The whole solution, in the bulk and in the boundary layer, can be deduced from the above expression for the pressure. For the temperature, it is

$$LT_b^{in}(x,s) = \frac{\frac{2}{3}\gamma_{0,\mathrm{IG}}\sqrt{D_c}\sqrt{\frac{\gamma_{0,\mathrm{IG}}-1}{\gamma_{0,\mathrm{IG}}}}\tilde{\tau}^{-\frac{3}{4}}\sqrt{s}\cosh\left[\frac{s}{c_{s0}}(1-x)\right]}{\sinh\left(\frac{s}{c_{s0}}\right) + \frac{2}{3}\gamma_{0,\mathrm{IG}}\sqrt{D_c}\sqrt{\frac{\gamma_{0,\mathrm{IG}}-1}{\gamma_{0,\mathrm{IG}}}}\tilde{\tau}^{-\frac{3}{4}}\sqrt{s}\cosh\left(\frac{s}{c_{s0}}\right)} \times Lf(s). \tag{4.140}$$

It is not possible to invert Eq. (4.140) using the residue theorem because of the presence of the sum $\sinh(x) + \cosh(y)$, which makes it impossible to analytically calculate its poles. Numerical inversion of the Laplace transform is thus necessary in order to determine the inner description in real space. However, the solution of Eq. (4.140) permits a closer approach to the critical point than is allowed using the classical theory of the piston effect (f.o.d.), which does not hold when the fluid gets closer than $\varepsilon^{\frac{2}{3}}$ to the critical point. According to the matched asymptotic description technique, the close vicinity of the critical point is described by the s.o.d, which is given by the asymptotic behavior of the inner description [see Eq. (4.140)] when $\tilde{\tau} \to 0$. This asymptotic behavior is studied in the next section.

4.4.3 Can Temperature Equilibrate at the Speed of Sound?

4.4.3.1 The Close Vicinity of the Critical Point

The asymptotic behavior of Eq. (4.139) for $\tilde{\tau} \to 0$ is

$$\lim_{\tilde{\tau}\to 0} Lp_b^{in}(s) = Lp_b^{s.o.d.} = \frac{3}{2}Lf(s)\left[1 - \frac{1}{\gamma_{0,IG}}\frac{1}{\sqrt{D_c}}\sqrt{\frac{\gamma_{0,IG}}{\gamma_{0,IG}-1}} \times \frac{\sinh\left(\frac{s}{c_{s0}}\right)}{\sqrt{s}\cosh\left(\frac{s}{c_{s0}}\right)}\tilde{\tau}^{\frac{3}{4}} + o\left(\tilde{\tau}^{+\frac{3}{4}}\right)\right].$$

Here the superscript $s.o.d.$ stands for second outer description. The whole solution derived from the previous equation for pressure can be written as a function of ε and τ as follows:

- In the thermal boundary layer:

$$T_{bl}^{s.o.d.} = 1 + \tau + \alpha Lf(s) + O\left(\alpha\frac{\tau^{\frac{3}{4}}}{\epsilon^{\frac{1}{2}}}\right)$$

$$p_{bl}^{s.o.d.} = \frac{3}{2}(1+\tau) - \frac{9}{8} + \alpha\frac{3}{2}Lf(s) + O\left(\alpha\frac{\tau^{\frac{3}{4}}}{\epsilon^{\frac{1}{2}}}\right)$$

$$\rho_{bl}^{s.o.d.} = 1 + \frac{\alpha}{\epsilon^{\frac{1}{2}}\tau^{\frac{1}{4}}}Lf(s)\frac{1}{\gamma_{0,IG}}\frac{1}{\sqrt{D_c}}\sqrt{\frac{\gamma_{0,IG}}{\gamma_{0,IG}-1}}\frac{\sinh\left(\frac{s}{c_{s0}}\right)}{\sqrt{s}\cosh\left(\frac{s}{c_{s0}}\right)} \tag{4.141}$$

$$\times \exp\left(-\frac{\sqrt{s}}{\sqrt{D_c}}\frac{x}{\epsilon^{\frac{1}{2}}\tau^{\frac{1}{4}}}\right) \tag{4.142}$$

$$+ o\left(\frac{\alpha}{\epsilon^{\frac{1}{2}}\tau^{\frac{1}{4}}}\right)$$

$$u_{bl}^{s.o.d.} = \alpha\frac{3}{2}\frac{1}{\gamma_{0,IG}}\sqrt{\frac{\gamma_{0,IG}}{\gamma_{0,IG}-1}}Lf(s)\frac{\sinh\left(\frac{s}{c_{s0}}\right)}{\cosh\left(\frac{s}{c_{s0}}\right)}$$

$$\times\left[1 - \exp\left(-\frac{\sqrt{s}}{\sqrt{D_c}}\frac{x}{\epsilon^{\frac{1}{2}}\tau^{\frac{1}{4}}}\right)\right] + o(\alpha). \tag{4.143}$$

- In the bulk:

$$
\begin{aligned}
T_b^{s.o.d.} &= 1+\tau+\alpha L f(s)\frac{\cosh\left[\frac{s}{c_{s0}}(1-x)\right]}{\cosh\left(\frac{s}{c_{s0}}\right)} + O\left(\alpha\frac{\tau^{\frac{3}{4}}}{\epsilon^{\frac{1}{2}}}\right)\\
p_b^{s.o.d.} &= \frac{3}{2}(1+\tau)-\frac{9}{8}+\alpha\frac{3}{2}Lf(s)\frac{\cosh\left[\frac{s}{c_{s0}}(1-x)\right]}{\cosh\left(\frac{s}{c_{s0}}\right)} + O\left(\alpha\frac{\tau^{\frac{3}{4}}}{\epsilon^{\frac{1}{2}}}\right)\\
\rho_b^{s.o.d.} &= 1+\alpha\frac{2}{3}\frac{1}{\gamma_{0,\mathrm{iG}}-1}Lf(s)\frac{\cosh\left[\frac{s}{c_0}(1-x)\right]}{\cosh\left(\frac{s}{c_{s0}}\right)} + o(\alpha)\\
u_b^{s.o.d.} &= \alpha\frac{3}{2}\frac{1}{\gamma_{0,\mathrm{IG}}}\sqrt{\frac{\gamma_{0,\mathrm{IG}}}{\gamma_{0,\mathrm{IG}}-1}}Lf(s)\frac{\sinh\left[\frac{s}{c_{s0}}(1-x)\right]}{\cosh\left(\frac{s}{c_{s0}}\right)} + o(\alpha).
\end{aligned}
\tag{4.144}
$$

4.4.3.2 Acoustic Saturation of the Piston Effect

Note that solutions of Eqs. (4.141) and (4.144) no longer depend on the vicinity of the critical point. Furthermore, the first-order temperature perturbation in the boundary layer is homogeneous and equal to the value at the wall. This indicates that, when closer to the critical point than $\varepsilon^{\frac{2}{3}}$, the fluid temperature equilibrates on the acoustic timescale.

The driving force for the evolution (heat diffusion in the thermal boundary layer) only appears at the second order (which is $\tau^{\frac{3}{4}}/\varepsilon^{\frac{1}{2}}$ smaller than the first order).

Accordingly, the acoustic timescale appears to be the shortest temperature equilibration timescale. The s.o.d. that was not reachable continuously by the piston-effect theory corresponds to the saturation of the piston effect: when the fluid approaches the critical point, it first crosses the limit for which the piston effect and the acoustic timescale are of the same order of magnitude.

Then, if the fluid gets closer to the critical point than this limit, the temperature equilibration time stops decreasing. The bulk response tends continuously to a regime in which this temperature equilibration time no longer depends on the distance to the critical point but is entirely driven by acoustics.

This particular behavior may prevent the use of acoustic filtering procedures to solve the hydrodynamic equations near the critical point in the saturated regime. Such procedures are commonly used when solving low Mach number compressible flows for which acoustic waves are small perturbations that can be filtered.

We will show in Chap. 5 that acoustic waves are perturbations in this unsaturated acoustic regime. Acoustic waves can even be ignored or filtered depending on the required accuracy.

The bulk temperature in the saturated regime has been calculated for the following heating function:

$$f(t) = \frac{t}{t_0}.$$

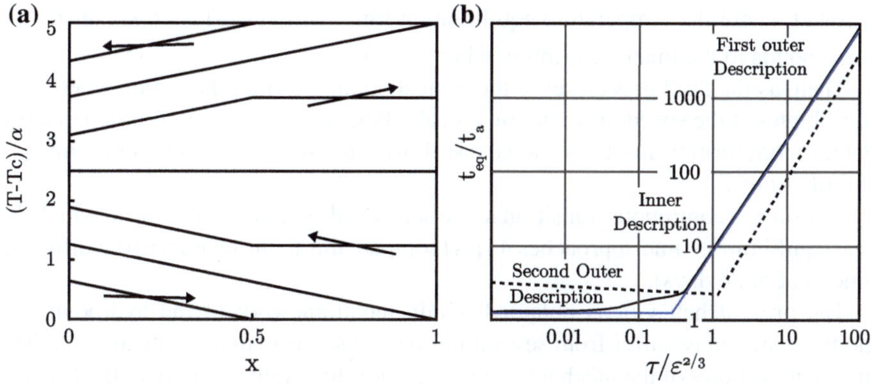

Fig. 4.14 Acoustic saturation. **a** Reduced bulk temperature (arbitrary units). *Arrows* indicate the instantaneous direction of propagation. **b** Reduced temperature equilibration time

The solution can be written as follows:

$$T_b^{sat} = 1 + \tau + \frac{1}{t_0}\left\{t + \frac{8}{\pi^2 c_{s0}}\sum_{n=1}^{+\infty}\frac{(-1)^n}{(2n-1)^2}\cos\left[\left(n-\frac{1}{2}\right)\pi(1-x)\right]\right.$$
$$\left.\times \sin\left[\left(n-\frac{1}{2}\right)\pi c_{s0} t\right]\right\}.$$

Figure 4.14a displays the temperature in the bulk for times extending from $t = 0$ to $t = \frac{4t_a}{c_{s0}}$ in steps of $\frac{1}{2}\frac{t_a}{c_{s0}}$ (which corresponds to steps of 22 μs in a 10 mm container filled with a model critical fluid that has the parameters given in Table 3.1). $\frac{t_a}{c_{s0}}$ represents the time it takes for an acoustic wave to cross the sample.

The instantaneous directions of propagation of the acoustic waves are indicated by small arrows. The strength of the acoustic wave (the ratio, at a given location, of the temperatures before and after the wave passage) is equal, to first order, to the temperature increase at the wall.

The reflection of the compression wave at the insulated wall can even result in a bulk temperature that is higher than the temperature at the boundary. This is not the case when the wall is isothermal (this case is treated in Sect. 5.3).

The evolution of the temperature equilibration time is plotted in Fig. 4.14b as a function of $\tilde{\tau} = \frac{\tau}{\varepsilon^{\frac{2}{3}}}$. The equilibration time is defined as the time necessary for the temperature at $x = 0.5$ to reach half of the value of the temperature at $x = 0$ (see Fig. 4.14a). The branch corresponding to the f.o.d. was calculated on the basis of the solution found in [29].

The inner description was computed by numerically inverting the Laplace transform of Eq. (4.140). The increasing branch of the equilibration time, for $\tau \geq \varepsilon^{\frac{2}{3}}$, corresponds to the piston-effect speeding up. When τ decreases, a crossover region is found where the temperature equilibration time saturates. This crossover region corresponds to the inner description where $\tau \sim \varepsilon^{\frac{2}{3}}$.

Then, as the fluid gets closer to the critical point, we enter the region where heat equilibrates at the speed of sound; this is called the second outer description (s.o.d.). For the model critical fluid considered just above, this crossover occurs approximately 2.4 mK from T_c.

It ensures a continuous transition to the acoustically saturated regime in which the heat equilibration time approaches 44 μs (compare this to the characteristic diffusion time of about 6 days).

The present behavior is obtained within the framework of the hydrodynamic model, which may suffer from several limitations so close to the critical point. The first is the validity of the mechanical model; that is to say, the validity of the Navier–Stokes equation in a region of diverging critical density fluctuations. The second limitation derives from the behavior of the bulk viscosity coefficient, which strongly diverges [5, 6, 12, 14, 17, 18, 21]. This divergence has never been clearly observed experimentally, presumably because of a extremely small amplitude prefactor. The third limitation is the presence of the critical adsorption layers at the walls [3, 10, 22], which is not accounted for. A fourth limitation is associated with the finite heat conductivity of the wall, which leads to a bottleneck effect [19]; see Sect. 8.3.3.

References

1. Abramowitz M, Stegun IA (eds) (1964) Handbook of mathematical functions. National Bureau of Standards, Washington
2. Bailly D, Zappoli B (2000) Hydrodynamic theory of density relaxation in near-critical fluids. Phys Rev E 62(2):2353–2368
3. Blümel S, Findenegg GH (1985) Critical adsorption of a pure fluid on a graphite substrate. Phys Rev Lett 54:447–450
4. Boukari H, Briggs ME, Shaumeyer JN, Gammon RW (1990) Critical speeding up observed. Phys Rev Lett 65(21):2654–2657
5. Carles P (1998) The effect of bulk viscosity on temperature relaxation near the critical point. Phys Fluids 10:2164–2176
6. Carles P, Dadzie K (2005) Two typical time scales of the piston effect. Phys Rev E 71:066310
7. Churchill RV (1972) Operational mathematics. McGraw-Hill Book Company, New York
8. Darrozes JS (1972) The method of matched asymptotic expansions applied to problems involving two singular perturbation parameters. Fluid Dyn Trans 6:72
9. Ferrell RA, Hao H (1993) Adiabatic temperature changes in a one-component fluid near the liquid–vapor critical point. Phys A 197(1–2):23–46
10. Fisher ME, de Gennes PG (1978) C R Acad Sci Paris B 287:207
11. Garrabos Y, Bonetti M, Beysens D, Perrot F, Fröhlich T, Carlès P, Zappoli B (1998) Relaxation of a supercritical fluid after a heat pulse in the absence of gravity effects: theory and experiments. Phys Rev E 57(5):5665–5681
12. Kadanoff LP, Swift J (1968) Transport coefficients near the liquid-gas critical point. Phys Rev 166:89
13. Kassoy DR (1979) The response of a confined gas to a thermal disturbance. i. Slow transients. SIAM J Appl Math 36:3

14. Kawasaki K (1970) Sound attenuation and dispersion near the liquid-gas critical point. Phys Rev A 1:1750
15. Kevorkian J, Cole JD (1981) Perturbation methods in applied mathematics. Springer, New York
16. Miura Y, Honda K, Kawai J, Matsumoto M, Sakurai M, Yoshihara S, Ohnishi M, Kobayashi H, Takenouchi T, Ishikawa M (2004) Need research. In: 2nd International symposium on physical sciences in space, Toronto, ON, Canada, 23–27 May 2004
17. Onuki A (1997) Bulk viscosity near the critical point. J Phys Soc Jpn 66:511–513
18. Onuki A (2002) Phase transition dynamics. Cambridge University Press, Cambridge
19. Onuki A, Ferrell RA (1990) Adiabatic heating effect near the gas-liquid critical point. Phys A 164(2):245–264
20. Onuki A, Hao H, Ferrell RA (1990) Fast adiabatic equilibration in a single-component fluid near the liquid–vapor critical point. Phys Rev A 41(4):2256–2259
21. Quentrec B (1979) A new analysis of sound propagation near the critical point of xenon. J Phys Lett Paris 40(13):257–261
22. Thomms M, Findenegg GH, Lewandowski H (1994) Critical adsorption of SF6 on a finelly divided graphite sustrate. Ber Bunsenges Phys Chem 98:477–481
23. Van Dyke M (1964) Perturbation methods in fluid mechanics. Academic Press, New York
24. Wilkinson RA, Zimmerli GA, Hao H, Moldover MR, Berg RF, Johnson WL, Ferrell RA, Gammon RW (1998) Equilibration near the liquid–vapor critical point in microgravity. Phys Rev E 57(1):436–448
25. Zappoli B (1992) The response of a nearly supercritical pure fluid to a thermal disturbance. Phys Fluids A 4:1040–1048
26. Zappoli B, Bailly D (1990) Transport in a confined compressible fluid under timedependent volumetric heat source. Phys Fluids A 2:1771–1779
27. Zappoli B, Carlès P (1996) Acoustic saturation of the critical speeding up. Phys D: Nonlinear Phenom 89:381–394
28. Zappoli B, Carlès P (1995) The thermo-acoustic nature of the critical speeding up. Eur J Mech B 14:41–65
29. Zappoli B, Durand-Daubin A (1994) Heat and mass transport in a near supercritical fluid. Phys Fluids 6(5):1929–1936

Chapter 5
Heat Transfer

Abstract The theory of heat propagation from a heat pulse is presented in the mean-field framework of the one-dimensional model. It is used to interpret the phenomena observed in three-dimensional experiments performed under microgravity for similar heating conditions. A heat flux is supplied at the heated wall, instead of an imposed parietal temperature. Such heat flux conditions (that mimics the heat source) are more adapted to the experimental investigations. A constant temperature is then imposed at the other boundary (that mimics the thermostated cell). As the piston-effect velocity is expected to be proportional to the heat flux at the wall, the response velocity can be checked by tuning the flux.

5.1 Theory of the Cooling Piston Effect

5.1.1 The Model and the Asymptotic Approach

The 1-D model is considered and solved analytically, as in [24]. The heating wall is at $x = 0$, and the boundary at $x = 1$ corresponds to the thermostated wall. In a first step, the walls have zero thickness and the thermostated one is then considered to be perfectly conducting. The basic equations that describe the fluid motion are Eqs. (3.31)–(3.34). The same critical fluid model (see Table 3.1) is also used in this analytical approach. However, the linearized equation of state is refined here (when applied to particular experiments) by considering the universal Ising values of the critical exponents (see Table 1.2), without changing the main body of the analytical work done so far. In a second step, a numerical solution of this 1-D model (see also [11]) is given in Sect. 5.4 of this chapter in order to account for the finite thickness and the finite thermal conductivity of the walls.

The thermal boundary condition that describes the surface thermal power $\mathcal{P}(t)$ supplied at the boundary at $x = 0$ by the heating device (like a thermistor in a self-heating pulse mode) is expressed in a dimensional form, homogeneous to a heat flux as follows:

B. Zappoli et al., *Heat Transfers and Related Effects in Supercritical Fluids*,
Fluid Mechanics and Its Applications 108, DOI: 10.1007/978-94-017-9187-8_5,

$$-\Lambda T_x(x=0,t) = \mathcal{P}(t) \quad \text{at } x=0, \tag{5.1}$$

where the thermal conductivity Λ is defined by Eq. (3.13). In the analytical approach, the heating power $\mathcal{P}$ is supplied to a wall of unit area during a finite time period t_0; i.e., $\mathcal{P}=0$ for $t<0$, $\mathcal{P}=\mathcal{P}(t)$ for $0\leq t\leq t_0$ and $\mathcal{P}=0$ for $t>t_0$. The cutoff time $t_0<t_D$ is smaller than the characteristic heat diffusion time $t_D=\frac{L^2}{D_T}$ with the cell length $L=1$ in arbitrary units. The thermal boundary condition at $x=1$ is equal to the initial temperature T_0 of the thermostat:

$$T(x=1,t)=T_0. \tag{5.2}$$

The initial equilibrium pressure of the fluid at critical density $\rho_0=\rho_c$ is p_0. The boundary conditions for the velocity field are zero velocity at both boundaries. Writing Eq. (5.1) in a nondimensional form, and considering the temperature dependence of Λ given by Eq. (3.14), the boundary condition becomes

$$-\left(1+\Lambda_c\tau^{-\frac{1}{2}}\right)T_x(x=0,t)=\varphi\mathcal{P}(t) \text{ and } T(x=1,t)=T_0, \tag{5.3}$$

where

$$\varphi=\frac{L}{\Lambda_{b0}^{\mathrm{MF}}T_c}\mathcal{P}_{\max} \tag{5.4}$$

gives the order of magnitude of the heat flux at the wall. $\mathcal{P}_{\max}$ is the maximum value of $\mathcal{P}(t)$. In the boundary conditions of Eq. (5.3), φ represents the heat flux expressed in $\mathcal{P}_{\max}$ units, and this remains on the order of 1 for any heating period.

5.1.2 The Acoustic Period

5.1.2.1 The Thermal Boundary Layer

In the same way as previously described in Chap. 4, we start by studying the acoustic time period. Again, the problem is singular. The variables need to be rescaled to fit with the steep temperature variations in the diffusion boundary layer, which have a characteristic size (denoted δ) that is linked to the vanishing heat diffusivity. This order of magnitude must be carefully obtained by performing an order of magnitude analysis of the Navier–Stokes equations. This analysis is the same as in Chap. 4, except for the temperature. Its order of magnitude must be deduced from the order of magnitude of the heat flux at the thermistor boundary, which is the driving force. The temperature can be expanded as

$$\widetilde{T}^a(z,t)=1+\tau+\eta_T T_{bl}^a(z,t)+o(\eta_T),$$

where η_T is the order of magnitude of the temperature perturbation. The boundary condition of Eq. (5.3) leads to the following nondimensional equation:

$$-\left(1+\Lambda_c^*\tau^{-\frac{1}{2}}\right)\frac{\eta_T}{\delta}T^a_{bl\,z}(z=0,t)=\varphi\mathcal{P}(t). \tag{5.5}$$

From this relation we can directly deduce the order of magnitude of the temperature perturbation at the boundary:

$$\eta_T=\frac{\delta\varphi}{\left(1+\Lambda_c^*\tau^{-\frac{1}{2}}\right)}.$$

We assume that the critical point is so close that the background properties become negligible, so that

$$\left(1+\Lambda_c^*\tau^{-\frac{1}{2}}\right)\approx\Lambda_c\tau^{-\frac{1}{2}}.$$

The order of magnitude of the temperature increase at the boundary is then

$$\eta_T\equiv\alpha=\frac{\tau^{-\frac{1}{2}}\delta\varphi}{\Lambda_c^*}. \tag{5.6}$$

Here, the value of α depends on the heat flux (through φ), whereas it was directly dependent on the reduced temperature increase in Chap. 4, where the temperature change was imposed at the wall $x=0$. When we look for the orders of magnitude of the hydrodynamics variables, the same reasoning as used in Chap. 4 leads to the following expansions in the thermal boundary layer on the acoustic timescale:

$$\begin{aligned}
\widetilde{T}^a(z,t)&=1+\tau+\varepsilon^{\frac{1}{2}}\tau^{\frac{3}{4}}\varphi T^a_{bl}(z,t)+o(\varepsilon^{\frac{1}{2}}\tau^{\frac{3}{4}}\varphi) && \varepsilon^{\frac{1}{2}}\tau^{\frac{3}{4}}\varphi\ll 1\\
\widetilde{\rho}^a(z,t)&=1+\varepsilon^{\frac{1}{2}}\tau^{-\frac{1}{4}}\varphi\rho^a_{bl}(z,t)+o(\varepsilon^{\frac{1}{2}}\tau^{-\frac{1}{4}}\varphi) && \varepsilon^{\frac{1}{2}}\tau^{-\frac{1}{4}}\varphi\ll 1\\
\widetilde{u}^a(z,t)&=0+\varepsilon\varphi u^a_{bl}(z,t)+o(\varepsilon\varphi) && \varepsilon\varphi\ll 1\\
\widetilde{p}^a(z,t)&=\tfrac{3}{2}(1+\tau)-\tfrac{9}{8}+\varepsilon\varphi p^a_{bl}(z,t)+o(\varepsilon\varphi) && \varepsilon\varphi\ll 1.
\end{aligned} \tag{5.7}$$

Here too, we note that the order of magnitude of the density perturbation is $\frac{1}{\tau}$ times larger than the temperature perturbation because of the divergence (as τ^{-1}) of the isothermal compressibility. The boundary layer thickness is found to be $\delta=\varepsilon^{\frac{1}{2}}\tau^{\frac{1}{4}}$ because the thermal diffusivity decreases as $\tau^{-\frac{1}{2}}$. The system of equations is formally the same as the system of Eqs. (4.34)–(4.36). However, the boundary condition at $x=0$ is now replaced with

$$\left.\frac{\partial T^a_{bl}}{\partial z}\right|_{z=0}(z=0,t)=-\mathcal{P}(t). \tag{5.8}$$

This system reduces to a diffusion equation for temperature, which is written

$$T^a_{bl\,t} = D_c T^a_{bl\,zz}, \tag{5.9}$$

where D_c is defined by Eq. (3.30). The boundary condition at $x = 0$ is given by Eq. (5.8), and that at $x = 1$ is replaced with the requirement that $\lim_{z\to\infty} T^a_{bl}(z, t) = 0$. As in Sect. 4.4.2, Eq. (5.9) is solved by the Laplace transform,

$$sLT^a_{bl}(z, s) = D_c LT^a_{bl\,zz},$$

with the boundary conditions

$$LT^a_{bl}(0, s) = -L\mathcal{P}(s)$$
$$\lim_{z\to\infty} LT^a_{bl}(z, s) = 0$$

and the initial condition $LT^a_{bl}(z, 0) = 0$. The solution of this ordinary differential equation of the variable z is a straightforward task, and yields

$$LT^a_{bl}(z, s) = \frac{L\mathcal{P}(s)}{\sqrt{\frac{s}{D_c}}} e^{-z\sqrt{\frac{s}{D_c}}}. \tag{5.10}$$

The Laplace transform of the velocity in the boundary layer is obtained from the momentum equation, and can be written as

$$Lu^a_{bl}(z, s) = \frac{2}{3} D_c \mathcal{P}(s) \left[1 - e^{-z\sqrt{\frac{s}{D_c}}}\right]. \tag{5.11}$$

It appears that the velocity at the edge of the thermal boundary layer ($z \to \infty$) is nonvanishing and on the order of $\varepsilon\varphi$, which indicates that, here too, there is a mass flux entering the bulk phase from the boundary layer. It is also proportional to the heat flux at the wall, since we can use the above equation to check that

$$\lim_{z\to\infty} Lu^a_{bl}(z, s) = \frac{2}{3}\mathcal{P}(s). \tag{5.12}$$

This is consistent with the energy equation written in the boundary layer [see Eq. (4.46)]. As expected, the piston-effect velocity can thus be changed as needed by changing the heat flux at the boundary. The other solutions for density and pressure can be found in [24].

5.1.2.2 The Bulk Phase

Reasoning similar to that employed in Chap. 4 would lead to perturbations of the hydrodynamic variables of amplitude $\varepsilon\varphi$ in the bulk phase. The same system of linearized equations as Eqs. (4.50)–(4.53) reduces to a linear propagation equation. This equation describes the compression waves generated by mass addition at the edge of the thermal boundary layer. The amplitude of the corresponding acoustic field is $\varepsilon\varphi$. In Laplace space, this equation can be written as

$$s^2 Lu^a_{b\,tt}(x,s) - c^2_{s0} Lu^a_{b\,xx}(x,s) = 0,$$

where c^2_{s0} is defined by Eq. (4.56). The following boundary conditions are deduced from the matching condition with the boundary layer region:

$$Lu^a_b(0,s) = \frac{2}{3} D_c \mathcal{P}(s). \tag{5.13}$$

From the zero velocity condition at $x = 1$, we obtain

$$Lu^a_b(1,s) = 0.$$

Note that the acoustic phenomenon that occurs in the bulk is purely isentropic to first order. Thermal diffusion is negligible in the bulk by definition; the viscous dissipation only matters at higher orders, since the velocity gradient is too weak. More generally, a one-dimensional flow always lacks viscous dissipation unless flow discontinuities are present (such as shock waves).

In Laplace space, the solutions for the velocity and for the temperature are

$$Lu^a_b(x,s) = \frac{2}{3} D_c \mathcal{P}(s) \frac{\sinh\left[\frac{s}{c_{s0}}(1-x)\right]}{\sinh(\frac{s}{c_{s0}})} \tag{5.14}$$

$$LT^a_b(x,s) = \frac{4\gamma_{0,IG} c_{s0}}{9} D_c \mathcal{P}(s) \frac{\cosh\left[\frac{s}{c_{s0}}(1-x)\right]}{\sinh(\frac{s}{c_{s0}})}. \tag{5.15}$$

These solutions describe compression waves that flash back and forth between the thermostated wall of the cell and the heating wall. If the heat flux is constant (i.e., $\mathcal{P}(s) = \frac{1}{s}$), the solutions in real space are

$$u^a_b(x,t) = \frac{2}{3} D_c \left[(1-x) - \frac{2}{\pi} \sum_{n=1}^{\infty} \frac{1}{n} \sin(n\pi x)\cos(n\pi c_{s0} t) \right]$$

$$T^a_b(x,t) = \frac{4\gamma_{0,IG} c_{s0}}{9} D_c \left[c_{s0} t + \frac{2}{\pi} \sum_{n=1}^{\infty} \frac{1}{n} \cos(n\pi x)\sin(n\pi c_{s0} t) \right]. \tag{5.16}$$

The above solution for the temperature corresponds to a linear compression wave. The traveling front reproduces the Heaviside step for the heat flux at $t = 0$. When the wave reaches a given location, it increases the local temperature by $\frac{4\gamma_{0,\mathrm{IG}} c_{s0}}{9} D_c$. As the wave flashes back and forth, the temperature of the medium increases over time. The evolution of the bulk temperature is given by the term $c_{s0} t$.

5.1.3 Piston-Effect Time Period

5.1.3.1 The Bulk Phase

The temperature in the boundary layer is on the order of $\varepsilon^{\frac{1}{2}} \tau^{\frac{3}{4}} \varphi$, while it is on the order of $\varepsilon \varphi$ in the bulk (see Sect. 5.1.2). At the outer edge of the thermal boundary layer, the temperature is thus zero at the order $\varepsilon^{\frac{1}{2}} \tau^{\frac{3}{4}} \varphi$. In the bulk, the temperature at $x = 0$ is a constant multiplied by $\varepsilon \varphi$, which is much smaller than $\varepsilon^{\frac{1}{2}} \tau^{\frac{3}{4}} \varphi$, and is thus zero at this order. The asymptotic matching between the bulk and boundary layer temperature is thus fulfilled. However, this hierarchy is not conserved as time goes by. Indeed, from Eq. (5.10), the temperature in the boundary layer, expressed in real space, is the convolution of the heating power with $\frac{1}{\sqrt{t}}$ (the inverse Laplace transform of $\frac{\sqrt{\pi}}{\sqrt{s}}$). From Eq. (5.15), the bulk temperature is the convolution product of the heating power with a constant. The temperature in the bulk thus increases $\sqrt{t}$ times faster than the boundary layer temperature. In order for the matching condition [similar to Eq. (4.62)] to be fulfilled again, the bulk temperature and the boundary layer temperature must be of the same order of magnitude. Thus, time must be rescaled as follows:

$$\varepsilon^{\frac{1}{2}} \varphi \tau^{\frac{3}{4}} T^a_{bl}(z, \frac{\Theta}{\sigma(\varepsilon, \tau)}) \equiv \varepsilon \varphi T^a_b(x, \frac{\Theta}{\sigma(\varepsilon, \tau)})$$

or

$$\varepsilon^{\frac{1}{2}} \varphi \tau^{\frac{3}{4}} = \varepsilon \varphi \sqrt{\frac{\Theta}{\sigma(\varepsilon, \tau)}}.$$

From the above equation, the characteristic timescale at which the piston effect acts is given by

$$\Theta = \frac{\varepsilon}{\tau^{\frac{3}{2}}} t.$$

For a constant heating flux, Eq. (5.16) for T and u display two important features:

- The order of magnitude of the velocity remains constant over time, in contrast to the temperature step case, where it goes to zero
- The temperature is the additive superposition of an increasing linear function of time and an acoustic term that remains bounded over time

The matching approach indicates that the solution on the piston-effect timescale can be written, according to the method of multiple timescales [12], as the sum of a slowly varying dominant mode and a rapid perturbative mode:

$$
\begin{aligned}
\bar{T}^{pe}(x,\Theta) &= 1+\tau+\tau^{\frac{3}{2}}\varphi T_b^{pe}(x,\Theta)+\varepsilon\varphi T_A(x,t,\Theta)+o(\varepsilon\varphi)\\
\bar{\rho}^{pe}(x,\Theta) &= 1+\tau^{\frac{3}{2}}\varphi\rho_b^{pe}(x,\Theta)+\varepsilon\varphi\rho_A(x,t,\Theta)+o(\varepsilon\varphi)\\
\bar{p}^{pe}(x,\Theta) &= \frac{3}{2}(1+\tau)-\frac{9}{8}+\tau^{\frac{3}{2}}\varphi p_b^{pe}(x,\Theta)+\varepsilon\varphi p_A(x,t,\Theta)+o(\varepsilon\varphi)\\
\bar{u}^{pe}(x,\Theta) &= 0+0+\varepsilon\varphi u_A(x,t,\Theta)+o(\varepsilon\varphi).
\end{aligned}
\tag{5.17}
$$

In the above equations, the subscript A corresponds to the rapid perturbative modes that depend on both the "fast" timescale (t variable) and a "slow" one (Θ variable). We thus use the method of matched asymptotic expansions to solve the singularity in space, and the method of multiple timescales to take into account the loss of validity of the acoustic solution at large times. This allows a deeper analysis of the interplay between the acoustic phenomena and the piston effect. Taking into account the fact that the variables t and Θ are independent, the following derivation rule holds for any thermodynamic variable $F(x,t,\Theta)$:

$$
\frac{\partial F}{\partial t}=\left.\frac{\partial F}{\partial t}\right|_{\Theta,x}+\left.\frac{\varepsilon}{\tau^{\frac{3}{2}}}\frac{\partial F}{\partial \Theta}\right|_{t,x},
$$

and, by incorporating the expansion of Eq. (5.17) into the basic Navier–Stokes equations given in Sect. 3.2.1, we obtain the following set of equations:

$$
\rho_{b\,\Theta}^{pe}+\rho_{b\,x}^{pe}=-u_{A\,x}
\tag{5.18}
$$

$$
\begin{aligned}
p_{b\,x}^{pe}&=0\\
p_{A\,x}&=-\gamma_{0,\mathrm{IG}}u_{A\,t}
\end{aligned}
\tag{5.19}
$$

$$
\begin{aligned}
\frac{3}{2}T_b^{pe}&=p_b^{pe}\\
\frac{3}{2}T_A&=p_A
\end{aligned}
\tag{5.20}
$$

$$
T_{b\,\Theta}^{pe}+T_{A\,t}=-\frac{3}{2}\left(\gamma_{0,\mathrm{IG}}-1\right)u_{A\,x},
\tag{5.21}
$$

with the initial conditions

$$
T_b^{pe}(x,0)=u_b^{pe}(x,0)=\rho_b^{pe}(x,0)=u_b^{pe}(x,0)=T_A=p_A=\rho_A=u_A=0.
\tag{5.22}
$$

At this step in the analytical approach, the system of Eqs. (5.18)–(5.21) has no boundary conditions. They will be obtained by performing boundary layer analysis (similar to the one made in Chap. 4): first at $x = 0$, and second at the thermostated wall at $x = 1$. Two distinct boundary layers are thus introduced: the heating boundary layer (HBL) located at the heating wall at $x = 0$, and the cooling boundary layer (CBL) at the thermostated cell boundary at $x = 1$. The boundary conditions for the bulk equations are replaced by the matching conditions with the HBL and the CBL:

$$T_b^{pe}(0,\Theta) = \lim_{z\to\infty} T_{hbl}^{pe}(z,\Theta)$$

$$T_b^{pe}(1,\Theta) = \lim_{z^*\to\infty} T_{cbl}^{pe}(z^*,\Theta) \tag{5.23}$$

$$p_b^{pe}(\Theta) = \lim_{z\to\infty} p_{hbl}^{pe}(z,\Theta) \tag{5.24}$$

$$p_b^{pe}(\Theta) = \lim_{z^*\to\infty} p_{cbl}^{pe}(z^*,\Theta), \tag{5.25}$$

where the subscripts *hbl* and *cbl* refer, respectively, to the heating boundary layer and the cooling boundary layer. The CBL boundary layer variable is defined as

$$z^* = \frac{1-x}{\delta},$$

where δ is the boundary layer thickness. Therefore, we first need to solve the boundary layer equations.

5.1.3.2 Thermal Boundary Layers

Following similar reasoning to that employed in Chap. 4, the variables are expanded as follows in the two boundary layers:

$$\begin{bmatrix} \tilde{T}_{hbl}^{pe}(z,\Theta) \\ \tilde{p}_{hbl}^{pe}(z,\Theta) \\ \tilde{T}_{cbl}^{pe}(z^*,\Theta) \\ \tilde{p}_{cbl}^{pe}(z^*,\Theta) \end{bmatrix} = \begin{bmatrix} 1+\tau \\ \frac{3}{2}(1+\tau)-\frac{9}{8} \\ 1+\tau \\ \frac{3}{2}(1+\tau)-\frac{9}{8} \end{bmatrix} + \tau^{\frac{3}{2}}\varphi \begin{bmatrix} T_{hbl}^{pe}(z,\Theta) \\ p_{hbl}^{pe}(z,\Theta) \\ T_{cbl}^{pe}(z^*,\Theta) \\ p_{cbl}^{pe}(z^*,\Theta) \end{bmatrix} + o(\tau^{\frac{3}{2}}\varphi)$$

$$\begin{bmatrix} \tilde{\rho}_{hbl}^{pe}(z,\Theta) \\ \tilde{u}_{hbl}^{pe}(z,\Theta) \end{bmatrix} = \begin{bmatrix} 1 \\ 0 \end{bmatrix} + \tau^{\frac{1}{2}}\varphi \begin{bmatrix} \rho_{hbl}^{pe}(z,\Theta) \\ u_{hbl}^{pe}(z,\Theta) \end{bmatrix}$$

$$\begin{bmatrix} \tilde{\rho}_{cbl}^{pe}(z^*,\Theta) \\ \tilde{u}_{cbl}^{pe}(z^*,\Theta) \end{bmatrix} = \begin{bmatrix} 1 \\ 0 \end{bmatrix} + \varepsilon\varphi \begin{bmatrix} \rho_{cbl}^{pe}(z^*,\Theta) \\ u_{cbl}^{pe}(z^*,\Theta) \end{bmatrix}.$$

By inserting the above expansions into the Navier–Stokes equations, we obtain

$$\rho^{pe}_{hbl,\Theta} + u^{pe}_{hbl\,z} = 0$$
$$\rho^{pe}_{cbl,\Theta} + u^{pe}_{cbl\,z^*} = 0 \tag{5.26}$$

$$p^{pe}_{hbl\,z} = p^{pe}_{cbl\,z^*} = 0 \tag{5.27}$$

$$\frac{3}{2}T^{pe}_{hbl} + \frac{9}{4}\rho^{pe}_{hbl} = p^{pe}_{hbl}$$
$$\frac{3}{2}T^{pe}_{cbl} + \frac{9}{4}\rho^{pe}_{cbl} = p^{pe}_{cbl} \tag{5.28}$$

$$\frac{3}{2}u^{pe}_{hbl\,z} - D_c T^{pe}_{hbl\,zz} = 0$$
$$\frac{3}{2}u^{pe}_{cbl\,z^*} - D_c T^{pe}_{cbl\,z^*z^*} = 0, \tag{5.29}$$

with the initial conditions

$$T^{pe}_{hbl}(z,0) = u^{pe}_{hbl}(z,0) = \rho^{pe}_{hbl}(z,0) = p^{pe}_{hbl}(z,0) = 0$$
$$T^{pe}_{cbl}(z^*,0) = u^{pe}_{cbl}(z^*,0) = \rho^{pe}_{cbl}(z^*,0) = p^{pe}_{cbl}(z^*,0) = 0$$

and the boundary conditions:

$$T^{pe}_{hbl}\big|_z (z=0,\Theta) = -\mathcal{P}(\Theta)$$
$$T^{pe}_{cbl}(z^*=0,\Theta) = 0$$
$$u^{pe}_{hbl}(z=0,\Theta) = 0$$
$$u^{pe}_{cbl}(z^*=0,\Theta) = 0.$$

Since the driving force for the evolution is the heat flux at $x = 0$, the solution is first sought in the hot boundary layer region. The pressure is homogeneous throughout the sample, and can be replaced everywhere by its value to first order $p_0(\Theta)$. Equation (5.28) then give, after some rearranging,

$$\left[T^{pe}_{hbl} - \frac{2}{3}p_0(\Theta)\right]_\Theta - D_c\left[T^{pe}_{hbl} - \frac{2}{3}p_0(\Theta)\right]_{zz} = 0.$$

The solution to this diffusion equation can be written in Laplace space as

$$LT^{pe}_{hbl}(z,q) = \frac{L\mathcal{P}(q)}{\sqrt{\frac{q}{D_c}}} e^{-z\sqrt{\frac{q}{D_c}}} + \frac{2}{3}Lp_0(q),$$

where q is the Laplace variable associated with the piston-effect time variable Θ. From the continuity equation, taking into account the fact that pressure is homogeneous, we obtain

$$Lu_{hbl}^{pe}(z,q) = \frac{2}{3} D_c L\mathcal{P}(q) \left[1 - e^{-\sqrt{\frac{q}{D_c}}z}\right].$$

A first matching condition can be deduced from the above equation:

$$Lu_A(0,q,s) = \lim_{z\to\infty} Lu_{hbl}^{pe}(z,q) = \frac{2}{3} D_c L\mathcal{P}(q). \tag{5.30}$$

Now considering the cold boundary layer, the equations also reduce to a diffusion equation:

$$\left[T_{cbl}^{pe} - \frac{2}{3}p_0(\Theta)\right]_\Theta - D_c \left[T_{cbl}^{pe} - \frac{2}{3}p_0(\Theta)\right]_{z^*z^*} = 0,$$

with the boundary condition

$$\left.\left(T_{cbl}^{pe} - \frac{2}{3}p_0(\Theta)\right)\right|_{z^*=0} = -\frac{3}{2}p_0(\Theta),$$

which has the following solution in Laplace space:

$$LT_{cbl}^{pe}(z^*,q) - \frac{2}{3}Lp_0(q) = -\frac{2}{3}L\mathcal{P}(q)e^{-z^*\sqrt{\frac{q}{D_c}}}$$
$$Lu_{cbl}^{pe}(z^*,q) = \frac{4}{9}\sqrt{D_c}\sqrt{q}Lp_0(q)\left[1 - e^{-z^*\sqrt{\frac{q}{D_c}}}\right].$$

Equation (5.30) gives the second matching condition for u_A:

$$Lu_A(1,q,s) = \lim_{z^*\to\infty} Lu_{hbl}^{pe}(z^*,q) = \frac{4}{9}\sqrt{D_c}\sqrt{q}Lp_0(q). \tag{5.31}$$

Equation (5.20), which describe the motion in the bulk phase, reduce to a propagation equation for the velocity field:

$$u_A(x,t,\Theta)_{tt} - c_{s0}^2 u_A(x,t,\Theta)_{xx} = 0, \tag{5.32}$$

which can be written in Laplace space as follows:

$$q^2 Lu_A(x,s,\Theta) - c_{s0}^2 Lu_A(x,s,\Theta)_{xx} = 0. \tag{5.33}$$

The boundary conditions deduced from the matching conditions of Eqs. (5.30) and (5.31) are

$$Lu_A(0,q) = \frac{2}{3}\frac{D_c}{s}L\mathcal{P}(q)$$
$$Lu_A(1,q) = \frac{4}{9}\frac{\sqrt{D_c}}{s}\sqrt{q}Lp_0(q). \tag{5.34}$$

The velocity $u_A(0,\Theta)$ corresponds to a mass addition to the bulk, which derives from the expansion of the fluid within the heating boundary layer. The velocity $u_A(1,\Theta)$ corresponds to the mass depletion of the bulk phase due to the contraction of the fluid located within the cold boundary layer. The latter corresponds to the opposite phenomenon to that occurring in the hot boundary layer [note the opposite signs of the diffusion terms in Eq. (5.29)]. This contraction also occurs in ideal gases, and explains a number of transport anomalies observed during crystal growth from the vapor phase [23]. However, its magnitude is much smaller than that for supercritical fluids.

When the acoustic wave emitted at the hot boundary layer reaches $x = 1$, the CBL forms, and the contraction of the fluid it contains gives rise to a nonzero boundary condition for the bulk velocity. Thus, the latter cannot exist before the sound wave emitted at the hot boundary layer reaches $x = 1$, and the boundary condition at $x = 1$ must be shifted by an acoustic time. This can be written as

$$Lu_A(1,q,s) = e^{-as} \times \frac{4}{9}\frac{\sqrt{D_c}}{s}\sqrt{q}Lp_0(q), \tag{5.35}$$

where $e^{-as}\frac{4}{9\sqrt{D_c}}\sqrt{q}L\mathcal{P}(q)$ is the Laplace transform of

$$H_{[0]}\left(\Theta - \frac{\varepsilon}{c_{s0}}\tau^{-\frac{3}{2}}\right)u_A\left(1, \Theta - \frac{\varepsilon}{c_{s0}}\tau^{-\frac{3}{2}}t\right).$$

The solution of Eq. (5.33) with boundary conditions of Eqs. (5.34) and (5.35) in physical space is

$$u_A(x,\Theta,t) = u_A(0,\Theta) - [u_A(0,\Theta) - u_A(1,\Theta)]$$
$$\times\left[x + \frac{2}{\pi}\sum_{n=1}^{\infty}\frac{1}{n}\sin(n\pi x)\cos(n\pi c_{s0}t)\right].$$

This solution clearly shows that the velocity field depends on two time variables: a "fast" variable t, which describes the acoustic phenomena, superposed onto a "slow" variable Θ, which describes the mechanisms occurring on the piston-effect timescale. $u_A(0,\Theta)$ and $u_A(1,\Theta)$ are the boundary conditions that drive the acoustic process in the bulk. They correspond, respectively, to a continuous compression wave that is emitted at the HBL and to a continuous expansion wave that is emitted at the CBL. The latter causes a cooling of the bulk phase by adiabatic expansion. This is the *cooling piston effect*, which will be shown to play a crucial role in heat transfer in supercritical fluids. From Eq. (5.20) for p_{Ax}, T_b^{pe} and T_A^{pe}, and the energy equation, it follows that

$$p_0(\Theta) = \frac{9}{4}(\gamma_{0,\mathrm{IG}} - 1) \int_0^{\Theta} [u_A(0,v) - u_A(1,v)]\, dv,$$

where v is an auxiliary variable specific to this equation. With a Laplace transform that gives:

$$qLp_0(q) = \frac{9}{4}(\gamma_{0,\mathrm{IG}} - 1)\left[\frac{2}{3}D_c L\mathcal{P}(q) - \frac{4}{9}\sqrt{D_c}\sqrt{q}Lp_0(q)\right].$$

This equation gives the explicit form of the homogeneous temperature in the bulk:

$$LT_b^{pe} = (\gamma_{0,\mathrm{IG}} - 1)D_c \frac{L\mathcal{P}(q)}{\sqrt{q}\left[\sqrt{q} + (\gamma_{0,\mathrm{IG}} - 1)\sqrt{D_c}\right]}.$$

The latter gives, by inverting the Laplace transform and invoking the convolution theorem, the following form of the homogeneous temperature in the bulk:

$$T_b^{pe}(\Theta) = (\gamma_{0,\mathrm{IG}} - 1)D_c\mathcal{P}(\Theta) \otimes \exp\left[(\gamma_{0,\mathrm{IG}} - 1)^2 D_c\Theta\right] \mathrm{erfc}\left((\gamma_{0,\mathrm{IG}} - 1)\sqrt{D_c\Theta}\right) \tag{5.36}$$

Here, $\otimes$ denotes the convolution product.

5.1.3.3 Complete Solution in the Physical Space

By merging the results obtained in the previous paragraphs, we obtain the following two-timescale solutions for the perturbation variables:

- In the hot boundary layer:

$$T_{hbl}^{pe}(z,\Theta) = \mathcal{P}(\Theta) \otimes \left[\frac{\sqrt{D_c}}{\sqrt{\pi}\sqrt{\Theta}} \exp\left(\frac{z^2}{4D_c\Theta}\right) + (\gamma_{0,\mathrm{IG}} - 1)\, D_c \right. \tag{5.37}$$
$$\left. \times \exp\left((\gamma_{0,\mathrm{IG}} - 1)^2 D_c\Theta\right) \mathrm{erfc}\left((\gamma_{0,\mathrm{IG}} - 1)\right)\sqrt{D_c}\sqrt{\Theta}\right]$$

$$\tilde{u}_{hbl}^{pe} \equiv u_{hbl}^{pe} = \frac{3}{2}\mathcal{P}(\Theta) \otimes \left[\delta(\Theta) - \frac{z\exp\left(-\frac{z^2}{4D_c\Theta}\right)}{2\sqrt{\pi}\sqrt{D_c}\Theta^{\frac{3}{2}}}\right].$$

- In the bulk phase:

$$T_b^{pe}(\Theta) = (\gamma_{0,\mathrm{IG}} - 1)D_c\mathcal{P}(\Theta) \otimes \exp\left[(\gamma_{0,\mathrm{IG}} - 1)^2 D_c\Theta\right]$$
$$\times\, erfc\left((\gamma_{0,\mathrm{IG}} - 1)\sqrt{D_c}\sqrt{\Theta}\right) \tag{5.38}$$

$$
\begin{aligned}
T_A(x, \Theta, t) &= \frac{4\gamma_{0,\mathrm{IG}} c_{s0}}{3\pi} \left[u_A(0, \Theta) - u_A(1, \Theta)\right] \sum_{n=1}^{\infty} \frac{1}{n} \sin(n\pi x) \cos(n\pi c_{s0} t) \\
u_A(x, \Theta, t) &= u_A(0, \Theta) - \left[u_A(0, \Theta) - u_A(1, \Theta)\right] \\
&\quad \times \left\{ x + \frac{2}{\pi} \sum_{n=1}^{\infty} \frac{1}{n} \sin(n\pi x) \cos(n\pi c_{s0} t) \right\}.
\end{aligned} \tag{5.39}
$$

- In the cooling boundary layer:

$$
\begin{aligned}
T_{cbl}^{pe}(z^*, \Theta) &= \mathcal{P}(\Theta) \otimes (\gamma_{0,\mathrm{IG}} - 1) D_c \\
&\quad \times \left\{ \exp\left((\gamma_{0,\mathrm{IG}} - 1)^2 \sqrt{D_c \Theta}\right) \operatorname{erfc}\left((\gamma_{0,\mathrm{IG}} - 1) \times \sqrt{D_c} \sqrt{\Theta}\right) \right. \\
&\quad - \exp\left((\gamma_{0,\mathrm{IG}} - 1) z^*\right) \exp\left((\gamma_{0,\mathrm{IG}} - 1)^2 D_c \Theta\right) \\
&\quad \left. \times \operatorname{erfc}\left[(\gamma_{0,\mathrm{IG}} - 1) \sqrt{D_c} \sqrt{\Theta} + \left(\frac{z^*}{2\sqrt{D_c}\sqrt{\Theta}} \right) \right] \right\}
\end{aligned} \tag{5.40}
$$

$$
\begin{aligned}
u_{cbl}^{pe}(z^*, \Theta) &= \mathcal{P}(\Theta) \otimes \frac{3}{2} (\gamma_{0,\mathrm{IG}} - 1) D_c \sqrt{D_c} \\
&\quad \times \left[\frac{1}{\sqrt{\pi}\sqrt{\Theta}} \left(1 - \exp\left(\frac{z^{*2}}{4 D_c \Theta} \right) \right) \right. \\
&\quad + (\gamma_{0,\mathrm{IG}} - 1) \sqrt{D_c} \exp\left((\gamma_{0,\mathrm{IG}} - 1) z^*\right) \times \exp\left((\gamma_{0,\mathrm{IG}} - 1)^2 D_c \Theta\right) \\
&\quad \times \operatorname{erf} c \left((\gamma_{0,\mathrm{IG}} - 1) \sqrt{D_c} \sqrt{\Theta} + \frac{z^*}{2\sqrt{D_c}\sqrt{\Theta}} \right) - (\gamma_{0,\mathrm{IG}} - 1) \sqrt{D_c} \\
&\quad \left. \exp\left((\gamma_{0,\mathrm{IG}} - 1)^2 D_c \Theta\right) \times \operatorname{erf} c \left((\gamma_{0,\mathrm{IG}} - 1) \sqrt{D_c} \sqrt{\Theta}\right) \right].
\end{aligned}
$$

Taking into account the asymptotic behaviors of the velocity in the HBL and CBL for large values of z and z^*, we obtain the following values of $u_A(0, \Theta)$ and $u_A(1, \Theta)$:

$$
\begin{aligned}
u_A(0, \Theta) &= \frac{3}{2} D_c \mathcal{P}(\Theta) \\
u_A(1, \Theta) &= \mathcal{P}(\Theta) \otimes \frac{3}{2} (\gamma_{0,\mathrm{IG}} - 1) D_c \sqrt{D_c} \\
&\quad \times \left\{ \frac{1}{\sqrt{\pi}\sqrt{\Theta}} - (\gamma_{0,\mathrm{IG}} - 1) \sqrt{D_c} \exp\left[(\gamma_{0,\mathrm{IG}} - 1)^2 D_c \Theta \right] \right. \\
&\quad \left. \times \operatorname{erf} c \left[(\gamma_{0,\mathrm{IG}} - 1) \sqrt{D_c} \sqrt{\Theta} \right] \right\}
\end{aligned}
$$

5.1.3.4 Solution for a Typical Heat Flux

The process where heat is supplied to the fluid as an imposed heat flux clearly illustrates the thermoacoustic nature of the piston effect. As a matter of fact, Eqs. (5.37), (5.39), and (5.40) for the temperature can show that acoustic waves are emitted by both the heating boundary layer at $x = 0$ and the cooling boundary layer at $x = 1$. However, the solution for the temperature field is calculated explicitly for the following typical heating function:

$$\mathcal{P}(\Theta) = 1 - H_{[0]}(\Theta - \Theta_0), \tag{5.41}$$

where $H_{[0]}(\Theta)$ is the Heaviside step function, which is equal to 0 for $(\Theta - \Theta_0) < 0$ and equal to 1 for $(\Theta - \Theta_0) > 0$. It must be noted that the cutoff time on the Θ scale depends on the distance to the critical point for fixed t_0, since $\Theta_0 = \frac{\varepsilon}{\tau^{\frac{3}{2}}} t_0$. We now introduce the following useful quantities:

$$\begin{aligned}
h(\Theta) &= \frac{1}{(\gamma_{0,\mathrm{IG}} - 1)^2 D_c} \exp\left[(\gamma_{0,\mathrm{IG}} - 1)^2 D_c \Theta\right] \operatorname{erf} c\left[(\gamma_{0,\mathrm{IG}} - 1)\sqrt{D_c}\sqrt{\Theta}\right] \\
&\quad - \frac{1}{(\gamma_{0,\mathrm{IG}} - 1)^2 D_c} + \frac{1}{(\gamma_{0,\mathrm{IG}} - 1)\sqrt{D_c}\sqrt{\pi}}\sqrt{\Theta} \\
h^*(z^*, \Theta) &= \exp\left[(\gamma_{0,\mathrm{IG}} - 1)^2 D_c \Theta + (\gamma_{0,\mathrm{IG}} - 1) z^*\right] \\
&\quad \times \operatorname{erf} c\left((\gamma_{0,\mathrm{IG}} - 1)\sqrt{D_c}\sqrt{\Theta} + \frac{z^*}{2\sqrt{D_c}\sqrt{\Theta}}\right) \\
&\quad - \operatorname{erf} c\left(\frac{z}{2\sqrt{D_c}\sqrt{\Theta}}\right) + 1 \\
&\quad - \exp\left[(\gamma_{0,\mathrm{IG}} - 1)^2 D_c \Theta\right] \operatorname{erf} c\,(\gamma_{0,\mathrm{IG}} - 1)\sqrt{D_c}\,\sqrt{\Theta} \\
h_1^*(z^*, \Theta) &= \frac{1}{(\gamma_{0,\mathrm{IG}} - 1)^2 D_c} \left\{\exp\left[(\gamma_{0,\mathrm{IG}} - 1)^2 \sqrt{D_c}\Theta + (\gamma_{0,\mathrm{IG}} - 1) z^*\right]\right. \\
&\quad \times \operatorname{erf} c\left((\gamma_{0,\mathrm{IG}} - 1)\sqrt{D_c}\sqrt{\Theta} + \frac{z^*}{2\sqrt{D_c}\sqrt{\Theta}}\right) \\
&\quad \left. - \left(1 + (\gamma_{0,\mathrm{IG}} - 1) z^*\right) \operatorname{erf} c\left(\frac{z^*}{2\sqrt{D_c}\sqrt{\Theta}}\right)\right\} \\
&\quad + \frac{2}{(\gamma_{0,\mathrm{IG}} - 1)\sqrt{D_c}\sqrt{\pi}}\sqrt{\Theta} \exp\left(-\frac{z^{*2}}{4 D_c \Theta}\right)
\end{aligned}$$

and

$$\begin{aligned}
u_A(0, \Theta) &= \frac{3}{2}\sqrt{D_c}\left[1 - H_{[0]}(\Theta - \Theta_0)\right] \\
u_A(1, \Theta) &= \frac{3}{2}\sqrt{D_c}\left\{1 - \exp\left[(\gamma_{0,\mathrm{IG}} - 1)^2 D_c \Theta\right] \operatorname{erf} c\left((\gamma_{0,\mathrm{IG}} - 1)\sqrt{D_c}\sqrt{\Theta}\right)\right. \\
&\quad - H_{[0]}(\Theta - \Theta_0)\left[1 - \exp\left[(\gamma_{0,\mathrm{IG}} - 1)^2 D_c (\Theta - \Theta_0)\right]\right. \\
&\quad \left.\left. \times \operatorname{erf} c\left((\gamma_{0,\mathrm{IG}} - 1)\sqrt{D_c}\sqrt{(\Theta - \Theta_0)}\right)\right]\right\}.
\end{aligned}$$

Therefore, the solutions of Eqs. (5.37)–(5.40) can be written as follows:

- In the heating boundary layer:

$$\begin{aligned} T^{pe}_{hbl}(z,\Theta) &= 2\sqrt{D_c}\left[\sqrt{\Theta}\, i\,\mathrm{erfc}\left(\frac{z}{2\sqrt{D_c}\sqrt{\Theta}}\right) - H_{[0]}(\Theta-\Theta_0)\right.\\ &\qquad\left.\times\sqrt{(\Theta-\Theta_0)}\, i\,\mathrm{erf}\,c\left(\frac{z}{2\sqrt{D_c}\sqrt{(\Theta-\Theta_0)}}\right)\right]\\ &\quad+\frac{\gamma_0}{\mathrm{Pr}_0}\left[h(\Theta)-H_{[0]}(\Theta-\Theta_0)h(\Theta-\Theta_0)\right]\\ \rho^{pe}_{hbl}(z,\Theta) &= \frac{4}{3}\sqrt{D_c}\left[\sqrt{\Theta}\, i\,\mathrm{erfc}\left(\frac{z}{2\sqrt{D_c}\sqrt{\Theta}}\right)\right.\\ &\qquad\left.-H_{[0]}(\Theta-\Theta_0)\sqrt{(\Theta-\Theta_0)}\, i\,\mathrm{erfc}\left(\frac{z}{2\sqrt{D_c}\sqrt{(\Theta-\Theta_0)}}\right)\right]\\ u^{pe}_{hbl}(z,\Theta) &= \frac{3}{2}\sqrt{D_c}\left[\left\{1-\mathrm{erfc}\left(\frac{z}{2\sqrt{D_c}\sqrt{\Theta}}\right)\right\}\right.\\ &\qquad\left.-H_{[0]}(\Theta-\Theta_0)\sqrt{(\Theta-\Theta_0)}\left\{1-\mathrm{erf}\,c\left(\frac{z}{2\sqrt{D_c}\sqrt{(\Theta-\Theta_0)}}\right)\right\}\right]. \end{aligned} \tag{5.42}$$

- In the bulk phase:

$$\begin{aligned} T^{pe}_b(\Theta) &= \frac{\gamma_{0,\mathrm{IG}}}{\mathrm{Pr}_0}\left[h(\Theta)-H_{[0]}(\Theta-\Theta_0)h(\Theta-\Theta_0)\right]\\ T_A(x,\Theta,t) &= \frac{4\gamma_{0,\mathrm{IG}}}{3\pi}c_{s0}\left[u_A(0,\Theta)-u_A(1,\Theta)\right]\\ &\quad\times\sum_{n=1}^{\infty}\frac{1}{n}\sin(n\pi x)\cos(n\pi c_{s0}t)\\ u_A(x,\Theta,t) &= u_A(0,\Theta)-\left[u_A(0,\Theta)-u_A(1,\Theta)\right]\\ &\quad\times\left\{x+\frac{2}{\pi}\sum_{n=1}^{\infty}\frac{1}{n}\sin(n\pi x)\cos(n\pi c_{s0}t)\right\}\\ p_0(\Theta) &= \frac{3}{2}T^{pe}_{hbl}(\Theta)\ \text{and}\ p_A(x,\Theta,t)=\frac{3}{2}T_A(x,\Theta,t)\\ \rho^{pe}_b(\Theta) &= \frac{2}{3(\gamma_{0,\mathrm{IG}}-1)}T^{pe}_b(\Theta)\\ \text{and}\ \rho_A(x,\Theta,t) &= \frac{2}{3(\gamma_{0,\mathrm{IG}}-1)}T_A(x,\Theta,t). \end{aligned} \tag{5.43}$$

- In the cooling boundary layer:

$$
\begin{aligned}
T^{pe}_{cbl}(z,\Theta) &= \frac{\gamma_{0,\mathrm{IG}}}{\mathrm{Pr}_0}\left\{\left[h(\Theta) - H_{[0]}(\Theta-\Theta_0)h(\Theta-\Theta_0)\right]\right.\\
&\qquad\left. - \left[h^*_1(z^*,\Theta) - H_{[0]}(\Theta-\Theta_0)h^*_1(z^*,\Theta-\Theta_0)\right]\right\}\\
\rho^{pe}_{cbl}(z,\Theta) &= \frac{3}{2}\frac{\gamma_{0,\mathrm{IG}}}{\mathrm{Pr}_0}\left[h^*(z^*,\Theta)\right.\\
&\qquad\left. - H_{[0]}(\Theta-\Theta_0)\, h^*(z^*,\Theta-\Theta_0)\right]\\
u^{pe}_{cbl}(z,\Theta) &= \frac{3}{2}\sqrt{D_c}\left[h^*(z^*,\Theta) - H_{[0]}(\Theta-\Theta_0)h^*(\Theta-\Theta_0)\right].
\end{aligned}
\tag{5.44}
$$

5.2 The Fourth Heat Transfer Mechanism

The uniformly valid description in space is obtained by applying the additive composition procedure (see Chap. 4) to solutions of Eqs. (5.42), (5.43), and (5.44). On the piston-effect timescale, the diffusion effects are confined to a very thin boundary layer of thickness τ. However, the bulk is subjected to acoustic compression and expansion waves that are emitted at the heating and cooling edges of the boundary layer, respectively. As mentioned in the previous section, this velocity field is directly linked to the expansion or contraction of the fluid contained within the HBL or the CBL. The complete two-timescale solution for the temperature is computed with the same set of parameters as those in Chap. 4.

This solution is plotted in Fig. 5.1 for the Heaviside step heating function. We can clearly see that the temperature increase is due to the acoustic waves that flash back and forth in the cavity; each time the wave passes a given location, the temperature increases by an amount proportional to the difference between the velocities at the edge of the HBL and CBL (i.e., to the difference between the heating and cooling fluxes). This proportionality reflects the resulting combination of the heating compression wave and cooling expansion wave.

We will see in the following that this cooling piston effect is an important process in heat transfer (see Chap. 9) and hydrodynamic instabilities (see Sect. 14.3) in the presence of the Earth's gravitational field.

5.2.1 *The Piston Effect as a Thermal Short Circuit*

As mentioned in the next section, the piston effect can be considered a fourth temperature equilibration mechanism. However, since the unheated walls are insulated, no net heat transfer through the sample is possible. In the present case, however, the wall at $x = 1$ is kept at the initial fluid temperature. When the bulk is heated by the piston effect, a cooling boundary layer appears near the thermostated wall. By a process similar to that occurring at the heated wall, a positive velocity field is generated by the contraction of the fluid present in this cooled layer. This contraction provokes

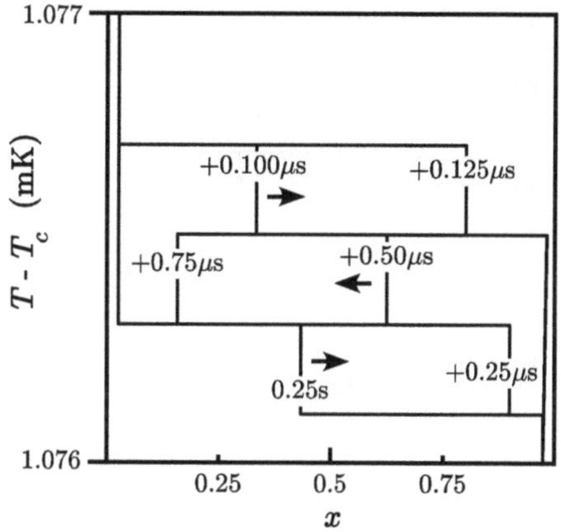

Fig. 5.1 The thermoacoustic nature of the piston effect as shown by the asymptotic solution. The thermal field is plotted every 25 μs. The compression wave emitted at the edge of the thermal diffusion boundary layer (*left wall*) provokes a traveling increase in the bulk temperature

backward-traveling expansion waves. As a result of the cooling piston effect, net transport of heat is thus possible. The mean temperature in the bulk, as given by the slowly varying component of the temperature, is calculated in detail in [24]. The explicit calculation is based on the near-supercritical fluid model, as introduced in Sect. 3.2.2. The fluid, at 1 K above T_c, is confined in a sample cell with a typical length of 12 mm. A constant heat flux of 1 W m^{-2} is applied for 0.5 s. According to Eq. (5.6), the temperature rise at the boundary is on the order of 10 mK. The other parameters are equal to those given in Chap. 4.

Figure 5.2a represents the mean temperature perturbation in the bulk phase after the heating period (for $\Theta \geq \Theta_0$). This temperature perturbation is homogeneous, which is a characteristic of the piston-effect heating. The boundary layers appear as very thin regions near the two boundaries of the sample cell. The important feature is that the fluid transfers heat from one side to the other without any bulk dissipation: the entropy production is confined to the boundary layers. Viscous dissipation, which never appears in the approximate equations, is of higher order (i.e., much smaller) throughout in the sample. The bulk fluid thus behaves as a thermal short circuit, and the heat transfer dissipation becomes independent of the sample length. However, this is only valid on the piston-effect timescale, when the thermal boundary layer thickness remains much smaller than the sample length. As the heat diffusion is very slow when approaching the critical point, the process can last until the characteristic heat diffusion time is reached (which is several hours at 1 K from the critical point for a 1 cm sample).

When the heat flux is turned off, the heating piston effect stops quasi-immediately. The only time lag is that of the propagation of the compression waves through the heating boundary layer, which occurs on a timescale much shorter than the acoustic time. The mean temperature decreases due to thermoacoustic expansion of the bulk as the cooling piston effect remains effective. Density profiles are plotted in Fig. 5.2b. The density profile shows that once the heat flux has stopped, the slope at $x = 0$ is zero, like the slope of the temperature profile (the pressure gradient is zero throughout the domain). In contrast, the density profile at $x = 1$ is characteristic of the presence

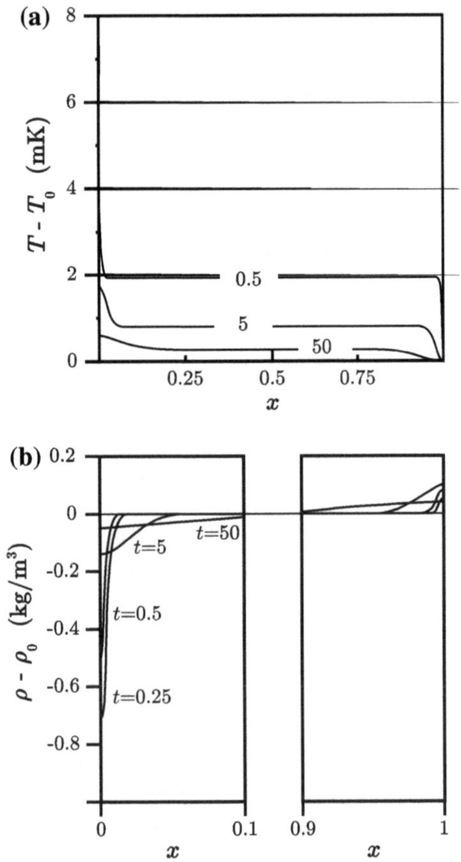

Fig. 5.2 Evolution on the piston-effect timescale: **a** temperature; **b** density. Time is in seconds and the distances are made nondimensional by normalizing to to the sample cell length

of a nonzero heat flux. The cooling piston effect, which is proportional to the heat flux at the wall, becomes weaker and weaker as the bulk temperature decreases. The evolution becomes increasingly diffusive everywhere. The time variable needs to be rescaled to describe the very long heat diffusion period. This rescaling can be performed as in Chap. 4.

Considering what we have discussed in this section, it is now worth addressing the question of whether the piston effect can be considered a fourth heat-transfer mechanism, in addition to the classical processes of conduction, radiation and convection.

5.2.2 The Piston Effect: A New Mode of Energy Transport?

In the previous chapter we saw that, when the heated sample fluid is isolated at $x = 1$, piston-effect heating equilibrates the bulk fluid temperature very quickly through isentropic compression. Now, if the fluid is confined between hot and cold walls, we saw above that energy can be transferred through the sample due to a combination of heating and cooling piston effects. However, the temperature change remains homogeneous in the bulk fluid due to an isentropic process too. This leads to a paradoxical situation, since a heat flux is classically the result of a temperature gradient in an isenthalpic process. Therefore, we will analyze the details of the process in the following, paying special attention to confirming that the mean-field description of the singular behavior of the characteristic time of these (heating and/or cooling) piston effects also leads to an accurate understanding of the true Ising singular behavior expected in real critical fluids.

The specific entropy s of the fluid evolves according to the following dimensional equation using the physical variables [see also Eq. (21.82)] :

$$\rho T\frac{ds}{dt} = \boldsymbol{\tau} : \mathbf{D} + \nabla.[\Lambda \nabla T], \tag{5.45}$$

where $\boldsymbol{\tau} : \mathbf{D}$ is the dissipation function [see also Eq. (21.74)], i.e., the twice-contracted product of the stress tensor $\boldsymbol{\tau}$ and the deformation rate tensor $\mathbf{D}$. A small temperature variation dT can be decomposed into a function of small variations in entropy S and pressure p:

$$dT = \frac{T}{c_{pc}}dS + \left(1 - \frac{1}{\gamma_{0c}}\right)\left(\frac{\partial T}{\partial p}\right)_S dp, \tag{5.46}$$

where the subscript extension c refers to the critical divergences of the specific heats at constant pressure $c_{pc} \propto \tau^{-\gamma}$ and constant volume $c_{Vc} \propto \tau^{-\alpha}$, and its related ratio $\gamma_{0c} = \frac{c_{pc}}{c_{Vc}} \propto \tau^{-\gamma-\alpha}$ (with $\gamma \simeq 1.24$ and $\alpha \simeq 0.11$; see Eqs. (1.76), (1.82) and Table 1.3). The 1-D asymptotic analyses developed in the previous chapters have shown that, during the evolution of the fluid as governed by the piston effect, (i) the pressure remains homogeneous when the timescale is longer than the acoustic time, and (ii) viscous dissipation is only a higher-order phenomenon. As a consequence, Eq. (5.45) leads (to first order) to the thermal diffusion equation at constant pressure:

$$\frac{dS}{dt} = \frac{\Lambda}{\rho c_p}\Delta S = D_T \Delta S, \tag{5.47}$$

in which the thermal diffusion coefficient $D_T \propto \tau^{\gamma - Y_\Lambda}$ goes to zero at the critical point (with $\gamma - Y_\Lambda = \nu + Y_\eta \simeq 0.67$, $\nu \simeq 0.63$, $Y_\eta \simeq 0.04$ and $Y_\Lambda \simeq 0.57$; see Table 1.5).

Thus, despite the rapid temperature equilibration of the bulk, the heat transport is led by diffusion, which remains very slow, and most of the thermal energy sent into the fluid remains inside the thermal boundary layers (see Fig. 5.3). As a consequence, the piston effect should not be considered a new mode of heat transport.

However, let us consider the configuration of Fig. 5.3, where a near 2-D sample located in an interferometer is heated at its left wall and kept at a constant temperature at its right wall. A hot diffusive boundary layer (of thickness δ_1) develops at the left

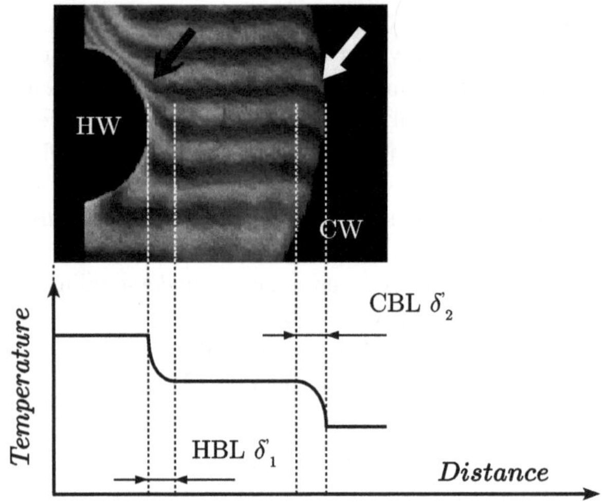

Fig. 5.3 Optical interferometer cell filled with SF_6 confined between two conductive walls a distance L apart. The temperature of the "hot" wall is maintained constant (HW, $T = T_c + 100$ mK, $T_c = 45.564$ K), while that of the "cold" wall has just been decreased by 10 mK (CW, $T = T_c + 90$ mK). The fluid is experiencing microgravity conditions; the *black arrow* indicates the HBL (thickness δ_1), and the *white arrow* the CBL (thickness δ_2)

wall (the density drops, as indicated by the bending of the interferometer fringes), while a cold boundary layer (of thickness δ_2) simultaneously forms on the right-hand side (the density increases, as indicated by the bending of the interferometer fringes). In-between, the temperature is homogeneous (as is the density, as the interferometer fringes are rectilinear). The effective dimensions of the sample for heat transport are thus reduced to $\delta_1 + \delta_2$. However, it is important to note that the acceleration of the heat transport by this effect is transient; the boundary layers ultimately reach the dimensions of the cell and the stationary state remains diffusive. The bulk fluid is a kind of thermal short circuit, which seems to contradict the resulting Eq. (5.47). This is, nevertheless, only an apparent contradiction, as the heat transport is performed by the two (hot and cold) boundary layers, which are connected by the bulk fluid that acts as a thermal short circuit. The thermal gradient thus increases from $\frac{\Delta T}{L}$ to $\frac{\Delta T}{\delta_1+\delta_2}$, which correspondingly increases and accelerates the heat transport.

5.2.3 Transfer Function of a Thermostated Cell

In a mean-field approach, now we consider the temperature increase in the bulk—limited to the component due to the slow, piston-effect timescale—that results from the interplay between the heating and cooling piston effects. It is defined by Eq. (5.17) for the temperature and given by Eq. (5.38). It can be written explicitly as

$$\overline{T}^{pe} - T_0 = \tau^{\frac{3}{2}}\varphi\left[\mathcal{P}(\Theta) \otimes (\gamma_{0,\mathrm{IG}} - 1)D_c e^{(\gamma_{0,\mathrm{IG}}-1)^2 D_c \Theta}\,\mathrm{erfc}\left((\gamma_{0,\mathrm{IG}} - 1)\sqrt{D_c}\sqrt{\Theta}\right)\right] + o(\tau^{\frac{3}{2}}\varphi),$$

where φ is the nondimensional parameter that characterizes the heat pulse as defined in Eq. (5.4). Introducing the physical variables and time t, we obtain

$$\overline{T}^{pe} - T_0 = T_c\tau^{\frac{3}{2}}\varphi\left[\mathcal{P}(t) \otimes (\gamma_{0,\mathrm{IG}} - 1)D_c e^{\frac{t}{t_{PE}}}\,\mathrm{erfc}\left(\sqrt{\frac{t}{t_{PE}}}\right)\right]\frac{\varepsilon}{\tau^{\frac{3}{2}}}$$
$$= \frac{1}{\rho_c L c_V}\mathcal{P}(t) \otimes e^{\frac{t}{t_{PE}}}\,\mathrm{erf\,c}\left(\sqrt{\frac{t}{t_{PE}}}\right).$$

Here, $t_{PE} = \frac{t_a}{(\gamma_{0,\mathrm{IG}}-1)D_c}\tau^{\frac{3}{2}}$ is the characteristic piston-effect time, and $\mathcal{P}(t)$ is the boundary heat flux per unit surface and unit time. We can also define

$$\Pi(t) = A_H \times \mathcal{P}(t)$$

as the instantaneous input heat power supplied to the fluid, where A_H is the heating surface area in contact with the fluid. Since $\rho_c LS$ is the total mass m of fluid provided, we obtain

$$\overline{T}^{pe}(t) - T_0 = \Pi(t) \otimes \frac{e^{\frac{t}{t_{PE}}}\,\mathrm{erfc}\left(\sqrt{\frac{t}{t_{PE}}}\right)}{mc_V}. \tag{5.48}$$

The function $e^{\frac{t}{t_{PE}}}\,\mathrm{erfc}\left(\sqrt{\frac{t}{t_{PE}}}\right)$ is the transfer function of the cell if the power supplied to the fluid through the heating surface is considered the entry function and the temperature increase in the bulk is thus the output function.

Different heating powers with the same total energy have been considered, as shown in Fig. 5.4a. The corresponding convolution products for the bulk temperature are given in Fig. 5.4b. The experimental value of the total energy will be given later, in Chap. 9. It corresponds to the value $\varphi = 10^{-2}$ and to a cutoff time of $t = 0.5$ s. Figure 5.4b shows that the maximum value of the temperature in the bulk and the time taken to reach it depend on the shape of the heat flux. However, the shape of the temperature response for longer times is almost independent of the shape of the heat flux, and depends only on the total energy it has received.

5.2.4 Energy Balance and Piston-Effect Efficiency

In the analytical approach, the piston effect was shown to homogenize temperature on a short timescale. The next step is to determine its efficiency from the energy balance. Let us consider a 1-D fluid sample that is initially at $T = T_0 > T_c$ and $\rho = \rho_c$, into which a heat flux $P(t)$ is sent at $x = 0$. This fluid sample is thermally insulated at

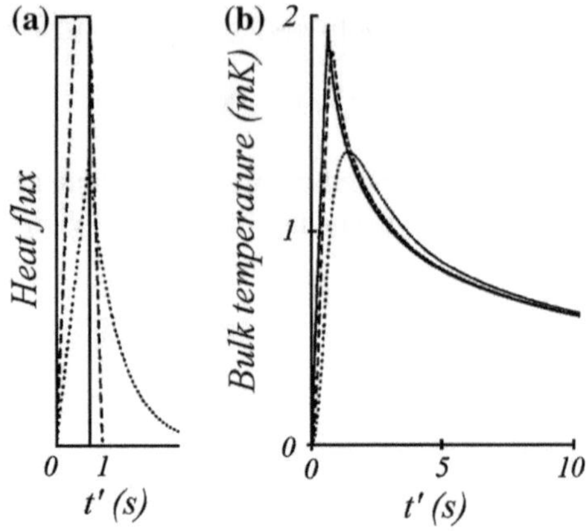

Fig. 5.4 **a** Different heat flux profiles at the wall with the same total energy. **b** Temperature evolution for the different heat fluxes; this shows that the temperature relaxation profile for longer times is near-independent of the heating time profile

$x = 1$ (in arbitrary length units). The same calculation as used in the previous section with $u_A(1, \Theta) = 0$ leads to the following value for the bulk temperature increase, ignoring the acoustic perturbations:

$$\overline{T}^{pe}(t) - T_0 = \Pi(t) \otimes \frac{1}{mc_V}.$$

In other words, if $Q(t)$ is the total energy supplied to the fluid up to time t, the temperature in the bulk can be written as

$$\overline{T}^{pe}(t) - T_0 = \frac{Q(t)}{mc_V}. \tag{5.49}$$

The same calculations for the mean density in the bulk phase give

$$\overline{\rho}^{pe}(t) - \rho_c = \frac{2}{3}\frac{\rho_c}{\gamma_{0,\mathrm{IG}} - 1}\frac{1}{T_0}\frac{Q(t)}{mc_V}. \tag{5.50}$$

From Eq. (3.11), which is valid for a van der Waals fluid, the variation in the total bulk energy due to the temperature and density variations is

$$E - E_0 = LS\left[c_V\left(T - T_0\right) - a\left(\rho - \rho_c\right)\right].$$

E_0 is the initial total energy of the fluid volume LS at the initial conditions $\{T_0, p_0, \varrho_c\}$. Introducing the above Eqs. (5.49) and (5.50) for $T - T_0$ and $\rho - \rho_c$, respectively, we obtain

$$E\left(t\right) - E_0 = Q\left(t\right)\left[1 - \frac{2a}{3\left(c_{p,\mathrm{IG}} - c_V\right)}\frac{\rho_c}{T_c}\right]$$

We recall that the c_V notation in the analytical approach of Chap. 4 means $c_V = c_{V,\mathrm{IG}}$ in the mean field approximation. Taking into account the values of a expressed as a function of the critical coordinates of the van der Waals fluid [see Eq. (1.12)] and $c_{p,\mathrm{IG}} - c_{V,\mathrm{IG}} = r$, the above equation can be written as

$$E(t) - E_0 = \frac{1}{4} Q(t),$$

which allows us to define the energy yield ζ as

$$\zeta = \frac{E(t) - E_0}{Q(t)} = \frac{1}{4} \tag{5.51}$$

for a van der Waals fluid. Only a small fraction ($\frac{1}{4}$) of the energy brought to the fluid is thus transmitted to the bulk phase, although the temperature increase of the bulk fluid is the same as that obtained as if all of the energy were transferred to the bulk by diffusion. In other words, as the internal energy of the van der Waals fluid also depends on density (energy decreases when density increases), less energy is needed to reach a given temperature by isentropic compression than by diffusion. The temperature efficiency of the piston effect in terms of increasing the bulk temperature is thus

$$\zeta^{-1} = 4 \tag{5.52}$$

and constant however the fluid temperature distance to its critical temperature.

5.2.5 Generalization to Real Fluids

The nondimensional model presented in Chap. 4, just like the results obtained in Sects. 5.2.3 and 5.2.4, are based on the van der Waals equation of state and a mean-field approach to the fluid behavior. In the following, we provide a calculation based on a more realistic estimate for the equation of state and the transport properties, and focus especially on gaining a better knowledge of the variations in the characteristic parameters of the piston effect when $\tau \to 0$. Let us consider the true velocity of sound c_s, defined using Eq. (1.84), i.e.,

$$c_s^2 = \frac{1}{\rho \kappa_s} = \frac{1}{\rho \kappa_T} + \frac{T}{\rho^2 c_V} \left(\frac{\partial p}{\partial T} \right)_\rho^2 . \tag{5.53}$$

As $\kappa_T \sim \tau^{-1.24}$ while $c_V \sim \tau^{-0.11}$ along the critical isochore (see Table 1.3 and [20] for example), the first term on the right-hand side of Eq. (5.53) can be ignored when $\tau \to 0$, and the sound velocity goes to zero as $c_s \sim \tau^{0.055}$ for $\rho = \rho_c$. To account for this small divergence, renormalizing the dimensionless scale time $\frac{\sqrt{\gamma_{0,\mathrm{IG}} r T}}{L} t$ defined relative to the acoustic time [see Eq. (3.9)] introduces the following change in the time variable:

$$t \Rightarrow \sqrt{\frac{c_{V,\mathrm{IG}}}{c_V}}t.$$

Accordingly, the governing nondimensional equations become

$$\begin{aligned}
\sqrt{\frac{c_{V,\mathrm{IG}}}{c_V}}\rho_t + (\rho u)_x &= 0 \\
\sqrt{\frac{c_{V,\mathrm{IG}}}{c_V}}\rho\left(u_t + u u_x\right) &= -\frac{1}{\gamma_{0,\mathrm{IG}}}P_x + \epsilon\frac{4}{3}u_{xx} \\
\sqrt{\frac{c_{V,\mathrm{IG}}}{c_V}}\rho T_t + \frac{c_{V,IG}}{c_V}\rho T_x &= -(\gamma_{0,\mathrm{IG}} - 1)\left(\frac{\partial p}{\partial T}\right)_\rho (T_0, \rho_c)u_x \\
&\quad + \epsilon\left[\frac{\gamma_{0,\mathrm{IG}}}{\mathrm{Pr}_0}\frac{\Lambda}{\Lambda_{b0}}T_{xx} + \left(\frac{\Lambda}{\Lambda_{b0}}\right)_x T_x\right] \\
&\quad + \epsilon\gamma_{0,\mathrm{IG}}\left(\gamma_{0,\mathrm{IG}} - 1\right)u_x^2,
\end{aligned} \tag{5.54}$$

where we have simultaneously re-expressed the main contributions of the physical reference parameters involved in the dimensional reduction scheme.

The nondimensional real equation of state can always be used in its linearized form, obtained from a Taylor expansion of the nondimensional pressure at the conditions T_0, ρ_c; i.e.,

$$p = p_0 + \left(\frac{\partial p}{\partial T}\right)_\rho (T - T_0) + \left(\frac{\partial p}{\partial \rho}\right)_T (\rho - \rho_c).$$

The solution of the above equations for a real fluid can be written in a form similar to that obtained when considering the van der Waals fluid (see Sect. 3.1.2). The D_c-renormalized coefficient that appears is now

$$D_c \Rightarrow D_c\left(\frac{\partial p}{\partial T}\right)_\rho^{-2}.$$

The speed of the sound wave is no longer the speed of sound for the van der Waals gas c_{s0}, and the corresponding change is

$$c_{s0} = \sqrt{\gamma_{0,\mathrm{IG}} r T} \Rightarrow c_s = \left(\frac{\partial p}{\partial T}\right)_\rho \sqrt{\frac{\gamma_0 - 1}{\gamma_0}}.$$

where appears now the real fluid ratio $\gamma_0 = \frac{c_P}{c_V}$ of the specific heats. The same calculations as those performed in Sect. 4.2 lead to the following modified form for the piston-effect timescale:

$$\Theta \Rightarrow \varepsilon \frac{\Lambda}{\Lambda_{b0}} \left(\frac{c_{V,\mathrm{IG}}}{c_V} \right) \left(\frac{\partial p}{\partial T} \right)_\rho \frac{t}{t_a}.$$

We now return to the dimensional quantities. The expression for the increase in temperature in the bulk becomes

$$\overline{T}^{pe}(t) - T_0 = \mathcal{P}(t) \otimes \frac{e^{\frac{t}{t_{PE}}} \operatorname{erfc}\left(\sqrt{\frac{t}{t_{PE}}}\right)}{m c_V},$$

where t_{PE} can now be written as

$$t_{PE} = \frac{\rho_c^3}{T_c} L^2 \left(\frac{\partial T}{\partial p} \right)_\rho^2 \frac{c_V^2}{\Lambda} \left(\frac{\partial p}{\partial \rho} \right)_T. \tag{5.55}$$

If we consider the thermodynamic relation $c_p - c_V = \frac{T}{\rho^2} \left(\frac{\partial p}{\partial T} \right)_\rho \left(\frac{\partial \rho}{\partial T} \right)_p$ [see Eq. (21.48)], we get, after some manipulation,

$$t_{PE} = \frac{L^2}{D_T} \frac{1}{\gamma_0 (\gamma_0 - 1)}. \tag{5.56}$$

Introducing the diffusion time $t_{D_T} = \frac{L^2}{D_T}$, corresponding to Eq. (1.47), we obtain

$$t_{PE} = \frac{t_{D_T}}{\gamma_0 (\gamma_0 - 1)}. \tag{5.57}$$

With $t_{D_T} \sim \frac{1}{D_{T,c}} \sim \tau^{Y_\Lambda - \gamma} = \tau^{-0.67}$ [see Eq. (1.45)] and $\gamma_0 = \frac{c_p}{c_V} \cong \gamma_{0c} \sim \tau^{-\gamma+\alpha} = \tau^{-1.13}$ [see Eq. (1.82)] when $\tau \to 0$ in a real fluid case, we get

$$t_{PE} \sim \tau^{Y_\Lambda + \gamma - \alpha} = \tau^{1.59}. \tag{5.58}$$

For the value $t = t_{PE}$, the thermal penetration length is

$$\delta = \sqrt{D_T t_{PE}} \sim \tau^{\gamma - \alpha} = \tau^{1.13}. \tag{5.59}$$

Note that for $\tau \to 0$, Eqs. (5.57) and (5.59) are asymptotically equivalent to the approximate thermodynamic formulations $t_{eq} = \frac{t_{D_T}}{\gamma_{0c}^2}$ of Eq. (4.11) and $\delta = \sqrt{D_T t_{eq}}$ of Eq. (4.10).

The energy yield of Eq. (5.51) can be now written as

$$\zeta = \frac{p_0}{T_c} \left(\frac{\partial T}{\partial p} \right)_\rho \tag{5.60}$$

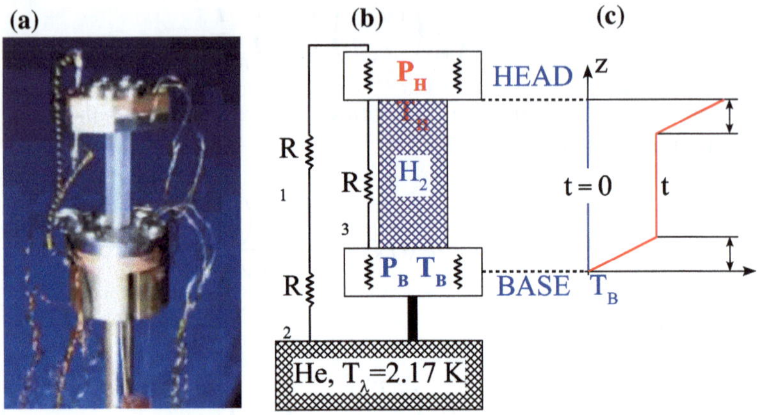

Fig. 5.5 Picture and sketch of the experimental heat pipe based on the piston effect. R_1 and R_2 are the thermal resistances of the copper threads that send energy to the heater in the head. R_3 is the thermal resistance of the PMMA cylinder that encloses the fluid (H_2). Because the thermal conductivity of copper varies a great deal in the range 2 − 40 K, R_1 corresponds to the value at 33 K and R_2 at 10 K. The sample is sealed with a thermal tap (ice flue)

For CO_2 and most gases, it appears that $\left(\frac{\partial p}{\partial T}\right)_\rho \approx 7\frac{p_0}{T_c}$, meaning that the temperature efficiency ζ^{-1} with which the piston effect increases the bulk temperature approaches the real value

$$\zeta^{-1} = \frac{Q(t)}{E(t) - E_0} \approx 7. \tag{5.61}$$

We recall that this temperature efficiency was 4 for a van der Waals gas [see Eq. (5.52)]. The above value obtained for a real critical fluid will be compared with experiment values in Chap. 9.

5.2.6 *Heat Pipes Based on the Piston Effect*

Since the bulk fluid is a thermal short circuit, it is natural to ask the question of whether compressible fluids near a critical point could be used to transport heat over long distances, just as heat pipes do. Heat pipes [3, 6] use the enthalpy of phase change (the latent heat L_H) during evaporation (hot end) to condensation (cold end) processes. The fluid transport is ensured by gravity-driven convection or capillarity. When the heat pipe fluid is close to its critical point, L_H vanishes, as does the yield.

Although a detailed study of the possibility of using the piston effect in near-critical fluids as a heat pipe principle is beyond the scope of this book, we will present some results [15] corresponding to the configuration of Fig. 5.5 here. The hydrogen sample is placed in a magnetic field gradient supplied by one end of a 10 T superconductive coil (see Sect. 23.3 for further details). The magnetic force is able to compensate for the weight of the sample for the value $\nabla\left(\frac{1}{2}\mathbf{B}^2\right) = -491\,\mathrm{T}^2\,\mathrm{m}^{-1}$ [22]. The hot (also called the head) and cold (also called the base) walls of the sample are made from two copper cylinders (Fig. 5.5). The head and base temperatures are measured using a Cernox^{TM} resistor, which consists of a composite material that exhibits a very small variation in resistance with magnetic field. The fluid is confined in a cylinder of length $L = 16.5\,\mathrm{mm}$, inner diameter 3 mm, and outer diameter 4 mm that is made of polymethylmethacrylate (PMMA). This material has a weak thermal conductivity ($\Lambda_{\mathrm{PMMA}} = 0.125\,\mathrm{Wm}^{-1}\mathrm{K}^{-1}$ at 33 K) and a low specific heat ($c_{p,\mathrm{PMMA}} = 180\,\mathrm{Jkg}^{-1}\mathrm{K}^{-1}$ at 33 K), thus ensuring near-adiabatic conditions. For a density of $\rho_{\mathrm{PMMA}} = 1.15 \times 10^3\,\mathrm{kgm}^{-3}$, the thermal diffusivity is $D_{T,\mathrm{PMMA}} = 6.04 \times 10^{-7}\,\mathrm{m}^2\mathrm{s}^{-1}$, corresponding to a typical diffusion time $t_{D,\mathrm{PMMA}} = L^2/D_{T,\mathrm{PMMA}} = 450\,\mathrm{s}$ along the length of the cylinder. The PMMA cylinder is sealed with stainless steel rings. Stainless steel is a thermal insulator at these low temperatures. The mass of the copper head is 2.37×10^{-3} kg. The base (cold part) is kept at a constant temperature T_B to within 30 μK by proportional–integral thermal regulation. The electric power P_B needed for the regulation is recorded. It is typically 3 mW when the head and the base are at the same temperature. Initially, the base, the head, the fluid and the cell walls are at the same temperature T_0. The temperature in the head is kept constant by supplying a constant power (∼5 mW) to compensate for the heat losses through the wires (see Fig. 5.5).

In order to study heat transmission through the sample, a heat flux is supplied at time $t = 0$ in the head by triggering a steady electric power $P_0 = 7.5\,\mathrm{mW}\left(\frac{Q_0}{A_H} \simeq 1\,\mathrm{kWm}^{-2}\right)$ (A_H is the heating area of the head). The heat transfer Q to the base is measured by the difference in electrical power needed to keep the base temperature constant:

$$Q = P_B\,(t < 0) - P_B\,(t > 0)\,. \tag{5.62}$$

Here, we report the heat transferred to the base in two situations: (i) an empty sample, and (ii) H_2 under near-critical conditions (Fig. 5.6). The result is striking. In the empty cell, the heat transfer increases slowly, with a slope of zero at short times,

$$\left(\frac{dQ}{dt}\right)_{t=0} = 0. \tag{5.63}$$

This is due to the fact that the heat front spends a finite time diffusing inside the PMMA cylinder on the length scale L before reaching the base. In contrast, in the supercritical fluid, the heat transfer starts to increase with a finite slope. This can be understood from simple arguments. We make the highly simplifying assumption that the heat transfer is limited by the cold boundary layer thickness only. The temperature T_b in the bulk fluid therefore evolves at short times as [see Eq. (5.49)]

$$\frac{\partial T_b}{\partial t} = \left(1 - \frac{c_V}{c_p}\right)\frac{Q_0}{\rho_c S L c_V}, \tag{5.64}$$

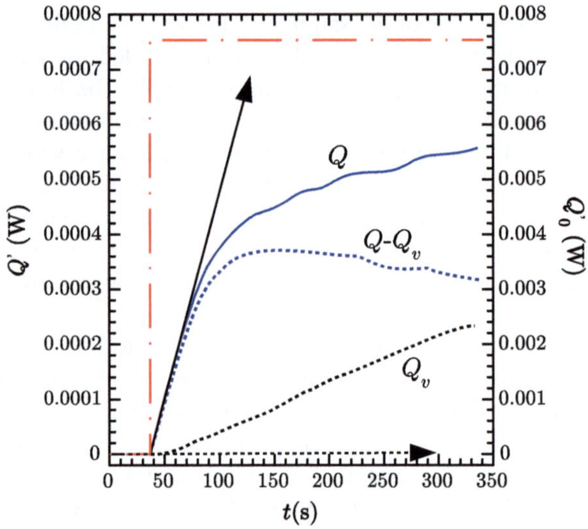

Fig. 5.6 Heat transport measured experimentally in H_2 (*solid line*) at near-critical density and $T_c + 0.1K(T_c = 33K)$. The diffusion time is $t_{D,H_2} = 2.2 \times 10^5$ s here and the piston-effect time is $t_{PE} = 3$ s. Initially, a constant power ($Q_0 = 7.5$ mW or 1 kW m^{-2}) is supplied to the head (*dashed curve*). Q is the power received by the base. The heat conduction by the sample walls when the sample is empty (Q_e, *broken line*) is also shown for comparison. The diffusion time in PMMA is $t_{D,\mathrm{PMMA}} = 450$ s. The difference $Q - Q_e$, which qualitatively represents the heat transfer by H_2 alone, is also plotted. The *arrows* represent the behavior at short times for H_2 (*solid arrow*) and empty (*dotted arrow*) cells

where S is the sample's cross-sectional area. The heat flux transferred to the base thus becomes

$$Q = \Lambda_{H_2} \frac{T_b - T_0}{\delta} A_H, \tag{5.65}$$

where Λ_{H_2} is the thermal conductivity of H_2. As $\delta = (D_T t)^{\frac{1}{2}}$, Eq. (5.65) yields

$$\frac{dQ}{dt} = \Lambda_{H_2} \left(1 - \frac{c_V}{c_p}\right) \frac{Q_0}{\rho_c L c_V D_T^{1/2}} t^{-\frac{1}{2}} \propto \left(\frac{t}{t_{PE}}\right)^{-\frac{1}{2}}, \tag{5.66}$$
$$\text{with } \left(\frac{dQ}{dt}\right)_{t=0} \to \infty,$$

where $t_{PE} = 3$ s is the characteristic piston-effect time. The slope at small times should therefore be infinite. However, it is finite in experiments because (i) the measurement implies an integration constant of several seconds and (ii) the kinetics are limited by the heat capacity of the head. It is also interesting to note that the prefactor in Eq. (5.66) varies with temperature as

$$\frac{dQ}{dt} \sim \tau^{Y_\eta/2+\nu/2+\alpha-\gamma} = \tau^{-0.8}. \tag{5.67}$$

When the temperature is closer to T_c, the transfer should then increase further. An increase is observed experimentally when T nears T_c [2].

However, we have only considered the short timescale (in Fig. 5.6, the piston-effect time is 3 s). The transport at long times is still driven by diffusion when the cooling and heating boundary layers merge. This will occur at very long times. Since diffusion is very slow, this final stage occurs only at times that can barely be reached experimentally. In Fig. 5.6, the diffusion time is $t_{D,H_2} = 2.2 \times 10^5$ s (i.e., 2.5 days). Therefore, heat conduction through the cell walls occurs instead, as the diffusion time in PMMA is $t_{D,\mathrm{PMMA}} = 450$ s. This is clearly visible in Fig. 5.6, where both the empty and H_2-filled samples show similar behavior at time ~150 s, a time much larger than the piston-effect time, and comparable to $t_{D,\mathrm{PMMA}}$.

5.2.7 Access to Isentropic Properties

As recalled in Chap. 21 (see Sect. 21.8), the internal energy balance [see Eq. (21.78)] is the sum of the energy involved in the work done by the pressure forces plus the viscous dissipation and the heat conduction. Under microgravity conditions, the velocity induced by the piston effect is very small, and the convective term in the material derivative can be neglected. On the other hand, the viscous dissipation is negligible for 1-D models. These approximations are validated by the asymptotic analysis carried out in Chap. 4. Therefore, the enthalpy balance equation given by Eq. (21.81), can be written

$$\rho c_p \frac{\partial T}{\partial t} = T\alpha_p \frac{\partial p}{\partial T} + \Lambda \Delta T. \tag{5.68}$$

Taking into account the fact that

$$\frac{T\alpha_p}{\rho c_p} = \left.\frac{\partial T}{\partial p}\right|_s = \left(1 - \gamma_0^{-1}\right) \left.\frac{\partial T}{\partial p}\right|_\rho,$$

Eq. (5.68) can be written as

$$\frac{\partial T}{\partial t} = \left(1 - \gamma_0^{-1}\right) \left.\frac{\partial T}{\partial p}\right|_\rho \frac{\partial p}{\partial T} + D_T \Delta T. \tag{5.69}$$

The first term on the right-hand side of Eq. (5.69) represents the piston-effect term that is proportional to the time derivative of the homogeneous bulk temperature $T_b(t)$. The time derivative of the pressure in the bulk is thus

$$\frac{dp}{dT} = p\beta_s \frac{dT_b(t)}{dt},$$

where $\beta_s = \frac{1}{p}\frac{\partial p}{\partial T}\Big|_s$ is an isentropic pressure coefficient.

If we introduce the deviation $\delta T_{BL}(x, t)$ of the boundary layer temperature from the bulk temperature, as defined by

$$T(x, t) = T_b(t) + \delta T_{BL}(x, t),$$

the second term in the right-hand side of Eq. (5.68) is the time derivative of the temperature deviation, so

$$\frac{d\left[\delta T_{BL}(x, t)\right]}{dt} = D_T \Delta T.$$

This term becomes very small near the critical point. As the ratio γ_0^{-1} decreases strongly whereas $\frac{\partial T}{\partial p}\Big|_\rho$ remains finite, $\frac{\partial T}{\partial p}\Big|_s$ remains on the order of $\frac{\partial T}{\partial p}\Big|_\rho$ and is thus finite. This is why the relative contribution of the work done by pressure forces to the temperature dynamics becomes prominent when nearing the critical point.

The density dynamics can be found by following a similar reasoning. The density field is expressed as a function of the independent variables s and P:

$$\frac{d\rho}{dt} = \frac{\partial \rho}{\partial s}\Big|_p \frac{ds}{dt} + \frac{\partial \rho}{\partial p}\Big|_s \frac{dp}{dt}.$$

Taking into account the definition of the isentropic expansion coefficient α_s,

$$\alpha_s = -\frac{1}{\rho}\frac{\partial \rho}{\partial T}\Big|_s = \gamma_0^{-1}\alpha_p,$$

we find, by replacing $\frac{dp}{dT}$ from Eq. (5.69), that

$$\frac{d\rho}{dt} = -\rho\alpha_s\left[\frac{dT}{dt} - \gamma_0 D_T \Delta T\right]$$

Introducing the temperature deviation, we obtain

$$\frac{d\rho}{dt} = -\rho\alpha_s \frac{dT_b(t)}{dt} - \rho\alpha_p \frac{d\left[\delta T_{BL}(x, t)\right]}{dt}. \tag{5.70}$$

Equation (5.70) is the relation between the density and the temperature dynamics. Simultaneous measurements of pressure and temperature in the bulk phase can thus provide the isentropic properties of the fluid, such as α_s or β_s (see [4, 5, 13]).

5.3 Inversion of the Reflection Rules of Acoustic Waves

Thermal adaptation layers along thermostated boundaries play a prominent role when approaching the critical point [14, 25, 27]. These layers are responsible for the strong thermoacoustic coupling that drives the thermal kinetics of the bulk phase. In what follows, other behaviors are investigated in the close vicinity of the critical point when

$$\tau \ll \varepsilon^{\frac{2}{3}}.$$

5.3.1 Numerical Simulation of Acoustic Wave Reflection at a Thermostated Wall

The set of boundary conditions that must be considered for the equations of motion is an important aspect of the description of low Mach number compressible fluid flows. In particular, these boundary conditions show how acoustic waves reflect at the walls, when described in terms of characteristic curves of the governing differential equations. A numerical simulation of the reflection process has been performed for different initial distances to the critical point. The details of this analysis can be found in [25].

5.3.1.1 The Numerical Simulation

The system under consideration is a 1-D slab-like container that is heated at its left boundary. The right boundary is considered to be either thermostated or insulated. The initial temperature is $T_c + 0.3$ mK, corresponding to $\tau \simeq 10^{-5}$. The temperature of the left wall increases at the nondimensional rate α [see Eq. (3.36)]. We focus here on the reflection process for the elementary incident compression wave emitted at the heated boundary. We explore the temperature and velocity perturbations at different times around $\frac{1}{c_{s0}}$. The temperature and velocity perturbations are plotted in Fig. 5.7 for thermostated (at the initial temperature) or insulated boundary conditions.

The temperature rise enables the propagation of the incident compression wave to be visualized. In the thermostated case (see Fig. 5.7a1), when the wavefront reaches the right boundary, a very thin adaptation layer forms in which a strong negative temperature gradient is expected. Due to the high fluid expandability, density strongly increases in this layer. This results in a positive velocity field at the edge of the thermal boundary layer. At the bulk scale, the velocity is nonzero at $x = 1$ (the fluid flows towards the right boundary; see Fig. 5.7b1). The incident compression wave does not see a closed end, but an outflow from an open cavity. It thus reflects as an expansion wave to fulfill the condition at the edge of the adaptation layer. This change in the nature of the reflected wave originates in the presence of a very thin adaptation layer near the wall. The contraction of the fluid in it modifies the boundary conditions of

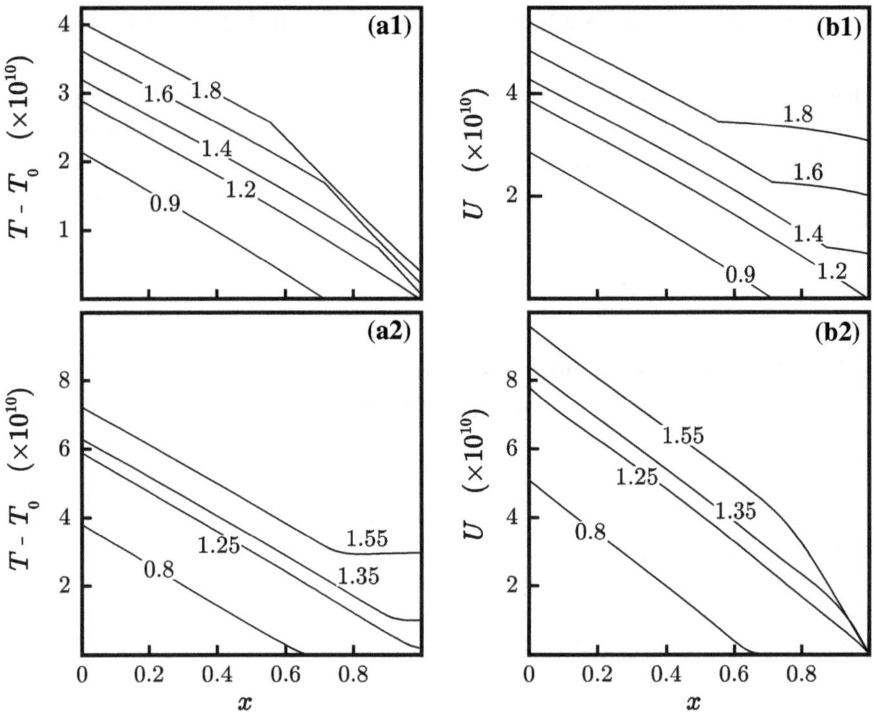

Fig. 5.7 Temperature (**a**) and velocity (**b**) during the reflection of an acoustic compression wave at a thermostated (**1**) or an insulated (**2**) wall at the bulk scale. The velocity is expressed in units of the velocity of sound

the propagation in the bulk. The boundary condition for the velocity in the bulk is zero at the wall when no significant adaptation layer is present (in ideal gases for example). In contrast, it is the fluid contraction velocity at the edge of the adaptation layer when such a layer is significant.

In order to check the role of the adaptation layer, let us consider the case of an insulated wall at $x = 1$. In this case, no adaptation layer should exist, even in near-critical fluids. Figure 5.7a2 shows that the thermal adaptation layer has indeed disappeared (there is no longer any significant temperature gradient at the wall). Accordingly, no contraction of the fluid is expected near the boundary: the bulk velocity tends to zero at $x = 1$ (see Fig. 5.7b2). The classical reflection rules are retrieved: the compression wave reflects as a compression wave to ensure that the boundary conditions of zero velocity are fulfilled. In the reflected wave, the temperature has now increased (see Fig. 5.7a2).

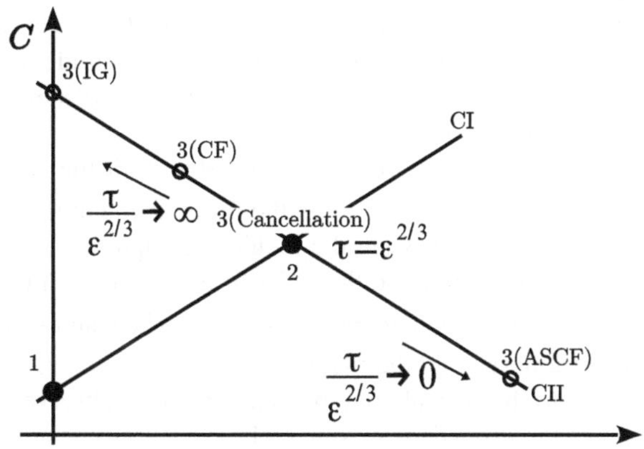

Fig. 5.8 Interpretation of the reflection process (for the case of an incident compression wave) within the framework of the theory of characteristics. *IG* ideal gas, *CF* critical fluid, *ASCF* acoustically saturated critical fluid

5.3.1.2 Scaling Analysis

A convenient way to analyze the reflection mechanism is to use the classical (u, c_s) diagram of the theory of characteristics [19] to solve hyperbolic problems. In this diagram, the x-axis represents the fluid velocity u, while the y-axis represents the velocity of sound c_s, which is proportional to $\sqrt{T}$. According to this theory, all possible states (labeled 2 in Fig. 5.8) of the fluid behind an incident elementary wave traveling to the right (labeled family C_I in the figure) are located on a straight line passing through the unperturbed initial state, labeled 1 in the figure, $(u_1 = 0, c_{s1})$. The interpretation of the reflection process with the help of the theory of characteristics as illustrated in Fig. 5.8 only applies to elementary waves; i.e., to waves that produce infinitely small changes in the fluid properties during an infinitely short timeframe. The strength, c_{s2}/c_{s1}, of the wave thus gives state 2 (u_2, c_{s2}). This strength depends in our case on the boundary heating conditions. Let us consider the case where an incident compression wave reaches the right wall (see Fig. 5.8). It reflects as a left-traveling wave (family C_{II}). The state of the fluid behind the reflected wave (denoted 3) is located on another straight line symmetrical to the first one; this line passes through 2 and is such that $u_3 = u_{BL}$, in order to fulfill the boundary condition imposed by the adaptation layer (see Fig. 5.8). For a van der Waals gas far from the critical point that exhibits a low isothermal compressibility, or for an ideal gas, the thermal adaptation layer contracts weakly and thus $u_{BL} \sim 0$ (point 3 on Fig. 5.8). The compression wave is thus reflected as a compression wave since $c_{s3} \sim \sqrt{T_3} > c_{s2} \sim \sqrt{T_2}$. This situation has been reported in the numerical simulations of [25].

As the critical point is neared, u_{BL} increases so that the amplitude of the reflected wave decreases. At some distance from the critical point, $u_{BL} = u_2$: the velocity generated by the contraction of the thermal adaptation layer is equal to the velocity behind the incident wave. A reflection wave is no longer needed to satisfy boundary conditions. Closer to the critical point, $u_{BL} > u_2$, the incident wave sees an outflow and reflects as an expansion wave. From the order of magnitude analysis given in Chap. 4, it is possible to find the order of magnitude of the distance to the critical point where this phenomenon occurs.

The order of magnitude of the velocity associated with the contraction of the boundary layer is $\alpha^* \varepsilon^{\frac{1}{2}} / \tau^{0.75}$, where α^* is the order of magnitude of the temperature drop in the boundary layer when the incident compression wave reaches the thermostated boundary. α^* is equal to the order of magnitude of the velocity of the fluid behind the incident wave (i.e., $u_2 = \alpha \varepsilon^{\frac{1}{2}} / \tau^{0.75}$). The result is that the order of magnitude of the outflow seen by the incident wave is $u_{BL} = \alpha \varepsilon / \tau^{\frac{3}{2}}$. This allows us to find the distance to the critical point beyond which we observe the inversion of the reflection rules of the acoustic waves, namely

$$\alpha \varepsilon / \tau^{\frac{3}{2}} = \alpha \varepsilon^{\frac{1}{2}} / \tau^{0.75};$$

that is to say,

$$\tau = \varepsilon^{\frac{2}{3}}.$$

Remarkably, this condition corresponds exactly to the crossover to acoustic saturation (see Sect. 5.3). In a 10 mm container filled with the CO_2-like model gas (see Sect. 3.1.2), this value is on the order of 2.8 mK. Note that this value is valid within the linear approach to the density response [see Eq. (4.29)]. In addition, this theory only considers slow variations in the temperature at the boundary, such that the adaptation layer has time to contract. In a faster process (higher frequency), the contraction might not have enough time to take place. The fluid velocity won't be large enough to interact with the reflection process, so we anticipate that inversion won't be observed.

5.3.2 Asymptotic Analysis of the Reflection Process in the Crossover Regime

For the sake of convenience, let us consider the situation where a compression wave is produced by supplying a heat flux to the left wall. The opposite wall is maintained at the initial temperature. A boundary layer is introduced at the right wall to match the bulk and wall temperatures. This latter is the cooling boundary layer (previously denoted CBL, see Sect. 5.1.3), whereas the boundary layer at $x = 0$ is the heating boundary layer (HBL). The approximate equations in the HBL and in the bulk are written within the framework of an inner description. The analysis is formally the same as in Sect. 4.4.2.

5.3.2.1 Heating Boundary Layer

The boundary condition at $z = 0$ accounts for an imposed heat flux:

$$\left[T_{hbl}^{in}(z,t) - \frac{2}{3}p_{hbl}^{in}(z,t)\right]_z = \sqrt{\widetilde{\tau}}\mathcal{P}(t) \quad \text{at} \quad z = 0, \tag{5.71}$$

where $\mathcal{P}(t)$ is the power deposited at the boundary. Note that the equation has been written as a function of the inner variable $\widetilde{\tau}$. The solution in the HBL can be written in Laplace space as

$$LT_{hbl}^{in}(z,s) - \frac{2}{3}Lp_{hbl}^{in}(z,s) = \frac{L\mathcal{P}(s)\sqrt{D_c}\widetilde{\tau}^{0.75}}{\sqrt{s}} \exp\left(-\widetilde{\tau}^{-0.25}\sqrt{s}\frac{z}{\sqrt{D_c}}\right),$$

which leads to the velocity in the CBL:

$$Lu_{hbl}^{in}(z,t) = \frac{2}{3}D_c L\mathcal{P}(s)\left[1 - \exp\left(-\widetilde{\tau}^{-0.25}\sqrt{s}\frac{z}{\sqrt{D_c}}\right)\right].$$

Invoking the matching condition between the bulk and the boundary layer regions, namely

$$\lim_{z\to\infty} Lu_{hbl}^{in}(z,s) = Lu_b^{in}(0,s),$$

gives the first boundary condition for the governing equation for the fluid motion in the bulk:

$$Lu_b^{in}(0,s) = \frac{2}{3}D_c L\mathcal{P}(s). \tag{5.72}$$

5.3.2.2 The Bulk

The approximate equations in the bulk [see Eq. (4.133)] reduce to a propagation equation. The boundary conditions obtained from the matching conditions between the HBL and the CBL are

$$u_b^{in}(0,.t) = \lim_{z\to\infty} u_{hbl}^{in}(z,t)$$
$$u_b^{in}(1,t) = \lim_{z\to\infty} u_{hbl}^{in}(z^*,t).$$

The general solution in the bulk is

$$u_b^{in}(x,s) = Lu_b^{in}(1,s)\frac{\sinh\left(\frac{sx}{c_{s0}}\right)}{\sinh\left(\frac{s}{c_{s0}}\right)} + Lu_b^{in}(0,s)\frac{\sinh\left(\frac{\sinh(1-x)}{c_{s0}}\right)}{\sinh\left(\frac{s}{c_{s0}}\right)}, \tag{5.73}$$

where $u_b^{in}(1,s)$ and $u_b^{in}(0,s)$ are given by the matching conditions with the boundary layers.

5.3.2.3 Cooling Boundary Layer

The equations describing the CBL are the same as those describing the HBL [see Eq. (4.133)], but the variable z is replaced by $z^* = 1 - z$:

$$\left[T_{cbl}^{in}(z^*,t) - \frac{2}{3}p_{cbl}^{in}(t)\right]_t = \sqrt{\tilde{\tau}}D_c\left[T_{cbl}^{in}(z^*,t) - \frac{2}{3}p_{cbl}^{in}(t)\right]_{z^*z^*},$$

with the boundary conditions

$$T_b^{in}(z^*=0,t) = 0$$

$$\lim_{z^*\to\infty}\left[T_{cbl}^{in}(z^*,t) - \frac{2}{3}p_{cbl}^{in}(t)\right] = 0$$

(the latter condition is deduced from the matching conditions with the bulk equations). The solution in the CBL in Laplace space is

$$LT_{cbl}^{in}(z^*,s) - \frac{2}{3}Lp_{cbl}^{in}(s) = -Lp_{cbl}^{in}(s)\exp\left(-\tilde{\tau}^{-0.25}\sqrt{s}\frac{z^*}{\sqrt{D_c}}\right).$$

This solution permits us to compute the fluid velocity generated by the CBL:

$$\lim_{z^*\to\infty} Lu_{cbl}^{in}(z^*,s) = Lu_b^{in}(1,s) = \frac{4}{9}\tilde{\tau}^{-0.75}\sqrt{D_c}\sqrt{s}Lp_{cbl}^{in}(s) \tag{5.74}$$

Equations (5.73), (5.72), and (5.74) give the velocity in the bulk phase as a function of the remaining unknown pressure term $Lp_{cbl}^{in}(s)$:

$$u_b^{in}(x,s) = \frac{4}{9}\tilde{\tau}^{-0.75}\sqrt{D_c}\sqrt{s}Lp_{cbl}^{in}(s)\frac{\sinh\left(\frac{sx}{c_{s0}}\right)}{\sinh\left(\frac{s}{c_{s0}}\right)} + \frac{2}{3}D_cL\mathcal{P}(s)\frac{\sinh\left(\frac{s(1-x)}{c_{s0}}\right)}{\sinh\left(\frac{s}{c_{s0}}\right)}. \tag{5.75}$$

From the momentum equation in the bulk, namely

$$p_{b\,x}^{in} = -\gamma_{0,\mathrm{IG}}u_{b\,t}^{in},$$

we obtain

$$Lp_b^{in}(x,s) = \frac{4}{9}\widetilde{\tau}^{-0.75}\sqrt{D_c}\sqrt{s}Lp_b^{in}(s)\frac{\cosh\left(\frac{sx}{c_{s0}}\right)}{\sinh\left(\frac{s}{c_{s0}}\right)} + \frac{2}{3}D_c L\mathcal{P}(s)\frac{\cosh\left(\frac{s(1-x)}{c_{s0}}\right)}{\sinh\left(\frac{s}{c_{s0}}\right)}.$$

Then, if we add the matching condition, namely

$$\lim_{x\to 1} Lp_b^{in}(x,s) = Lp_{cbl}^{in}(s),$$

the pressure term is found to be

$$Lp_b^{in}(s) = -\frac{\frac{2}{3}\gamma_{0,\mathrm{IG}}c_{s0}\widetilde{\tau}^{-0.75}D_c L\mathcal{P}(s)}{\sqrt{D_c}\widetilde{\tau}^{-0.75}\sinh(\frac{s}{c_{s0}}) - \frac{4}{9}\gamma_{0,\mathrm{IG}}c_{s0}\sqrt{s}\cosh(\frac{s}{c_{s0}})}. \tag{5.76}$$

The velocity in the bulk is obtained by substituting the previous equation into Eq. (5.75):

$$\begin{aligned} Lu_b^{in}(x,s) &= \tfrac{2}{3}D_c L\mathcal{P}(s) \\ &\times \left\{\frac{\widetilde{\tau}^{-0.75}\sinh\left(\frac{s(1-x)}{c_{s0}}\right) + \frac{4}{9}\gamma_{0,\mathrm{IG}}c_{s0}\sqrt{D_c}\sqrt{s}\cosh\left(\frac{s(1-x)}{c_{s0}}\right)}{\widetilde{\tau}^{-0.75}\sinh\left(\frac{s}{c_{s0}}\right) + \frac{4}{9}\gamma_{0,\mathrm{IG}}c_{s0}\sqrt{D_c}\sqrt{s}\cosh\left(\frac{s}{c_{s0}}\right)}\right\}. \end{aligned} \tag{5.77}$$

5.3.2.4 The Wave Reflection Process

Taking the limit $\widetilde{\tau} \to 0$ in Eqs. (5.77) and (5.76) gives the bulk velocity and pressure fields in the saturated regime (superscript *sat*):

$$Lu_b^{sat}(x,s) = \frac{2}{3}D_c L\mathcal{P}(s)\frac{\cosh\left(\frac{s(1-x)}{c_{s0}}\right)}{\cosh\left(\frac{s}{c_{s0}}\right)} \tag{5.78}$$

$$Lp_b^{sat}(s) = \frac{2}{3}\gamma_{0,\mathrm{IG}}c_{s0}D_c L\mathcal{P}(s)\frac{\sinh\left(\frac{s(1-x)}{c_{s0}}\right)}{\cosh\left(\frac{s}{c_{s0}}\right)}. \tag{5.79}$$

We consider a constant heat flux at the wall ($L\mathcal{P}(s) = \frac{1}{s}$) as a test case. The Laplace transforms of Eqs. (5.78) and (5.79) are

$$u_b^{sat}(x,t) = \frac{2}{3}D_c\left\{1 - \frac{4}{\pi}\sum_{k=1}^{\infty}\frac{1}{2k-1}\sin\left[(k-\frac{1}{2})\pi x\right]\cos\left[(k-\frac{1}{2})\pi c_{s0}t\right]\right\}$$

$$p_b^{sat}(x,t) = \frac{2}{3}\gamma_{0,\mathrm{IG}}c_{s0}D_c\left\{\frac{4}{\pi}\sum_{k=1}^{\infty}\frac{1}{2k-1}\cos\left[(k-\frac{1}{2})\pi x\right]\sin\left[(k-\frac{1}{2})\pi c s_0 t\right]\right\}.$$

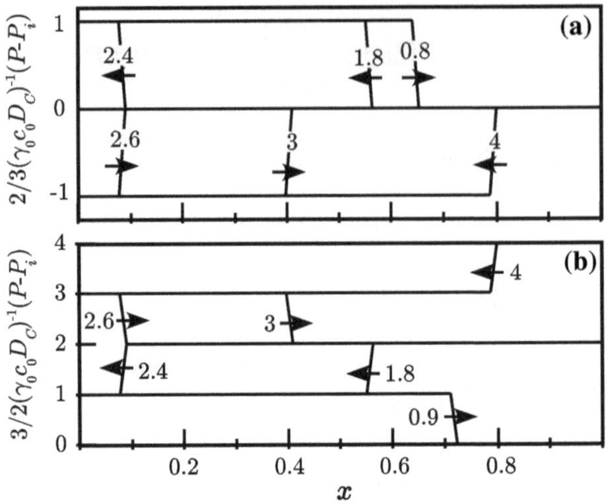

Fig. 5.9 Reflection process of a compression wave: pressure in the saturated regime (**a**) and pressure in the unsaturated (piston-effect) regime (**b**). *Arrows* indicate the direction of propagation; time is counted in units of $\frac{1}{c_{s0}}$

The pressure field given by the above expression is plotted in Fig. 5.9a. The compression wave reflects, as expected, as an expansion wave. In the limit $\tilde{\tau} \to \infty$; that is to say, in the unsaturated (piston-effect) regime, the pressure generated by a constant heat flux can be written as follows in physical space:

$$p_b^{sat}(x,t) = \frac{2}{3}\gamma_{0,\mathrm{IG}} c_{s0} D_c \left\{ c_{s0}t + \frac{2}{\pi}\sum_{k=1}^{\infty}\frac{1}{k}\cos\left[k\pi x\right]\sin\left[k\pi c_{s0}t\right] \right\}.$$

The corresponding pressure plot given in Fig. 5.9b confirms that the reflected wave is a compression wave. Note that the time-averaged pressure is constant in the saturated regime. Since temperature and pressure are linked by $T_b^{sat}(x,t) = \frac{2}{3}p_b^{sat}(x,t)$, the fluid temperature remains constant on average. The next section confirms that energy is transferred at the speed of sound and is not stored in the fluid.

5.3.2.5 Energy is Transferred at the Speed of Sound

The velocity in the CBL becomes, in Laplace space,

$$Lu_{cbl}^{sat}(z^*,s) = \frac{2}{3}D_c\left[1 - \exp\left(-\tilde{\tau}^{-0.75}\sqrt{s}\frac{(1-z^*)}{\sqrt{D_c}}\right)\right],$$

and in physical space,

$$u_{cbl}^{sat}(z^*, t) = \frac{2}{3} D_c \operatorname{erf} c \left(\frac{\widetilde{\tau}^{-0.25} \frac{(1-z^*)}{\sqrt{D_c}}}{2\sqrt{t}} \right).$$

As the velocity at the edge of the boundary layer is proportional to the heat power at the wall $\mathcal{P}(x = 0, t)$,

$$\mathcal{P}((x = 0, t) \propto u_b^{sat}(x = 0, t) = \frac{2}{3} D_c$$
$$\mathcal{P}((x = 1, t) :\propto u_b^{sat}(x = 1, t) = 0$$

for $0 < t < \frac{1}{c_{s0}}$, and

$$\mathcal{P}(x = 0, t) \propto u_b^{sat}(x = 0, t) = \frac{2}{3} D_c$$
$$\mathcal{P}((x = 1, t) : \propto u_b^{sat}(x = 1, t) = -\frac{4}{3} D_c$$

for $\frac{1}{c_{s0}} < t < \frac{2}{c_{s0}}$.

The energy supplied to the fluid between times 0 and $\frac{2}{c_{s0}}$ is $\frac{2}{3} D_c \times \frac{2}{c_{s0}}$. It is transferred to the left wall between time $\frac{1}{c_{s0}}$ and $\frac{2}{c_{s0}}$ with double the power ($\frac{4}{3} D_c \times \frac{1}{c_{s0}}$). The energy supplied to the fluid at the heated wall during the time it takes for a wave to cross the fluid and then return is thus all deposited at the cooling boundary at the same time: heat is transported from the heating boundary to the cooling boundary at the speed of sound.

5.3.2.6 The Reflection of a Heat Pulse

We consider the following heat pulse function at the left boundary:

$$\mathcal{P}(x = 0, t) = 1 \text{ for } t \leq t_0$$
$$\mathcal{P}(x = 0, t) = 0 \text{ for } t > t_0.$$

Its Laplace transform is

$$L\mathcal{P}(x = 0, s) = \frac{1 - e^{-t_0 s}}{s}.$$

Using the latter in Eq. (5.77), and inverting the Laplace transforms of the respective limits $\widetilde{\tau} \to 0$ and $\widetilde{\tau} \to \infty$, we obtain the fluid velocity in the saturated (superscript *sat*) and unsaturated (piston-effect, superscript *pe*) regimes, namely

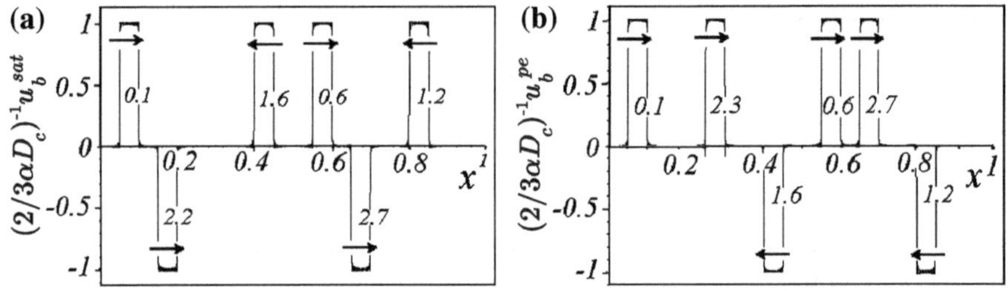

Fig. 5.10 Reflection of a pulse of finite extension: velocity in the saturated regime (**a**) and in the unsaturated (piston-effect) regime (**b**). *Arrows* indicate the direction of propagation; time is counted in units of $\frac{1}{c_{s0}}$

$$
\begin{aligned}
u_b^{sat}(x,t) &= -\frac{8}{3\pi} D_c \sum_{k=1}^{\infty} \frac{1}{2k-1} \sin\left[(k-\frac{1}{2})\pi x\right] \\
&\times \left\{\cos\left[(k-\frac{1}{2})\pi c_{s0} t\right] - \cos\left((k-\frac{1}{2})\pi c_{s0}(t-t_0)\right)\right\} \\
u_b^{pe}(x,t) &= -\frac{4}{3\pi} D_c \sum_{k=1}^{\infty} \frac{1}{k} \sin(k\pi x)\left\{\cos(k\pi c_{s0} t) - \cos(k\pi c_{s0}(t-t_0))\right\}.
\end{aligned}
$$

The corresponding profiles are given in Fig. 5.10.

The detailed reflection process at the crossover has not yet been explored for pulse heating. The inverse Laplace transform cannot be derived analytically because no isolated poles can be found in Eqs. (5.76) and (5.77) (a numerical inversion would be necessary).

5.4 Wall Conductivity Effects

Until now, ideal thermal boundary conditions have been imposed. In real situations, perfect temperature steps or heat pulses are quite difficult to achieve. The issue of the general impact of the finite conductivity and finite thickness of the container walls has been addressed in [7, 11]. In the following, we consider the 1-D numerical simulation performed in [11], which best represents the 3-D experimental situation described below.

The properties of the solids are chosen to mimic those of the experimental device (using the ALICE facility onboard the MIR space station; for details, see Sect. 23.1.1 in Chap. 23). The left heating wall (called S_1) is made of a 0.5 mm thick layer of zirconia (Zr). The right wall (called S_2) is made of a 5 mm thick layer of beryllium copper (Be-Cu) alloy. The fluid layer (thickness $L = 10$ mm) in-between consists of CO_2 ($\rho = \rho_c$, $T > T_c$).

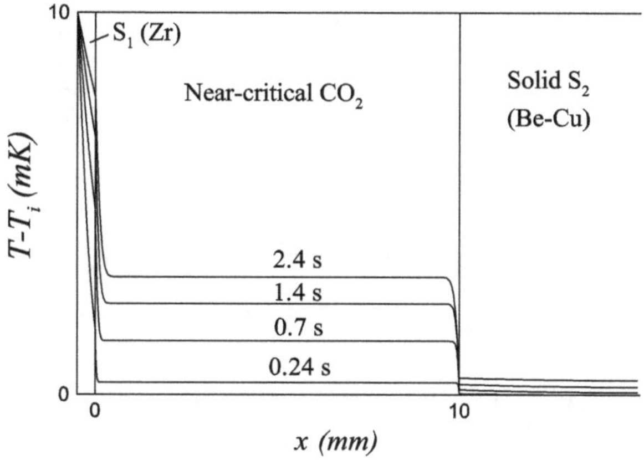

Fig. 5.11 1-D numerical simulation of the temperature profile for a fluid initially at $T_i = T_c + 1\,\mathrm{K}$. At $t = 0$, a temperature step of 10 mK is imposed at the external boundary of the *left heating wall* S_1. The *right wall* S_2 is insulated at its *outer right edge*. The fluid thickness corresponds to $0 \leq x \leq L$ (= 10 mm)

The simulations of the fluid phase are based on the numerical solution of the 1-D hydrodynamic equations given in Sect. 3.2. The heat transfer in the walls is modeled by solving the heat diffusion equation in the walls coupled with the energy equation in the fluid.

Temperature and fluxes are continuous at the interface between the walls and the fluid. A temperature step of 10 mK is imposed from S_1 at 1 K above the critical temperature. S_2 is insulated at its outer edge.

5.4.1 Insulated Wall Case: Cooling Piston Effect

Calculations are performed on the piston-effect timescale [7, 16, 17, 24]. The uniform temperature rise in the bulk fluid occurred quickly. It is apparent from Fig. 5.11 that the temperature increase at the fluid–S_2 contact ($x = L$) as well as in the solid S_2 remains weak compared to that of the bulk fluid.

The condition imposed on the fluid by the Be-Cu alloy resembles that of a thermostat. The small difference from a real thermostat at $x = L$ (labeled isoTh) can be estimated from Fig. 5.12. The latter displays heat losses from the fluid versus time for different materials (including Be-Cu) and the isoTh case.

The decision to plot heat losses instead of the temperature deviation is justified by the fact that the piston-effect heating rate is ruled by the heat fluxes exchanged [24, 26].

In addition to the two curves showing the heat flux at $x = L$ when $T = T_i$ (thermostat, in the absence of S_2) and when S_2 is made of Be-Cu, Fig. 5.12 plots the results obtained for two other materials (Pyrex glass and epoxy) that are usually considered insulators.

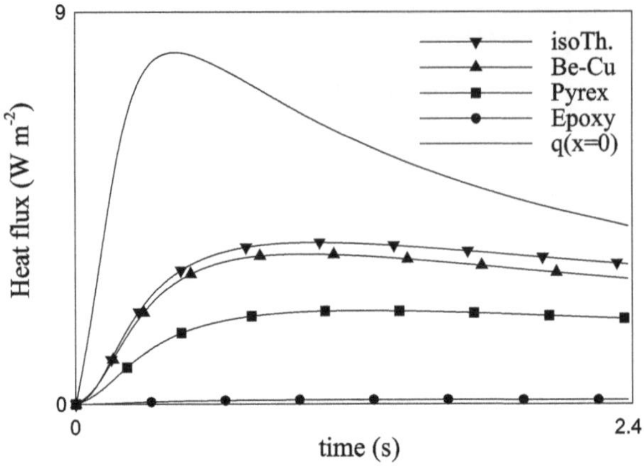

Fig. 5.12 Heat fluxes at $x = 0$ and at $x = L$ (fluid–solid contact) for different materials. The *curve isoTh* corresponds to the isothermal condition at $x = L$

It is clear that a Pyrex wall is inadequate for ensuring insulation in the present transient situation, since heat losses are non-negligible compared to the injected fluxes (dashed line), and even comparable with those occurring when S_2 is thermostated. The insulating epoxy appears to be more appropriate for ensuring adiabatic conditions.

These results show that, in most experiments devoted to near-critical fluid dynamics under local heating (or cooling), two antagonistic piston effects coexist, since heat losses drive a cooling piston effect at the same time as heat injection drives a heating one.

There are only two exceptions to such a situation: that encountered in [21], where the wall temperature of a spherical container is uniformly monitored from the outside (no cooling piston effect), and that concerning the relaxation after a local short heat pulse; in this case, the authors of [9] studied the cooling caused by the (quasi) thermostated walls of the container (no heating piston effect).

The question of whether a unique (heating or cooling) piston effect is present in an experiment is important for characterizing the thermal behavior of a sample when approaching T_c. When heating is performed in a thermostated cell, the amplitude of the bulk temperature decreases near T_c because of the increasing amplitude of the cooling effect.

In such a case, less energy is stored in the bulk fluid, which tends ultimately to behave like a thermal short circuit [24].

5.4.2 *Effect of Heat Diffusion in the Walls*

In Fig. 5.11, a thermal gradient inside the wall S_1 can be discerned. The latter modifies the thermal boundary conditions at the fluid–solid interface. In order to study the

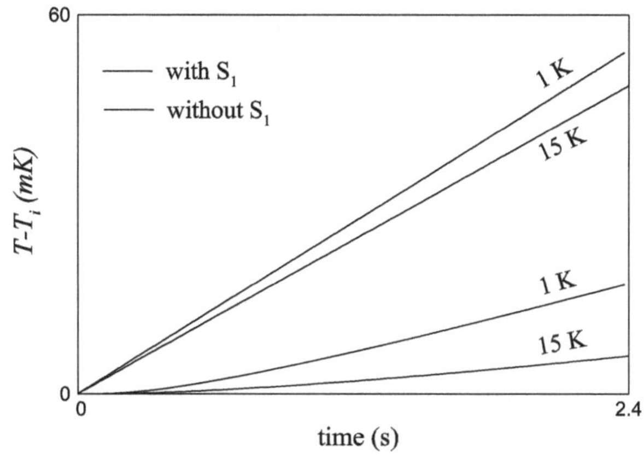

Fig. 5.13 Temperature evolution at $x = \frac{L}{2}$ when the fluid is initially set to either $T_c + 1$ K or $T_c + 15$ K. The unheated wall at $x = L$ is adiabatic. The case "with S_1" (*solid curves*) corresponds to imposing the heat flux at $x = -5$ mm, at the outer edge of S_1. The case "without S_1" (*dashed curves*) corresponds to imposing the heat flux directly at $x = 0$

phenomenon, computations at $T_c + 1$ K and $T_c + 15$ K were performed in which heating was delivered at a constant flux (50 W m^{-2}), a condition that is easier to meet in experiments than a constant temperature step.

The heat flux was imposed at the outer edge of S_1 at $x = -5$ mm or directly at $x = 0$. In the following, we consider the adiabatic or thermostated boundary condition at $x = L$.

5.4.2.1 The Adiabatic Case

Figure 5.13 displays the results.

The temperature evolution is recorded at the center of the fluid ($x = \frac{L}{2}$). Note that the heating rate $\frac{dT}{dt}$ is proportional to the heat flux [see Eq. (5.49)]. The differences between the results obtained "with" or "without" S_1 are significant and increase with time. From Eq. (5.49), the case "without S_1" should be independent of $T - T_c$ on the piston-effect timescale in the present van der Waals approximation.

However, a very weak variation with $T - T_c$ is observed at the same time, as shown in Fig. 5.13. We attribute this variation to the thermal property changes in the hot boundary layer. During the heating period with S_1, the heat flux at $x = 0$ never reaches the imposed value (50 W m^{-2}) at $x = -5$ mm, as illustrated in Fig. 5.14a. S_1 stores energy and thus damps the heat flux absorbed by the fluid. A $T - T_c$ dependence is observed, which is analyzed below.

It is instructive to plot the temperature variations at the interface $x = 0$ (see Fig. 5.14b). At constant energy (i.e., at the same time), the fact that the temperature at the solid–fluid interface is smaller at 1 K than at 15 K above T_c shows that the solid stores less energy closer to T_c.

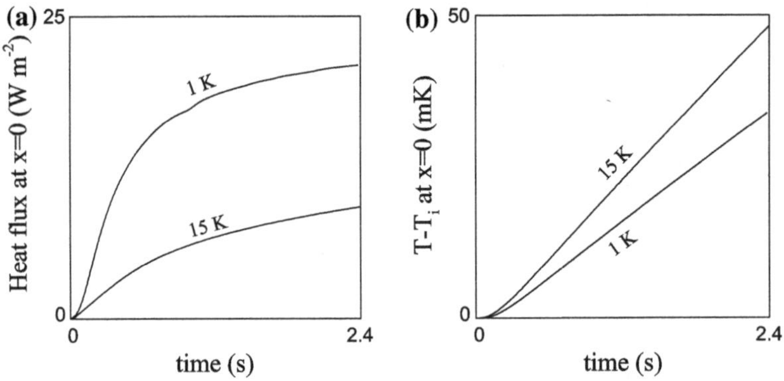

Fig. 5.14 Heat flux (**a**) and temperature (**b**) at $x = 0$ for the adiabatic case at $x = L$

More energy is passed to the fluid boundary layer (Fig. 5.14a). The decrease in the temperature of the solid from 15 K to 1 K above T_c is thus the result of the increase in the fluid's c_p as T goes to T_c.

Nevertheless, we observe in Fig. 5.14b that, at any given time, the boundary layer temperature decreases as the temperature of the solid above T_c decreases. If the behavior of the fluid's c_p explains why the solid tends to store less energy near T_c, the thermal conductivity Λ of the fluid (which also increases near T_c) will govern the heat flux at the interface.

As a matter of fact, Λ is about three times larger (in our model) at $T_c + 1$ K than at $T_c + 15$ K. We observe in Fig. 5.14a that the increase in the heat flux is on the same order.

Figure 5.13 shows that the thermal evolution of a near-critical fluid is mainly affected by the "finite" thermal properties of the container walls. The latter can be of the same order of magnitude as the thermal fluid properties that vary with $T - T_c$. These questions are discussed by Ferrell and Hao [7] (see also Sect. 9.1.4).

5.4.2.2 The Thermostated Case

The temperature evolutions for the same heating and $T - T_c$ conditions as in Fig. 5.13 are reported in Fig. 5.15.

The effect of the presence of S_1 is still important, even though it appears to be somewhat weaker at 1 K above T_c than in the adiabatic case. Note that the temperature behavior without S_1, where the temperature rise is *smaller* closer to T_c, is different from that in the adiabatic case.

This lowering is due to the presence of a cooling piston effect associated with the cold boundary layer at S_2. However, this is not valid when the effect of S_1 is taken into account. The temperature rise is then *larger* near T_c. It is striking that the

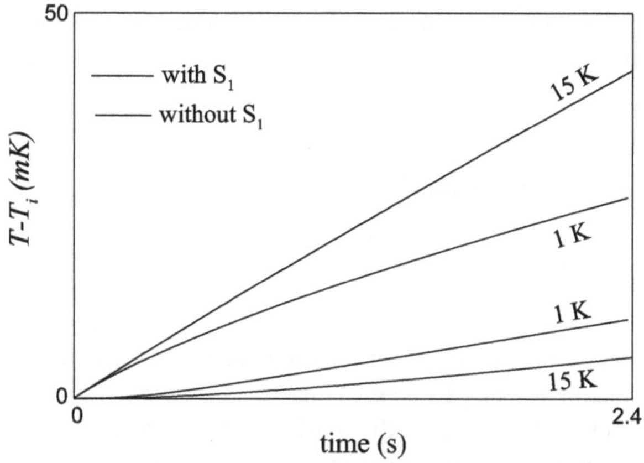

Fig. 5.15 Bulk temperature evolution at $x = \frac{L}{2}$ when the fluid is initially set to either $T_c + 1$ K or $T_c + 15$ K. The unheated wall at $x = L$ is thermostated. The case "with S_1" (*solid curves*) corresponds to imposing the heat flux at $x = -5$ mm, at the outer edge of S_1; the case "without S_1" (*dashed curves*) BZ: done corresponds to imposing the heat flux directly at $x = 0$ (as in Fig. 5.13)

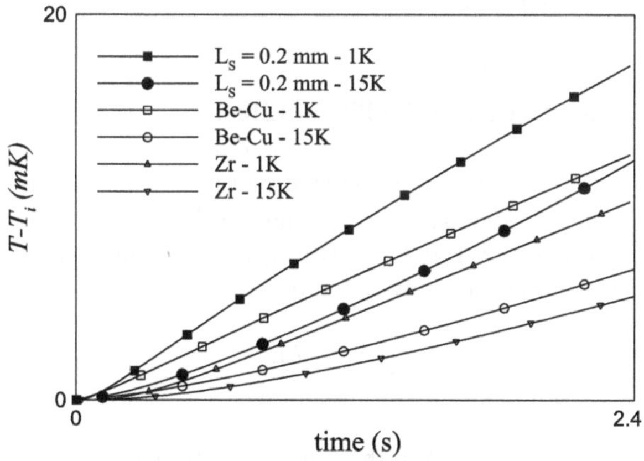

Fig. 5.16 Bulk temperature evolution at $x = \frac{L}{2}$ when the fluid is initially set to either $T_c + 1$ K or $T_c + 15$ K for different S_1 materials and thicknesses (L_s). The unheated wall at $x = L$ is thermostated

presence of S_1 inverts the temperature behavior. In order to interpret this behavior, let us consider first the process at $x = 0$.

The heat flux injected in the fluid is nearly the same in both the adiabatic and thermostated cases, as it is controlled by the hot boundary layer in the fluid. In Fig. 5.13a, the heat flux increases near T_c. Since the latter drives the piston effect, it is not surprising that we obtain a larger bulk temperature rise at $T_c + 1$ K than at $T_c + 15$ K.

To conclude this short review of the influence of boundary effects, Fig. 5.16 depicts the results obtained when S_1 is made thinner or constructed from Be-Cu and Zr.

As expected from the above discussion, the bulk temperature rise is always larger at $T_c + 1$ K than at $T_c + 15$ K.

5.5 Piston Effect Visualized

5.5.1 *Velocity of the Piston*

The "piston" that compresses the fluid is visualized in the experiments below, which also enable the piston velocity to be measured.

As already discussed in Chap. 4, the border of the thermal boundary layer is convected during piston-effect thermalization and this materializes the "piston." The piston velocity is generally quite small [see Eq. (4.46)]: for a heat pulse of 0.5 mW mm^{-2} it is on the order of a few μm s^{-1}, which is quite difficult to detect.

However, under certain circumstances, when heat is produced in a cavity, the fluid velocity can become very fast and give rise to detectable convection that looks like a "jet" (see Ref. [8] for details).

5.5.1.1 A Cavity to Strengthen the Piston Effect

In order to determine the velocity of the HBL, let us consider the following model. The heat source is a slit (volume V, section area A, length L) in which a mass m of fluid is contained initially (see Fig. 5.17). During heating, this mass is instantaneously heated at constant volume and shows a temperature rise of

$$\delta T_V = \frac{\delta E}{m c_V}. \tag{5.80}$$

Here, δT_V is the temperature rise of the fluid element that has received a portion of energy δE and remains at a constant (slit) volume V, and thus has the specific heat c_V. The average fluid density $\rho = m/V$ remains constant. The temperature increase δT_V induces a pressure rise

$$\delta p = \left(\frac{\partial p}{\partial T}\right)_\rho \delta T_V. \tag{5.81}$$

In the following step, we assume that the pressure relaxes, without dissipating to its initial value, through an isentropic (i.e., adiabatic) expansion. The density change due to the pressure decrease is

$$\delta \rho = \left(\frac{\partial \rho}{\partial p}\right)_s (-\delta p). \tag{5.82}$$

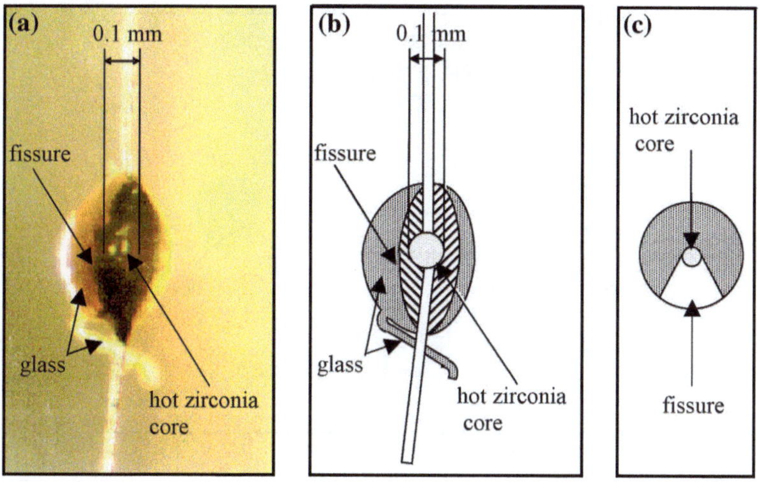

Fig. 5.17 **a** Photo of the Th2 thermistor with the fissure and the central metal oxide sphere. **b**, **c** Sketches of the thermistor

Straightforward application of the transformations for thermodynamic derivatives given in Chap. 21 (see Sect. 21.3.2), results in

$$\left(\frac{\partial \rho}{\partial p}\right)_s = -\frac{c_V}{c_p}\left(\frac{\partial \rho}{\partial T}\right)_p \left(\frac{\partial T}{\partial p}\right)_\rho . \tag{5.83}$$

By making use of Eq. (5.80), the density variation due to expansion becomes

$$\delta\rho = \frac{1}{c_p}\left(\frac{\partial \rho}{\partial T}\right)_p \frac{\delta E}{m} . \tag{5.84}$$

We note that the final fluid state after this process is identical to the one after the process of heating at constant pressure, which directly includes the thermal expansion

$$\delta T_p = \frac{\delta E}{m\, c_p} , \tag{5.85}$$

where again we obtain Eq. (5.84):

$$\delta\rho = \left(\frac{\partial \rho}{\partial T}\right)_p \delta T_p = \frac{1}{c_p}\left(\frac{\partial \rho}{\partial T}\right)_p \frac{\delta E}{m} . \tag{5.86}$$

In order to determine the flow velocity u_0 of the expansion of the volume heated with power $\dot{q} = \delta E/\delta t$, we write the variation in the volume element δV by assuming a one-dimensional expansion δL with constant cross-sectional area A:

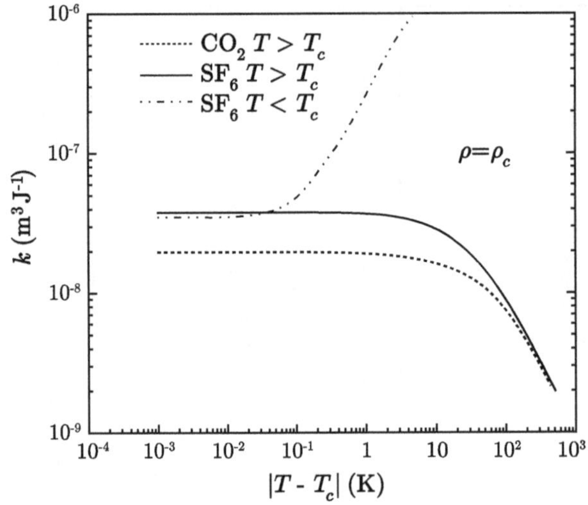

Fig. 5.18 The parameter $k = \frac{\beta_p}{\rho c_p}$ is plotted for CO_2 and SF_6 as a function of temperature along the critical isochore and for the gas phase below T_c. Near T_c, this parameter is constant. For high temperatures, the parameter is seen to decrease as the fluid tends to the ideal gas state. Below T_c, the value for the SF_6 gas phase increases for $T_c - T > 0.1$ K. Near T_c, the ratio of the k values for SF_6 and CO_2 is nearly 2

$$\delta V = A\delta L = -\frac{m}{\rho^2}\delta\rho. \tag{5.87}$$

Combining Eqs. (5.84) and (5.87), we obtain the final expression for the expansion velocity $u_0 = \frac{\delta L}{\delta t}$:

$$u_0 = k\frac{\dot{q}}{A}, \tag{5.88}$$

where

$$k = \frac{\beta_p}{\rho c_p}.$$

According to Eq. (5.88), u_0 is proportional to the injected heat flux $\frac{\dot{q}}{A}$. This relation is derived from the general nondimensional form of Eq. (4.46), as discussed in Ref. [24].

In Fig. 5.18, the factor k is plotted for both CO_2 and SF_6 with respect to the temperature along the critical isochore according to the equations of state [1, 10, 18]. It is clear that this factor does not vary strongly near the critical point because the divergences of c_p and β_p compensate for each other. In effect, near T_c, where $c_p \gg c_V$, Eq. (5.88) reduces to

$$u_0 = \frac{1}{T}\left(\frac{\partial T}{\partial p}\right)_\rho \frac{\dot{q}}{A}, \tag{5.89}$$

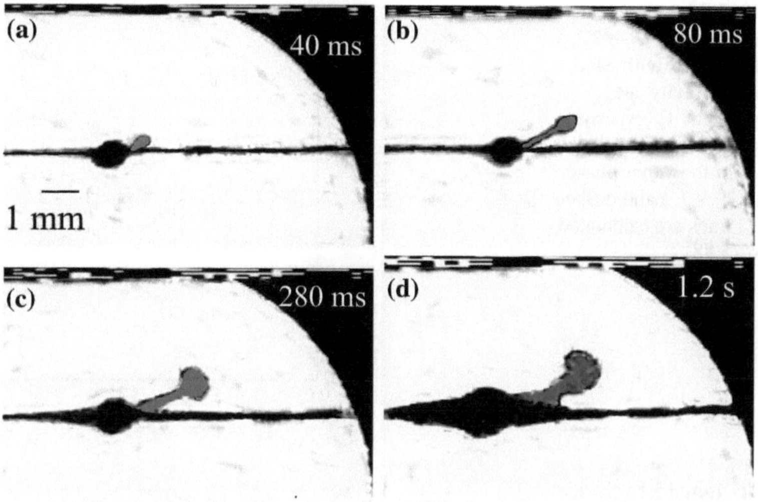

Fig. 5.19 The evolution of a jet after a 40 ms heat pulse in SF_6 under weightless conditions at $T_c + 2.15$ K. Time is 40 ms (**a**), 80 ms (**b**), 280 ms (**c**), and 1200 ms (**d**) after the start of the pulse. For clarity, regions around the thermistor and jet have been erased and the inside of the jet has been colored *gray*

where the prefactor $\frac{1}{T}\left(\frac{\partial T}{\partial p}\right)_\rho$ is independent of temperature near T_c [9]. For SF_6, the coefficient k is nearly twice as large as that for CO_2 in the single-phase region above T_c (Fig. 5.18). For a given heat flux, a larger velocity is thus obtained with SF_6.

5.5.1.2 Visualization

A strong convection velocity is observed in the following experimental configuration. A thermistor, Th2 (diameter 0.26 mm, time constant 10 ms), is located in the center of a thermostated cylindrical sample like that already used to detect the piston effect (for details, see Chaps. 9 and 23).

It is heated by an electric current during a time Δt (40 ms) with the heating power ranging from 0 to 30 mW. The optical cell (diameter $D = 11.6$ mm, thickness $e_f = 6.79$ mm) is located in one arm of a Twyman–Green interferometer (see Figs. 23.3 and 23.4) that allows the density inhomogeneities, averaged over the cell thickness, to be detected.

Another thermistor (Th2), located 5 mm from Th1, measures the temperature of the bulk fluid. After the heat pulse, the temperature of the heated thermistor can also be measured.

The thermistor Th1 shows a fissure or crevasse at its surface, and the temperature of the fluid in the fissure increases locally much more than the temperature of the

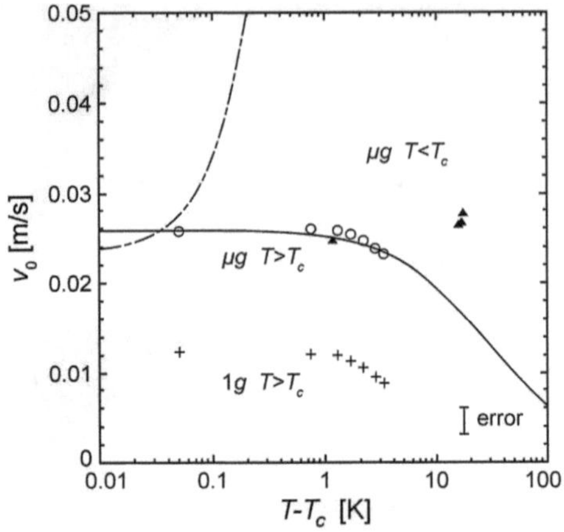

Fig. 5.20 Initial jet velocity u_0 with respect to $|T - T_c|$. Results obtained with SF_6 under microgravity are compared with 1g experiments. Below T_c, the velocity increases in the vapor phase. The full ($T > T_c$) and dashed ($T < T_c$) lines are calculated from Eq. (5.88)

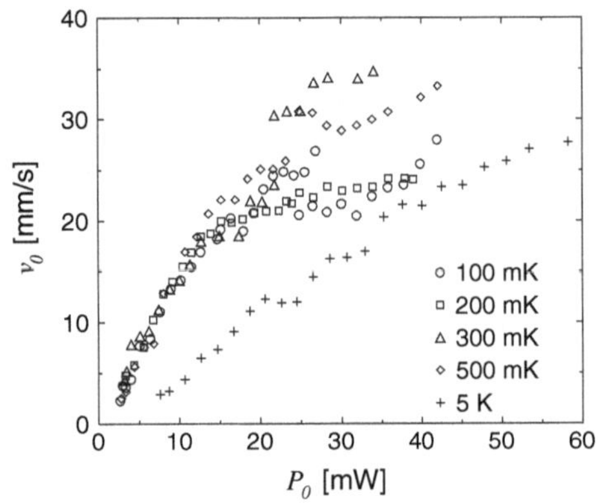

Fig. 5.21 Initial jet velocity u_0 for CO_2 at critical density and at different temperature differences from T_c, with respect to the electric heating power P_0

other part of the HBL around the thermistor. The edge of the HBL thus reaches a large velocity and looks like a jet. This is especially true at the very start of the heating.

However, as time passes, such a fluid column loses its speed because of the viscous friction and the lateral heat diffusion. The evolution of a jet for a heat pulse at $T = T_c + 2.15$ K is shown in Fig. 5.20a–d.

The initial jet velocity u_0, as determined based on the average value for the first 40 ms of heating, is plotted in Fig. 5.20 versus the initial temperature. It is clear that the evolution of the jet is not sensitive to the temperature very close to the critical point in agreement with Eq. (5.89), and decreases far from T_c according to Eq. (5.88) and Fig. 5.18.

In Fig. 5.21, u_0 is plotted for various temperature differences from T_c as a function of the electric power rate $P_0 \approx Q$ (the relation between P_0 and Q is detailed in [9] and Chap. 23). Across a large range of heating powers, the velocity is found to be proportional to the electric power supplied when the latter is not too high, in accordance with Eq. (5.88). For the highest values of electric power, the velocity is seen to saturate. This saturation is the effect of increasing energy dissipation by heat diffusion from the "hot" jet fluid into the colder bulk fluid (see Ref. [8]).

A jet is also observed as an effect of the Earth's gravitational field (Fig. 5.20). However, the velocity values are approximately 50 % smaller than those obtained under reduced gravity. That said, gravity-induced convection does disturb the expansion process of the fluid, even on such a small timescale (for a detailed discussion, see [8]).

References

1. Abbaci A, Sengers JV (1990) Technical report BN1111. University of Maryland, College Park
2. Beysens D, Chatain D, Evesque P, Garrabos Y (2008) Nucleation and growth of a bubble pattern under vibrations in weightlessness. Europhys Lett 82:36003
3. Beysens D, Chatain D, Nikolayev VS, Ouazzani J, Garrabos Y (2010) Phys Rev E 82:061126
4. de Bruijn R, van Diest RJJ, Michels AC (1996) Isochoric specific heat c_v in the near-critical region of SF_6 from isentropic thermalization in microgravity. In: Viviani A (ed) Proceedings of the 2nd European symposium on fluids in space. Congressi srl, Roma, pp 334–338
5. de Bruijn R (1999) Heat transfer in a critical fluid under microgravity conditions: a spacelab experiment. PhD thesis, University of Amsterdam, Amsterdam, p 197
6. Faghri A (1995) Heat pipe science and technology. Taylor and Francis, New York
7. Ferrell RA, Hao H (1993) Adiabatic temperature changes in a one-component fluid near the liquid–vapor critical point. Phys A 197(1–2):23–46
8. Frohlich T, Beysens D, Garrabos Y (2006) Piston-effect-induced thermal jets in near-critical fluids. Phys Rev E 74(4):046307
9. Garrabos Y, Bonetti M, Beysens D, Perrot F, Fröhlich T, Carlès P, Zappoli B (1998) Relaxation of a supercritical fluid after a heat pulse in the absence of gravity effects: theory and experiments. Phys Rev E 57(5):5665–5681
10. Jin GX (1993) PhD thesis, University of Maryland, College Park
11. Jounet A, Zappoli B, Mojtabi A (2000) Rapid thermal relaxation in near-critical fluids and critical speeding up: discrepancies caused by boundary effects. Phys Rev Lett 84(15):3224–3227
12. Kevorkian J, Cole JD (1981) Perturbation methods in applied mathematics. Springer, New York
13. Michels AC, de Bruijn R, Karapantsios TD, van Diest RJJ, van den Berg HR, van Deenen B, Sakonidou EP, Wakeham WA, Trusler JPM, Louis A, Papadaki M, Straub J (1995) Adiabatic compressive heating of critical fluids under microgravity conditions. In: The 30th 1995 national heat transfer conference, vol 3. Portland, OR, USA, 6–8 August 1995
14. Miura Y, Yoshihara S, Ohnishi M, Honda K, Matsumoto M, Kawai J, Ishikawa M, Kobayashi H, Onuki A (2006) High-speed observation of the piston effect near the gas-liquid critical point. Phys Rev E 74:010101(R)
15. Nikolayev VS, Beysens D (1999) Boiling crisis and non-equilibrium drying transition. Europhys Lett 47:345–351
16. Onuki A, Ferrell RA (1990) Adiabatic heating effect near the gas–liquid critical point. Phys A 164(2):245–264
17. Onuki A, Hao H, Ferrell RA (1990) Fast adiabatic equilibration in a single-component fluid near the liquid–vapor critical point. Phys Rev A 41(4):2256–2259
18. Sengers JV (1994) Effects of critical fluctuations on the thermodynamic and transport properties of supercritical fluids. In: Supercritical fluids: fundamentals for application, Chap. 9, Kluwer, Dordrecht, pp 231–272

19. Shapiro AH (1954) The dynamics and thermodynamics of compressible fluid flow. Wiley, New York
20. Stanley HE (1971) Introduction to phase transitions and critical phenomena. Clarendon, Oxford
21. Straub J, Eicher L (1995) Density and temperature relaxation in the two-phase region near the critical point of a pure fluid. Phys Rev Lett 75(8):1554–1557
22. Wunenburger R, Chatain D, Garrabos Y, Beysens D (2000) Magnetic compensation of gravity forces in (p-)hydrogen near its critical point: application to weightless conditions. Phys Rev E 62(1):469–476
23. Zappoli B, Bailly D (1990) Transport in a confined compressible fluid under time-dependent volumetric heat source. Phys Fluids A 2:1771–1779
24. Zappoli B, Carlès P (1995) The thermo-acoustic nature of the critical speeding up. Eur J Mech B 14:41–65
25. Zappoli B, Carlès P, Amiroudine S, Ouazzani J (1995) Inversion of acoustic waves reflection rules in near critical pure fluids. Phys Fluids 7:2283–2287
26. Zappoli B, Amiroudine S, Gauthier S (1999) Rayleigh–Taylor instability in near-critical pure fluids. Int J Thermophys 20:257–265
27. Zhang P, Shen B (2009) Thermoacoustic wave propagation and reflection near the liquid–gas critical point. Phys Rev E 79:060103(R)

Part II
Heat Transfer Experiments Performed in Weightless Conditions

Chapter 6
Introduction to Heat Transfer Experiments Performed in Weightless Conditions

We will start with a brief review of the experimental status of the field when space experiments were first carried out to investigate critical point phenomena (see for example [10]). At that time (the late 1980s), the piston effect was not even suspected. These experiments were performed in a gravity-free environment to get rid of (i) hydrostatic pressure (the cell becomes stratified on Earth under its own weight due to the divergence of κ_T, and thus cannot reach the critical point) and (ii) buoyancy during thermalization and/or phase separation (as already discussed in Sect. 1.8). Due to the critical slowing down of the thermal diffusivity, scientists expected very long equilibration times: on the order of days or more (depending on the sample fluid size).

However, surprisingly, in a sounding rocket experiment (see [12]), the fluid temperature was seen to very quickly follow the cell body temperature. We now know that this fast thermalization was due to the piston effect. This effect had already been predicted in ideal gases. In such a gas, although the isothermal compressibility κ_T is not very large, an acoustic wave can be generated within a confined volume when the boundary temperature is suddenly changed [9]. However, the corresponding adiabatic fluid motion cannot thermalize the gas faster than heat diffusion [11]. In contrast, near the gas–liquid critical point where κ_T diverges, the adiabatic fluid motion can become more efficient than diffusion, and this gives rise to the piston-effect mechanism described theoretically in Chaps. 4 and 5 (see also [14]).

Early experiments were not designed to describe and probe the mechanism of the piston effect in particular. They therefore suffered from some limitations and complications that made the results difficult to analyze:

- On Earth, it is quite difficult to separate out the effects of the piston effect and convection [3]. The high value of $\left(\frac{\partial\rho}{\partial T}\right)_p$ results in large density gradients, even in the presence of minute thermal gradients. Also, as D_T tends to zero, the Rayleigh number diverges and the fluid then becomes extremely unstable (see Sect. 1.8).

B. Zappoli et al., *Heat Transfers and Related Effects in Supercritical Fluids*, Fluid Mechanics and Its Applications 108, DOI: 10.1007/978-94-017-9187-8_6,

This leads to strong convective motion, even for rather small temperature perturbations. Thermalization occurs by convective motion, and this competes with and hides the piston effect. As a result, the fast heat transport on the ground was attributed to convective transport before the piston effect was suspected. Gravity-driven convection can also thin the thermal boundary layers, thus increasing the heat flux. The relative contributions of the different processes involved in thermalization are therefore quite difficult to estimate [4, 6, 8]. To remove any ambiguity, piston-effect dynamics should be investigated free of the effects of gravity.

- Under reduced gravity, the final equilibration time is still governed by diffusion (see Sect. 4.3). Cell homogenization is thus difficult or impossible to attain very near the critical point during the available experimental time [6, 8]. The density inhomogeneities resulting from the piston effect itself relax with the heat diffusion time, which diverges (see Table 1.5 and [1, 2, 6, 8, 13]). Checking for equilibrium thus necessitates a special cell design to avoid dead volumes and allow adequate light-based testing (Schlieren diffraction, light interferometry, etc.).
- When experiments are performed in the ultimate temperature range—very close to T_c (within a few mK from T_c), the transient temperature and heat flux at the cell walls are difficult to control due to the imperfect nature of the cell materials [6]. An ideal temperature step of 1 mK is quite difficult to achieve at the cell wall when comparing the large equilibration times of common materials with the characteristic piston-effect times.
- There is also a general limitation on the comparison of 1-D models with real 3-D experiments.

As a result, most of the observations that had been reported by the late 1980s were qualitative in nature, although there was the observation that temperature equilibration was much faster than expected for a purely heat-diffusive process.

Weightless experiments were therefore designed in order to circumvent the above problems and obtain quantitative information on the piston-effect mechanism. Thus, in the first two chapters of Part II (Chaps. 7 and 8), which focuses on heat-transfer experiments, we will introduce the main aspects of the experimental test cells that have been used to study heat and mass transfer phenomena in supercritical and near-critical fluids. Both chapters are complemented by the brief review of the hardware given in Chap. 23. Controlled heat source subsystems were used to heat the fluid; for example, a small thermistor was operated while immersed in the fluid [7], or a thin conductive layer deposited on a plane substrate was employed [5]. A comparison of 1-D models with these real 3-D experiments is provided in Chap. 8. A careful analysis of the heat transferred through the fluid is then performed in Chap. 9 in order to probe the validity of the piston-effect mechanism of heat and mass transfer in a one-phase critical fluid cell ($T > T_c$) through comparison with the results yielded by the 1-D model obtained in Chap. 5. Note that data obtained at temperatures far from the critical point are required to adjust the characteristic thermal timescales to the material constraints. The last chapter (Chap. 10) of this part presents heat and mass transfer experiments in near-critical two-phase fluids ($T < T_c$) induced by similar temperature changes under microgravity.

References

1. Bonetti M, Perrot F, Beysens D, Garrabos Y (1994) Fast thermalization in supercritical fluids. Phys Rev E 49(6):R4779–R4782
2. Bonetti M, Perrot F, Beysens D, Garrabos Y (1995) Fast adiabatic heating and temperature relaxation in near-critical fluids under zero gravity. Int J Thermophys 16:1059–1067
3. Boukari H, Briggs ME, Shaumeyer JN, Gammon RW (1990) Critical speeding up observed. Phys Rev Lett 65(21):2654–2657
4. Boukari H, Shaumeyer JN, Briggs ME, Gammon RW (1990) Critical speeding up in pure fluids. Phys Rev A 41(4):2260–2263
5. de Bruijn R, van Diest RJJ, Karapantsios TD, Michels AC, Wakeham WA, Trusler JPM (1997) Heat transfer in pure critical fluids surrounded by finitely conducting boundaries in microgravity. Phys A 242(1–2):119–140
6. Garrabos Y, Le Neindre B, Guenoun P, Perrot F, Beysens D (1993) Transport of heat and mass in near-critical fluids. Microgravity Sci Technol 6:110–118
7. Garrabos Y, Lecoutre-Chabot C, Hegseth J, Nikolayev VS, Beysens D, Delville J-P (2001) Gas spreading on a heated wall wetted by liquid. Phys Rev E 64(5):051602
8. Guenoun P, Khalil B, Beysens D, Garrabos Y, Kammoun F, Le Neindre B, Zappoli B (1993) Thermal cycle around the critical point of carbon dioxide under reduced gravity. Phys Rev E 47(3):1531–1540
9. Kassoy DR (1979) The response of a confined gas to a thermal disturbance. i. Slow transients. SIAM J Appl Math 36:3
10. Moldover MR, Sengers JV, Gammon RW, Hocken RJ (1979) Gravity effects in fluids near the gas–liquid critical point. Rev Mod Phys 51(1):79–99
11. Radhwan AM, Kassoy DR (1984) The response of a confined gas to a thermal disturbance: rapid boundary heating. J Eng Math 18:133–156
12. Straub J, Eicher L, Haupt A (1995) Dynamic temperature propagation in a pure fluid near its critical point observed under microgravity during the German spacelab mission d-2. Phys Rev E 51(6):5556–5563
13. Zappoli B, Carlès P (1995) The thermo-acoustic nature of the critical speeding up. Eur J Mech B 14:41–65
14. Zappoli B, Bailly D (1990) Response of a solid–gas growth interface to an increase in temperature. Int J Thermophys 33(9):1839–1847

Chapter 7
Quick Overview of Some Test Cells

Abstract The test cells used in the studies of the CO_2 and SF_6 critical point phenomena in microgravity conditions are presented. Some disk-shapped cells correspond to optical observations in transmission or interferometry, while a complex test cell use a specific design that combines transmission and interferometry observations. A blind spherical cell is designed to measure the specific heat at constant volume. These cells also includes (internal and/or external) heaters that exhibit Joule effect- or electrostriction-induced thermal phenomena. All these cells used during the MIR missions and the Space Suttle missions are appropriate for studying the temperature equilibration and the piston effect mechanism during pulse heating, as well as the long time behavior of the relaxation of density inhomogeneities, together with the measurements of the thermal diffusivity coefficient and the isochoric specific heat. More details on the hardware facilities and space missions can be found in the last chapter of appendices part.

7.1 Test Cells for Space Experiments

7.1.1 Test Cells Developed by Garrabos and Coworkers

The test cells developed by Garrabos and coworkers [7–9] were filled with critical CO_2 and critical SF_6 and widely used in the next chapters to provide experimental demonstrations of the relevance of the analytical developments given in Chaps. 4 and 5. Two typical pictures of the (direct observation and interferometry) cells are shown in Fig. 23.3 of Chap. 23. These small, disk-shaped cells that have a typical mass of $\sim$100 g and unsupported (glued) windows for optical use in transmission or interferometry can be characterized by their aspect ratios (thickness/diameter), which are typically in the range $\sim\frac{1}{2}$–$\sim\frac{1}{10}$, and their fast thermal responses, which are typically $\sim$1 s or less. As a representative example of the different optical cells developed by Garrabos et al. we briefly show the main components

B. Zappoli et al., *Heat Transfers and Related Effects in Supercritical Fluids*, Fluid Mechanics and Its Applications 108, DOI: 10.1007/978-94-017-9187-8_7,

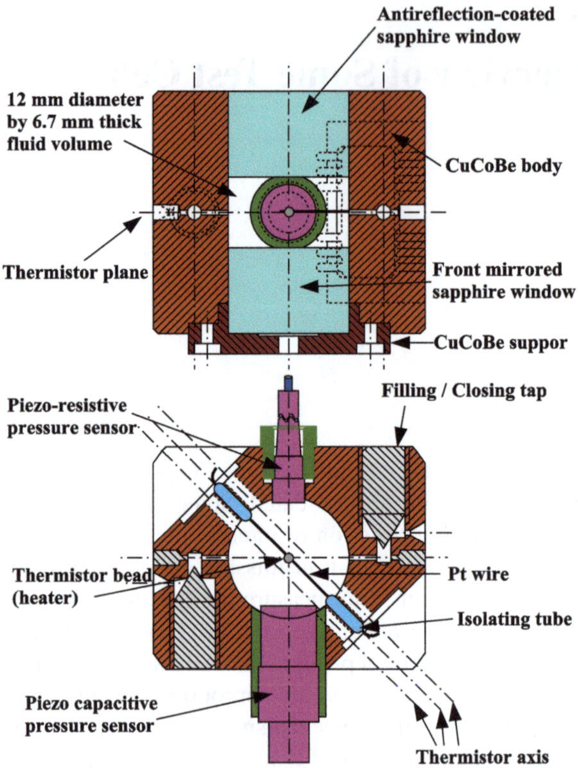

Fig. 7.1 The experimental cell developed by Garrabos and coworkers. *Top* cross section along the optical observation; *bottom* cross section perpendicular to the optical observation

of the interferometer cell. This cell was used to perform a detailed experimental study of the heat and mass transfers between a small thermistor heater located in the central part of the fluid chamber. We will report the corresponding experimental results in Chap. 9.

The experimental cell is filled with CO_2 (from Air Liquide; purity is better than 99.998 %) to critical density (467.9 kg m^{-3}) with an accuracy of better than 0.1 %. The body cell (Fig. 7.1a) is a parallelepipedic block with outer dimensions of $17 \times 24 \times 27$ mm^3 made of copper–cobalt–beryllium alloy. The fluid is enclosed in a cylinder 11.6 mm in diameter (denoted D) and 6.7 mm in thickness (denoted e), which is sealed at each end by sapphire windows.

The thermistors TH1 (the heat source) and TH2 (the bulk temperature measurement) are directly immersed in the fluid at equal distances from the two windows. The separation between their centers (4.6 mm) compares well with the thermistor–wall distance (3.5 mm). The thermistors (Thermometric B35 PB 103 F-A, with a time constant of 100 ms and a maximum power rating of 35 mW) are nearly spherical in shape and 0.9 mm in diameter. Each thermistor is sealed onto a thin

(0.1 mm diameter) wire of a platinum alloy, and the whole is encapsulated in a shock-resistant glass coating. A more detailed description of the cells will be given when required below.

7.1.2 Test Cells Developed by Wilkinson and Coworkers

The test cells developed by Wilkinson et al. [18] are conventional disk-shaped cells with supported windows, as shown in Figs. 7.2 and 7.4, leading to a larger size and a larger mass than in the previous glued-window design of 7.1. We briefly describe the interferometric measurements of density changes associated with the late stage of thermal equilibrium in SF_6 performed by Wilkinson et al., which led to accurate measurements of the thermal diffusion coefficient. Indeed, the interferometer test cell is designed to asymmetrically heat a SF_6 sample cell filled to critical density (Fig. 7.2). The fluid is contained in a coin-shaped volume $V_F = 75\,\text{mm}^3$ bounded by two cylindrical sapphire windows and by a circular copper spacer. The inner volume is a cylinder 10.0 mm in diameter and 0.957 mm in axial thickness. A foil heater is cemented to a 25° sector of the spacer's outside diameter. A copper block that is diametrically opposed from the heater connects the spacer thermally to the cell's aluminum flanges. This orientation of the heater and the heat sink leads to asymmetric heating of the spacer and thus the sample.

When the heater is switched off, the cell's copper and sapphire walls can be considered near-isothermal boundaries for the fluid sample. Let us estimate the main characteristics of the wall properties and the sample geometry that drive the late stages of the equilibration process. The total area of the fluid in thermal contact with the walls has the approximate value

$$A_F = A_{windows} + A_{spacer} + A_{tube} \approx 200\,\text{mm}^2,$$

$$\text{where} \quad \begin{cases} A_{windows} = 157 \ \text{mm}^2 \\ A_{spacer} \ \ = 30.1 \ \text{mm}^2 \\ A_{tube} \ \ \ \ = 12 \ \ \ \text{mm}^2 \end{cases}$$

are the respective areas of the two windows, the spacer, and the filling tube in contact with the fluid. The ratio $\frac{A_F}{V_F} = 2.63\,\text{mm}^{-1}$ leads to a characteristic length of 0.38 mm, which is comparable to the half-thickness ($L_0 = 0.48$ mm) of the sample cell. This geometry justifies a 1-D model where the fluid is contained in a cylinder with isothermal walls of radius $R = 5$ mm and height $2L_0$, with $L_0 \ll R$ (aspect ratio $\frac{L_0}{R} \approx 0.1$). The largest corresponding time constant is $t_D = \frac{4L_0^2}{\pi^2 D_T}$.

During the space experiments, the density distribution within the sample is disturbed from equilibrium by applying an electrical pulse to the cell heater. The pulse temporarily heats the cell body asymmetrically, thereby producing a nonequilibrium density distribution. After inducing such density gradients, interferometer images are used to observe the subsequent evolution of the density distribution (see for

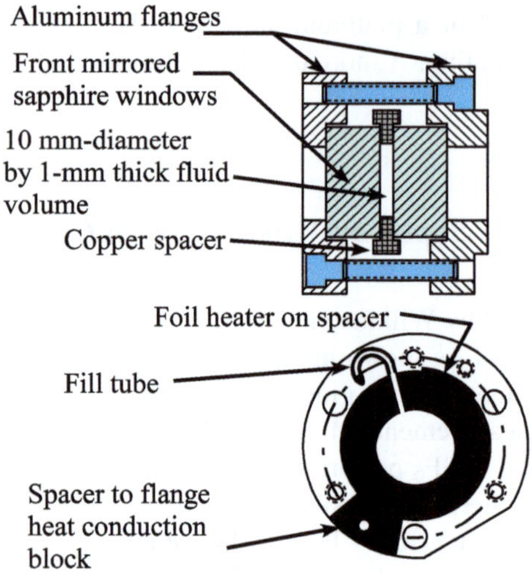

Fig. 7.2 Two views of the interferometer test cell operated by Wilkinson and coworkers (see Ref. [18]) on ESA's CPF during the IML-2 mission. The test cell is filled with SF_6 at critical density. The test cell has an axial thickness of 0.96 mm, which was chosen in order to limit the longest relaxation time to a few hours and to reduce the effect of the sample's large turbidity near T_c on the interferometry images. A specific foil heater is cemented to a 25° sector of the spacer's outside diameter. Diametrically opposite to the heater, a copper block connects the spacer thermally to the cell's aluminum flanges

example Fig. 7.3 from [18]). The analysis assumes that the observed density deviations are proportional to the temperature deviations within the sample (this assumption requires very low values for density and temperature deviations in comparison to the critical density and to the temperature difference from the critical point, respectively). Measurements performed late in the density gradient relaxation, where the time dependence is expected to be exponential, confirm our understanding of the late-stage diffusive equilibration near the liquid–vapor critical point. Analysis in the temperature range $1.4\,\text{mK} \leq T - T_c \leq 198.2\,\text{mK}$ supports diffusive relaxation very near to T_c. This results in measurements of D_T as close to T_c as $T - T_c = 1.4\,\text{mK}$ [18].

A similar test cell for observation in transmission was also built by Wilkinson and coworkers [20, 21] on ESA's CPF to study the density changes due to an electric field (electrostriction effect).

Schematic diagrams of the sample cell are shown in Fig. 7.4. The main parts of the coin-shaped cell, which is also studied by interferometry, are similar to those of the previous test cell: SF_6 fluid is confined between two flat quartz windows and by a washer-shaped stainless-steel spacer.

The spacer defines a cylindrical volume with an inner diameter of 12 mm and a thickness of 2 mm. Approximately a quarter of this volume is occupied by a 1 mm thick semi-circular quartz disk attached to a window. The disk divides the cell into two parts.

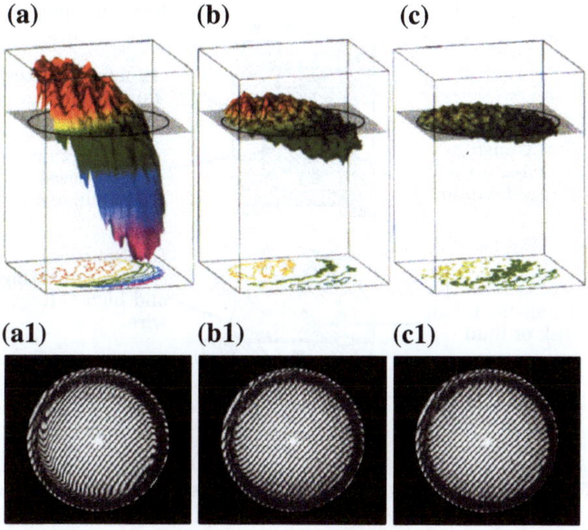

Fig. 7.3 A sequence of interferograms (lower parts **a**1, **b**1, **c**1) and the corresponding net phase maps (upper parts **a**, **b**, **c**) for the density disturbance provoked by a 0.28 J heating pulse of 240 s at $T - T_c = 8.5$ mK. The heater location corresponds to the *lower left boundary of the interferograms* and the *lower right boundary of the phase maps*. The net phase was proportional to the deviation of the local density from equilibrium. Each phase map is represented by both a surface and a contour map. Zero net phase is indicated by the *plane surrounding the surface* and the *heavy green contour line*. The *dark circle in the plane* represents the cell's boundary. **a** Start of the analysis ($t = 0$), which corresponds to the end of the heating pulse. The contour interval is equal to the 0.2 fringe. **b** Approximately one time constant later ($t = t_D = 42$ min). The contour interval is equal to the 0.2 fringe. **c** $t = t_D = 84$ min. The contour interval is equal to the 0.1 fringe (from [18])

The cavity near the filling tube is 1 mm thick, while the other cavity is 2 mm thick. A 0.125 mm diameter wire passes through the 2 mm thick section of the sample parallel to the edge of the quartz disk about 1.8 mm away from it (see Fig. 7.4).

7.1.3 Test Cell Developed by Michels and Coworkers

The test cell developed by Michels et al.[11] where simultaneous interferometric observations and thermal measurements are performed is shown in Fig. 7.5.

This cell illustrates the importance of accounting for complex geometry and material properties when studying the propagation of a planar thermal stepwise disturbance into a homogeneous sample. It consists of two interconnected cylindrical chambers with a total fluid volume of approximately $V_F = 6\,\text{cm}^3$. The cell is filled with SF_6 at critical density ($m_F = 4.44$ g). Along the lower optical axis, a smaller chamber enables light-scattering measurements to be made at discrete angles between 22° and 90°, together with direct observations.

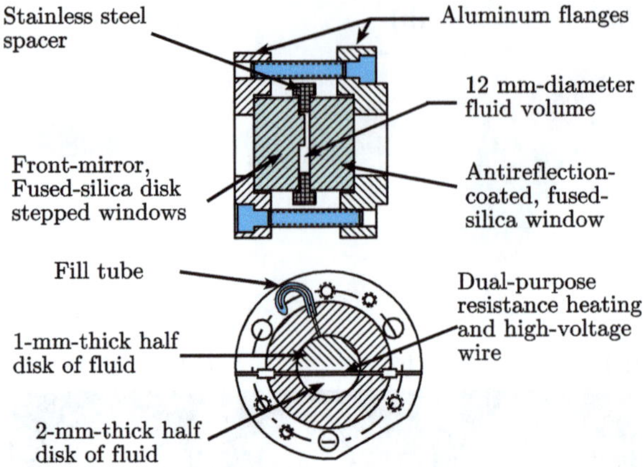

Fig. 7.4 Schematic diagrams of the sample cell (see text). The fluid-filled volume is 12 mm in diameter. Half of this region is 2 mm thick; the other half is 1 mm thick. The fine wire that passes through the fluid was charged to 500 V to generate the electric field (from Ref. [22])

In the upper part, the larger chamber accommodates a mirror to form one arm of a Twyman–Green interferometer. The core of the interferometer chamber is a heating plate made of a very thin gold layer deposited onto a flat quartz substrate, which was cut from a synthetic quartz cylinder 20 mm in diameter and 16 mm high. The heating plate is located 1 mm from the axis of symmetry and is parallel to it.

The field of view of the interferometer permits observations of the fluid up to a distance of 3 mm perpendicular to the heater. The gold layer is deposited by vaporization. Its area is $14 \times 18\,\text{mm}^2$ and it is 20 nm thick. The measured total resistance of the gold layer, R_H, is $3.6\,\Omega$. Heat is generated by sending an electric current through this plate resistor by using the current source subsystem (CSS) of CPF. The current ranges from 0 to 1 A in steps of 20 μA. During the space experiment, 66 heating runs were performed. The power delivered by the plate heater was varied between 10 μW and 180 μW, while the pulse duration was varied between 60 s and 300 s. Due to the large value of the plate heater area ($A_H \approxeq 252\,\text{mm}^2$), the delivered surface power did not exceed $0.7\,\text{W m}^{-2}$. Density changes in the fluid adjacent to the heating layer are detected using the interferometer images.

A 4 mm thick piece of synthetic quartz is glued to the bottom of the heater substrate by molecular contact. It is coated with silver by vaporization so that it can be used as a mirror. One of the two beams that builds the interferogram enters the fluid parallel to the gold layer and is reflected by this mirror such that it is made to pass along the heater a second time and leaves the fluid through the entrance window.

Given that a 16 mm path must be traversed twice, the requirement for the heater to be parallel to the optical axis and for the mirror to be perpendicular to it ensures that the beam of laser light must cover a fluid layer smaller than the resolution of the interferometer image, which is on the order of 10 μm. The entrance window is 8 mm thick, and both of its surfaces are comparably flat to that of the mirror (smaller

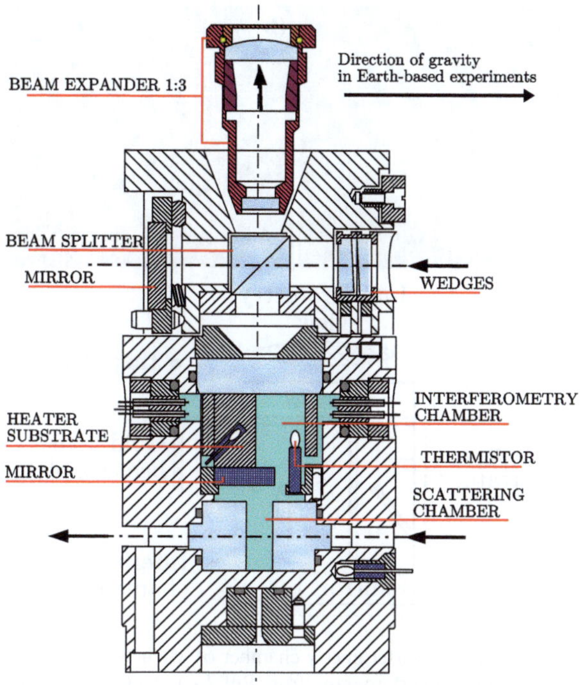

Fig. 7.5 Spaceflight test cell operated by Michels and coworkers on CPF during the 1994 IML-2 mission (from Ref. [4]). To study heat transfer in critical pure fluids, the test cell is filled with SF_6 at critical density. The core of the test cell is the plate heater, which is made of a very thin layer of gold deposited on a flat quartz substrate. A constant heating power is generated by running an electric current through this layer. The density changes in the fluid adjacent to the plate heating boundary surface are monitored using the interferometer images

than one tenth of a wavelength). Inside the interferometry chamber, the temperature is measured by two YSI precision thermistors (time response is about 3 s). One is in the fluid and one lies behind the heater in a mounting hole inside the quartz substrate. The fluid thermistor is located approximately 9 mm from the heater. The substrate thermistor is located 4 mm from the heater. A third thermistor is located in the SCU and acts as an independent monitor of the SCU temperature. Thermistors enable temperature rises to be measured with a sensitivity of 10 μK.

Space experiments that investigate heat transfer by imposing a plane thermal disturbance on the fluid have been performed in this interferometer chamber. However, the cell has a complex geometry and is made of different materials. It is difficult to determine both the thermal transport properties and the surface area of the whole cell. Michels et al. identified the five component materials that contributed the most to heat losses and had the largest heat exchange surface areas. In spite of this complexity, the phenomenology of the piston-effect mechanism was unambiguously observed and described.

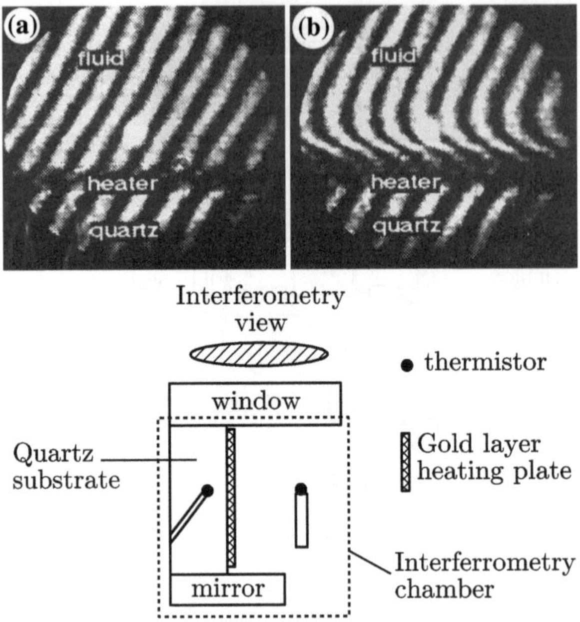

Fig. 7.6 Schematic view of the interferometry chamber of Michels et al SF_6 test cell that shows the construction of the resulting interferogram (from Ref. [4]). A magnification factor of 3 is used, which leads to a field of view 4 mm in diameter. *Top* interferometer fringe patterns of the one-phase domain($T > T_c$), **a** for complete equilibrium in the sample cell before the heating pulse starting at $t = 0$ s, and **b** at time $t = 57$ s after the onset of heating. The spatial extent (here on the order of 1 mm) of a diffusive boundary layer is clearly demonstrated by the bending of the fringes. The homogeneous density variations in the bulk are difficult to observe by the naked eye because changes are certainly less than 1 % of the original fringe pattern. However, small density variations were confirmed by the correlated temperature variations measured by the thermistor that was located in the fluid at about 9 mm from the heater plate

Their operating conditions were a constant heat power of 130 μW delivered by the plate heater for 60 s (corresponding to 7.8 mJ of dissipated total energy).

The interferometer images clearly show a hot diffusive thermal boundary layer (characterized by curved fringes) that develops close to the heater (see Fig. 7.6).

The spatial extent of the HBL depends on the initial temperature difference from T_c, but is always small (less than 1 mm) compared to the typical size of the remaining fluid part (on the order of a centimeter). The bulk fluid away from the developing boundary layer appears to be homogeneous (fringes are parallel). Density variations are difficult to observe because they are too small due to the low efficiency of the heater (a significant amount of the energy delivered to the resistive layer is absorbed by the quartz substrate). However, the adiabatic effect can also be simultaneously measured by the thermistor located in the fluid 9 mm from the heater (see Fig. 7.7).

Other interesting results are the temperature evolutions for heating runs at different $(T - T_c)$ (see Fig. 7.7): The thermistor responses are essentially synchronous with

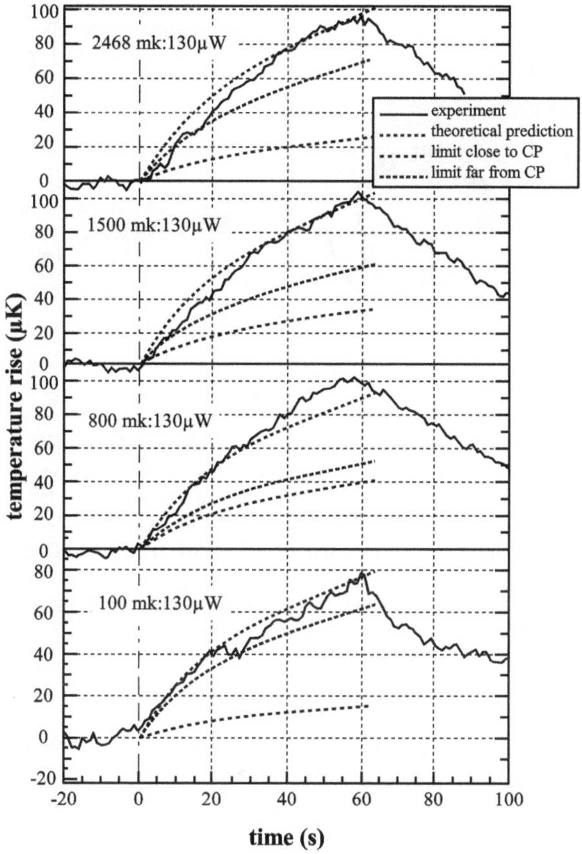

Fig. 7.7 Comparison of local bulk temperature evolutions at various temperatures far from the critical temperature ($T - T_c \geq 100\,\text{mK}$) during heat pulses of constant power ($q_H = 130\,\mu\text{W}$). *Solid lines* indicate temperatures measured by the thermistor immersed in the fluid 9 mm from the heater. The *dashed lines* are the theoretical predictions using exact formulation (*upper lines*) and approximate formulations valid close to T_c (*median lines*) and far from T_c (*lower lines*) (see [4])

the onset of the heating period (which corresponds to zero on the x-axis), a signature of the piston effect which is not found when dealing with thermal conduction only. Moreover, there is noticeable agreement with theoretical results obtained using a 1-D model [4], as shown by the upper dashed line (see also Eq. 5.48 in Sect. 5.2.3) for each temperature.

More detailed comparisons with theoretical predictions can be found in [4, 6, 11, 17]. Their analyses, as a function of $T - T_c$, provide quantitative data for the determination of three important parameters:

1. The characteristic time t_{PE} of the piston effect [6, 11]
2. The thermal diffusivity coefficient D_T [17]

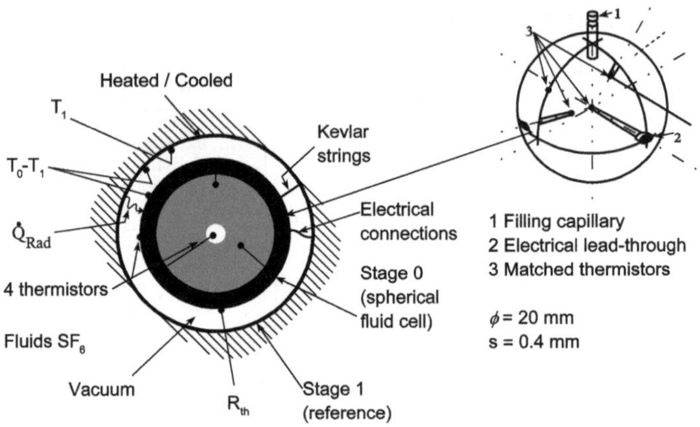

Fig. 7.8 Setup for the scanning radiation calorimeter operated by Straub et al. on the HPT-HYDRA experiment in 1993, during German Spacelab Mission D-2. The spherical cell (stage 0) is heated/cooled by radiative heat exchange with the surrounding shell (stage 1). To minimize heat transport by conduction, the thermostat is positioned under vacuum. The connections between the stages are achieved using Kevlar strings and special electrical contacts. The specific heat behavior is determined by measurements of $T_0 - T_1$ and T_1, which are related to the thermal resistance R_{th} of the test cell. The sample cell, which is filled with SF_6 at critical density, has an outer diameter of 20 mm and a wall thickness of 0.4 mm. Four thermistors are used to measure the temperature distribution in the cell

3. The heat capacity at constant volume c_V, obtained from adiabatic measurements accounting for the piston effect [5].

7.1.4 Test Cell Developed by Straub and Coworkers

During the D-2 mission, a scanning radiation calorimeter (see Figs. 7.8, 7.9) developed by Straub et al.[13, 14], was used to measure the isochoric heat capacity of SF_6 during heating and cooling runs [15]. A specific sphere-shaped cell shown schematically in Fig. 7.8 was developed for these measurements. In this cell, the loss of local information due to the absence of optical diagnoses is compensated for by the spherical, well-shaped design of the calorimeter cell. The high symmetry of this design permits precise 1-D modeling of the thermal processes. The spherical cell is made of copper produced by an electrolytic coating process, and has an inner diameter of 19.4 mm and a thickness of 0.35 mm. The sample mass, the cell volume, and the (heating/cooling) wall area in contact with the fluid are: $m_F = 2.773$ g, $V_F = 3.7626\,\text{cm}^3$, and $A_F = 11.7\,\text{cm}^2$, respectively. The cell is equipped with four thermistors (0.35 mm in diameter, 100 ms time constant, 1.66 Hz sampling rate). The sample cell is heated (stage 0 in Fig. 7.8) and then passively cooled, mainly by radiative exchange with the surrounding reference shell (stage 1 in Fig. 7.8). Only 10 % of the total heat exchange is due to heat conduction through the electrical

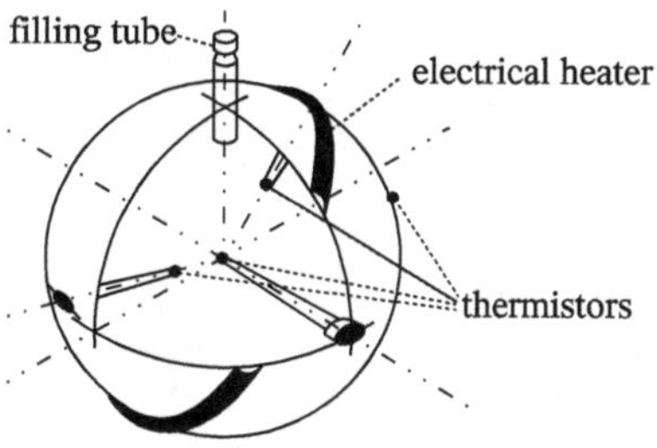

Fig. 7.9 Sketch of the calorimeter sample cell (made of copper with a gold layer inside and outside) which shows the heater constructed by winding a wire onto the outer surface of the cell along a great circle. The thermal conductivity of the adhesive is 1 $W\,m^{-1}\,K^{-1}$. Because of the thermal resistance of the adhesive, the measured temperature response was delayed by about 2–3 s, and at the end of the pulse the temperature continues to increase for the same time lag

connections of the cell shell. Different cooling/heating ramps were performed with rates $\left(\frac{dT_0}{dt}\right)$ ranging between $-0.065\,K\,h^{-1}$ and $-1.3\,K\,h^{-1}$. The specific heat measurements were not considered here despite the great interest in this fundamental quantity for determining the thermodynamic behavior of a real Iisng-like system near its liquid-gas critical point. The detailed analyses of these measurements can be found in [1, 2, 10].

The calorimeter was also used to perform experiments [15, 16] on temperature equilibration, which confirmed its fast dynamics. In these experiments, the spherical cell was heated by a 3.85 mW electric pulse for 10 s. A total of 39 different temperatures in the range $0.03\,K < |T - T_c| < 5.25\,K$ were investigated, and the temperature response was measured with the four thermistors. One thermistor measures the temperature of the outer surface of the spherical cell, while the other three others measure the fluid temperature at three radial positions: 3.2, 6.0, and 8.4 mm from the wall (see Fig. 7.8). Each heat pulse increases the temperature of the whole system (cell + fluid) between 8 and 15 K , depending on $T - T_c$.

The heater is a round wire that is pasted onto the outer surface of the cell along a great circle (see Fig. 7.9).

The thermistor measuring the wall temperature is mounted a distance of about 5 mm from the heater. Here, the measured temperature was found to be equal to the calculated wall mean temperature to within 0.1 mK. At the end of the heat pulse, the maximum temperature difference in the cell wall was estimated to be less than 1 mK [16]. Accordingly, the whole wall of the cell was treated as an isothermal heating area. The important geometrical parameter of this spherical heating problem is then the ratio of the heating area A_H to the total sample volume, $\frac{A_H}{V} = \frac{3}{R} = 0.3125\,mm^{-1}$. Since $A_H \equiv A_F$, this cell provides a higher ratio than any other cell (Wilkinson's, Michel's, Garrabos's, ...) with the same critical fluid mass. In addition, the spherical symmetry of this sample cell enables close comparison with a 1-D spherical numerical model. Moreover, the results [15, 16] obtained with this ideally shaped (but *blind*) cell become of great interest if we note that the total fluid volume

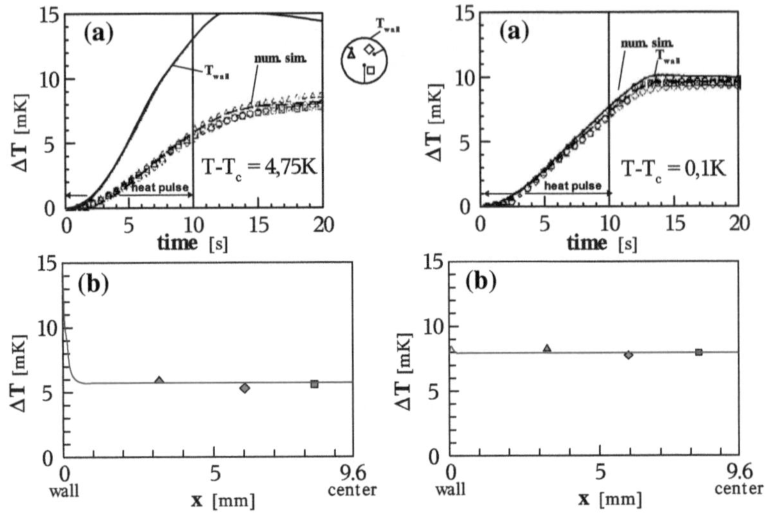

Fig. 7.10 *Top* Temperature responses in the fluid (*triangles, squares, diamonds*) and at the sample wall (*solid line*) during and after the heat pulse at $T - T_c = 4.75$ K (*Left* (**a**)) and at $T - T_c = 0.1$ K (*Right* (**a**)), as compared to the results from the numerical simulation (*dashed lines*). *Bottom* Calculated temperature distribution (*solid line*) at the end of the heat pulse at $T - T_c = 4.75$ K (*Left* (**b**)) and at $T - T_c = 0.1$ K (*Right* (**b**)). The measured temperatures at this time are indicated by the same symbols

and thus the total mass of SF_6 are of the same order of magnitude as those for the cell of Michels et al.

Figure 7.10a shows the temperature response during and after the 10 s heat pulse at initial temperatures of $T - T_c = 4.75$ K (left part) and 0.1 K (right part). The solid line represents the measured temperature T_W at the cell wall, and the data are the bulk temperatures measured at different thermistor positions. The dashed line shows the results of numerical calculations based on 1-D spherical modeling when the fluid velocity is neglected. The spatial coordinate x is measured in the radial direction, with $x = 0$ at the cell wall and $x = R$ at the center. In the numerical calculations, the measured temperature at the wall was used as an input boundary condition. This avoids any influence of the initial time delay (about 3 s) due to the finite thermal resistance of the cell-plus-fluid ensemble.

Note that at $T - T_c = 4.75$ K (see the left parts of Fig. 7.10a and b), the three temperatures in the fluid at the positions $x = 3.2, 6.0$, and 8.4 mm are almost equal, while a noticeable temperature difference has developed in a boundary layer between the wall and the bulk. The calculated temperature distribution in the radial direction after the 10 s heat pulse is also plotted in Fig. 7.10. The calculated thermal penetration length is less than 0.5 mm. The uniform temperature increase in the bulk is caused by isentropic compression due to the piston effect as a consequence of the expanding fluid in the hot boundary layer. The calculated curve follows the measured values, which confirms the above hypothesis: isentropic temperature propagation and the physical validity of the 1-D numerical model based on near-ideal spherical geometry and near-perfect thermal insulation.

At $T - T_c = 0.1$ K (see the right parts of Fig. 7.10a, b), there is almost no observable temperature difference between the wall and the bulk fluid. The calculated temperature evolution fits the data very well. After 10 s, the calculated temperature difference between the wall and the bulk is only 0.2 mK and the thickness of the boundary layer is estimated to be 0.2 mm, as shown in Fig. 7.10b. As mentioned by the authors, this behavior arises because the thermal contact between the cell wall and SF_6 is modified when T goes to T_c due to the diverging thermophysical properties of the fluid. Practically speaking, as T_c is approached, the temperature difference between the wall and the bulk fluid becomes so small that, within the limits of measurement uncertainty, this difference cannot be observed. Such a behavior very close to T_c where the thermal nature of the fluid dominates the heat exchanged at the wall–fluid contact is called the (thermal) *bottleneck effect* and is discussed further in section.

In spite of this bottleneck effect, which mainly leads to a drastic change in the dynamics of the heat exchange very close to T_c, it remains possible to demonstrate that the amplitude of the fast temperature response in the bulk depends only on the amount of energy delivered to the fluid during the heat pulse (see Eqs. (4.90) and (4.76) in Chap. 4). For example, after 13 s at $T - T_c = 0.1$ K in Fig. 7.10 (right part), the maximum temperature increase of the wall $\Delta T_W \simeq 9.5$ mK is only slightly larger than the maximum temperature increase in the bulk, $\Delta T_{bulk} \simeq 9$ mK . For the copper cell, the corresponding energy increase was $\Delta Q_{cell} = C_{cell} \Delta T_W \simeq 14$ mJ (with $C_{cell} = m_{cell} c_{p,\,copper} = 1.46\,\mathrm{J\,K^{-1}}$ [15]), while for the fluid, the equivalent quantity was $\Delta Q_F = m_F c_V \Delta T_{bulk} = 32$ mJ (with $m_F = 2.773\,g$; $c_V = 1290\,\mathrm{J\,kg^{-1}\,K^{-1}}$ [15]). The total energy increase of the calorimeter was then $Q_{tot} = \Delta Q_{cell} + \Delta Q_F \simeq 46$ mJ, which should be compared with the total energy 38.5 mJ provided by the heater. This difference can be attributed to the uncertainties in both the injected power and temperature measurements.

Similarly, at $T - T_c = 4.75$ K (see Fig. 7.10), after 13 s, the temperature increases for the wall and the bulk are $\Delta T_W = 15$ mK, and $\Delta T_{bulk} = 8.0$ mK, respectively. The corresponding amounts of energy are then $\Delta Q_{cell} \simeq 22$ mJ and $\Delta Q_F = 19$ mJ (with $c_V = 880\,\mathrm{J\,kg^{-1}\,K^{-1}}$ [15]), leading to $Q_{tot} \simeq 41$ mJ, a value that compares well with the one above at $T - T_c = 0.1$ K. These estimations implicitly assume that the heating surface is not the wire but the complete sphere. The 1-D equivalent model of the heating sphere must preserve the ratio between the total sample volume and the heated area. This ratio is a key characteristic length for estimating the bulk fluid temperature response in a 1-D model. Here we have benefited from the "calorimeter nature" (i.e., there is no external energy dissipation) of Straub et al. cell: the heat losses in the calorimeter cell were estimated to be lower than 10 % by Straub and coworkers. Therefore, adiabatic compression of the bulk occurs near-isentropically and causes an instant temperature change in the bulk that is proportional to the total heat injected into the fluid, as shown by Eq. (5.36) in Sect. 5.1.3, and as also expected in [3, 12, 19].

In comparison, the heating conditions used by Michels et al. (see Sect. 7.1.3) appear to be significantly different from Straub et al. For example, in Fig. 7.7, at the end of the 60 s heating period at $T - T_c = 0.1$ K, the measured temperature increase in the bulk is $\Delta T_{bulk} = 75\,\mu$K. This value corresponds to a supplied energy $Q_F = m_F c_V \Delta T_{bulk} = 430\,\mu$J, which represents only 5.5 % of the total heat delivered by the plate heater ($Q_{H,tot} = 7.8$ mJ). In this case, the very small temperature rise cannot be explained by either the difference in the heated fluid masses (factor of about 1.6), or the difference in the thermal energy supplied (factor of about 5), or even the energy partition between the substrate and the fluid (about 15 % for this temperature difference from T_c). The remaining energy difference (on the order of 300–400 μJ) reflects the contribution of the cold piston effect that takes place at the thermostated wall boundaries. The total fluid–wall (cold) area can be estimated as $A_F \approx 2.1\,\mathrm{cm}^2$, while the calculated heating (hot) area ($A_H \approx 0.25\,\mathrm{cm}^2$) is only 12 % of the fluid boundary surface. The "conductimeter" character of the cell (i.e., implying a noticeable temperature gradient between the heater part and the thermostated part of the cell) leads to a real heat transfer mechanism via bulk temperature equilibration through the (heating and cooling) piston effects. The difference from classical heat exchange by a pure conduction mechanism is demonstrated here by the initial homogeneous temperature increase of the bulk that induces the simultaneous formation of a cold boundary layer at the thermostated walls. During short transient times, a new heat transfer mode is observed where two spatially disconnected fluid layers (the hot and cold nonhomogeneous boundary layers) generate two piston effects resulting from the balance between the compression and the expansion of the bulk fluid (as shown from the 1-D model analysis given in Sect. 5.2). However, the complex geometry of Michels et al. cell and the difficulties involved in precisely controlling the heat flux directed into the fluid remain the main factors limiting the quantitative description of the heat transport.

References

1. Barmatz M, Zhong F, Shih A (2004) Need research. Int J Thermophys 25:1667
2. Barmatz M, Zhong F, Shih A (2005) Need research. Int J Thermophys 26:921
3. Boukari H, Shaumeyer JN, Briggs ME, Gammon RW (1990) Critical speeding up in pure fluids. Phys Rev A 41(4):2260–2263
4. de Bruijn R (1999) Heat transfer in a critical fluid under microgravity conditions: a spacelab experiment. PhD thesis, University of Amsterdam, Amsterdam, p 197
5. de Bruijn R, van Diest RJJ, Michels AC (1996) Isochoric specific heat c_v in the near-critical region of SF_6 from isentropic thermalization in microgravity. In: Viviani A (ed) Proceedings of the 2nd European symposium on fluids in space. Congressi srl, Roma, pp 334–338
6. de Bruijn R, van Diest RJJ, Karapantsios TD, Michels AC, Wakeham WA, Trusler JPM (1997) Heat transfer in pure critical fluids surrounded by finitely conducting boundaries in microgravity. Phys A 242(1–2):119–140
7. Garrabos Y, Perrot F, Beysens D (1991) Heat and mass transport in a hypercompressible fluid under zero gravity (invited paper). In: ESA (ed) Proceedings of the first European symposium on fluids in space, volume ESA SP-353. ESA, Paris, pp 357–388

8. Garrabos Y, Le Neindre B, Guenoun P, Khalil B, Beysens D (1992) Observation of spinodal decomposition in a hypercompressible fluid under reduced gravity. Europhys Lett 19:491
9. Garrabos Y, Le Neindre B, Guenoun P, Perrot F, Beysens D (1993) Transport of heat and mass in near-critical fluids. Microgravity Sci Technol 6:110–118
10. Haupt A, Straub J (1999) Evaluation of the isochoric heat capacity measurements at the critical isochore of SF_6 performed during the German spacelab mission d-2. Phys Rev E: Stat, Nonlinear, Soft Matter Phys 59(2):1795–1802
11. Michels AC, de Bruijn R, Karapantsios TD, van Diest RJJ, van den Berg HR, van Deenen B, Sakonidou EP, Wakeham WA, Trusler JPM, Louis A, Papadaki M, Straub J (1995). Adiabatic compressive heating of critical fluids under microgravity conditions. In: The 30th 1995 national heat transfer conference, vol 3. Portland, OR, USA, 6–8 August 1995
12. Onuki A, Hao H, Ferrell RA (1990) Fast adiabatic equilibration in a single-component fluid near the liquid-vapor critical point. Phys Rev A 41(4):2256–2259
13. Straub J, Nitsche K (1993) Isochoric heat capacity c_V at the critical point of SF_6 under micro- and earth-gravity: results of the German spacelab mission d1. Fluid Phase Equilib 88:183–208
14. Straub J, Haupt A, Nitsche K (1993) Radiation calorimeter for heating and cooling ramps used for hysteresis measurements at phase transition. Fluid Phase Equilib 88:123–135
15. Straub J, Eicher L, Haupt A (1995) Dynamic temperature propagation in a pure fluid near its critical point observed under microgravity during the German spacelab mission d-2. Phys Rev E 51(6):5556–5563
16. Straub J, Eicher L, Haupt A (1995) The process of heat and mass transport at the critical point of pure fluids. Int J Thermophys 16:1051–1058
17. Wakeham WA, Trusler JPM, de Bruijn R, van Diest RJJ, Karapantsios TD, Michels AC (1997) Heat transport near the critical region of a pure fluid in microgravity. In: Viviani A (ed) The 2nd European symposium on fluids in space. Congressi srl, Rome
18. Wilkinson RA, Zimmerli GA, Hao H, Moldover MR, Berg RF, Johnson WL, Ferrell RA, Gammon RW (1998) Equilibration near the liquid-vapor critical point in microgravity. Phys Rev E 57(1):436–448
19. Zappoli B, Bailly D, Garrabos Y, Le Neindre B, Guenoun P, Beysens D (1990) Anomalous heat transport by the piston effect in supercritical fluids under zero gravity. Phys Rev A 41(4):2264–2267
20. Zhong F, Horst M (1995) Density equilibration near the liquid-vapor critical point of a pure fluid: single phase $t\ t_c$. Phys Rev E 51(4):3223–3241
21. Zhong F, Meyer H (1999) Heat transport in a pure fluid near the critical point: steady state and relaxation dynamics. J Low Temp Phys 114:231–255
22. Zimmerli GA, Wilkinson RA, Ferrell RA, Moldover MR (1999) Electrostriction of a near-critical fluid in microgravity. Phys Rev E 59(5):5862–5869

Chapter 8
The Pancake Test Cell

Abstract A detailed analysis of the design and materials that are used in an ideal pancake cell is presented. Indeed, most of the studies have been carried out in disk-shaped cells, like "pancake" cells, where a thin fluid layer is confined between two flat parallel transparent windows of large thickness, needed to support the high internal pressure of the confined fluid. The main characteristic parameters of length and time governing the fluid thermal behavior during the pancake cell heating are defined.

8.1 Pancake Cell Design

The schematic design of a pancake cell is given in Fig. 8.1. The fluid layer of thickness e_F is confined between two flat parallel transparent windows of thickness e_W and external diameter D_W separated by a spacer of internal thickness e_F and internal diameter D. A simplified optical channel assumes that the aspect ratio $\frac{\text{thickness}}{\text{diameter}} = 1$ for the window (i.e., $D_W = D$ and $e_W = D$). Generally, the whole optical field of view is D. In the following, a typical value of $D = 10$ mm is considered, with e_F ranging from 1 to10 mm. The total length is then $2D + e_F$ in the direction of observation. The relative variations in the window-fluid area and cell body-fluid area depend on the aspect ratio $r_A = \frac{e_F}{D}$ of the fluid sample. Three typical values, $r_A = \frac{1}{10}$, $\frac{1}{2}$, and 1, will be considered in the following.

The body of the pancake cell is assumed to be a parallelepipedic block of external dimensions $L_{body} = l_{body} = 2D + e_f$ and $h_{body} = 2D$ (see Fig. 8.1), which makes it easier to estimate its mass, heat capacity and thermal response time. Therefore, only the thermal characteristics of two materials, those of the windows and body, are needed. Anticipating the list of materials of interest (given below), it is worth noting that sapphire and fused silica exhibit large differences in their thermal properties, leading to significant differences in heat exchange through the optical window. Similarly, choosing oxygen-free copper (OFHC) rather than stainless steel has a large impact on the thermal response of the cell body. Moreover, modeling the thermal

B. Zappoli et al., *Heat Transfers and Related Effects in Supercritical Fluids*, Fluid Mechanics and Its Applications 108, DOI: 10.1007/978-94-017-9187-8_8,

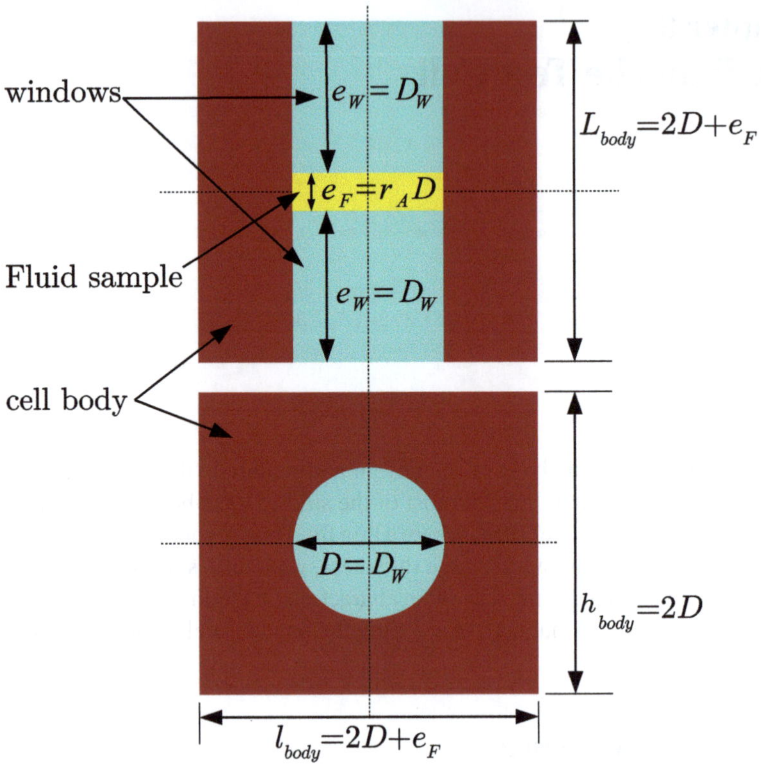

Fig. 8.1 Characteristic dimensions of the pancake cell

response of the whole cell over several decades can be achieved by considering air or water as the test fluid, in addition to near-critical SF_6 and CO_2.

The following discussion shows that a single effective length can characterize the thermal properties of a given pancake cell.

8.2 Characteristic Length of an Ideal Pancake Cell

The simplest definition of a characteristic length L comes from the 1-D theoretical approach of Chaps. 4 and 5, which relates L to $\frac{A_{acting}}{V_F}$. Here, A_{acting} is the heating or cooling area and V_F is the total volume of the critical fluid. Recall that this 1-D approach is only valid for ideal walls with infinite conductivity. In this situation, the characteristic timescale $t_{PE} = \frac{L^2}{D_T \gamma_{0c}(\gamma_{0c}-1)} = \frac{t_D}{\gamma_{0c}(\gamma_{0c}-1)}$ (see Eq. 5.57 in Sect. 5.2.5) of the piston effect can also be interpreted as the time $t_{eq} = \frac{\delta^2}{D_T} = \frac{t_D}{\gamma_0^2}$ (see Eq. 4.11 in the Introduction of Chap. 4) needed to transfer, from a boundary layer of size $\delta \simeq \frac{L}{\gamma_0}$, an amount of energy that can adiabatically heat the remaining bulk fluid of size $L-\delta$. Both of these times exhibit the same asymptotic behavior when $\gamma_0 \gg 1$. Approaching T_c, the homogeneous bulk fluid of thickness $L-\delta$ acts during $t_0 \simeq t_{PE}$ as a diffusive-like but very thin layer, because its adiabatic thermal resistance $\left(\sim \frac{1}{L\rho_c c_V}\right)$ is significantly lower than the diffusive thermal resistance $\left(\sim \frac{\sqrt{D_T}}{\lambda} = \frac{1}{\sqrt{\lambda \rho_c c_p}}\right)$ of the boundary layer of thickness δ. There are, however, some ambiguities when considering a realistic conduction process in a 3-D sample. For instance, L can be the closest distance between acting and nonacting walls; L can be the largest distance between two walls; L can be the cell diameter; and so on. To overcome this difficulty, it is necessary to perform a dimensional analysis for either internal or external heating (see also Sect. 23.2 for pratical details).

8.2.1 External Heating (EH) Case

In the EH mode, the acting area corresponds to the total area A_F surrounding the fluid volume, and what matters in the estimation of the total energy of the adiabatic process is the total heat capacity of the fluid. The latter is proportional to the fluid volume at given ρ and c_V. Therefore, to maintain the dimensional analogy between a 1-D model and the experiment, it is necessary to keep the ratio $\frac{\text{fluid volume}}{\text{area}}$ of the surrounding wall constant. $L_{EH} = \frac{V}{A_F}$ is thus the natural characteristic length. Of course, the closer the resemblance of the actual cell to a 1-D geometry, the more accurate the above definition. This ideal situation corresponds to the following two limiting cases:

1. A very thin pancake cell ($L_{EH} \approx \frac{e_f}{2}$ for $\frac{e_f}{D} \ll 1$)
2. A spherical calorimeter ($L_{EH,sphere} = \frac{R}{3}$, where R is the cell radius)

8.2.2 Internal Heating (IH) Case

There are two main complications of the analysis of an experiment in the IH mode, where the area of the hot boundary layer is such as $A_{HBL} = A_{HB}$ with A_{HD} equal to experimental area of heater device.

- First, IH introduces an additional heat flow through the external walls, where the adiabatic condition is very difficult to achieve practically.
- Second, IH separates the area A_{HD} of the heater device from the total area A_F of the fluid in contact with the external walls.

At the nonadiabatic boundaries, the walls remain colder than the fluid and a cold boundary layer (CBL) of area $A_{CBL} \leq A_F$ develops. Adiabatic compression takes place, which cools down the bulk fluid. It is worth mentioning that, when the heating process stops, the temperature relaxation again involves $A_F = A_{CBL} + A_{HD}$. For the experimental situation, note that the connecting wires of the heater device lead to heat losses, increasing A_{CBL}.

All IH configurations correspond to $\frac{A_{HD}}{A_F}$ ratios between 0 (the case for a very small internal heater, such as a point-like thermistor source) and $\frac{1}{2}$ (the case for a large internal heater with a surface area that is close to that of the external cell walls). The value of $\frac{1}{2}$ also corresponds to the configuration of the theoretical 1-D model used in Chap. 5, where A_{HD} (at $x = 0$) $= A_{CBL}$ (at $x = L$).

A possible definition for the characteristic length would be $L_{IH} = \frac{V_{Cell}}{A_{HD}}$. However, if we consider the obvious condition $A_{HD} < A_{CBL}$, we obtain $L_{IH} = \infty$ for $\frac{A_{HD}}{A_F} = 0$ and $L_{IH} = 2\frac{V_{cell}}{A_F}$ for $\frac{A_{HD}}{A_F} = \frac{1}{2}$. Then L_{IH} would always be larger than the largest length involved in a 1-D conduction problem, which is unphysical.

There is no such difficulty if we instead choose the characteristic length $L_{IH} = \frac{V_{Cell}}{A_F}$. This characteristic length remains the same for both heating and cooling, as $A_{CBL} \approx A_F$ appears to be always active. This definition has another advantage in that it is the same for both the IH and EH modes, thus leading to the following general definition for a given cell made of perfect materials (infinite conductivity):

$$L_{IH} = L_{EH} = \frac{V_{Cell}}{A_F}. \tag{8.1}$$

8.2.3 Linear and Cylindrical 1-D Modeling (Ideal Cell)

The 1-D model of Chap. 5, as depicted in Fig. 8.2a, can be used to model the thermal behavior of an ideal pancake cell. To do so, it is necessary to define (i) the length and (ii) the heating conditions.

From the above analysis, the characteristic length can be written as

$$L = \frac{\text{cell volume}}{\text{acting}(\equiv \text{surrounding) wall area}} = \frac{V}{A_F} = \frac{e_f}{2}\left(\frac{1}{1 + 2\frac{e_f}{D}}\right). \tag{8.2}$$

If we now consider the internal heating conditions, the 1-D model imposes that the heating area (A_F) must be equal to the cooling area (A_W) (see Fig 8.2a).

When the effective experimental heating and cooling areas are quite different, a value for the heat flux that is different from the experimental one must be considered in order to properly describe the experiment.

This can be done through the use of a cylindrical configuration, for example, as discussed in [4] and shown in Fig. 8.2b. In Fig. 8.2, we assume that the ratio $\frac{A_{HBL}}{A_{CBL}}$ is chosen to be equal to the experimental ratio $\frac{A_{HD}}{A_F}$.

This modification is needed to obtain the experimental amplitudes with a 1-D model (Fig. 8.2a).

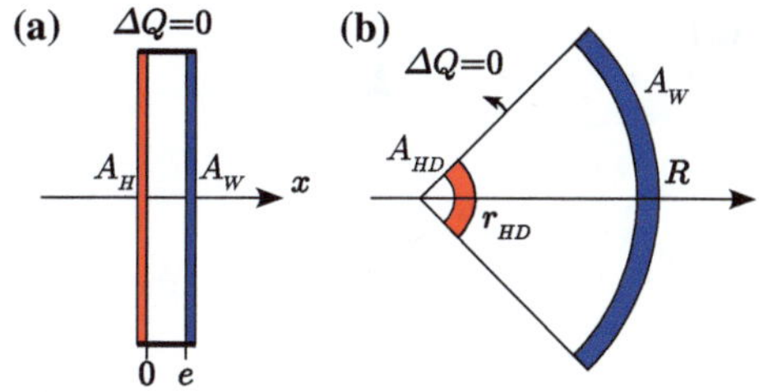

Fig. 8.2 1-D models of an ideal pancake cell with heating and cooling walls: **a** linear model and **b** cylindrical model

Recall that the characteristic length defined by Eq. (8.2) is only valid during the initial transient period such that $t \lesssim$ a few t_{PE}. For larger times ($t \approx t_D$), where the remaining heat transfer mechanism is pure diffusion, a new detailed re-analysis of the length scale problem is needed for each specific cell design (see, for example, [1, 5, 6]).

8.3 Thermal Properties of the Wall

8.3.1 Thermal Contact Between a Material and a Critical Fluid

Let us consider a wall-fluid interface in the 1-D model as shown in Fig. 8.3a. An effective thermal contact at $x = 0$ links two imperfect systems: at $x < 0$, the cell body is labeled with the subscript W, and has a finite thermal impedance Z_W (as defined just below) at a homogeneous initial temperature T_{W_∞}; at $x > 0$, a fluid with a thermal impedance Z is present at a homogeneous initial temperature $T_{F_i} = T_{W_\infty} - \Delta T$.

At $t = 0$, the cell body is placed in thermal contact with the sample fluid ($0 \leqslant x \leqslant L$). We consider a time such that the thermal penetration length δ_{th} in the fluid (thermal boundary) is small; i.e., $\left(\frac{\delta_{th}}{L}\right)^2 \ll 1$. During this initial transient, the effective thermal contact at $x = 0$ links two nonideal systems (considered to be semi-infinite on this timescale) at two different temperatures. It is well known that the interface temperature $T_{int}(x = 0)$ is independent of time, and can be written as

$$T_{int}(x=0) = \frac{Z T_{W_\infty} + Z_W T_{F_i}}{Z + Z_W}. \tag{8.3}$$

In Eq. (8.3), $Z = \frac{\sqrt{D_T}}{\lambda} = \frac{1}{\sqrt{\lambda \rho_c c_P}}$ and $Z_W = \frac{\sqrt{D_{T_W}}}{\lambda_W} = \frac{1}{\sqrt{\lambda_W \rho_W c_{P_W}}}$ are the thermal impedances of the fluid and the wall, respectively. Equation (8.3) results from the conservation of the heat flux per unit area at the boundary j_{WF} $(x = 0)$, which gives the resulting heat flux per area unit at the interface,

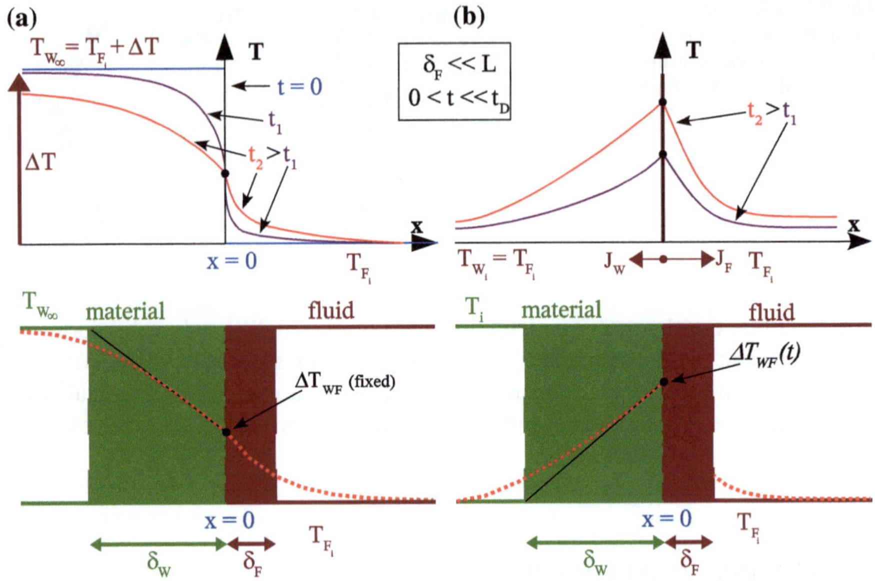

Fig. 8.3 Initial temperature profiles close to the thermal contact between two materials that exchange heat by an isobaric conduction process. **a** A constant temperature $T_{W_\infty} = T_{F_i} + \Delta T$ is imposed. **b** A constant heat flux j is imposed at the interface $x = 0$

$$j_{WF}(x=0) = \frac{1}{Z + Z_W} \frac{T_{W_\infty} - T_{F_i}}{\sqrt{\pi t}}. \tag{8.4}$$

From Eqs. (8.3) and (8.4), we note that a perfect material (i.e., a material with infinite conductivity or zero thermal impedance) imposes its temperature on any imperfect material (finite conductivity), which, in turn, drives the heating power.

A similar description holds for the initial conduction where a constant heat flux is delivered at the wall-fluid interface, as shown in Fig. 8.3b. Here, the total heat flux $J = j_F + j_W$ is divided between the fluid and the materials by the constant ratio r_{WF}, defined as follows:

$$r_{WF} = \frac{j_F}{j_W} = \frac{Z_W}{Z}. \tag{8.5}$$

We note from the above that the heat flux is captured by the perfect material with zero thermal impedance.

The configuration of Fig. 8.3a with $Z_W = 0$ corresponds to the theoretical wall boundary condition at $x = 0$ in the 1-D model of Chap. 4 when an ideal temperature step ΔT is applied at the wall. The configuration of Fig. 8.3b corresponds to the 1D model of Chap. 5 when a constant heat flux j is directed into the fluid (the wall is considered either to have zero heat capacity or to be infinitely thin).

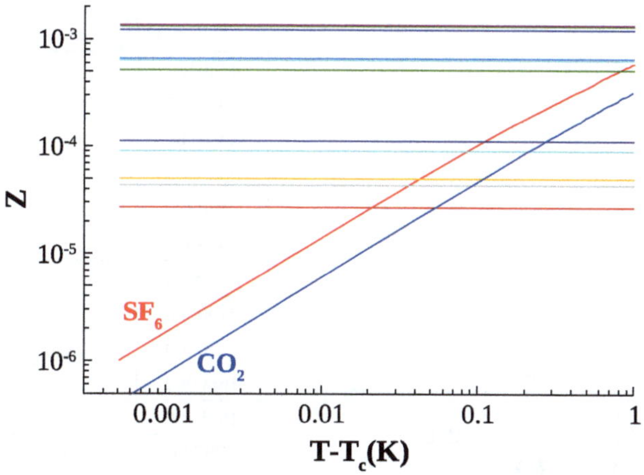

Z	
0.146119	air
0.00135147	PtFe
0.00132972	Vespel
0.00123064	Delrin
0.000662702	suprasil
0.000632563	eau
0.000515574	Zircone
0.000112867	StainlessSteel
0.0000908436	Saphir
0.0000498762	CuBe
0.0000431701	aluminium
0.0000272984	of hc

Fig. 8.4 Singular behaviors of the thermal impedances Z of SF_6 and CO_2 with respect to $T - T_c$, as compared with the thermal impedances of cell wall materials (log–log scale). The Z values (in SI units) of various materials are given in the table on the *right*

The next section compares the singular thermal behaviors of CO_2 and SF_6 as a function of $T - T_c$ with the (constant-like) thermal properties of selected wall materials.

8.3.2 Relative Thermal Properties of Materials and Fluids

A nonexhaustive Z-table for the materials involved in the design of the test cells is provided on the *right* of Fig. 8.4 also gives, for the sake of comparison, the (singular) thermal impedances of SF_6 and CO_2 at various $T - T_c$ values. For $T - T_c \gtrsim 3 - 5$ K, the value of Z for both fluids is larger than any Z_W value except for those of some highly isolating materials (e.g., technical organic polycarbonate PTFE). In contrast, close to T_c ($T - T_c \lesssim 20 - 50$ mK), the critical fluids become more conductive than any of the selected wall materials. In short, depending on the temperature, a critical fluid can take Z values that are either on the order of the most isolating materials or less than the most conductive materials.

The low value of Z close to T_c can be attributed to either the high heat capacity (heat accumulation) or the low heat diffusivity.

The behavior of the volumic heat capacity ρc_p is illustrated in Fig. 8.5 and demonstrates the heat-accumulative nature of the critical fluids in the temperature range of interest. The behavior of the thermal conductivity Λ is illustrated in Fig. 8.6. The thermal conductivity can become larger than those of some materials: the hypothesis of infinite wall conductivity can become problematic.

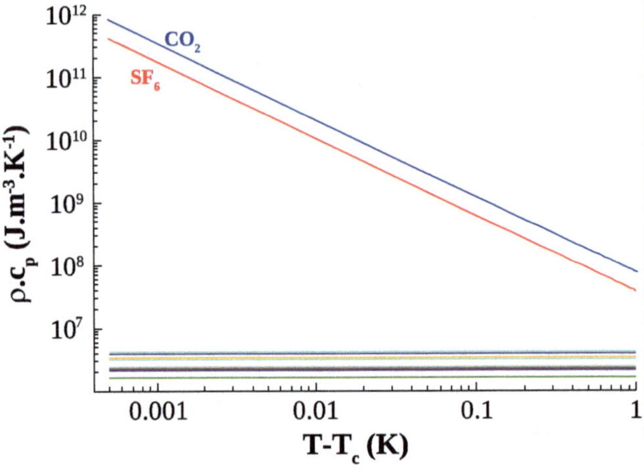

ρC$_p$	
4186182	eau
3925000	StainlessSteel
3458560	ofhc
3435800	CuBe
3188800	Saphir
2439000	aluminium
2280000	Zircone
2190000	PtFe
2130000	Delrin
1650000	suprasil
1615900	Vespel
1787.66	air

Fig. 8.5 Singular behaviors of the volumetric heat capacities ρc_p (J m^{-3} K^{-1}) of SF_6 and CO_2 as a function of $T - T_c$ (log–log scale), as compared to the volumetric heat capacities of selected wall materials. The corresponding (temperature-independent) values are given in the table on the *right*

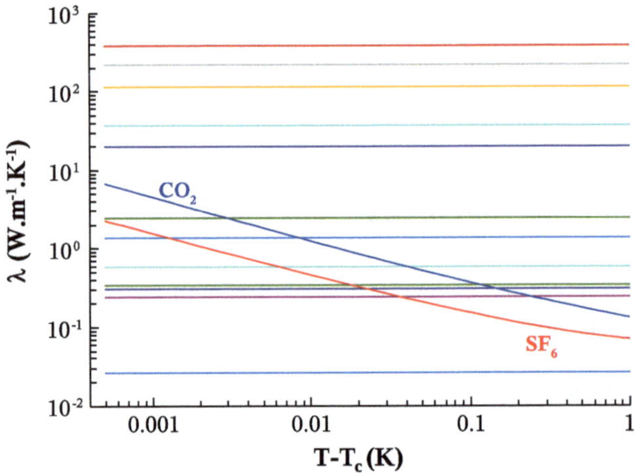

λ	
388	ofhc
220	aluminium
117	CuBe
38	Saphir
20	StainlessSteel
2.65	Zircone
1.38	suprasil
0.597	eau
0.35	Vespel
0.31	Delrin
0.25	PtFe
0.0262	air

Fig. 8.6 Singular behaviors of the thermal conductivities Λ (W m^{-1} K^{-1}) of SF_6 and CO_2 as a function of $T - T_c$ (log–log scale), as compared to the thermal conductivities of selected wall materials. The corresponding (temperature-independent) values are given in the table on the *right*

Recall that for perfectly conductive cell walls, the piston-effect timescale is expressed by [see Eq. (4.11)]

$$t_{PE} \simeq t_{eq} \simeq \frac{t_D}{\gamma_{0c}^2} \sim (T - T_c)^{-1.7} , \tag{8.6}$$

which goes to zero as the critical point is approached. Very close to T_c (typically $T - T_c \lesssim 10\,\text{mK}$), the critical fluids become more conductive than any of the selected wall materials, and the fluid impedance drops below that of any material, as illustrated in Fig. 8.4. A crossover then occurs. The equilibration regime becomes dominated by the wall conductivity. This is the so-called bottleneck regime ($\frac{Z_F(T-T_c)}{Z_{W,i}} \ll 1$), where the dynamics at short times are limited by the transient conduction process in the cell wall. As for the ideal regime dominated by infinite conductivity, a clear understanding of the conductivity bottleneck regime comes from extending the above derivation of the timescale t_{eq} of Eq. (8.6). Therefore, the new characteristic timescale must account for the thermal properties of the wall, as discussed in the next section.

8.3.3 Finite Conductivity Cell: The Bottleneck Regime

We investigate the piston-effect dynamics in the presence of cell walls with finite thermal impedance. As the critical point is approached, the thermal impedance of the fluid drops far below that of the cell walls, and there is a crossover beyond which the bulk fluid equilibration regime is dominated by the fluid conductivity ($\frac{Z}{Z_i} \ll 1$). The dynamics will be limited by the bottleneck conduction process in the cell wall.

We consider the 1-D configuration of Fig. 8.3a: a semi-infinite cell material ($x < 0$, subscript W, with a finite thermal impedance Z_W, at temperature $T_{W\infty}$) is in thermal contact with the sample fluid ($0 \lesssim x \ll L$), its thermal impedance Z is a function of ($T_{F_i} - T_c$), its initial temperature is T_{F_i}, and its boundary at $x = L$ is adiabatic. Very close to T_c (at $T_{F_i} - T_c \lesssim 100\,\text{mK}$), $Z < Z_W$; see Fig. 8.5. In this temperature range where $\frac{Z}{Z_W} < 1$, the fluid temperature and the heat current at $x = 0$ [expressed in Eqs. (8.3) and (8.4)] can be written as

$$T_{int}(x=0) \approx T_{F_i}\left[1 + \frac{Z}{Z_W}\left(\frac{T_{W\infty} - T_{F_i}}{T_{i_F}}\right)\right] \tag{8.7}$$

and

$$\dot{j}_{WF}(x=0) \approx \left(1 - \frac{Z}{Z_W}\right)\left[\frac{1}{Z_W}\left(\frac{T_{W\infty} - T_{F_i}}{\sqrt{\pi t}}\right)\right], \tag{8.8}$$

respectively. Equations (8.7) and (8.8), as well as Fig. 8.5, clearly show that the thermal properties of the wall material play a significant role in the temperature change at the wall. Its finite thermal impedance governs the exchange of heat flux with the fluid very close to T_c.

We follow a reasoning similar to Ferrell and Hao's [3] in order to obtain the new thermal equilibration time $t_{c,Bott}$ in the conductivity bottleneck regime. During $t_{c,Bott}$, the thermal penetration length in the fluid is $\delta_{th} = \sqrt{D_T t_{c,Bott}}$. Provided that the conditions $t_D \gg t_{c,Bott} \geq t_{PE}$ are satisfied during this initial transient, the interface temperature remains close to the bulk temperature due to the piston effect. However, the heat flux injected into the fluid boundary layer is now driven by Z_W^{-1}.

Moreover, Eqs. (8.7) and (8.8) show that the ratio $\frac{Z}{Z_W}$ drives the contribution. From the conservation of heat flux at the boundary, at any (small) time t, the mean temperature increase in the fluid boundary layer, $\frac{E_{BL_F}(t)}{l_{th,F}(t)\rho_c c_p}$, is directly proportional to the mean temperature increase in the wall material boundary layer, $\frac{E_{BL_W}(t)}{\ell_{th,W}(t)\rho_W c_{pW}}$. Here, $\ell_{th,F}(t) \approx \sqrt{D_T\, t} \ll L$ and $\ell_{th,W}(t) \approx \sqrt{D_{T_W}\, t}$ measure the thermal penetration lengths of the temperature inhomogeneities into the fluid and into the wall, respectively. At constant heat flux, the fluid temperature at $x = 0$ increases as $\sqrt{t}$ [see Eqs. (8.7) and (8.8)]. Thus, the bulk fluid temperature also increases as $\sqrt{t}$. For the present bottleneck regime, the temperature equilibration can only be obtained after the time needed for the wall boundary layer temperature $\frac{E_{BL_W}(t)}{\delta_W \rho_W c_{pW}}$ to reach the increase in bulk temperature has elapsed; i.e., when the time-dependent energy $E_{BL_W}(t)$ that has diffused into the wall boundary layer during $t_{c,Bott}$ has reached the energy E_b needed to increase T_{F_i} to T_{int}. We then infer that

$$\frac{E_{BL_W}}{\delta_W \rho_W c_{pW}} \approx \frac{E_b}{\delta_W \rho_W c_{pW}} \approx \frac{E_b}{L\rho_c c_V}.$$

Thus $\delta_W \approx L\frac{\rho_c c_V}{\rho_W c_{pW}}$. The value of $t_{c,Bott}$ can readily be written as $t_{c,Bott} = \frac{\delta_W^2}{D_{T_W}}$, where D_{T_W} is now the thermal diffusivity of the wall material. The new characteristic timescale of the piston effect in the fluid, modified by the conductivity bottleneck of the wall material, is then

$$t_{c,Bott} = \frac{L^2}{D_{T_W}}\left(\frac{\rho_c c_V}{\rho_W c_{pW}}\right)^2. \tag{8.9}$$

As expected, the new piston-effect time also depends on the properties of the wall material and varies with the fluid temperature as $c_V^2 \propto (T - T_c)^{-0.22}$. It is noticeable that the piston-effect time does not goes to zero anymore but instead slowly diverges. A detailed description of the crossover between both regimes is given in the next section. Note that it is important not to confuse this behavior with the acoustic saturation of the piston effect which occurs very close to T_c for a very different reason: the propagation of energy at the speed of sound c_s; see Sect. 4.4.3.

8.4 Bottleneck Crossover Behaviors for a Pancake Cell

A single expression for the characteristic time that characterizes the two limiting regimes of the piston effect can now be defined as

$$t_{PE} = \frac{\left(\ell_{eff}\right)^2}{D_T} \times \frac{1}{\gamma_0\,(\gamma_0 - 1)}, \tag{8.10}$$

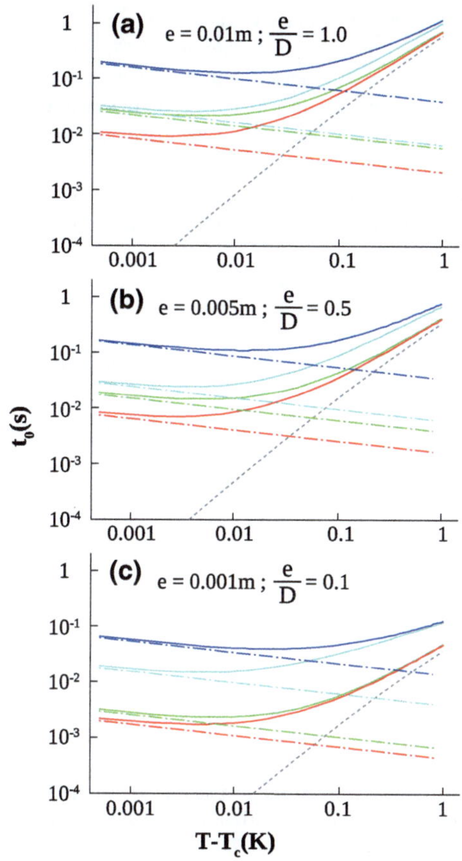

Case	Color	spacer	Windows
1	green	CuBe	Saphir
2	blue	StainlessSteel	suprasil
3	red	ofhc	Saphir
4	cyan	aluminium	suprasil

D = 0.01 m

Fig. 8.7 Calculated (log–log scale) crossovers of the characteristic piston-effect time as a function of $T - T_c$, when using a pancake cell filled with critical SF_6, for different materials (see the table in Fig. 8.4) and different aspect ratios: **a** $r_A = 1$; **b** $r_A = 0.5$; **c** $r_A = 0.1$

where we have introduced the effective characteristic length ℓ_{eff} of the sample cell, defined as follows:

$$\ell_{eff} = \frac{V}{\sum_{i=1}^{N} \left(\frac{Z}{Z+Z_i}\right) A_i}. \tag{8.11}$$

As in the previous definition of L in Eq. (8.2), ℓ_{eff} represents the ratio $\frac{\text{fluid volume}}{\text{effective acting area}}$, where the effective acting area A_{eff} is now a weighted sum over the surface area A_i of the N different wall segments i of thermal impedance Z_i.

The two limiting regimes observed above correspond to Eq. (8.10) with $Z_i \ll Z$ far from T_c and $Z_i \gg Z$ close to T_c, respectively. The temperature at which

Experiment	Color	Spacer 1	Spacer 2	% 2/1	Windows	D (m)	e (m)	L_{PE} (m)
ESEME	green	CuBe	CuBe	0	Saphir	0.0120	0.00670	0.001583
Wilkinson	pink	aluminium	aluminium	0	Saphir	0.0100	0.00200	0.000714
Straub	blue	ofhc	ofhc	0	ofhc	0.0192	0.01920	0.003200
Michels 1	red	suprasil	aluminium	52.5	suprasil	0.0206	0.01800	0.003276
Michels 2	red	suprasil	aluminium	30.0	suprasil	0.0200	0.00250	0.001000
Michels 3	red	suprasil	aluminium	100.0	suprasil	0.0600	0.00215	0.001003
Michels 4	gray	Z = 0.000333				0.0250	0.05120	0.001200

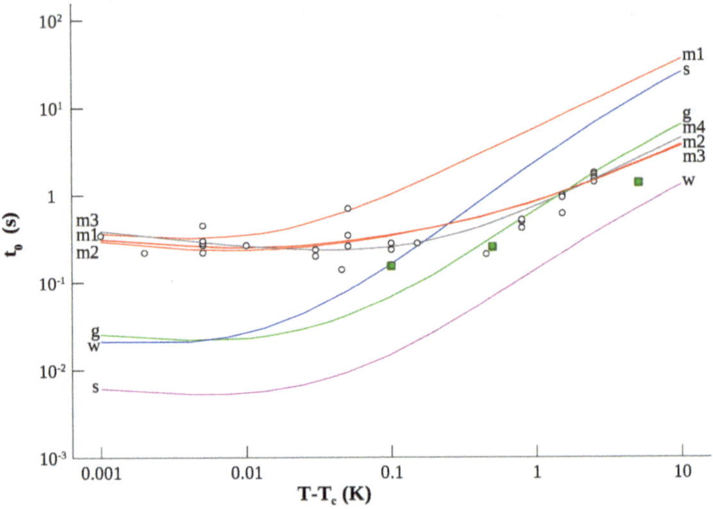

Fig. 8.8 Calculated (log–log scale) crossover behaviors of the characteristic piston-effect time as a function of $T - T_c$, obtained using pancake cell models of various experimental cells filled with critical SF_6 (for labels and colors see the inserted table and Sect. 7.1); *Open circles* From Michels et al 's cell (see Sect. 7.1.3 and Ref. [2]); *Full squares* From Garrabos and al's cell (see Sect. 7.1.1 and Ref. [4])

the crossover takes place is illustrated in Fig. 8.7 for different cell aspect ratios and materials (using SF_6 as the fluid to facilitate comparison with experiments; see Ref. [4]).

In addition, the pancake cell approximation for the test cells presented in Sects. 7.1.1 to 7.1.4 is provided in the table of Fig. 8.8. In spite of the crude approximation we have used for these complex 3-D cells, this figure shows that the technical constraints involved with such cell designs and cell monitoring enable the crossover behavior to be observed in the temperature range 10–100 mK.

References

1. Berg RF (1993) Thermal equilibration near the critical point: effects due to three dimensions and gravity. Phys Rev E 48(3):1799–1805

2. de Bruijn R (1999) Heat transfer in a critical fluid under microgravity conditions: a spacelab experiment. PhD thesis, University of Amsterdam, Amsterdam, p 197
3. Ferrell RA, Hao H (1993) Adiabatic temperature changes in a one-component fluid near the liquid–vapor critical point. Phys A 197(1–2):23–46
4. Garrabos Y, Dejoan A, Lecoutre-Chabot C, Beysens D, Nikolayev VS, Wunenburger R (2001) Piston effect in a supercritical fluid sample cell: a phenomenological approach of the mechanisms. J Phys IV 11(Pr6):23–34
5. Wilkinson RA (1998) Density relaxation of liquid–vapor critical fluids in earth's gravity. Int J Thermophys 19:1175–1183
6. Wilkinson RA, Zimmerli GA, Hao H, Moldover MR, Berg RF, Johnson WL, Ferrell RA, Gammon RW (1998) Equilibration near the liquid–vapor critical point in microgravity. Phys Rev E 57(1):436–448

Chapter 9
Heat Transport by the Piston Effect: Experiments

Abstract Heat transfer experiments are presented where heat pulses are produced by a point-like thermistor at the center of a thermostated cell. This configuration allows the theoretical analyses of the piston effect mechanism to be tested. It is observed, first, a hot boundary layer, developing at the heat source, which shows large coupled density-temperature inhomogeneities. This part relaxes by a diffusive process, whose density and temperature relaxations are slowed down close to the critical point. During heating, the dynamics of the bulk fluid part, which remains uniform in temperature and density, is accelerated near the critical point and governed by the characteristic time of the piston effect. At the thermostated walls, a slightly cooler boundary layer forms, simultaneously. It cools down the bulk by also a piston effect mechanism. The final sample cell relaxation to the temperature and density equilibration is governed by the time of the thermal diffusion, which corresponds to the slowest mechanism. Comparison with a one-dimensional model shown good agreement with experimental results when the characteristic length of the three-dimensional sample cell is obtained from a pancake cell model. A brief illustration of the situation in the presence of gravity is also given.

9.1 The 1-D Model

In this section we will summarize the results of the 1-D model (see Chap. 5) of a CO_2 sample that is assumed to behave as a van der Waals gas, with *ad hoc* modifications to account for the real equation of state (see Sect. 5.2.5).

The space variable x covers the domain $x = \{0, L\}$, where L is the fluid thickness. The sample is initially at equilibrium temperature T_i and pressure p_i on the critical isochore (density $\rho = \rho_c$) in the vicinity of the critical point ($T_i \approx T_c$). At time $t = 0$, a heating power $\mathcal{P}(t)$ is sent from the cell wall at $x = 0$ for the duration of the total heating period Δt_H. The temperature regulation of the cell wall at $x = L$ corresponds to the condition $T(x = L, t) = T_i$. The fluid is described by a linearized equation of

B. Zappoli et al., *Heat Transfers and Related Effects in Supercritical Fluids*, Fluid Mechanics and Its Applications 108, DOI: 10.1007/978-94-017-9187-8_9,

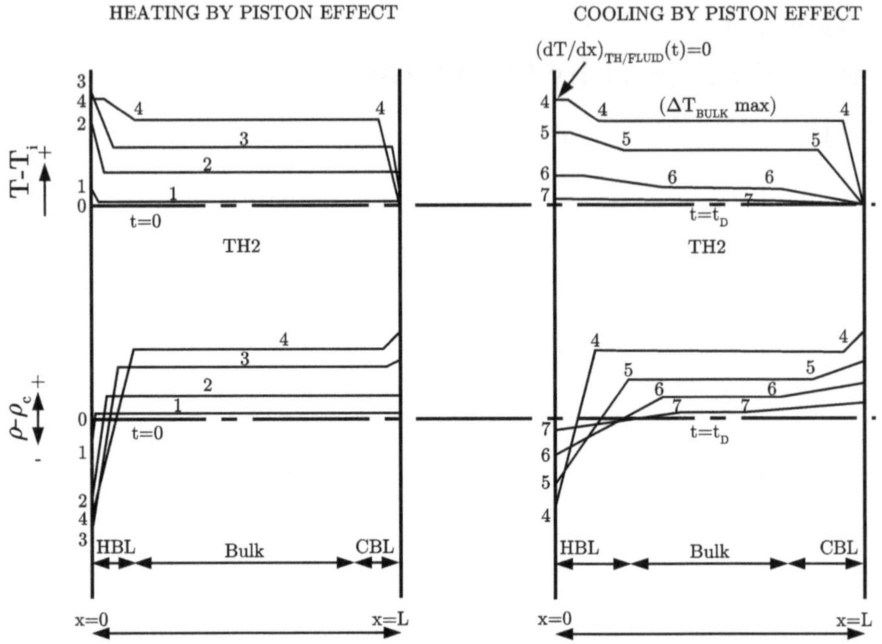

Fig. 9.1 Schematic behavior of the relaxation in (*upper*) temperature ($T - T_i$) and (*lower*) density ($\rho - \rho_c$) in a 1-D cell model after a heat pulse is released at the $x = 0$ wall. Time is referenced to the beginning of the pulse (*zero horizontale line*). The heat pulse duration (ΔT_H) corresponds to the period of time during which the heat flux at the wall differs from zero, and ΔT_{bulk} reaches a maximum at $t = \Delta T_H$

state, and its behavior obeys the 1-D compressible unsteady Navier-Stokes equations for Newtonian fluids. The corresponding equations were solved analytically by means of the singular perturbation method given in Chap. 5. The development of the heating and cooling boundary layers, as explained in Sect. 5.2, drives the bulk temperature and density changes shown in Fig. 5.2. A typical experimental sequence is depicted in Fig. 9.1.

9.1.1 Thermalization by a Heat Pulse

When a heat pulse $\mathcal{P}(t)$ is released at the wall at $x = 0$, the temperature response $T_b(t)$ in the bulk is found as the convolution product ($\otimes$) of $\mathcal{P}(t)$ with a transfer function $F(t)$ (see Sect. 5.2.3):

$$T_b(t) - T_i = \left(\frac{\mathcal{P}(t)}{mc_V}\right) \otimes F(t), \tag{9.1}$$

Table 9.1 Thermal parameters for CO_2 calculated at various experimental temperatures

$T_i - T_c$ (K)	$\left(\frac{\partial P}{\partial T}\right)_\rho$ (MPa K^{-1})	$\left(\frac{\partial P}{\partial \rho}\right)_T$ (N m kg^{-1})	c_V (J kg^{-1} K^{-1})	γ_{0c}	$10^9 D_T$ (m^2 s^{-1})	t_D (s)	t_{PE} (s)	t_c^b (s)
16.8	0.17	7340.9	1081	6.6	18.3	123	3.92	3.4
15.9	0.17	6882.4	1090	6.9	17.5	129	3.71	3.3
0.8	0.17	194.2	1740	120.7	1.62	1389	0.087	0.095
0.6	0.17	136.2	1812	164.7	1.33	1692	0.0631	0.0627
0.4	0.17	83.6	1923	252.3	0.995	2261	0.0358	0.0357
0.2	0.17	36.0	2140	525.5	0.593	3794	0.0138	0.0138

The temperature dependence of pressure along the critical isochore is considered to be linear (see *column 2*). Critical parameters of CO_2: $T_c = 304.13$ K, $p_c = 7.376$ MPa, $\rho_c = 467.8$ kg m^{-3} (see Table 1.2). Characteristic times in *columns*7–9 are calculated for the value $L = 1.5 \times 10^{-3}$ m, as estimated from Eq. (9.11)

where m is the total mass of the fluid. $F(t)$ can be written as:

$$F(t) = \exp\left(\frac{t}{t_{PE}}\right) \operatorname{erf} c \left(\frac{t}{t_{PE}}\right)^{\frac{1}{2}}, \tag{9.2}$$

where the bulk temperature dynamics appear to be dictated by the characteristic time t_{PE} of the piston effect (see Eq. 5.56):

$$t_{PE} = \frac{t_D}{\gamma_{0c}\left(\gamma_{0c} - 1\right)}. \tag{9.3}$$

The experiments were performed in the ALICE facility (see Chap. 23.1) during the July 1992 Cassiopeia mission onboard the Russian MIR station, using heating pulses with particular durations produced by a point-like heat source (a bead thermistor) located at the center of a disk-shaped, thermostated CO_2 sample cell. More details of the experiments are given in Ref. [5]. In the experimental temperature range, the diffusion time $t_D = \frac{L^2}{D_T}$ and the piston-effect time of Eq. (9.3) take the values given in Table 9.1, with $L = 1.5$ mm, as estimated from Eq. (9.11) (see Sect. 9.1.3 below). The values in Table 9.1 show that the piston-effect time $t_{PE} < 0.1$ s represents a large reduction in the temperature equilibration for $T_i - T_c < 1$ K, and is on the order of 3.7 s at $T_i - T_c \simeq 16$ K.

An interesting feature of Eq. (9.1) is the fact that for values of t that are much larger than Δt_H, $T_b - T_i$ tends to behave like $F(t)$ regardless of the time history of $\mathcal{P}(t)$. This is illustrated in Fig. 9.2b, where $T_b - T_i$ is calculated for three different heat pulse shapes $\mathcal{P}(t)$ with the same total energy Q and different values of the heating period Δt_H (see Fig. 5.4a). The first shape is a Dirac function $\mathcal{P}(t) = Q\delta(t)$, which corresponds to $\Delta t_H \ll t_{PE}$. It is used below in Sect. 9.1.4 to compare with the Ferrell and Hao model. The two other heat pulses in Fig. 5.4a have finite durations that are comparable with the experimental values $\Delta t_H \simeq 2-3$ s (with $Q \simeq 8$ mJ; see Sect. 9.2.5). These latter two heat pulses correspond to $\Delta t_H \gtrsim t_{PE}$ in the experimental temperature range. Therefore, Q is the energy effectively

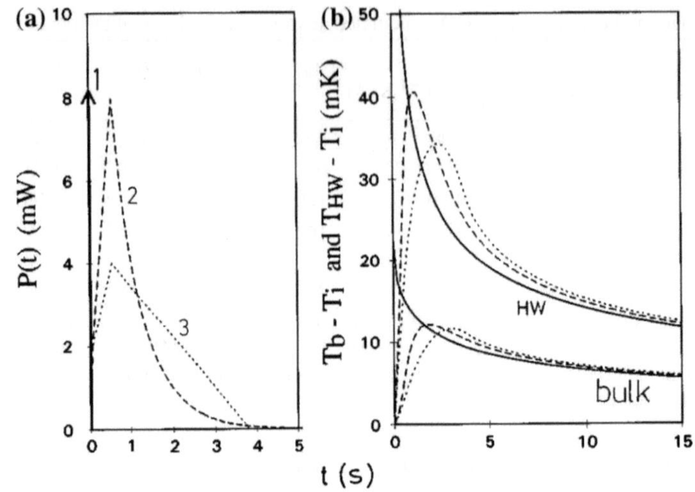

Fig. 9.2 **a** Different heat flux profiles at the wall with the same total energy. **b** Temperature evolutions for the different heat fluxes. This shows that the temperature relaxation profile for long times is nearly independent of the heating time profile

transmitted to the fluid at the end of the heat pulse. We then observe in Fig. 5.4b that the curves representing the bulk temperature modifications tend to merge after 10–15 s ($\simeq 5\,\Delta t_H$), whatever the exact shape of the heating pulse during the initial period. In particular, we note that the bulk temperature relaxation is not greatly affected by the effective time at which $\mathcal{P}(t)$ reaches its maximum ($\simeq 0.5$ s; on the order of the electrical pulse duration: see Sect. 9.2.5). This result will be a great help when adapting the model to the experimental conditions.

Any thermodynamic variable in the boundary layers can also be written in the form of the convolution product with the corresponding transfer function. In particular, the temperature at the hot wall $T_{HW}(x = 0, t)$ can be expressed as

$$T_{HW} = T_b(t) + \left(\frac{\mathcal{P}(t)}{L\rho_c c_P S_{HW}}\right) \otimes H(t), \tag{9.4}$$

where S_{HW} is the hot wall area. The transfer function $H(t)$ is

$$H(t) = \frac{1}{\sqrt{t}}\left(\frac{t_D}{t}\right)^{\frac{1}{2}}, \tag{9.5}$$

where t is only scaled using the diffusive time t_D. This result reflects the diffusive nature of the heat transfer in the hot boundary layer. A similar analysis to the one performed in Chap. 5 for the bulk temperature would show that the wall temperature very quickly becomes independent of the time distribution of the heating power for a given total energy supplied at the boundary.

Another interesting result of this model is the fluid velocity u, which reaches its maximum value u_{HBL} at the edge of the hot boundary layer [see Eq. (4.46)]. Taking into account Eq. (4.46) and the generalization to real fluids in Sect. 5.2.5, u_{HBL} can be written in the following dimensional form:

$$u_{HBL} = \frac{1}{T_c}\left(\frac{\partial T}{\partial p}\right)_\rho \frac{\mathcal{P}(t)}{S_{HW}} \simeq \frac{\mathcal{P}(t)}{7p_c S_{HW}}, \tag{9.6}$$

where we have used the approximation $\left(\frac{\partial T}{\partial p}\right)_\rho \simeq \frac{T_c}{7p_c}$ (see Table 9.1). Heating by the piston effect is thus only effective as long as $\mathcal{P}(t)$ differs from zero; i.e., only during Δt_H. As an important consequence, it is easy to estimate the order of magnitude of the expected maximum fluid velocity; for example, $u_{HBL} \simeq 20\,\mu\mathrm{m\,s}^{-1}$ under the experimental conditions where a total energy $Q \simeq 8$ mJ is supplied to CO_2 ($p_c \simeq 7.3$ MPa) by applying a constant mean power value $\bar{\mathcal{P}} = \langle\mathcal{P}(t)\rangle \simeq 2.7$ mJ for $\Delta t_H \simeq 3$ s through an external surface $S_{HW} \simeq 3\,\mathrm{mm}^2$ of a small spheroidal thermistor. This also supports the idea that the expanding hot boundary layer acts as an effective piston, since the fluid velocity at the frontier with the bulk is proportional to the heat flux (heat power per unit area) supplied to the fluid. This mechanism where the fluid velocity can become large enough to produce jets was addressed in detail in Sect. 5.5.1.

Equation (9.6) is also valid for the cold boundary layer. The corresponding velocity u_{CBL} is proportional to the heat flux through the cold wall (of surface area S_{CW}). However, the cooling caused by the piston effect remains effective in this case as long as $T_b(t) - T_i$ differs from zero; i.e., during the whole relaxation time period, on the order of t_D.

9.1.2 Energy Yield and Efficiency

The bulk temperature behavior is a result of the competition between heating and cooling at the walls. It is possible to calculate the energy balance resulting only from the heating piston effect, where the equations are linearized. This corresponds to considering the cold wall to be adiabatic during Δt_H.

At any time during the duration of heating, the energy that is supplied to the whole fluid $E(t)$ is split into the energy in the hot boundary layer and the energy $\Delta E(t)$ in the bulk. The ratio $\zeta = \frac{\Delta E(t)}{E(t)}$ is the energy efficiency of the piston effect, and this can be calculated from the model (see Chap. 5) provided the temperature and density profiles in the cell are known at all times. The resulting energy balance between the bulk and the hot boundary layer leads to the following result:

$$\zeta = \frac{\Delta E(t)}{E(t)} = \frac{p_i}{T_i}\left(\frac{\partial T}{\partial p}\right)_\rho, \tag{9.7}$$

where the yield ζ is independent of time. It varies only slightly with $(T_i - T_c)$. For a van der Waals fluid,

$$\zeta = \frac{1}{4}, \tag{9.8}$$

while, for real fluids (see Table 9.1),

$$\zeta \sim \frac{1}{7}. \tag{9.9}$$

This energy efficiency is low, which means that most of the energy remains stored in the boundary layer.

Simultaneously, the cold boundary layer develops slowly and induces cooling with the same energy efficiency $\zeta = \frac{\Delta E'(t)}{E'(t)}$ (the prime denotes the energies involved in the cooling process). Therefore, the total energy $\Delta E_b(t)$ in the bulk is the balance between the energy transferred by the piston effect from the hot wall, in other words $\Delta E(t)$, and the energy lost through the piston effect at the cold wall, in other words $\Delta E'(t)$. The energy effectively transmitted to the bulk $\Delta E_b(t)$ is therefore

$$\Delta E_b(t) = \Delta E(t) - \Delta E'(t) = \zeta \left[E(t) - E'(t)\right]. \tag{9.10}$$

The remaining energy is stored in the boundary layers.

Note that, at any time and according to Eq. (9.1), the bulk temperature is

$$\Delta T_b(t) = \frac{E(t) - E'(t)}{mc_V},$$

where $E(t) - E'(t)$ is the total amount of energy stored in the fluid at time t. $\Delta T_b(t)$ is precisely the temperature that would have been reached if all this energy had diffused into the bulk. As far as temperature equilibration is concerned, the piston effect is thus very efficient. This remark seems to conflict with the fact that the energy efficiency is low, with only $\zeta = \frac{1}{7}$ of the total energy sent into the bulk. This can be explained by the fact that the internal energy of a near-critical fluid depends significantly on density (see, for example, Eq. (4.6) for the van der Waals fluid case), and here the bulk density is changed due to the small flux of matter at the HBL border [see Eq. (9.6)].

9.1.3 Adaptation of the Length Scales to a 3-D Experiment

When $\mathcal{P}(t)$ is known, the estimation of the length scale L is critical to the correct adaptation of the 1-D model to 3-D experiments. Using Eq. (8.1) of Sect. 8.2 and the cell dimensions given below (Sect. 9.2.1), we obtain

$$L = \frac{\text{cell volume}}{\text{cold wall area}} = 1.5\,\text{mm}. \tag{9.11}$$

In the following we will choose to adopt this value. Confidence in this choice is enhanced by the good agreement between the model and the experimental data.

Note that this definition of the characteristic length leads to a modification of the model when the hot and cold wall areas are different (see Sect. 8.2.3), as is the case in this experiment. The amplitudes of the local temperature and density in the hot boundary layer must be magnified by the ratio $\frac{S'_{CW}}{S'_{HW}} \simeq 80$ from the results of the linear 1-D model where $\frac{S_{CW}}{S_{HW}} = 1$. This peculiar result is detailed below in our analysis of the temperature relaxation of the hot boundary layer.

9.1.4 Comparison with the Ferrell and Hao Model

Ferrell and Hao [3] have considered the cooling relaxation of an adiabatic fluid after it has been heated by a heat pulse approximated by a Dirac function, $\mathcal{P}(t) = Q\,\delta(t)$. They found the following relation for the bulk temperature evolution:

$$T_b(t) - T_i = \left(\frac{Q}{mc_V}\right) F_{FH}(t), \tag{9.12}$$

where

$$F_{FH}(t) = \exp\left[\frac{t}{t_0}\right] \operatorname{erfc}\left[\frac{t}{t_0}\right]^{\frac{1}{2}}. \tag{9.13}$$

Equation (9.13) can be compared to Eq. (9.2), where $t_0 = t_{PE}\left(\frac{\gamma_{0c}}{\gamma_{0c}-1}\right) \simeq t_{PE}$. The Ferrell and Hao model corresponds to the impulse response of our model when $\Delta t_H \ll t_{PE}$. We note that the approach used in [3] does not give the temperature behavior at the heating wall, since the initial value of this temperature (at $t = 0$) should theoretically be infinite for a Dirac heat pulse ($\Delta t_H = 0$). From Fig. 9.2, these models become strictly equivalent at long times when approaching the critical point.

9.2 Experimental Setup for ALICE

Some aspects of the experimental setup for ALICE have been given in Sect. 23.1.1; see also [1]. Here, we report additional details concerning the geometrical arrangement and the characteristic thermal responses of the different containers and materials. In addition, we will adequately describe the evolution of the energy supplied to the fluid, taking into account the losses.

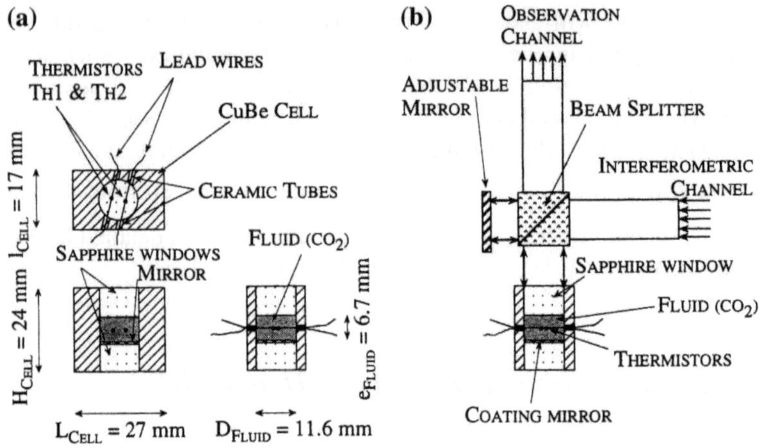

Fig. 9.3 Experimental setup for the study of thermal relaxation in supercritical fluids: **a** interferometer cell; **b** schematic design of the Twyman–Green interferometer

9.2.1 The Interferometer Cell

A picture of the experimental cell is shown in Fig. 23.3b and its detailed design is given in Sect. 7.1.1. The cell is filled with CO_2 (from Air Liquide; purity is better than 99.998 %) at a critical density of 467.9 kg m^{-3} to an accuracy of better than 0.1 %. The cell body is a parallelepipedic block of external dimensions $17 \times 24 \times 27$ mm^3 made of copper–beryllium alloy. The fluid is enclosed in a cylinder 11.6 mm in diameter and 6.7 mm thick (hereafter denoted e), which is closed at each end with sapphire windows.

Recall that the cell is located close to the center of symmetry of the SCU (sample cell unit), which is made of electrolytic copper (see for detail Sect. 23.1.1 and Fig. 23.2). The SCU is a cylinder (60 mm in external diameter, 115 mm long) that can be cooled and heated by means of Peltier elements located at its top and bottom. The SCU is the central part of a high-precision thermostat with a temperature stability that is better than 50 μK and temperature gradients of less than 100 μK cm^{-1}. The temperature is controlled and measured by two stabilized YSI 44900 thermistors (Yellow Springs Instrument Co.). One thermistor, denoted YSI_{mes} here, is located close to the fluid cell and is used to determine the cell temperature. The other thermistor is positioned close to the bottom Peltier elements and is used for temperature control.

9.2.2 Bulk Density Measurements

The cell is a part of a Twyman–Green interferometer (see Fig. 9.3) located in the SCU. The interference pattern is detected by a CCD video camera working at 25 fps.

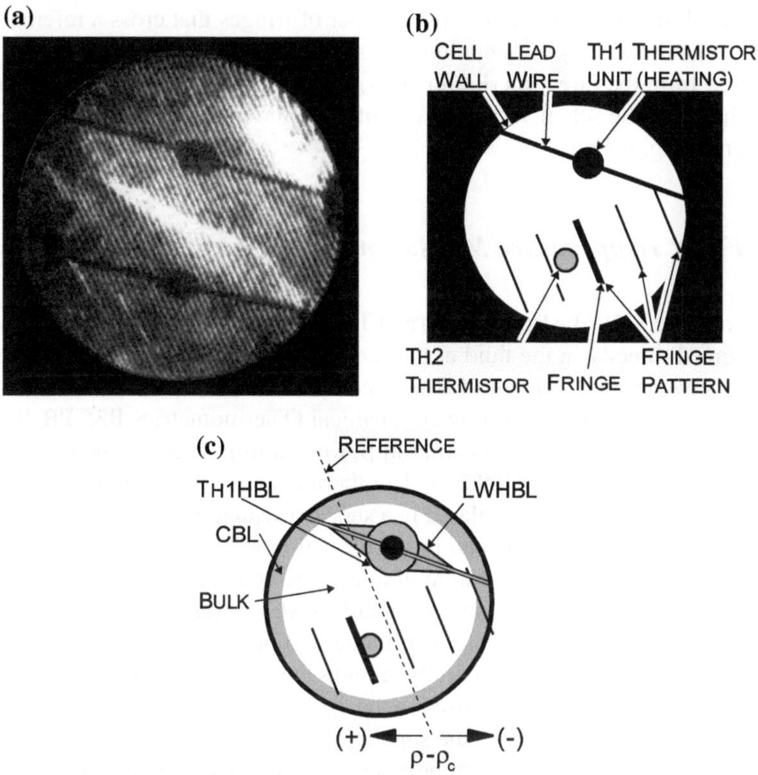

Fig. 9.4 **a** View of the interferometer cell at equilibrium temperature $T_i = T_c + 16.8\,\mathrm{K}$. The fringes are parallel lines; the illumination is not uniform because of mirror defects. **b** Schematic representation of (**a**) with an arbitrary reference line parallel to the fringes. **c** Same as (**b**) during the heat pulse supplied by the TH1 thermistor (see Figs. 9.8 and 9.12). In the selected areas (labeled BULK, TH1HBL, LWHBL and CBL), we observe a shift or distortion of the fringe pattern. These modifications of the pattern from the direction of the reference line (to the left or right) are related to the signs (+ or −, respectively) of the relative changes in density from the initial critical density

The interference pattern provides quantitative measurements of the fluid refractive index and thus of its density. At each equilibrium temperature, the pattern in the sample cell exhibits about 40 parallel straight fringes, corresponding to a homogeneous density within the pattern. A reference pattern is given in Fig. 9.4a (for $T_i = T_c + 16.8\,\mathrm{K}$) and this is schematized in Fig. 9.4b for practical analysis, as illustrated in Fig. 9.4c.

Using $n_C = 1.108$ for the refractive index of CO_2 (see Table 1.6), $\rho_C = 467.9\,\mathrm{kg\,m^{-3}}$, $e = 6.7\,\mathrm{mm}$ and $\lambda = 633\,\mathrm{nm}$, Eq. (23.5) (see Sect. 23.1.3) gives

$$\frac{\Delta\rho}{\rho} \approxeq \frac{\lambda}{2\,(n-1)\,e}\delta f \approx 0.044\,\%\ \text{per fringe.} \qquad (9.14)$$

The fringe shift is then counted as the number of fringes that cross a reference line parallel to the initial fringe orientation. The sign (+ or −) of the density variation can be inferred from the direction (left or right) that the pattern shifts in during heating [1], as shown in the schematic representations of the interferometric images of the cell given in Fig. 9.4b, c.

9.2.3 Fluid Temperature Measurements

The thermistors TH1 (the heat source) and TH2 (the bulk temperature measurement) are immersed directly in the fluid at the same distance from the two windows. The distance between their centers (4.6 mm) compares well with the thermistor–wall distance (3.5 mm). The thermistors are identical (Thermometrics B35 PB 103 F-A, 100 ms time constant, 35 mW maximum power rating). They are nearly spherical in shape with a diameter of 0.9 mm. The thermistor is sealed onto a thin wire of platinum alloy and then encapsulated in a shock-resistant glass coating.

Temperature measurements can be performed in a 2 K range around $T_c + 16$ K and $T_c + 0.5$ K to within a resolution of a few mK and at a sampling rate of 25 s^{-1} or 0.1 s^{-1}. Temperature data cannot be obtained while TH1 is activated for heating.

Each thermistor in the measurement mode is supplied with a 500 mV regulated voltage. The dissipated power generally remains lower than 4.5 μW, a value obtained when the higher sampling rate is used in the $T_c + 16$ K temperature range. From the technical data sheet provided by the supplier, the corresponding maximum temperature elevation can be estimated to be around 3 mK. This value has been indirectly confirmed by a careful analysis of the temperature measurements, and has therefore been subtracted from the values measured at the higher sampling rate. No temperature correction was needed at the lower sampling rate, where the average dissipated power is on the order of 2×10^{-2} μW.

9.2.4 Thermal Characteristics of the Container

Representing the SCU as an ensemble of spherical containers all with the same volume-to-surface ratio provides the characteristic lengths listed in column 5 of Table 9.2. This analogy is suggested by the near-spherical shape of the internal heating thermistor TH1. The typical thermal response $t_{R,X}$ of each container (labeled X) corresponds to the attainment of 63 % of its final equilibrium temperature. The $t_{R,X}$ values are listed in Table 9.2. Note that these values differ from the 1-D diffusive times $t_{D,X}$ by a factor of π^{-2}. At this point, we will make three remarks:

1. In order to reduce the thermal time constants of the cell, the cell body is made of copper–beryllium alloy and the windows are made of sapphire, which leads to $t_{R,CELL} \approx 1$ s. The condition ($t_{PE} > t_{R,CELL}$) that must be fulfilled to be able to study the dynamics of cooling is achieved for $T - T_c > 1$ K (see Table 9.1).

2. The thermal response of the lead wires affects the dynamics of pulse heating. The leads couple the thermistor inside the fluid, where heat is produced, to the SCU environment, where the temperature is stabilized. In this case, the analogy with a spherical problem is not relevant anymore. The length scale (labeled with an asterisk) reported in column 5 of Table 9.2 corresponds to the distance from the thermistor to the external surface of the cell. The corresponding time constant is $t_{R,LW} \approx 2$ s.
3. The central location of the cell in the SCU ensures thermal equivalence between the Peltier elements when they react to the heat production in the cell. We note that the temperature of the cell, as measured by the YSI_{mes} thermistor, exhibits a time constant of 2 s, which is comparable to the characteristic thermal time of the SCU.

From the above analysis, it is possible to infer a realistic time $t_{YSI} \simeq (2 \pm 1)$ s at which a modification of the fluid temperature at the central location of the fluid sample can be detected at the location of the YSI_{mes} thermistor.

9.2.5 Heat Pulse Characteristics

When TH1 is used as a heater, a stabilized voltage of 10 V is delivered during $\Delta t_{elec} = 480$ ms through a temperature-stabilized resistor of 274 Ω in series with the thermistor. The dissipated power is a function of the TH1 thermistor resistance R_{TH1}, which varies with the local temperature T. The power produced inside TH1 is thus a function of time and the temperature difference from T_c. The typical range is 5–35 mW in the self-heating mode of the thermistor unit.

Since experimental information cannot be obtained on Earth, where the thermal exchange is modified by convection during the heat pulse (see Sect. 9.3.1), it is necessary to develop a model for the TH1 heating behavior. A schematic representation of the energy partition is given in Fig. 9.5. It appears that only 50 % of the total energy that is produced by the Joule effect is transferred to the hot boundary layer around the thermistor bead. Some of the remaining energy corresponds to the self-heating mode of the thermistor bead, while another fraction of that energy is lost through conduction (mainly through the lead wires) during the time $t_{R,LW}$. Thus, it appears to be more realistic to say that TH1 can heat CO_2 during an effective heat pulse of duration $\Delta t_H \approx \Delta t_{elec} + t_{R,LW} \approx 2.5$ s.

A numerical simulation of the conduction problem can be performed by assuming a finite spherical geometry for the thermistor bead. The central thermistor part of TH1 is assimilated into a heat source sphere (S1) of diameter 0.2 mm, to which electrical power is delivered at constant voltage U.

The electric power is calculated from the characteristic law $R_{TH1}(T)$ of the B35 thermistor bead. The glass coating that forms the thermistor bead is accounted for by considering a second sphere (S2) of outer diameter 0.9 mm enclosing S1. Both S1 and S2 are considered to be made of identical glass with the same thermal properties as commercial zirconia (Table 9.2). This "heater" model is similar to the real thermistor,

Table 9.2 Thermal parameters for the various containers (X) involved in temperature relaxation experiments

X (material)	Geometry	Dimensions (mm)	Volume $\left(\mathrm{mm}^3\right)$	Equivalent sphere diameter (mm)	Effective mass (g)	$t_{R,X} \sim \left(\frac{\Phi}{2\pi}\right)^2 D_T - 1$ (s)
SCU (1)	Cylinder	60×115	325×10^{-6}	72.0	2500	1.2
Cell (2)	Parallelepipedic	$27 \times 17 \times 24$	11×10^{-6}	21.8	74	0.35
Windows (5) ($\times 2$)	Cylinder	11.6×8.5	1.8×10^{-6}	10.3	7.1	0.22
Thermistors (3) ($\times 2$)	Sphere	0.9	0.38×10^{-9}	0.9	2×10^{-3}	0.079
Lead wires (4) ($\times 2$)	Cylinder	0.1×5.25	82.5×10^{-12}	11 ($\sim$2H)(*)	1.8×10^{-3}	$1 - 2$ (*)

Material	ρ $\left(\mathrm{kg\,m^{-3}}\right)$	c_p $\left(\mathrm{J\,kg^{-1}\,K^{-1}}\right)$	Λ $\left(\mathrm{W\,m^{-1}\,K^{-1}}\right)$	$10^4\,D_T$ $\left(\mathrm{m^2\,s^{-1}}\right)$
Electrolytic copper	8950	385	388	1.1
Cu–Be Alloy	8200	419	117	0.34
Zirconia	5700	400	2.65	0.0116
Platinum alloy	21400	134	68.8	0.24
Sapphire	3996	800	38	0.12

The time constants $t_{R,X}$ were evaluated using a spherical geometry model for each container, except for the lead wires (*) (see text). The lower table gives the thermal properties of the materials

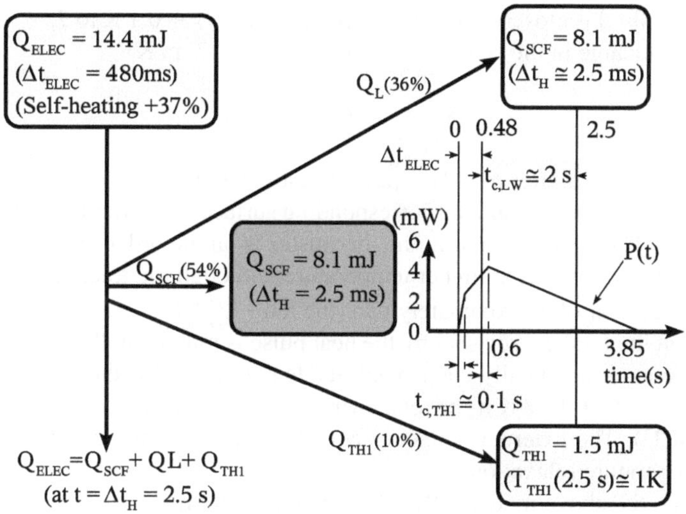

Fig. 9.5 Partition of the energy (in the absence of convection) corresponding to supplying 480 ms of electricity to the heating thermistor TH1. The fluid temperature is $T = T_c + 16.8$ K. $P(t)$ is the time profile of the heat flux delivered to the critical fluid from the TH1 external surface (see text)

as shown by the calculated thermal time constants: 63 % of the maximum temperature rise occurs in 79 ms (Table 9.2), a value close to the 100 ms value given by the manufacturer. Moreover, the temperature increase due to the power dissipated on the glass material during the higher rate temperature measurements is 3 mK, a value which agrees with the observations. In the simulation, S1 and S2 are located at the center of an infinite medium with spherical symmetry that is at rest. The medium's thermal properties correspond to those of CO_2 at a homogeneous density $\rho = \rho_c$. The effect of the leads is not considered in this model. Time discretisation for numerical calculations is 25 ms.

It appears that, in this simulation, the distance to the critical point is a key parameter for describing the thermistor behavior. When the fluid is initially at $T_c + 16.8$ K, the temperature change at the center of S1 follows (t in s, T in K)

$$\Delta T(t) \approx T_i + 18.5t^{\frac{1}{2}}.$$

The thermistor resistance changes accordingly. After $t = 480$ ms, the total energy supplied to the core of the thermistor can be estimated as $Q_{elec} \approx 14.4$ mJ (of which 37 % goes to the self-heating of the thermistor during the same period), and the external surface temperature change on S2 is close to 5 K. Typically, the energy needed to obtain this temperature increase of 5 K is 6.3 mJ, which represents 46 % of Q_{elec}. This result shows that just over 50 % of the total energy is dissipated into the hot boundary layer surrounding the thermistor.

When the fluid is closer to the critical point (range $T_c + 0.1$ K to $T_c + 1.0$ K), the temperature change in the center of S1 follows (t in s, T in K)

$$\Delta T(t) \approx T_i + 12.0t^{\frac{1}{2}}.$$

The energy supplied after a 480 ms pulse is estimated to be $Q_{elec} \approx 7.4$ mJ (26 % of it goes to self-heating) and the corresponding surface temperature change in S2 is close to 2 K. The energy stored in the thermistor is on the order of 3 mJ, leading to the same relative ratio as at higher temperature, although the total energy has been divided by a factor of approximately two (the ratio $\frac{R_{TH1}(48\,°\mathrm{C})}{R_{TH1}(32\,°\mathrm{C})} \simeq \frac{1}{2}$).

Note that the total energy sent by the heat pulse is rather small, which makes the SCU temperature practically unchanged (the temperature rise should be lower than 15 μK at $T_i = T_c + 16.8$ K if the total energy of 14.4 mJ is injected instantaneously into the SCU). The external boundary conditions of the fluid are then isothermal during temperature relaxation.

However, the above model suffers from two limitations:

1. It only considers heat transfer into the hot boundary layer, neglecting the energy that is transferred into the bulk by the piston-effect mechanism. However, this latter contribution remains small due to the low yield of the piston-effect mechanism (see Sect. 9.1.2).
2. It ignores the heat losses through the wires. These losses increase during the self-heating of the thermistor bead.

Despite these two limitations, and although a precise evaluation of the coupling between the self-heating mode and heat losses is difficult, an order of magnitude can be estimated for the respective fractions of energy supplied to the fluid and lost by the wires after their typical thermal response time: 2 s. For example, let us consider CO_2 at $T_c + 16.8$ K. We assume a linear temperature decrease from 5 K, the (above-estimated) temperature at the end of the electrical pulse, to 1 K (an arbitrary choice). We also assume that the temperature inhomogeneity along the wire axis is linear with a spatial extent of 1 mm and thus find that 7 mJ is lost, with a corresponding uncertainty of 0.5 mJ. It appears that this amount is comparable to the (above-calculated) energy (6.3 mJ) needed to heat the external surface of the thermobead at 5 K.

Based on the estimated heat pulse duration ($\Delta t_H \approx \Delta t_{elec} + t_{C,LW} \approx 2.5$ s; see above), the total heat energy Q_{elec} (=14.4 mJ) can be separated into two main contributions, the energy for TH1 and the wires, $Q_{TH1} + Q_L$ (=4.8+1.5 mJ) and the energy supplied to the fluid, Q_{SCF} (=8.1 mJ), as illustrated in Fig. 9.5. Q_{TH1} then corresponds to the heat needed to produce the temperature increase (+1 K) of the thermistor bead after 2.5 s, while Q_L is the energy lost by solid conduction through the lead wires during 2 s. Q_{SCF} is the energy that diffuses into the hot boundary layer during this heating period. During the typical time responses ($t_{C,LW}$) of the lead wires and the SCU ($t_{C,SCU}$)—approximately 2 s after the end of the electrical pulse, TH1 still heats the fluid. This is why, when TH1 works in a self-heating mode, the estimated heat pulse duration $\Delta t_H \approx \Delta t_{elec} + t_{C,LW} \approx 2.5$ s is longer than the electric pulse duration.

A more realistic time profile for the power $\mathcal{P}(t)$ as estimated from microgravity experiments (see Sect. 9.4.1.2 and Fig. 9.13) is also reported in Fig. 9.5. These results (i) properly account for the role of the thermistor time constant $t_{C,TH1}$ ($\approx$100 ms) in the predominant initial heating, and (ii) adjust the heating period to the period where the bulk fluid temperature increases.

9.3 Results

9.3.1 Heat Pulse Under Earth's Gravity

For the sake of comparison with the microgravity experiments, the thermal evolution of the heated CO_2 cell under Earth's gravity (1g) (see Figs. 9.6 and 9.7) was studied.

When $t < \Delta t_{elec}$ (see Fig. 9.6a and b), a plume develops within 100 ms after the beginning of the pulse. This value is on the order of the thermistor time constant. The energy produced during that time is as low as 2 mJ. Convection settles in the fluid during the electrical excitation, extends along the lead wires, and prevents any stable thermal boundary layer from forming around the thermistor (see Fig. 9.6b).

When $t > \Delta t_{elec}$, convection stops (see Fig. 9.6c) and the fluid cell relaxes to equilibrium (see Fig. 9.6d). The temperature relaxation at TH1 is reported in Fig. 9.7 for two typical temperature differences from the critical point ($T_c + 16.8$ K and $T_c + 0.2$ K).

The temperature after the electric pulse at T_c+16.8 K (5 K, extrapolated to 480 ms) is approximately twice as large as that at $T_i = T_c + 0.2$ K (2.5 K, extrapolated to 480 ms). This is the same ratio seen for the electric power, as calculated from the resistance values [RTH1($T_i = T_c + 0.2$ K)$\approx$2 RTH1($T_i = T_c + 16.8$ K)]. The temperature relaxation increases as T_c is approached.

Therefore, on Earth, the gravitational flow greatly modifies the heat transfer. In addition to the expected piston effect, the fast relaxation of TH1 is associated with the viscous relaxation of convective flows that also increase the heat flux between the thermistor and the thermostat, lowering the hot boundary layer temperature.

In contrast to TH1, the temperature modification in the bulk is too small to be detected by TH2 and in the SCU by YSI_{mes}. The fringe pattern is only disturbed by the plume, and recovers its equilibrium a few seconds after Δt_{ELEC}.

9.3.2 Interferometric Observations Under Microgravity

The evolutions of typical patterns at $T_c + 16.8$ K and $T_c + 800$ mK under reduced gravity are reported in Fig. 9.8. These different videoprints correspond to the schematic

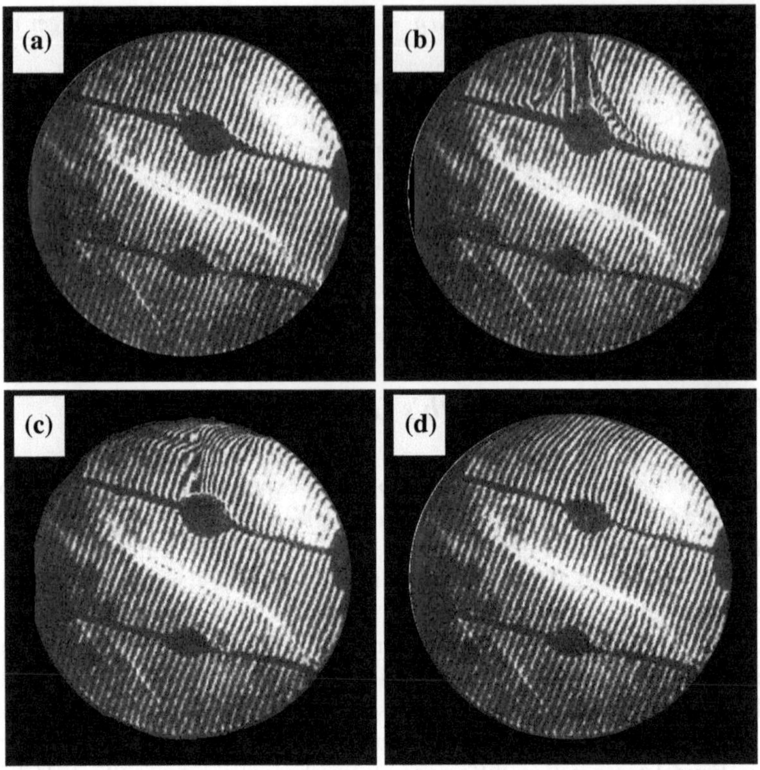

Fig. 9.6 Heat pulse at the equilibrium temperature $T_i = T_c + 16.8$ K under terrestrial gravitational acceleration ($1g$). Gravity is directed downwards. **a** $t = 80$ ms after the beginning of the pulse; **b** $t = 360$ ms; **c** $t = 1$ s; **d** $t = 4$ s. Note that the convective plume clearly develops at $t = 120$ ms after the beginning of the pulse. At $t = 4$ s there are no noticeable density inhomogeneities, which indicates a fast relaxation process

drawings given in Fig. 9.9, where selected areas are defined in Fig. 9.4c. Time is relative to the beginning of the pulse.

When $t < \Delta t_H$ (heating period), a HBL develops around TH1 (TH1HBL area in Fig. 9.4c). Its detailed evolution is discussed in [1]. The main features of the phenomenon are as follows. The layer appears as a black region, nearly spherical in shape, which shows a well-defined border (Fig. 9.8a and a′). The fact that the HBL is optically dark is the result of large density gradients associated with large temperature differences (the pressure is spatially uniform at that timescale). In Fig. 9.8a, there is a progressive extension of the dark region along the lead axis (LWHBL area in Fig. 9.4c), which corresponds to heat conduction through the lead wires between TH1 and the SCU.

A striking observation is the simultaneous shift under weightlessness of the whole fringe pattern, with the fringes remaining straight and parallel as indicated in the schematic drawing in Fig. 9.9. This bulk phenomenon, which coincides with the

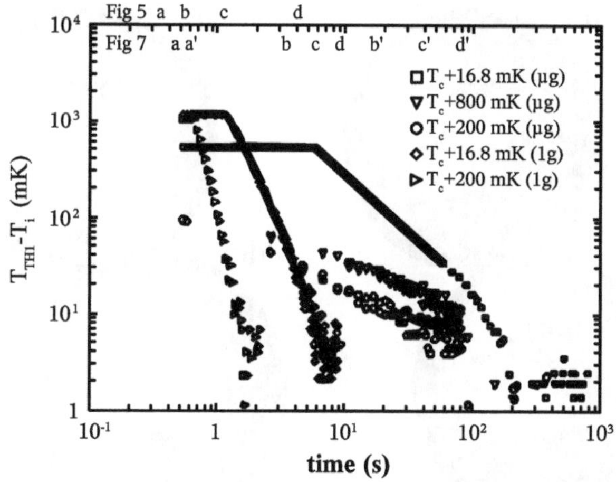

Fig. 9.7 Temperature relaxation of the heating thermistor TH1 at different equilibrium initial temperatures after a heat pulse (480 ms of electrical excitation) under $1g$ and microgravity conditions. At $T_i = T_c + 16.8$ K, the temperature measurement at short times is limited by the technically available temperature range (see text). *Top*: time reference for the videoprint sequence given in Fig. 9.6a–d ($1g$ conditions at $T_i = T_c + 16.8$ K), in Fig. 9.8a–d (microgravity conditions at $T_i = T_c + 16.8$ K) and in Fig. 9.8a′–d′ (microgravity conditions at $T_i = T_c + 800$ mK)

growth of the HBL, is the signature of the piston effect [1]. It corresponds to a spatially uniform density change throughout the entire cell. During the heating period, the pattern shifts to the left, (see Fig. 9.9(b) to (d)) corresponding to a uniform density increase (see Fig. 9.8c). The process is inverted during the cooling period (see Fig. 9.9(e) to (h)).

When $t > \Delta t_H$ (cooling period), the HBL continues to develop, but at the same time it fades out so that the fringes reappear (see Fig. 9.8b–d and b′–d′). The bulk pattern slowly returns to equilibrium (parallel fringes). This relaxation mechanism lasts at least 100 s and appears to be the longest timescale of the phenomenon. The longer relaxation time for the equilibration of the cell at $T_c + 800$ mK compared to that at $T_c + 16.8$ K is a consequence of critical slowing down.

In the following we detail the first few seconds after the end of the electrical excitation period.

9.3.2.1 The HBL Around TH1

This HBL evolution is similar at all temperatures, and exhibits kinetics that slow down when approaching T_c. The layer continues to expand during the cooling period while simultaneously fading out (Fig. 9.8c, d and c′, d′).

In order to quantify the dimensions of the HBL, a simulation of the interferometric fringe pattern as 2-D Gaussian-distributed density inhomogeneities within a uniform medium was performed [2] (see Fig. 9.10). Density inhomogeneities on the order

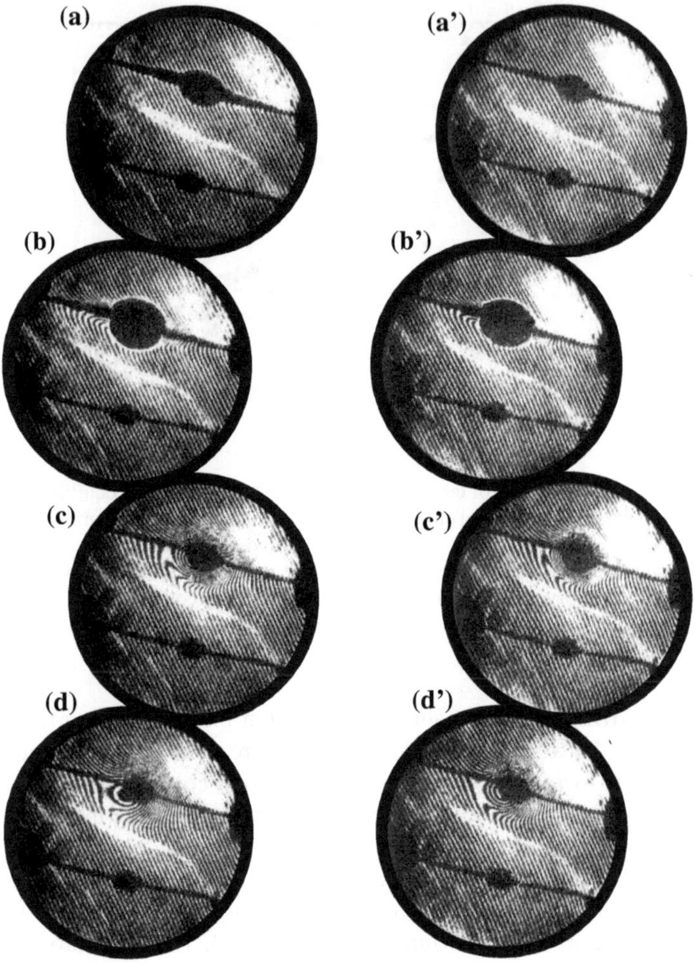

Fig. 9.8 Time sequence of the relaxation of density inhomogeneities after a 480 ms heat pulse (microgravity conditions). Note the hot boundary layer (TH1 HBL in Fig. 9.4c), which forms around the upper heating thermistor (the *black region* increases around TH1). Note also the fading out of this boundary layer and the reappearance of distorted fringes as time proceeds. *Left* initial fluid temperature $T_i = T_c + 16.8$ K; **a** 480 ms after the beginning of the heat pulse; **b** 3 s; **c** 8 s; **d** 12 s. *Right* initial fluid temperature $T_i = T + 800$ mK: **a**′ 480 ms after the beginning of the heat pulse, **b**′ 18 s; **c**′ 55 s; **d**′ 86 s. A time comparison of the relaxation of the HBL density inhomogeneities for the two different initial temperatures T_i demonstrates diffusive critical slowing down

of 1 % lower the fringe spacing by a factor of 4 and practically suppress the fringe visibility. We thus define the HBL border as the location of the last distorted fringe. We note the distortion of the fringes from the direction of the reference line to the right, which reflects the local decrease in density within the hot boundary layer (see Figs. 9.8c, d and 9.9h).

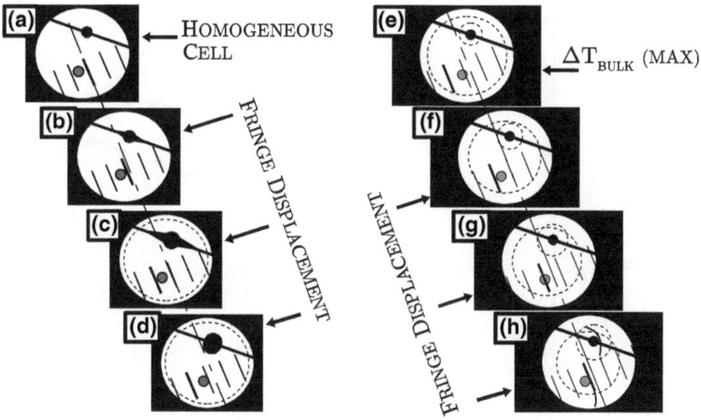

Fig. 9.9 Schematic representation of the time sequence of pulse heating ((**a**) to (**d**)) and subsequent thermal relaxation ((**e**) to (**h**)) (under microgravity). The main features of Fig. 9.8 are sketched. *Arrows* reverse of the direction of the parallel fringe displacements during pulse heating (**b**) to (**d**) and thermal relaxation (**f**) to (**h**). *Broken circles* borderline between the bulk and the HBL (small circle diameter) and the CBL (large circle diameter). The single fringe distorsions in the cold and hot boundary layers are oversimplified in (**h**) (see text)

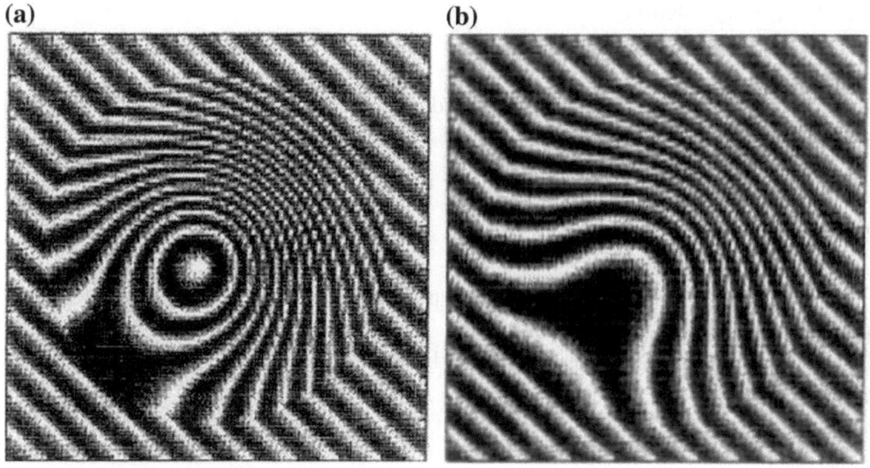

Fig. 9.10 Numerical simulation of the interference fringe pattern. Effect of a 2-D Gaussian-distributed refractive index inhomogeneity in a uniform refractive index medium (see [2]). Value of the central maximum amplitude: **a** $\frac{\delta n}{n} = 10^{-4}$, **b** $\frac{\delta n}{n} = 0.5 \times 10^{-4}$. Note the striking similarity of the fringe distortion to the interferometric pattern observed in Fig. 9.8d and d′ within the hot boundary layer (TH1HBL in Fig. 9.4c)

Figure 9.11 shows the HBL thickness $\delta_{HBL} = \frac{\delta - \delta_0}{2}$ as a function of time, where δ_0 and δ are the thermistor and HBL diameters, respectively. The diameter of the HBL is measured at $T_c + 16.8$ K and $T_c + 400$ mK along an axis perpendicular to the lead wire, where its perturbative role is minimized. The effective time exponent of

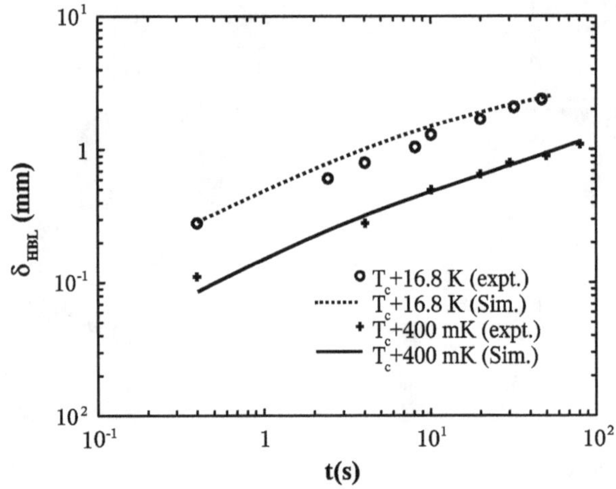

Fig. 9.11 Experimental thickness δ_{HBL}(expt.) of the hot boundary layer (TH1HBL in Fig. 9.4c) versus time during heating and cooling periods at two different equilibrium temperatures. *Open circles* $T_i = T_c + 16.8\,\mathrm{K}$; *crosses* $T_i = T_c + 400\,\mathrm{mK}$. Measurements (expt.) were performed after the end of the electrical excitation period (480 ms) of the TH1 thermistor. The numerical simulation results δ_{HBL}(Sim.) are based on a purely diffusive process (see [4]). *Dashed line* $T_i = T_c + 16.8\,\mathrm{mK}$; *solid line* $T_i = T_c + 400\,\mathrm{mK}$

the $\delta_{HBL}\,(t)$ growth law can be determined, and it has a value close to $\frac{1}{2}$. A numerical simulation reported in [1, 4] demonstrates the scaling by t_D (see Fig. 9.11).

9.3.2.2 The HBL Along the Leads

The hot boundary layer along the leads rapidly vanishes in a time that is nearly independent of T_i (see Fig. 9.9b, c and b′, c′). This phenomenon is due to the heat losses by conduction along the leads, which can be considered a (complicated) perturbative effect.

9.3.2.3 The Bulk

During the cooling period, the fringe pattern slowly returns to its initial position, with the fringes remaining straight and parallel. This change in the pattern shows that the bulk density uniformly returns to its initial value, and demonstrates the existence of a cooling piston effect. The typical evolution is on the order of tens of seconds.

The inversion in the fringe shift occurs after a time of around 3 s, which corresponds approximately to the heating time Δt_H estimated above. The temperature measurements in the bulk ($T_b = T_{TH2}$) confirm this interferometric observation.

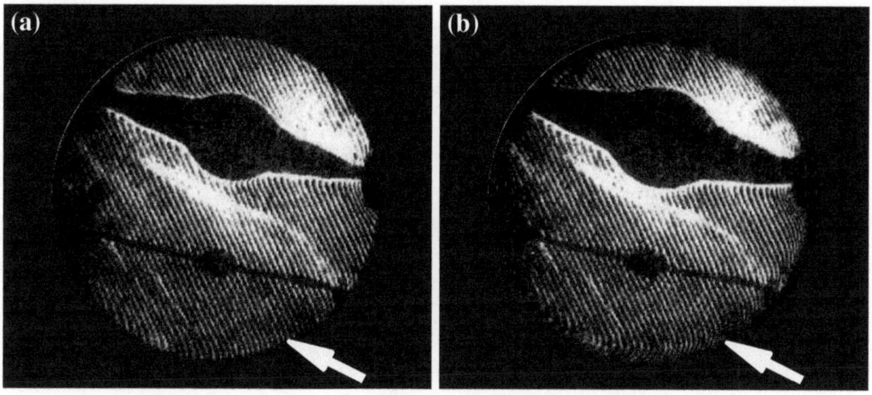

Fig. 9.12 Cold boundary layer formation during a continuous heat pulse (see the CBL region defined in Fig. 9.4c). **a** $t = 4$ s after the beginning of the pulse; **b** $t = 6$ s. The *arrows* point to the fringe deformation close to the cell walls (thermostated at the initial equilibrium temperature $T_i = T_c + 16.8$ mK, while the fringes remain straight. Due to the continuous heating at the location of TH1 and to the correlated continuous increase in the bulk temperature, this CBL corresponds to a significant increase in the local density, as shown by significant bending of the fringes to the left (see also Fig. 9.4c)

9.3.2.4 The CBL at the Cell Wall

Figure 9.12 shows the formation of a CBL under continuous heating [1]. These figures are a clear justification for the schematic representation of Fig. 9.7c. The CBL is demonstrated by the bending of the fringes at the cell wall, while fringes in the bulk remain straight. In contrast to the HBL observations, note the bending of the fringes from the reference line to the left, which reflects the increase in density within the CBL. However, such a layer is difficult to observe with this short pulse, as the corresponding density gradient and thickness are very small (the thickness increases but the density gradient decreases).

9.3.3 Temperature Behavior Within the HBL

The temperature relaxation of the heating thermistor TH1 is measured at different temperatures from $T_c + 16.8$ K to $T_c + 0.2$ K. Figure 9.13 shows a log–log plot of $T_{TH1} - T_i$ as a function of time for three typical temperature differences ($T_c + 16.8$ K, $T_c + 800$ mK and $T_c + 200$ mK). As the critical point is approached, the temperature of TH1 decreases noticeably (as in the terrestrial observations, although with a different time relaxation).

Two factors are responsible for the lowering of the HBL temperature produced by a constant energy heat pulse: first, the divergence of c_p, which enables the HBL

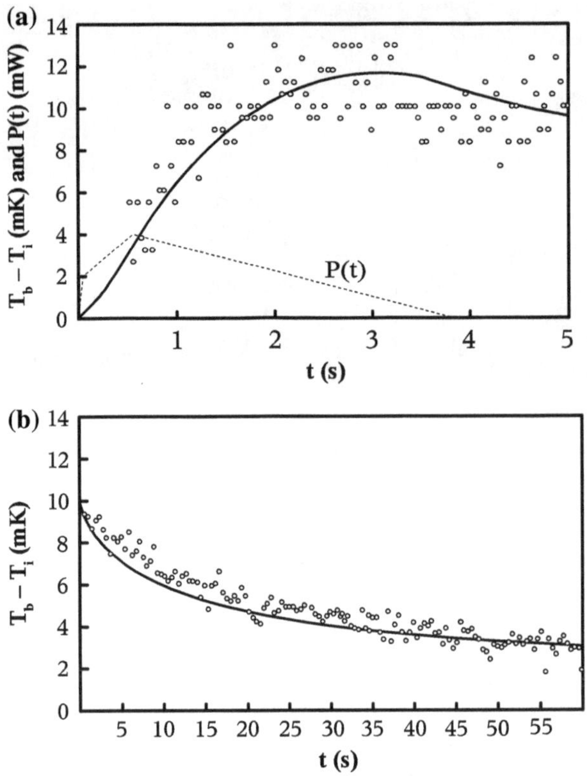

Fig. 9.13 Temperature relaxation ($T_b - T_i$) of the bulk fluid versus time as measured by thermistor TH2 (*open circles*) and as fitted by the model (*solid lines*). Initial equilibrium fluid temperature $T_i = T_c + 16.8$ K. **a** Evolution during the first 5 s after the beginning of the heat pulse; the time–power profile $\mathcal{P}(t)$ of the heat pulse is given by the broken line. **b** Evolution during the period 5–60 s

to store more energy; second, the critical enhancement of the thermal conductivity Λ, which favors energy transfer from the thermistor into the fluid.

9.3.4 Temperature Behavior in the Bulk Fluid

We mainly refer to the bulk temperature evolution at $T_c + 16.8$ K (see Fig. 9.8a–c) as measured by TH2, $T_b = T_{TH2}$ (with a correction at $t < 60$ s of 3 mK due to the thermistor self-heating). Time is counted from the beginning of the heat pulse.

After the electrical pulse has stopped, the temperature still increases. A maximum is reached at about 3 s (see Fig. 9.8a). This fast heating period is due to the development of the HBL during Δt_H (estimated above to be $\approx$2.5 s). The bulk fluid is heated uniformly, as shown by the shift in the whole fringe pattern while the fringes remain straight.

Cooling towards the thermostat temperature proceeds more slowly (see Fig. 9.8b). For the sake of comparison, we also report in Fig. 9.8c the HBL temperature as measured by TH1, $T_{HW} = T_{TH1}$ (log–log plot). Although the HBL is always hotter than the bulk, it does not appear to interfere with the bulk uniform cooling ensured by the CBL.

Similar results are obtained at $T_c + 15.9$ K and $T_c + 0.8$ K, but at this last temperature the bulk variation is significantly smaller, making a comparable analysis difficult. When $T_i < T_c + 0.8$ K, no temperature rise is detected by the TH2 thermistor. This decrease in the temperature change in the bulk is due to the divergence of c_V and to the increasing efficiency of the CBL: the fluid acts as a true thermal short circuit, as discussed in Sect. 5.2.1 (see Ref. [6]).

9.4 Quantitative Comparison with the 1-D Model

9.4.1 The Bulk Temperature Behavior at $T_c + 16.8$ K

9.4.1.1 Characteristic Length of the Model

As discussed in Sect. 8.2.3, knowledge of L (the characteristic length) and $\mathcal{P}(t)$ (the time dependence of the heating power) is needed to facilitate a comparison with experiments. The smallest length scale $L = 1.5$ mm of Eq. (9.11) corresponds to a characteristic time t_{PE} ($\approx$3.5 s), similar to the rough estimate above for Δt_H ($\approx$2.5 s).

Note that t_{PE} is larger than the typical equilibration response time of the thermostat ($\approx$2 s, see Table 9.2), which validates the boundary condition of a perfectly thermostated wall $[T(x = L) = T_i]$ used in the model. The total energy supplied to the fluid by the heating thermistor is $Q = 8.1$ mJ (see Sect. 9.2.5). Based on this value of Q, an estimated time profile for the heat flux $\mathcal{P}(t)$ can be proposed, and this takes the form of a rapid rise to a maximum value for 0.6 s (corresponding to the period during which electricity was supplied, added to the time response of the thermistor) followed by a slower decrease (corresponding to the period of heat loss through the leads).

The value of the maximum and the slope of the decrease are fixed by two criteria: the total energy provided, and the fact that the maximum resulting bulk temperature must occur around 3 s after the start of the heating period.

Analytical results obtained based on this heat flux are compared in the following with the bulk temperature relaxation at $T_c + 16.8$ K. Based on this analysis, the estimation of L is checked and then the time evolution of the heat flux directed into the fluid is defined.

Table 9.3 Fit results obtained for the bulk temperature relaxation ΔT_b measured by TH2 using Eq. (9.15), where A_0 and t_{PE} are the adjustable parameters

Time interval of the fit (s)	A_0(mK)	t_{PE} (s)
5–6	38.8	1.3
	(22.6)	4.4
10–60	149	0.085
	(22.6)	4.3

The values in parentheses were imposed on the fit (see text)

9.4.1.2 Fits to the Bulk Temperature Relaxation

After the end of the heat pulse, the bulk temperature increasingly tends to behave like the impulse response $F(t)$, whatever the time profile of the flux (see Fig. 9.2b and [6]). A sufficiently long time after the end of the heating period, the data can be fitted to [see Eq. (9.2)]

$$\Delta T_b = A_0 \exp\left[\frac{t}{t_{PE}}\right] \operatorname{erf} c \left[\frac{t}{t_{PE}}\right]^{\frac{1}{2}}, \tag{9.15}$$

where ΔT_b is the bulk temperature rise as measured by TH2. t_{PE} and $A_0 = \frac{Q}{mc_v}$ are adjustable parameters (Q is the total energy supplied at the hot wall). Fits using this relation were then performed in two different ranges, 5–60 s or 10–60 s. If the fits are similar, 5 s is a sufficiently long time for the fluid to "forget" the time history of the heat flux.

The results of these fits are summarized in Table 9.3. The amplitudes A_0 and t_{PE} show a large discrepancy between the ranges. This is due to the fact that, for sufficiently large values of t, ΔT_b tends to behave like $A_0\sqrt{\frac{t_{PE}}{t}}$. Accordingly, it becomes more and more difficult for a fitting algorithm to discriminate between the relative influence of A_0 and t_{PE}, even if the global fit is correct.

This point is confirmed by the calculation of the product $A_0\sqrt{t_{PE}}$ for each fit, which leads to similar results in both cases (the difference is less than 3 %). In order to obtain more accuracy for the value of t_{PE}, we now consider the estimate of the total energy sent into the fluid ($Q = 8.1$ mJ) and impose $A_0 = \frac{Q}{mc_v} = 22.6$ mK.

The fits performed with t_{PE} as the only free parameter give very similar values for both ranges, $t_{PE} \approx 4.3$ s (see Table 9.3). This value compares well with the value inferred from Eq. (9.11) ($t_{PE} = 3.5$ s with $L = 1.5$ mm), and shows the relevance of the ratio $\frac{\text{fluid volume}}{\text{cold wall area}}$ for adapting the 1-D model to the experiment.

Let us comment on the thermistor self-heating in the high-rate measurement mode. In order to estimate the temperature bias, some fits were made on the experimental data, based on the function $\frac{A_1}{\sqrt{t}}$ where A_1 is a free parameter. A_1 was calculated on experimental data ranging from 5 s to 300 s, with different self-heating corrections

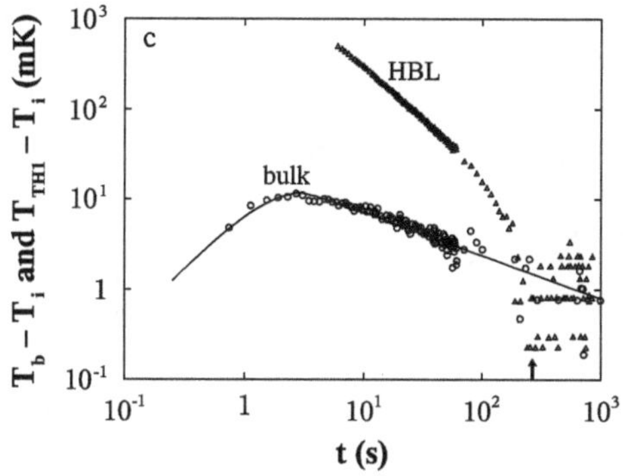

Fig. 9.14 Temperature relaxation ($T_b - T_i$) of the bulk fluid versus time as measured by thermistor TH2 (*open circles*) and as fitted by the model (*solid lines*). Initial equilibrium fluid temperature $T_i = T_c + 16.8$ K. A log–log plot is shown for the whole time range; the temperature relaxation of the hot boundary layer as measured by thermistor TH1 is also depicted *(open triangles)*. The *arrow* marks the typical time needed to fully restore density and temperature equilibria

applied between 5 s and 60 s. The best fit corresponds to the expected value of 3 mK, which independently confirms the manufacturer's specifications.

We now consider the value $L = 1.5$ mm, as set in the model, and return to $\mathcal{P}(t)$. The data of interest mainly relate to the temperature rise at early times (see Fig. 9.12a).

According to Fig. 9.5, the two parameters are Δt_H and the maximum power (at $t = 0.6$ s), with the value of 8.1 mJ taken as a constraint on the integrated total energy Q. The best fit is shown in Fig. 9.12a. This corresponds to $\Delta t_H = 3.8$ s, which is comparable to but slightly larger than the previous estimate (2.5 s) (Fig. 9.5).

9.4.2 HBL Relaxation at TH1

9.4.2.1 Critical Behavior of the HBL Temperature

The following approach is based on Eq. 9.4, focusing on the time dependence of $T_{HW}(x = 0)$ for the intermediate range $\Delta t_H < t < t_D$. For temperatures that are nearer to T_c than 1 K, $t_{PE} < \Delta t_H$. Using Eq. 9.4, we have determined that the late-time response of $\Delta T_{HW} = T_{HW} - T_i$ behaves asymptotically as

$$\Delta T_{HW}(x = 0) = \frac{A}{\sqrt{t}}. \tag{9.16}$$

The amplitude A is proportional to the total energy supplied to the fluid, but does not depend on the time history of the heat flux. If the same total energy is supplied to the fluid at each temperature T_i, Eq. 9.4 leads to

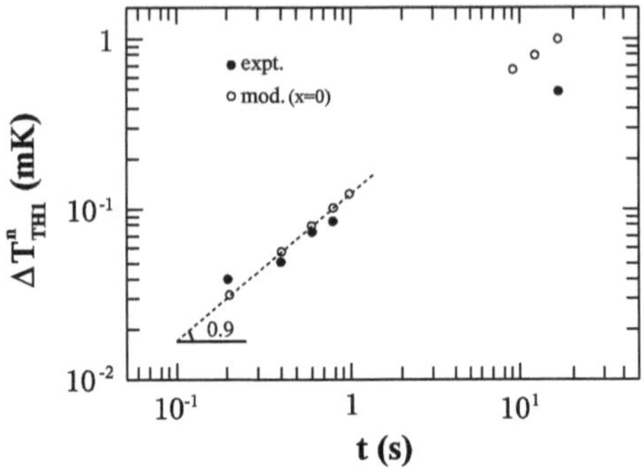

Fig. 9.15 Normalized hot wall temperature $\Delta T^n_{TH1}(x = 0, t = 10\,\text{s}, Q = 1\,\text{mJ})$ as a function of $(T_i - T_c)$ [see text and Eqs. (9.16) and (9.17)]. *Filled circles* TH1 measurements, *open circles* 1-D model

$$A \propto \frac{\sqrt{t_{PE}}}{c_V}, \tag{9.17}$$

which asymptotically vanishes as $(T_i - T_c)^{0.91}$.

After a long enough time (>6 s), the data measured by TH1 follows a $t^{-\frac{1}{2}}$ law (see the related discussion in Sect. 9.4.2.2). As a consequence, the comparison between the experimental data and the model is performed for $t = 10$ s.

The measured values are normalized to the same total energy (1 mJ) using the calculated power delivered by the Joule effect in the thermistor.

Moreover, to adapt the 3-D experiment to the model conditions, the temperature differences $T_{TH1} - T_i$ are also divided by the ratio of the area of the cold wall to that of the hot wall ($\frac{S_{CW}}{S_{HW}} \cong 80$, see Sect. 8.2.3). A comparison between these normalized experimental values ΔT^n_{TH1} $(x = 0, t = 10\text{ s}, Q = 1\text{ mJ}, S = S_{CW})$ and the corresponding calculated values from the model is shown in Fig. 9.15.

The quantitative agreement is noticeable considering the uncertainties attached to each estimate of the above quantities. The $(T_i - T_c)$ dependence of A with an exponent of 0.91 implicitly assumes that the critical exponent associated with D_T has an asymptotic value of 0.67. This assumption is incorrect in the temperature range of the measurements, as D_T exhibits a larger effective exponent in that range. The effective exponent of the temperature dependence of A is then lower than the expected value of 0.9, as observed in Fig. 9.15.

9.4.2.2 Diffusive Scaling of the HBL Temperature Relaxation

The form of the right-hand side of Eq. 9.4 suggests a normalized scaling for the relaxation of the hot wall temperature.

Figure 9.16 shows a log–log plot of the relaxation of the reduced HW temperature $\Delta T^* = \frac{mc_p[T_{HW}(t) - T_i]}{Q}$ versus the reduced time $t^* = \frac{t}{t_D}$ for the different equilibrium temperatures.

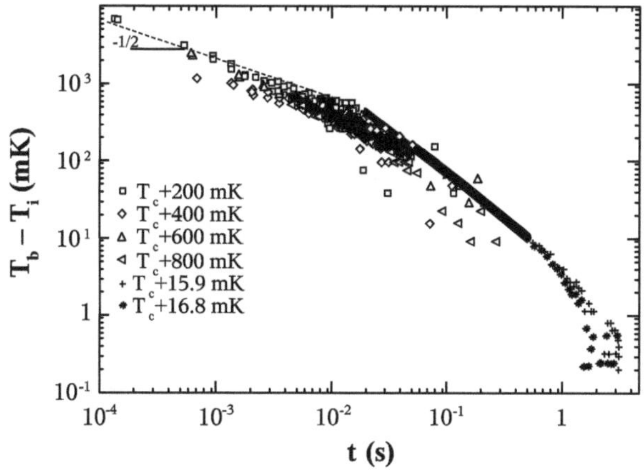

Fig. 9.16 Log–log plot of $\Delta T^* = \frac{mc_p[T_{HW}(t)-T_i]}{Q}$ versus the reduced time $t^* = \frac{t}{t_D}$. Near T_c, theory predicts an initial slope of $-\frac{1}{2}$ [see the text and Eqs. (9.4) and (9.5)]. $T_{HW}(t)$ is the relaxation temperature of the hot boundary layer (TH1HBL in Fig. 9.4c), as measured by the thermistor TH1

Here, m is the total mass of the fluid. All data collapse to a single curve that exhibits initial power law behavior of $(t^*)^{-\frac{1}{2}}$. Such a power law is found for diffusive processes that take place in semi-infinite media. For small values of t^*, the HBL relaxation is not influenced by the finite dimensions of the cell. We note that when $t^* \cong 1$, $\Delta T^* \cong 1$.

This indicates that, for $t \cong t_D$, the energy initially transmitted into the HBL has diffused into the bulk. This again confirms the diffusive character of the HBL relaxation.

References

1. Bonetti M, Perrot F, Beysens D, Garrabos Y (1994) Fast thermalization in supercritical fluids. Phys Rev E 49(6):R4779–R4782
2. Bonetti M, Perrot F, Beysens D, Garrabos Y (1995) Fast adiabatic heating and temperature relaxation in near-critical fluids under zero gravity. Int J Thermophys 16:1059–1067
3. Ferrell RA, Hao H (1993) Adiabatic temperature changes in a one-component fluid near the liquid–vapor critical point. Phys A 197(1–2):23–46
4. Frohlich T, Bouquet S, Bonetti M, Garrabos Y, Beysens D (1995) Self-similar solutions of the non-linear diffusion equation and application to near-critical fluids. Phys A 218(3–4): 419–436
5. Garrabos Y, Bonetti M, Beysens D, Perrot F, Fröhlich T, Carlès P, Zappoli B (1998) Relaxation of a supercritical fluid after a heat pulse in the absence of gravity effects: theory and experiments. Phys Rev E 57(5):5665–5681
6. Zappoli B, Carlès P (1995) The thermo-acoustic nature of the critical speeding up. Eur J Mech B 14:41–65

References

Chapter 10
Coexisting Liquid–Vapor Phases

Abstract Heat and mass transfer induced by temperature changes is studied in near-critical, two-phase fluids in weightlessness. Two situations are considered. (i) Slight heating from the sample walls in a two-phase sample, without phase change. It results a striking overheating of the vapour phase, which becomes hotter than the heating wall in a seemingly violation of the Second principle of thermodynamics. The effect is more pronounced to within a certain distance from the critical point. (ii) Boiling, where heat transfer is associated with phase change. This process is used in many types of heat transfer technologies as usually larger amount of heat can be transferred than in a one-phase fluid. Evaporation, which causes a recoil force at the liquid–vapor–solid contact line can dewet the liquid phase from the wall, inducing the redoubtable boiling crisis where the substrate temperature rises enormously and can eventually melt. Weightless experiments and critical slowing down allow the details of the process to be studied in details and compared with the recoil force theory.

The nonequilibrium behavior of near-critical two-phase fluids is complex to describe, since the fluid is inhomogeneous and a phase change can also occur. In addition, interface motion and deformation caused by the thermal perturbations can modify the phase distribution within the fluid and influence the heat transfer itself. From a practical point of view, two-phase heat transfer associated with a phase change—called boiling (see [8])—is involved in many heat-transfer technologies, as it usually allows a larger amount of heat to be transferred than in a one-phase fluid. In a boiling process, we expect that evaporation will dewet the liquid phase from the wall, creating a liquid–gas–solid contact line.

Under microgravity, there is an absence of the buoyancy that on Earth causes gas bubbles nucleating on a hot surface to rise, which aids boiling studies. The vicinity of the critical point enables the thermophysical properties of the fluid to be varied continuously by simply changing the temperature, thus taking advantage of the well-known universal power laws of critical phenomena and the complete wetting of one phase (liquid generally).

B. Zappoli et al., *Heat Transfers and Related Effects in Supercritical Fluids*,
Fluid Mechanics and Its Applications 108, DOI: 10.1007/978-94-017-9187-8_10,

In this chapter, we present experimental results on heat and mass transfer in near-critical two-phase fluids induced by temperature changes in microgravity. Two different kinds of experiments were performed to this end [33]:

1. In situ temperature measurements during temperature steps
2. An optical study of the deformation of the liquid–vapor interface when this interface is in contact with the heating walls of the sample cell

10.1 Experimental Arrangement

These results were obtained with the ALICE 2 facility onboard the Mir space station, using different fluid samples and different heating protocols, during the French/Russian Cassiopeia (1996), Pegasus (1998), and Perseus (1999) missions, as well as the French/American GMSF (1999) mission.

10.1.1 Thermal Control, Measurement, and Stimuli

A description of the ALICE 2 facility is given in Chap. 23 (see also [17]). The interferometer cell used for the two-phase studies is shown in Fig. 10.1. A YSI thermistor (rise time of 2 s, relative accuracy of 100 μK) imbedded in the sample cell unit (SCU) near the fluid volume was used to follow the temperature evolution of the cell walls (labeled T_W). The temperature measurements were performed with three in situ Thermometrics B10 thermistors (10 ms rise time, 500 μK accuracy, 0.2 mm in diameter). Two of these (labeled Th1 and Th2) were located close to the cell wall (at a distance of about 1 mm from it), and were always in the liquid phase. The temperatures measured by Th1 and Th2 are labeled T_{L1} and T_{L2}. The third thermistor (Th3) was mounted in the center of the cell, so that the gas bubble (of volume fraction $\frac{1}{2}$: cell filled at $\rho = \rho_c$) always surrounded Th3. The temperature measured by Th3 is labeled T_V. The measurement sampling rates were 25 Hz for T_{L1}, T_{L2} and T_V and 1 Hz for T_W during the first five minutes following a temperature quench, and then 0.1 Hz for the next 55 min.

10.1.2 Optics

The two-phase fluid sample was visualized by differential interferometry (using a Twyman–Green interferometer; see Sect. 23.1.3), as shown in Fig. 10.1a. Since the contact angle is zero near the critical point, the liquid–gas meniscus between the two parallel windows formed a semicircular interface in the plane perpendicular to the windows (see Fig. 10.1b). The interface appears dark in the images because the liquid–gas meniscus refracts the incident light away from the axis of the cell.

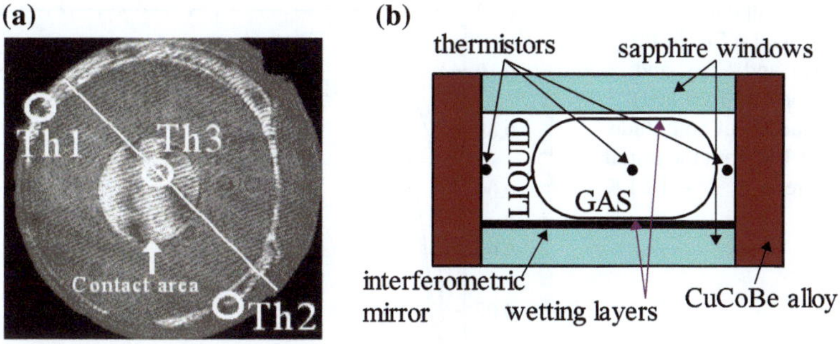

Fig. 10.1 **a** Interferometric image of the cell (with a thickness of 6.7 mm) used to perform in situ temperature measurements. *White circles* show the positions of the thermistors Th1, Th2 and Th3, whose measurements are labeled T_L^1, T_L^2 and T_V, respectively. **b** Schematic cross-section of the cell showing the gas and liquid phases

10.1.3 Samples

Both pure SF_6 and CO_2 were used. The required fluid density in the interferometric cells was achieved to high precision, within 0.2 %, by weighing and checking on the ground that the meniscus appeared in the middle of the cell after a temperature step below T_c [18]. The initial state of the fluid samples before heating was a flat bubble that was slightly constrained by the two windows, as drawn schematically in Fig. 10.1b.

10.1.4 Procedures

In the experiments dedicated to in situ temperature measurements, a series of rapid, positive temperature quenches of amplitude $\Delta T = T_f - T_i = 100\,\text{mK}$ and 50 mK were performed for initial temperatures T_i ranging from $T_c - 10.1$ K to $T_c - 0.1$ K. A quench of amplitude ΔT consisted of a sharp linear increase in T_W up to 80 % of ΔT in less than 10 s, followed by a smooth evolution up to the final temperature T_f.

10.2 Temperature Evolution in Liquid and Vapor

In this part, we will present striking temperature measurements where we observed that the vapor temperature was higher than that of the heating wall. The evolutions of the temperature rises at the cell wall $\delta T_W = T_W - T_i$, in the liquid $\delta T_{L1} = T_{L1} - T_i$, $\delta T_{L2} = T_{L2} - T_i$, and in the vapor $\delta T_V = T_V - T_i$, during a temperature quench of $+100$ mK from $T_i = T_c - 10.1$ K, are shown in Fig. 10.2.

Fig. 10.2 Temperature rises at the cell wall δT_W, in the liquid δT_{L1} and δT_{L2}, and in the vapor δT_V (see text), during a quench of amplitude $\Delta T = +0.1$ K from the initial temperature $T_i = T_c - 10.1$ K

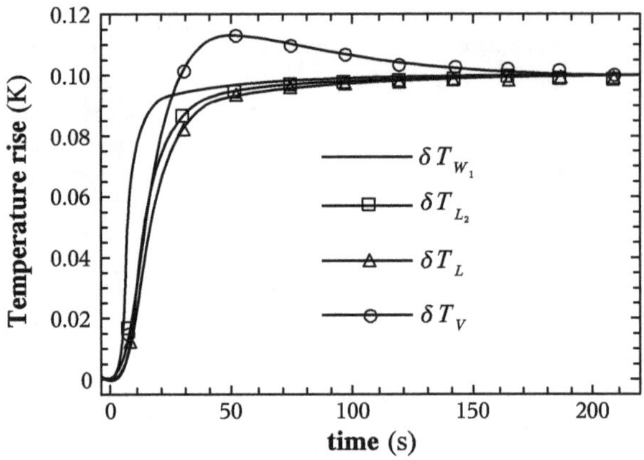

At the beginning of the quench ($t = 0$–25 s), all temperatures increase almost linearly over time, with T_{L1} and T_{L2} increasing less rapidly than T_V and remaining smaller than T_W. Before the end of the quench, the temperature in the vapor (T_V) reaches much higher than T_W. T_V then decreases slowly up to its final value, T_f. These observations and the vapor-overheating phenomenon can be explained as follows.

10.3 Initial Temperature Evolution

We first compare the initial temperature evolutions in liquid and vapor. During the adiabatic heat transfer process, the increase in pressure due to the expansion of the fluid heated at the boundary can be considered to be homogeneous within the sample (the timescales involved here are longer than the time of flight of sound waves).

During the quench, δT_W reaches 90 % of ΔT in less than 10 s, and its evolution is linear over time for at least 5 s (see Fig. 10.2). At the same time, the evolutions of δT_{L1}, δT_{L2} and δT_V are also found to be linear over time for at least 5 s (see Fig. 10.2). We calculate the slopes of the temperature rises in gas $\frac{d(\delta T_V)}{dt}$ and in liquid $\frac{d(\delta T_{L1})}{dt}$; $\frac{d(\delta T_{L2})}{dt}$ over 5 s. The quantities $s_1 = \left(\frac{\frac{d(\delta T_V)}{dt}}{\frac{d(\delta T_{L1})}{dt}}\right) - 1$ and $s_2 = \left(\frac{\frac{d(\delta T_V)}{dt}}{\frac{d(\delta T_{L2})}{dt}}\right) - 1$ are plotted as a function of $T_c - T_i$ in Fig. 10.3.

It was pointed out by Onuki and Ferrell [21] that, during the short period when the piston effect is effective, neither heat nor mass can be transferred between a vapor and a liquid that are initially coexisting, because the thermal diffusivity is weak near the critical point. Consequently, the liquid and vapor should behave as if they were independent, and the temperature increases in the liquid (δT_L) and the vapor (δT_V) should be homogeneous and isentropic. Therefore, in each bulk phase (here, "bulk" is the fluid region beyond the thermal diffusion layers, which is heated only by adiabatic compression), the ratio $\frac{\delta T_V}{\delta T_L}$ is expected to be

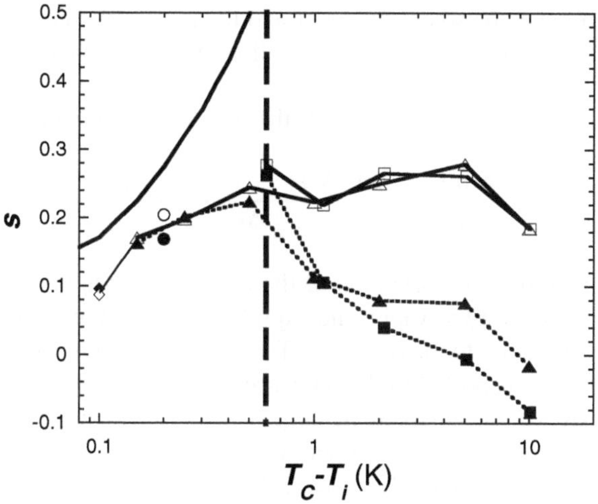

Fig. 10.3 The relative initial slopes $s_1 = \frac{\left(\frac{d(\delta T_V)}{dt}\right)_{t=0}}{\left(\frac{d(\delta T_{L1})}{dt}\right)_{t=0}} - 1$ (*solid symbols*; *dotted line* is a guide for the eye) and $s_2 = \frac{\left(\frac{d(\delta T_V)}{dt}\right)_{t=0}}{\left(\frac{d(\delta T_{L2})}{dt}\right)_{t=0}} - 1$ (*open symbols*; *solid line* is a guide for the eye) as a function of $T_c - T_i$. Quench amplitudes: +100 mK (*squares*), +50 mK (*triangles*), +25 mK (*diamonds*), +20 mK (*circles*). The *vertical dashed line* at 0.6 K separates the regions where Th1 and Th2 are (*left*) in the liquid bulk and (*right*) in the liquid hot boundary layer. *Bold solid curve*: $\frac{\left(\frac{\partial T}{\partial P}\right)_S^G}{\left(\frac{\partial T}{\partial P}\right)_S^L} - 1$ computed using the equation of state from [34]

$$1 + s = \frac{\delta T_V}{\delta T_L} = \frac{\left(\frac{d(\delta T_V)}{dt}\right)_{t=0}}{\left(\frac{d(\delta T_L)}{dt}\right)_{t=0}} = \frac{\left(\frac{\partial T}{\partial p}\right)_S^V}{\left(\frac{\partial T}{\partial p}\right)_S^L}, \tag{10.1}$$

where the subscript S means a derivative at constant entropy and the subscripts V and the L label the adiabatic derivative of T with respect to p in a one-phase vapor and a one-phase liquid. Since $\left(\frac{\partial T}{\partial P}\right)_S^V > \left(\frac{\partial T}{\partial P}\right)_S^L$, the vapor should then be more rapidly heated than the liquid, and this is actually observed.

In Fig. 10.3, the values $s = s_i = \frac{\left(\frac{d(\delta T_V)}{dt}\right)_{t=0}}{\left(\frac{d(\delta T_{Li})}{dt}\right)_{t=0}} - 1, i = \{1, 2\}$, as obtained from the measured temperature slopes, are related to the thermistors Th1 and Th2. The curves are compared to the theoretical prediction of Eq. (10.1) for $T_c - T < 0.6$ K, where temperatures are measured in the bulk fluid. Figure 10.3 shows that the data are smaller than the expected value from Eq. (10.1). This disagreement is presumably due to the relatively weak heat and mass exchange between the liquid and the vapor during the adiabatic compression. Indeed, during the piston-effect heat-transfer process, only a thin layer of fluid is able to permit liquid–vapor coexistence, and the temperature difference between the vapor and the bulk liquid causes heat and mass flow through the interface that acts to equilibrate the chemical potential near the interface.

According to Onuki's statement [22], the entropy conservation at the interface implies that

$$\frac{dm_V}{dt} = \frac{(\Lambda_V \nabla T_V - \Lambda_L \nabla T_L) .\mathbf{n}}{\Delta h}, \tag{10.2}$$

where $\frac{dm_V}{dt}$ is the rate of vaporization of the liquid, Δh is the latent heat of vaporization, Λ_V, and Λ_L are the vapor and liquid thermal conductivities, and $\mathbf{n}$ is the unit vector perpendicular to the interface. The contribution $(\Lambda_V \nabla T_V - \Lambda_L \nabla T_L) .\mathbf{n}$ represents the heat transported to the interface, which is negative at the start of adiabatic heating. Hence, the adiabatic heating process causes vapor to condense at the interface at the beginning of the temperature rise. This condensation was numerically observed by Zhong and Meyer [35]. Due to condensation, the vapor is less heated than the liquid, leading to a smaller vapor temperature growth rate than the one deduced from Eq. (10.1).

10.4 Vapor Overheating

We now analyze the vapor-overheating phenomenon. In order to compare the vapor overheating detected during quenches of different amplitudes, the temperature rise in the vapor δT_V is scaled to the temperature rise of the cell walls δT_W, and the behavior of its maximum $\left(\frac{\delta T_V}{\delta T_W}\right)_{MAX}$ is plotted as a function of $T_c - T_i$ in Fig. 10.4. This exhibits a maximum of 123 % around $T_c - T_i \simeq$ 5–7 K. The vanishing differences in thermophysical properties between vapor and liquid at the critical point imply that—asymptotically near the critical point—the vapor and the liquid show the same thermal response. On the other hand, far from the critical point, the efficiency of the adiabatic heating is reduced and the heat transfer is mainly diffusive: a situation that prevents the vapor from overheating. The existence of a maximum for the overheating of the gas at $T < T_c$ thus arises due to the competition between these two limiting behaviors.

In this cell, the vapor bubble is almost completely isolated from the thermostated cell walls by the liquid (the area of contact between vapor and windows is less than 6 % of the overall heating area). During the quench, the adiabatic heating process stops when T_L^{bulk} has reached T_W (there is no longer any temperature gradient at the boundary). Before the equilibration of T_L^{bulk} and T_W, the bulk vapor is heated more than the bulk liquid by the homogeneous pressure increase. Since the vapor bubble is not in contact with the heating wall, its temperature has no influence on the temperature gradient that drives the expansion of the liquid hot boundary layer (HBL). T_V^{bulk} can thus exhibit a large overshoot. The situation is different on Earth, where gravity causes the vapor to be in contact with the upper part of the cell. The HBL developing in the vapor stops expanding when the bulk vapor temperature

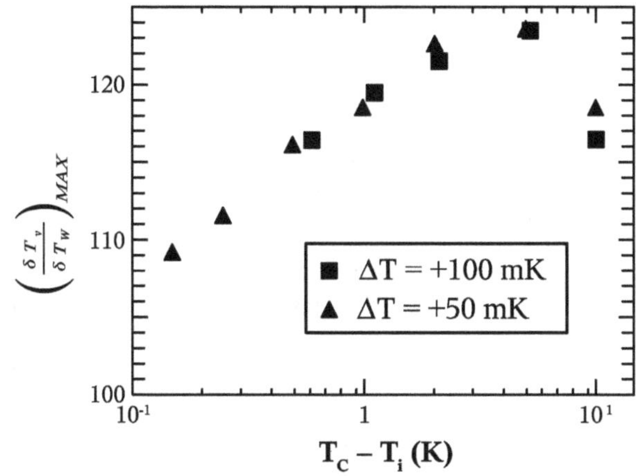

Fig. 10.4 Vapor overheating, as measured by $\left(\frac{\delta T_V}{\delta T_W}\right)_{MAX}$, as a function of $T_c - T_i$. *Full squares* +100 mK quenches. *Full triangles* +50 mK quenches. The occurrence of such a high level of overheating is due to the geometry of the liquid and vapor phases

reaches T_W. When T_V^{bulk} exceeds T_W, the vapor HBL may even contract (cooling piston effect) if the liquid HBL continues to expand in order to compensate for vapor overheating. Numerical simulations [32, 36] confirm this influence of the phase distribution on the temperature evolution.

10.5 Vapor Temperature Relaxation

The long-time thermal relaxation of T_V towards T_f after the maximum overheating follows the exponential law

$$T_V - T_f = \theta \exp\left(-\frac{t}{t_R}\right), \tag{10.3}$$

with an amplitude of θ and a relaxation time of t_R [33]. This analytical form is justified by the fact that, at long times, only the (exponential) relaxation of the slowest thermal mode can be observed [4]. The variation in t_R as a function of $T_c - T_i$ (see Fig. 10.5) follows the equation

$$t_R = t_0\,(T_c - T_i)^x\,, \tag{10.4}$$

with t_0 and x as free fitting parameters. The best fit value is $x = -0.92 \pm 0.05$, which appears to be comparable to the measured value (-0.89) of the effective critical exponent of the thermal diffusivity for both vapor and liquid [14]. We conclude that the relaxation of the temperature inhomogeneities generated by the adiabatic heat transfer is diffusive, although the temperature inhomogeneities are large. This is due

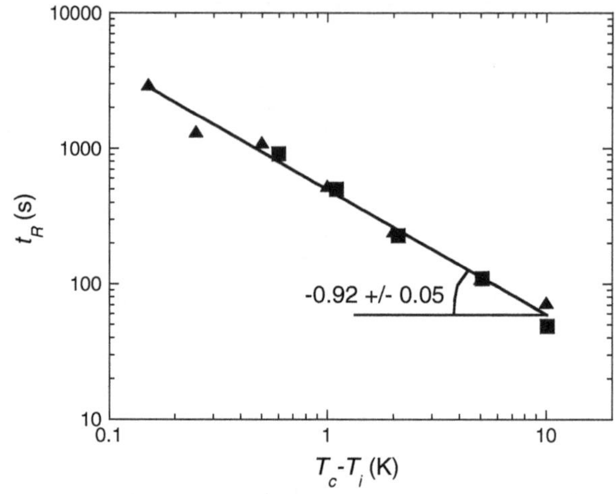

Fig. 10.5 Variation of the characteristic time t_R of the exponential relaxation of the gas temperature, as a function of $T_c - T_i$ (log–log plot). *Full squares* +100 mK quenches. *Full triangles* +50 mK quenches. *Continuous line* fit with Eq. (10.4)

to the thermal isolation of the gas bubble from the cell walls by the liquid layer. The overheating of the gas is not seen by the walls, so fast relaxation by the cold piston effect is prevented.

10.6 Evolution of the Contact Area

The thickness of the (liquid) wetting layers on Earth is very sensitive to temperature gradients [16] due to buoyancy. In the absence of gravity, the evolution of the fringes allows the evolution of the film thickness to be analyzed during the drying of the film, using the interferometer images shown in the lower part of Fig. 10.6 (see [31]). About ten fringes scroll until the drying is complete, which corresponds to an initial (equilibrium) thickness on the order of 15–20 μm [31].

We can also estimate the above film thickness at equilibrium. For length scales on the order of or larger than 0.1 μm, which is the typical scale of the phenomena detected here, the long-range van der Waals forces in the retarded regime are the only attractive forces [2]. In the liquid phase at a distance ζ from the wall, the energy of fluid–solid interaction per unit surface area is classically written as

$$E_{\mathrm{vdW}} = \frac{A}{3\zeta^3}, \tag{10.5}$$

where A is the Hamaker constant. The disjoining pressure Π is associated with this energy between the fluid and solid:

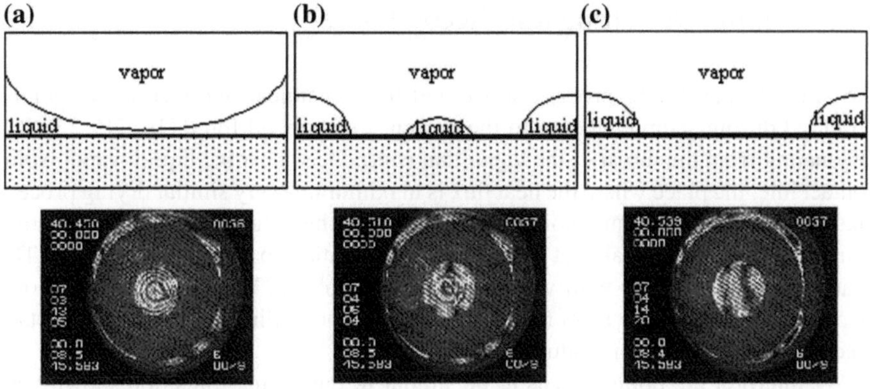

Fig. 10.6 *Top* Schematic sketch of the drying process for the contact region during heating of the two phase configurations of the cell, reported in the bottom part: **a** at the start of heating; **b** and **c** evolution during sample heating. *Bottom* Interferometer pictures of the SF_6 cell (6.7 mm thickness) at $T = T_c - 5\,\text{K}$

$$\Pi = -\frac{dE_{\text{vdW}}}{d\zeta} = \frac{A}{\zeta^4}. \tag{10.6}$$

For the sample configuration given in Fig. 10.1b, the pressure difference between the vapor and liquid is mainly due to the highest curvature induced by the sample thickness e; i.e.,

$$p_V - p_L \simeq \frac{2\sigma}{e}, \tag{10.7}$$

which should be equal to the disjoining pressure Π of Eq. (10.6) at equilibrium. In Eq. (10.7), σ is the liquid–vapor surface tension. Using Eqs. (10.6) and (10.7), the relation $\Pi = p_V - p_L$ leads to a typical thickness

$$\zeta \simeq \left(\frac{eA}{2\sigma}\right)^{\frac{1}{4}}. \tag{10.8}$$

Using the typical value $A \simeq 10^{-24}\,\text{N}\,\text{m}^2$ [7], we easily get an estimate of $\zeta \simeq 12\,\mu\text{m}$ for each wetting layer of the SF_6 sample shown in Fig. 10.1b. This estimate corresponds well to the measured 15–20 μm wetting layer between the window and the vapor.

Let us now analyze the film-drying process during the heating period. When the vapor is still colder than the wall, dewetting occurs as a ring (see Fig. 10.6b), with a wet liquid island in the center that eventually dries (see Fig. 10.6c). This drying is well known to people who deal with the ebullition process, and especially the boiling crisis, when the dry area spreads and a vapor film isolates the liquid from the heater [27]. A detailed analysis of the drying phenomena is provided below.

10.7 Boiling, Dewetting and Recoil Force

We attribute the drying process above and the dynamical interface deformations observed during sample heating to the presence of a recoil force [19, 20, 23, 26], originating from the thrust of the vapor that is produced near the liquid–vapor–solid contact line: the place where the heat flux is maximum. A very similar drying process takes place during the liquid boiling process at large heat flux. When surface heating is increased past a critical heat flux, a sudden transition to film boiling occurs. The heater becomes covered with vapor and may burn out [23, 26, 27]. This burnout, also called the departure from nucleate boiling or the boiling crisis, is an important practical problem in many industries.

We interpret the boiling crisis to be similar to the drying transition shown here [19], with the main difference being that, during boiling, the interface deformation is achieved by high vapor production, whereas near the critical point, the heat flux involved remains very small because of the vanishing latent heat. The boiling crisis is observed when the heat flux q_S from the solid heater exceeds a threshold value q_{CHF}, which is called the critical heat flux (CHF). The rapid formation of the vapor film on the heater surface sharply decreases the heat transfer and leads to local heater overheating.

10.7.1 The Recoil Force at the Triple Contact Line

The boiling crisis phenomenon is local [5]: it only depends strongly on the local values of the parameters in a very thin layer of liquid adjacent to the heating surface, and apparently does not induce any change in the flow. This layer is nearly quiescent because of the no-slip boundary condition for the fluid velocity at the heating surface. The most important parameter is the distribution of the local temperature. As a consequence of this local origin, the threshold depends strongly on the wetting properties of the heating surface. Numerous experiments [12, 15] show the general tendency: poor wetting of the heating surface by the liquid favors the boiling crisis and *vice versa*.

Experiments that have attempted to visualize [28, 30] the dry spots under the vapor bubbles on the heating surface show that a single dry spot suddenly begins to spread at the CHF. In [3, 24, 25], vapor recoil instability [23] is proposed as a reason for the departure from nucleate boiling. Although it is not clear how an instability can cause dry spots to spread, the authors show that the vapor recoil force can be important at large evaporation rates.

The recoil force originates in the uncompensated momentum of the vapor that is generated at the liquid–vapor interface during evaporation. In the reference frame of the liquid, the conservation of momentum implies that

$$\mathbf{p}_r + \eta_V(\mathbf{v}_V + \mathbf{v}_i) = 0, \tag{10.9}$$

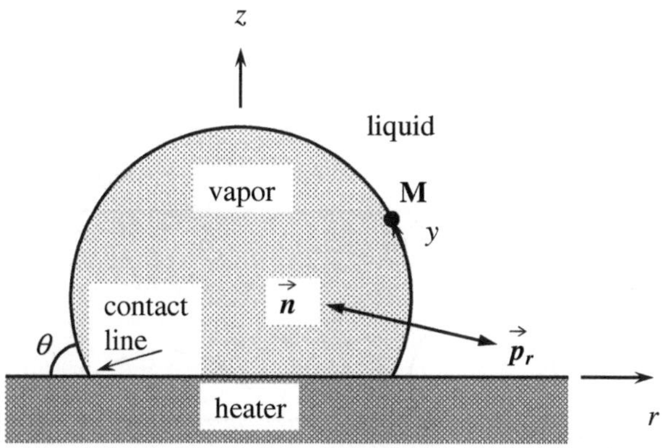

Fig. 10.7 Vapor bubble on the heating surface surrounded by liquid. The directions of the vectors $\mathbf{p}_r$ and $\mathbf{n}$ are shown, as well as the axes for the system of cylindrical coordinates

where $\mathbf{p}_r$ is the vapor recoil force per unit interface area, which varies across the vapor–liquid interface, η_V is the evaporated mass per unit time and unit interface area, $\mathbf{v}_i$ is the interface velocity, and $\mathbf{v}_V$ is the vapor velocity with respect to the interface. It is easy to establish that $\mathbf{v}_i = -\eta_V \rho_L^{-1}\mathbf{n}$, where $\mathbf{n}$ is a unit vector normal to the interface directed inside the vapor bubble (see Fig. 10.7). Mass conservation at the interface yields $\mathbf{v}_V = -\rho_L \rho_V^{-1}\mathbf{v}_i$, where ρ_L and ρ_V are the mass densities of the liquid and the vapor. Therefore, Eq. (10.9) implies that [23]

$$\mathbf{p}_r = -\eta_V^2 \left(\rho_V^{-1} - \rho_L^{-1}\right)\mathbf{n}. \tag{10.10}$$

The surface deformation caused by this force is important whenever the evaporation is strong. The rate of evaporation η_V can be related to the local heat flux across the interface q_L by the equality

$$q_L = H\eta_V, \tag{10.11}$$

where H is the latent heat of evaporation. Hereafter, we neglect heat conduction in the vapor with respect to the latent heat effect.

Below, we consider only the case of systems at high pressure, which is assumed to be comparable to the critical pressure for the sake of simplicity. The growth of the bubble is then slow enough to allow the surface tension to equilibrate its shape, and hydrodynamic forces can be neglected [27]. This allows the problem to be considered in the quasi-static approximation (However, vapor recoil remains the leading effect for any pressure).

The spreading of the dry spot looks similar to the spreading of a liquid that wets a solid. However, in the case of departure from nucleate boiling, it is *vapor* that seems to "wet" the solid. This never happens for a nonmetal liquid under equilibrium conditions (zero heat flux) on a perfectly clean and smooth metal surface [7], with the

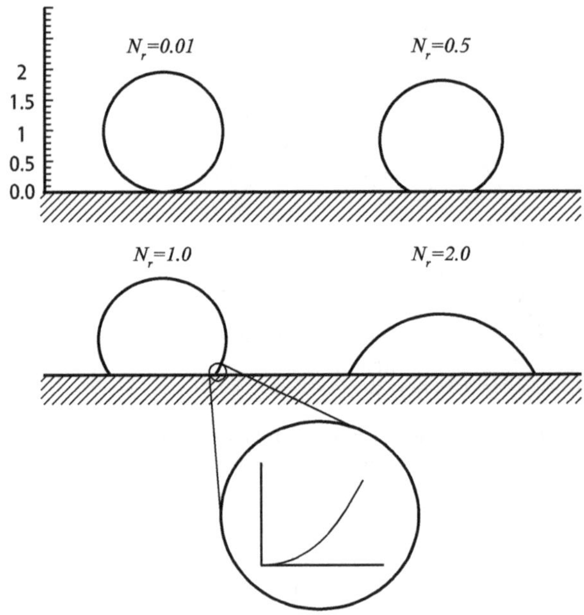

Fig. 10.8 Shape of the 2-D vapor bubble on the heating surface under the action of the vapor recoil force, as calculated for $a = 0.01$ (see text). The volume V is the same for all bubbles. The actual contact angle is zero, as demonstrated in the *inset*. The coordinates are scaled by $\left(\frac{3V}{4\pi}\right)^{\frac{1}{3}}$

finite contact angle possibly being due solely to the surface defects. A kind of drying transition occurs due to the vapor recoil force at some heat flux that we associate with the CHF. Using the quasi-static approximation, the variational approach [9] can be applied to analyze the shape of a vapor bubble just before the boiling crisis. The free energy of the system consists of two parts. The first part is conventional [9]:

$$U_1 = \sigma A + \sigma_{VS} A_{VS} + \sigma_{LS} A_{LS} - \lambda V, \tag{10.12}$$

where σ, σ_{VS}, and σ_{LS} are the surface tensions for the vapor–liquid, vapor–solid and liquid–solid interfaces, respectively; A_{VS} and A_{LS} are the corresponding interface areas; and A is the area of the vapor–liquid interface (Fig. 10.8). The last term in Eq. (10.12) reflects the fact that the shape of the bubble should be found for its given volume V, with λ being the Lagrange multiplier. The equation $\delta U_1 = 0$, where δU_1 is the energy change due to an infinitesimal displacement $\delta \mathbf{r}$ of A [9], yields the classical conditions for bubble equilibrium in the absence of external forces.

The second part, U_2, of the free energy accounts for the virtual work done by the external forces:

$$\delta U_2 = -\int_{(A)} \mathbf{p}_r \cdot \delta \mathbf{r} dA. \tag{10.13}$$

The minimization $\delta U_1 + \delta U_2 = 0$ of the total energy leads [9] to two equations. The first is the Laplace condition for local equilibrium at the interface:

$$K\sigma = \lambda + p_r, \tag{10.14}$$

where K is the local curvature of the bubble and $p_r = |\mathbf{p}_r|$. The second equation is $\cos\theta = c$, where $c = \frac{\sigma_{VS} - \sigma_{LS}}{\sigma}$ and θ is the contact angle (see Fig. 10.7). For the case $c > 1$ (as in the case of water on a metal surface), the second equation should be replaced with the condition $\theta = 0$.

Let us denote by y the distance along the bubble contour measured from the triple line to a given point M, as shown in Fig. 10.7. To find the bubble shape by solving Eq. (10.14), we need to know the vapor recoil as a function of y. In the following, we introduce a rough approximation to solve the very complicated problem of the heat exchange around the growing bubble. Saturated boiling is assumed. Thus, the vapor–liquid interface is maintained at the temperature T_s, the saturation temperature at constant system pressure. We also assume for simplicity that the thermal effect of convection can be taken into account by renormalizing the liquid's thermal conductivity.

To estimate how p_r varies near the contact line (when $y \to 0$), we consider the bubble to be two-dimensional, with the contact angle $\theta = \pi/2$ for the sake of simplicity. Since we are describing the heat exchange in a thin layer adjacent to the heating surface, we can imagine the bubble contour A to be a line Oy perpendicular to the Ox heater line. Then q_L can be obtained by solving a simple 2-D problem of unsteady heat conduction in a quarter plane $x, y > 0$, where the point $O(x = 0, y = 0)$ corresponds to the contact line. The boundary and the initial conditions for this problem can be written in the form

$$T_L|_{x=0} = T_s, \quad -\lambda_L \left.\frac{\partial T_L}{\partial y}\right|_{y=0} = q_S, \quad T_L|_{t=0} = T_s,$$

where $T_L(x, y, t)$ is the liquid's temperature, λ_L is the liquid's thermal conductivity, and q_S is the heat flux from the heating surface, which is assumed to be uniform for the case of a thin heating wall. The solution to this heat conduction problem can be written as

$$T_L = T_s + \frac{q_S}{\lambda_L}\sqrt{\frac{D_{TL}}{\pi}} \int_0^t \frac{dt}{\sqrt{t}} \left(\frac{x}{2\sqrt{D_{TL}t}}\right) \exp\left(-\frac{y^2}{4D_{TL}t}\right), \tag{10.15}$$

where D_{TL} is the liquid's thermal diffusivity. This solution implies the following expression for $q_L(y)$:

$$q_L = -\lambda_L \left.\frac{\partial T_L}{\partial x}\right|_{x=0} = -\frac{q_S}{\pi} E_1\left(\frac{y^2}{4\alpha_L t}\right). \tag{10.16}$$

The exponential integral $E_1(y)$ [1] decreases as $\exp(-y)/y$ as $y \to \infty$. Therefore, p_r, which is related to q_L via Eqs. (10.10) and (10.11), is a rapidly decreasing function of y. Note that $E_1(y)$ diverges logarithmically at the point $y = 0$. The divergence demonstrates that the evaporation is strongest in the vicinity of the contact line. Since this singularity is integrable, the heat flux through any finite part of the interface is finite.

In what follows, for any θ value, we will use the dependence $p_r(y)$ in the following form, which retains the main physical features:

$$p_r = -C \ln\left(\frac{y}{L}\right) \exp\left[-\left(\frac{y}{y_r}\right)^2\right], \tag{10.17}$$

where C is the amplitude of the recoil force and L is the length of the half-contour of the 2-D bubble. y_r is the characteristic length of the vapor recoil decay, which changes over time and is proportional to $\sqrt{D_{TL}t}$ (see Eq. 10.16). Meanwhile, the bubble grows, and its radius is also proportional to $\sqrt{D_{TL}t}$ during the late stages of its growth (see [27]). Therefore, y_r is proportional to the bubble size; this fact is taken care of by the expression $y_r = aL$, where a is the nondimensional fraction of the bubble surface on which the vapor recoil is important. From a physical point of view, y_r characterizes the width of the superheated layer of liquid, which is always less than the bubble size [27], so $a \ll 1$. This allows the upper limit for the integration to be infinity in the following nondimensional expression for the strength of the vapor recoil:

$$N_r = \frac{1}{\sigma}\int_0^\infty p_r \, dy. \tag{10.18}$$

The integration can be performed analytically, yielding the following relation between C and N_r: $N_r = CaL/(4\sigma)\sqrt{\pi}[\gamma + \log(4/a^2)]$, where $\gamma \simeq 0.577$ is Euler's number [1]. Using the cylindrical (r, z) system of coordinates (see Fig. 10.7), Eq. (10.14) can be written in parametric form as a system of three ordinary differential equations:

$$\frac{dr}{dy} = \cos u, \tag{10.19}$$

$$\frac{dz}{dy} = \sin u, \tag{10.20}$$

$$\frac{du}{dy} = -\frac{\sin u}{r} + \lambda + p_r(y), \tag{10.21}$$

with the boundary conditions $z = 0, u = \theta$ at $y = 0$ and $r = 0, u = \pi$ at $y = L$. A fourth condition is necessary to determine the unknown L. The mathematical problem is then closed by the equation that allows the Lagrange multiplier λ to be determined:

$$V = \pi \int_0^L r^2 \sin u dy, \tag{10.22}$$

where V is the given bubble volume.

In the following, the case $\theta = 0$ is considered (for example, water on a metal heater or near-critical fluids on any substrate). The solution of Eqs. (10.17)–(10.22) is presented in Fig. 10.8 for $a = 0.01$. Note that, although the actual contact angle is zero for all of the curves (see the inset in Fig. 10.8), the apparent contact angle grows with $N_r \propto q_S^2$. At small values of the heating flux q_S, the dry spot under the bubble corresponds to the size of the vapor bubble nucleation spot (which is assumed to be zero for the calculations corresponding to Fig. 10.8). The bubble grows with its contact line pinned at the defects on the surface of the heater, until the bubble departs due to the action of gravity or external hydrodynamic forces. The departure size of the bubble is small because the adhesion (which is proportional to the contact line length [29]) is small. At some value of q_S, the contact line can "*de-pin*" under the action of the vapor recoil before the bubble detaches from the heating surface. Figure 10.8 shows that the dry spot reaches a size equal to the bubble diameter at $N_r \sim 1$, a value that compares well with the estimate. The adhesion force increases as the dry spot increases, thus hindering the departure of the bubble. The departure time (i.e., the time interval during which the bubble is attached to the heating surface) grows sharply. We consider that this feedback is the cause of the departure from nucleate boiling, with this value of q_S being associated with q_{CHF}.

For poor wetting conditions ($\theta > 0$), a smaller value of the vapor recoil (and q_{CHF}) is needed to create a large dry spot. This explains the decrease in q_{CHF} with increasing contact angle and surface roughness [12].

10.7.2 Vapor Recoil in Near-Critical SF_6

10.7.2.1 Observations

Here, we consider results that were obtained using a sample of SF_6 at near-critical density (off by +0.25 %) [11], similar to the cell described above in Sect. 7.1.1, with a thickness of 1.664 mm. However, this cell is different in that (i) the cell is not in an interferometer and is observed via parallel light normal to the windows, and (ii) the cell is not equipped with thermistors. Since the windows were glued to the copper alloy wall, some of the glue is squeezed inside the cell. This glue forms a ring that blocks the transmission of light in a thin layer of the fluid adjacent to the copper wall, making it inaccessible for observations. Because of this glue layer, the windows may also be slightly tilted with respect to each other. A 10 mm diameter ring was engraved on one of the windows of each cell in order to calibrate the sizes of the images (see the upper images in Fig. 10.9a). As the bubble is pressed against the window, even a minute angle between them results in a steady force that pushes the bubble near to the wall.

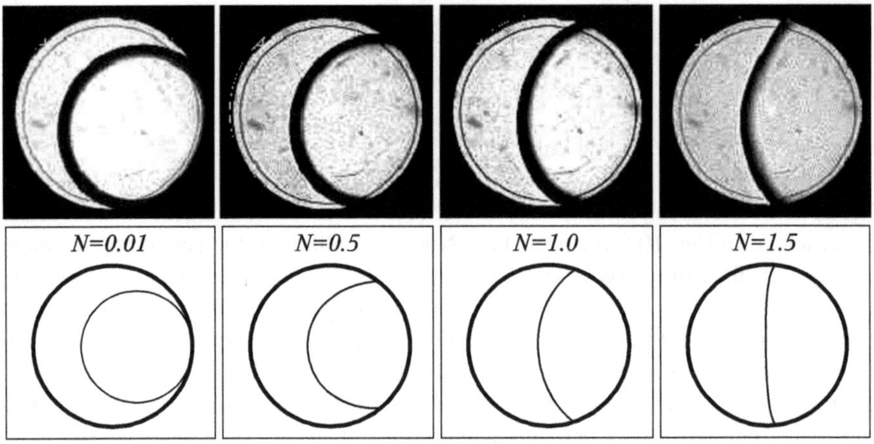

Fig. 10.9 *Top* evolution of the shape of the bubble during a temperature ramp of $\approx 7.2\,\text{mK}\,\text{s}^{-1}$. **a** $t = 0$ (initial equilibrium state). **b** $t = 338$ s. **c** $t = 780$ s. **d** $t = 900$ s. *Bottom* calculated bubble shape for different values of the nondimensional strength of vapor recoil, N, which goes to infinity at the critical point. Note that the actual contact angle is zero for all the curves

The cell is heated from room temperature almost linearly over time at a rate of $\approx 7.2\,\text{mK}\,\text{s}^{-1}$. The interface appears dark because the liquid–gas meniscus refracts the normally incident light away from the cell axis. After the temperature ramp has started, but still far from the critical temperature, the shape of the bubble changes. The contact area A_{VS} of the vapor with the solid (i.e., the copper wall) appears to increase. This increase is accompanied by an evident increase in the apparent contact angle; see the top images in Fig. 10.9b–d.

Note that temperature quenches (fast temperature jumps from one temperature to another) are also performed at various temperature differences from T_c. The typical value for a quench is $+0.1$ K. The images show bubble spreading, which is slight far from the critical point and stronger near the critical point. After each quench, as soon as the heating stops, the bubble interface begins to return to its initial form (see [10, 13] for more details).

10.7.2.2 Evolution of the Interface During Heating

The above experimental data show that the spreading gas and the associated interface deformation are caused by a nonequilibrium phenomenon. Marangoni convection due to a temperature change δT_i along the gas–liquid interface cannot be the source of this evolution, since (i) convections are not observed in the video films, and (ii) the interface must remain at constant (saturation) pressure throughout the process [10]. The bubble is thus deformed by the normal stress exerted on the interface by the recoil from the departing vapor.

The interface shape can be determined from the modified Laplace equation [see Eq. (10.14)]

$$\sigma K = \Delta p + p_r, \tag{10.23}$$

where Δp is a constant to be determined from the 3-D bubble volume. The 3-D local curvature K is equal to the sum of the 2-D local curvature (denoted $c_{\parallel}$) in the image plane and the 2-D local curvature (denoted $c_{\perp}$) in the perpendicular plane, as shown in Fig. 10.1b. The relatively low heat flow through the less conductive sapphire windows implies a small p_r near the contact line on the windows, as compared to the large value of Δp associated with the strong curvature $c_{\perp} \simeq \frac{2}{e}$ due to the small cell thickness e. Then, during heating, $K \approx c_{\parallel}$. The shape of the interface can thus be obtained from the 2-D equation

$$c_{\parallel}\sigma = \Delta p' + p_r(y), \tag{10.24}$$

where $\Delta p'$ is a constant determined by the 2-D bubble volume, and y is a coordinate that varies along the bubble contour in the image plane.

In order to find the distribution of the evaporation rate η_V at the interface, it is necessary to solve the entire heat transfer problem. Because the bulk temperature varies sharply in the boundary layer adjacent to the walls of the cell, and the interface temperature is constant, the largest portion of the mass transfer across the interface takes place near the triple contact line. Thus, η_V is large in the vicinity of the contact line. When $T \to T_c$, we assume that η_V has the following form:

$$n(y) = g(y)(T_c - T)^a, \tag{10.25}$$

i.e., η_V has the same local temperature behavior as T_c is approached. The rate of change in the mass m_V of the vapor bubble is defined using the following power law:

$$\frac{dm_V}{dt} = \int n(y)\mathrm{d}y \sim (T_c - T)^a, \tag{10.26}$$

where the integration was performed over the total vapor–liquid interface area. On the other hand,

$$\frac{dm_V}{dt} = \frac{d}{dt}(V\phi\rho_V), \tag{10.27}$$

where V is the cell volume and the vapor volume fraction ϕ is assumed to be $\phi = \frac{1}{2}$. Near the critical point, $\rho_V - \rho_c \sim (T_c - T)^{\beta}$ (with $\beta = 0.325$; see Table 1.2), so that $\frac{dm_V}{dt} \sim (T_c - T)^{\beta-1}\frac{dT}{dt}$ as $T \to T_c$ according to Eq. (10.27). Thus, Eq. (10.26) results in $a = \beta - 1$. Using Eq. (10.10) and $\sigma \sim (T_c - T)^{2\nu}$ (with $\nu = 0.63$; see Table 1.2), we find that the local change in curvature due to the vapor recoil scales as

$$c_{\parallel} \approx \frac{p_r}{\sigma} \sim (T_c - T)^{3\beta-2-2\nu}. \tag{10.28}$$

Because the exponent $3\beta - 2 - 2\nu \approx -2.3$ is very large, it even manifests itself far from the critical point, in agreement with experiment. In summary, as $T \to T_c$, the mass growth of the vapor follows the growth in its density (its volume remains constant), so that the diverging vapor production near the critical point drives a diverging recoil force.

The change in interface curvature has a striking effect on the bubble shape because it is not homogeneously distributed along the bubble interface. Since the evaporation is strongest near the heating wall, where the strongest temperature gradients form, both p_r and $c_{\parallel}$ increase strongly near this wall (i.e., near the triple contact line). Because the slope of the interface changes so abruptly near the contact line, the apparent contact angle is much larger than its actual value.

As $c_{\parallel}$ is proportional to the second derivative of the bubble profile, Eq. (10.24) is a differential equation with a boundary condition that is given by the actual contact angle [6]. This actual contact angle defines the first derivative (i.e., the slope) of the bubble profile at the solid wall. It is also specified by the interfacial tension balance, and must be zero near the critical point. This condition of a contact angle of zero gives a boundary condition for Eq. (10.24). In order to illustrate a possible solution of Eq. (10.24), we solved it using the same expression for $p_r(y)$ as in Eq. (10.17); i.e.,

$$p_r(y) \propto -N_r \ln\left(\frac{y}{L}\right) \exp\left[-\left(\frac{y}{0.1\,L}\right)^2\right], \tag{10.29}$$

where $y \in \{0, L\}$. $y = 0$ at the solid wall and L is the bubble half-contour. The numerical coefficient N_r in Eq. (10.29) can be determined from Eq. (10.18), where the upper integration limit can be replaced with infinity without any loss of accuracy. Although the expression (10.29) for the vapor recoil pressure is not rigorous, it contains the main physical features of the solution to the heat conduction problem: weak divergence at the contact line and rapid decay away from it. It was shown in [20] that the rigorous numerical solutions obtained far from the critical point follow this behavior. The result of this calculation is shown in Fig. 10.9. Since Eq. (10.28) implies

$$N_r \sim (T_c - T)^{-2.3} \to \infty \tag{10.30}$$

as $T \to T_c$, the increase in N_r mimics the approach to the critical point and qualitatively explains the observed shape of the vapor bubble (see Fig. 10.9). The increase of the apparent contact angle and the vapor–solid contact area A_{VS} can be seen in the Fig. 10.9. Very similar bubble spreading is observed far from T_c when boiling with a large heat flux [28, 30]. The main difference is that the large value of N_r is achieved by high vapor production, which can be realized by strong overheating rather than by critical effects.

It is well documented from experiments [27] that the CHF decreases rapidly when the fluid pressure p approaches the critical pressure p_c; i.e., when $T \to T_c$ in our constant volume system. This tendency was not well understood previously. However, the divergence of the factor N_r, as discussed above, helps us to understand it. We first note that the evaporation rate η_V scales as the applied heat flux q_S

and $N_r \sim q_S^2$, using Eqs. (10.10) and (10.18). By assuming that the boiling crisis at $q_S \simeq q_{\mathrm{CHF}}$ begins when N_r attains its critical value $N_{\mathrm{CHF}} \sim 1$, Eq. (10.30) yields

$$q_{\mathrm{CHF}} \sim (T_c - T)^{1+\nu-3\beta/2} \sim (T_c - T)^{1.15}. \tag{10.31}$$

The same exponent is also valid for the pressure scaling,

$$q_{\mathrm{CHF}} \sim (p_c - p)^{1.15}. \tag{10.32}$$

Equation (10.32) explains the observed tendency for $q_{\mathrm{CHF}} \to 0$ as $p \to p_c$.

Although the strict requirements for temperature stability and weightlessness lead to experimental difficulties when studying the boiling crisis in the near-critical region, they also present some important advantages. Only a very small heating rate (heat flux) is needed to reach the boiling crisis because q_{CHF} is very small. At such low heat fluxes, the bubble growth is extremely slow, due to the critical slowing down. The experiments are able to observe the spreading gas (i.e., the drying out that leads to the boiling crisis) for ~900 s (see Fig. 10.9 and [10, 13] for details). Such experiments not only permit excellent time resolution; they also allow the complicating effects of rapid fluid motion to be avoided.

References

1. Abramovitz M, Stegun IA (eds) (1972) Handbook of mathematical functions. National Bureau of Standards, Washington
2. Adamson AW, Gast AP (1997) Physical chemistry of surfaces. Wiley, New York
3. Avksentyuk BP, Ovchinnikov VV (1995) Two-phase flow modelling and experimentation. Edizioni ETS, Pisa
4. Behringer RP, Onuki A, Meyer H (1990) Thermal equilibration of fluids near the liquid-vapor critical point: ^{3}He and ^{3}He-^{4}He mixtures. J Low Temp Phys 81:71–102
5. Bricard P, Péturaud P, Delhaye J-M (1997) Understanding and modelling DNB in forced convective boiling: modelling of a mechanism based on nucleation site dryout. Multiph Sci Technol 9(4):329–379
6. Brown GR, Meyer H (1972) Study of the specific-heat singularity of He^3 near its critical point. Phys Rev A 6(1):364–377
7. de Gennes PG (1985) Wetting: statics and dynamics. Rev Mod Phys 57(3):827
8. Dhir VK (1998) Boiling heat transfer. Annu Rev Fluid Mech 30:365–401
9. Finn R (1986) Equilibrium capillary surfaces. Springer, New York
10. Garrabos Y, Lecoutre-Chabot C, Hegseth J, Nikolayev VS, Beysens D, Delville J-P (2001) Gas spreading on a heated wall wetted by liquid. Phys Rev E 64(5):051602
11. Garrabos Y, Lecoutre-Chabot C, Wunenburger R, Delville J P, Beysens D (1999) Critical boiling phenomena observed in microgravity. J Chem Phys 96(6):1066–1073
12. Hahne E, Grigull U (eds) (1977) Heat transfer in boiling. Hemisphere, Washington
13. Hegseth J, Garrabos Y, Nikolayev VS, Lecoutre-Chabot C, Wunenburger R, Beysens D (2002) Gas "wets" a solid wall in orbit. Int J Thermophys 23:89
14. Jany P, Straub J (1987) Thermal diffusivity of fluids in a broad region around the critical point. Int J Thermophys 8:165–180
15. Katto Y (1994) Critical heat flux. Int J Multiph Flow 20:53–90

16. Kayser RF, Schmidt JW, Moldover MR (1985) Wetting layers and dispersion forces for a fluid in contact with a vertical wall. Phys Rev Lett 54(7):707–710
17. Marcout R, Zwilling JF, Laherrere JM, Garrabos Y, Beysens D (1995) Alice 2, an advanced facility for the analysis of fluids close to their critical point in microgravity. Microgravity Q 5:162–170
18. Morteau C, Salzman M, Garrabos Y, Beysens D (1997) In: Viviani A (ed) The 2nd European symposium on fluids in space, Napoli, Italy, 22–26 April 1997 Congressi srl, Roma, pp 327–333
19. Nikolayev VS, Beysens D (1999) Boiling crisis and non-equilibrium drying transition. Europhys Lett 47:345–351
20. Nikolayev VS, Beysens D, Lagier G-L, Hegseth J (2001) Growth of a dry spot under a vapor bubble at high heat flux and high pressure. Int J Heat Mass Transf 44:3499–3511
21. Onuki A, Ferrell RA (1990) Adiabatic heating effect near the gas-liquid critical point. Phys A 164(2):245–264
22. Onuki A (1991) Sound propagation in phase-separating fluids. Phys Rev A 43(12):6740–6755
23. Palmer HJ (1976) The hydrodynamic stability of rapidly evaporating liquids at reduced pressure. J Fluid Mech 75:487–511
24. Pavlov PA, Liptchak AI (1992) Metastable phase states and kinetics of relaxation. Russian Academy of Sciences, Ural Division, Sverdlovsk, pp 119 (in Russian)
25. Sefiane K, Benielli D, Steinchen A (1998) A new mechanism for pool boiling crisis, recoil instability and contact angle influence. Coll Surf A 142:361–373
26. Straub J (1995) In: Rathe L, Walter H, Feuerbacher B (eds) The IXth European symposium on gravity-dependent phenomena in physical sciences. Springer, Berlin, pp 351–359
27. Tong LS (1997) Boiling heat transfer and two-phase flow, 2nd edn. Taylor and Francis, New York
28. Torikai K, Suzuki K, Yamaguchi M (1991) Study on contact area of pool boiling bubbles on a heating surface (observation of bubbles in transition boiling). JSME Int J Ser II 34:195–199
29. van Helden WGJ, van der Geld CWM, Boot PGM (1995) Forces on bubbles growing and detaching in flow along a vertical wall. Int J Heat Mass Transf 38:2075–2088
30. van Ouwerkerk HJ (1972) Burnout in pool boiling: the stability of boiling mechanisms. Int J Heat Mass Transf 15:25–28
31. Wunenburger R (2000) Transfert de Chaleur et Effet des Vibrations dans les Fluides Diphasiques au Voisinage du Point Critique. PhD thesis, Universite Bordeaux 1, Bordeaux
32. Wunenburger R, Garrabos Y, Lecoutre C, Beysens D, Hegseth J, Zhong F, Barmatz M (2002) Thermal response of a two-phase near-critical fluid in low gravity: strong gas overheating as due to a particular phase distribution. Int J Thermophys 23:103–115
33. Wunenburger R, Garrabos Y, Lecoutre-Chabot C, Beysens D, Hegseth J (2000) Thermalization of a two-phase fluid in low gravity: heat transferred from cold to hot. Phys Rev Lett 84(18):4100–4103
34. Wyczalkowska AK, Sengers JV (1999) Thermodynamic properties of sulfurhexafluoride in the critical region. J Chem Phys 111(4):1551
35. Zhong F, Meyer H (1999) Heat transport in a pure fluid near the critical point: steady state and relaxation dynamics. J Low Temp Phys 114:231–255
36. Zhong F, Meyer H (1996) Density equilibration near the liquid-vapor critical point of a pure fluid. ii. Coexisting phases for $t \leq t_c$. Phys Rev E 53(6):5935–5948

Part III
Effects of a Steady-State Acceleration Field

Chapter 11
Introduction to Effects of a Steady-State Acceleration Field

The two previous parts of this book provided theoretical and experimental results on the hydodynamic behavior of supercritical and near-critical fluids following a parietal thermal disturbance in the absence of any acceleration field, as it is well-known that motion in such hyperdilatable and hyperexpandable fluids is extremely sensistive to gravity (see Sect. 1.8). However, in the past few decades, considerable attention has been focused on hydrodynamic instabilities in supercritical fluid layers—mainly in relation to the situation where the temperature gradient is in the opposite direction to the Earth's gravitational acceleration, a situation that is referred to as the Rayleigh–Bénard configuration (see Sect. 1.8.3 in Chap. 1 and [1, 16] for more details). Aside from the pioneering work of Busse [8] in 1967, a number of experiments have been performed to explore organized fluid motion close to or in the turbulent regime [2, 5, 6, 9, 14, 15, 17, 18, 21] . Key problems have also been addressed theoretically. Among the results obtained, it is worth noting that the Nusselt number is now known to be related to other characteristic numbers, scaling theories, large-scale temperature fluctuations, pattern formation, spiral turbulence, and the random reversal of macroscopic flows. Readers should also refer to [7] for a review of the state of the art in Rayleigh–Bénard convection in the year 2000, and to Chaps. 4 and 5 of [12] (published in 2006 for the Henri Bénard Centenary Review). More recently, non-Oberbeck–Boussinesq effects in turbulent convection have been studied by Ahlers et al. [3, 4].

Based on this brief review of organized fluid motion in the presence of a steady-state acceleration field, we find that the hydrodynamic behaviors of supercritical fluids can be schematically divided into those arising from the temperature dependence of the transport coefficients and those that depend on the equation of state, while theoretical and experimental studies have mainly focused on organized fluid motion due to the temperature-dependent transport properties of supercritical fluids. However, the hydrodynamic effects that arise mainly due to the equation of state—or, more precisely, due to the high compressibility and high dilatability of

B. Zappoli et al., *Heat Transfers and Related Effects in Supercritical Fluids*, Fluid Mechanics and Its Applications 108, DOI: 10.1007/978-94-017-9187-8_11,

supercritical fluids—are also of great interest. Although a numerical solution is yet to be obtained with a reliable Navier–Stokes code written for a near-critical fluid, very little is known about the hydrodynamic behavior of hypercompressible (and then hyperexpandable) systems. Some theoretical work has been done in an attempt to obtain stability criteria for fluid layers subjected to adverse temperature gradients [10, 11]. Even the simpler situation of normally compressible non-Boussinesq fluids (such as ideal gases) is yet to be extensively studied from a hydrodynamic point of view. For example, Spradley and Churchill [20] have studied pressure-driven (or thermoacoustic) motion that is generated by the local expansion of normally compressible fluids. Paolucci [19] has studied convection in strongly differentially heated cavities filled with normally compressible media. In addition to these studies of normally compressible media, a number of technical studies performed in the early 1970 s were devoted to investigating the hydrodynamic behaviors of very compressible cryogenic fluids, such as supercritical oxygen, hydrogen or helium, which are known to be of great technological interest. Heinmiller [13] has numerically studied convective flows in cryogenic oxygen tanks in order to obtain a description of the pressure collapse which follows the fast mechanical destratification of a near-critical fluid stored under microgravity conditions. However, even though these early papers contained interesting features that are strongly linked to present ideas, it was not possible to identify the basic heat transport mechanisms due to the inadequacy of the computational resources available at that time. Therefore, in Part III of this book, we explore these particular effects, especially the responses of supercritical fluids to transient time-dependent boundary heating or the imposition of a steady-state acceleration field (we will consider the effects of a time-dependent acceleration field in Part IV). Obviously, since convection is a multidimensional problem, the 1-D model derived in Part I for $g = 0$ must be generalized to a 2-D or even 3-D configuration that accounts for the relative directions of the thermal gradient and acceleration field. To do this, we must develop modeling approaches and numerical simulations based on the techniques described in Chap. 22. Therefore, this part of the book is divided into three chapters. In the first (Chap. 12), we study the interaction between the piston effect and gravitational convection numerically for a square cavity filled with a one-phase critical fluid that undergoes parietal heating or point-source heating. The configuration where this square cavity is heated from below is then addressed in Chap. 13. Finally, experiments performed in a low-acceleration field are presented in Chap. 14.

References

1. Alhlers G (2006) Dynamics of spatio-temporal cellular structures, Chap 4. Experiments with Rayleigh–Bénard convection. Springer, New York, pp 95–102
2. Ahlers GG, Xu X (1999) Phys Rev Lett 86:3320
3. Ahlers G, Dressel B, Oh J, Pesch W (2009) Strong non Oberbeck-Boussinesq effects near the onset of convection in a fluid near its critical point. J Fluid Mech 642:15–48

4. Ahlers G, Araujo FF, Funfschilling D, Grossmann D, Lohse S (2007) Non-Oberbeck-Boussinesq effects in gazeous Rayleigh–Bénard convection. Phys Rev Lett 98:054501-1-4
5. Ashkenazi S, Steinberg V (1999) High Rayleigh number turbulent convection in a gas near the gas–liquid critical point. Phys Rev Lett 83(18):3641–3644
6. Assenheimer M, Steinberg V (1993) Rayleigh-Bénard convection near the gas–liquid critical point. Phys Rev Lett 70(25):3888–3891
7. Bodenschatz E, Pesch W, Ahlers G (2000) Recent developments in Rayleigh-Bénard convection. Annu Rev Fluid Mech 32:709–778
8. Busse FH (1967) The stability of finite amplitude cellular convection and its relation to an extremum principle. J Fluid Mech 30:625–649
9. Chavanne W (1997) PhD thesis, University of Journal Fourrier, Grenoble
10. Gitterman M, Steinberg VA (1970) Criteria of occurrence of free convection in a compressible viscous heat conducting fluid. J Appl Math Mech 34:305
11. Gitterman M, Steinberg VA (1972) Establishment of thermal equilibrium in a liquid near the critical point. High Temp (USSR) 10(3):565
12. Guyon E, Mutabazi I, Wesfreid JE (eds) (2006) Dynamics of spatio-temporal cellular structures Henri Bénard centenary review. Springer Tracts in Modern Physics. vol 207 of Springer, New York
13. Heinmiller PJ (1970) A numerical solution of the navier-stokes equations for a supercritical fluid thermodynamic analysis. Technical report (T.R.W. Rep. 17618-H080-RO-00) Houston
14. Kogan AB, Meyer H (2001) Heat transfer and convection onset in a compressible fluid: ^{3}He near the critical point. Phys Rev E 63(5):056310
15. Kogan AB, Murphy D, Meyer H (1999) Rayleigh-Bénard convection onset in a compressible fluid: ^{3}He near t_c. Phys Rev Lett 82(23):4635–4638
16. Manneville P (2006) Dynamics of spatio-temporal cellular structures Henri Bénard centenary review, Chap 3. Rayleigh-Bénard convection: thirty years of experimental, theoretical, and modeling work. Springer, New York, pp 41–65
17. Niemela JJ, Skrbek L, Sreenivasan KR, Donnelly RJ (2000) Turbulent convection at high rayleigh numbers. Nature 404:837
18. Niemela JJ, Skrbek K, Sreenivasan KR, Donnelly RJ (2001) The wind in confined thermal convection. J Fluid Mech 449:169–178
19. Paolucci S (1982) On the filtering of sound from the Navier-stokes equations (sand 82-8257). Technical report, Sandia National Laboratories, Albuquerque
20. Spradley LW, Churchill SW (1975) Pressure and buoyancy-driven thermal convection in a rectangular enclosure. J Fluid Mech 70:705–720
21. Xu X, Bajaj KMS, Ahlers G (2000) Heat transport in turbulent Rayleigh-Bénard convection. Phys Rev Lett 84(19):4357–4360

Chapter 12
Interaction Between the Piston Effect and Gravitational Convection

Abstract Strong gravitational instabilities are observed on the ground in heated near-critical fluids. However, the piston effect homogenizes temperature. The origin of strong convective instabilities in a near homogeneous pure fluid addresses the question posed by the evaluation of the interaction between the gravitational convection and the piston effect. After a brief one-dimensional model generalization to the two- or three-dimensional models subjected to a vertical steady state acceleration, the two-dimensional studies of the experimental heating are detailed for two configurations: a side-heated cavity and an immersed point heat source.

12.1 Governing Equations

The 3-D governing equations are the Navier–Stokes equations written in their general forms (see Chap. 21) when the fluid system is subjected to the Earth's gravitational field **g** (with a modulus that is expressed in units of g_0; see Sect. 1.7). The definitions of the nondimensional dependent variables are the same as in Sect. 21.5. From Eqs. (21.57), (21.61) and (21.78), the governing equations are then

$$\rho_t + \nabla \cdot (\rho \mathbf{v}) = 0$$

$$\rho_t + \nabla(\rho \mathbf{v} \otimes \mathbf{v}) = -\gamma_{0,\mathrm{IG}}^{-1} \mathbf{P} + \varepsilon \left[\Delta \mathbf{v} + \frac{1}{3} \nabla (\nabla \cdot \mathbf{v}) \right] + \frac{1}{\mathrm{Fr}} \rho \mathbf{g}$$

$$\rho T_t + \nabla(\rho T) = -(\gamma_{0,\mathrm{IG}} - 1)(p + a\rho^2) \nabla \cdot \mathbf{v} + \frac{\varepsilon}{\mathrm{Pr}_0} \nabla \left[\left\{ 1 + \Lambda (T-1)^{-0.5} \right\} \nabla T \right] + \varepsilon \gamma_{0,\mathrm{IG}} (\gamma_{0,\mathrm{IG}} - 1) \phi_D,$$

where the additional number

$$\mathrm{Fr} = \frac{c_{s0}}{L g_0}$$

B. Zappoli et al., *Heat Transfers and Related Effects in Supercritical Fluids*, Fluid Mechanics and Its Applications 108, DOI: 10.1007/978-94-017-9187-8_12,

is the Froude number, which characterizes the order of magnitude of a typical buoyancy force compared to a characteristic inertial force in the flow. Note that the velocity in the definition of the Froude number is the sound velocity. The van der Waals equation of state is written in the same way as in Sect. 3.1.1.

The initial conditions correspond to an isothermal near-critical fluid at rest and in thermodynamic equilibrium in the Earth's gravitational field. They account for the stratification due to the diverging isothermal compressibility. The stratification cannot be neglected, even for a 1 cm high sample cell. The weak barometric pressure gradient is enough to induce a non-negligible density inhomogeneity. The initial stratified density field is obtained by solving the hydrostatic equilibrium equations numerically:

$$-\frac{1}{\gamma_{0,\mathrm{IG}}}\frac{dp(y)}{dy} = \frac{1}{\mathrm{Fr}}\rho(y)g_0,$$

where y is the vertical axis parallel to the Earth's gravitational field and $p(y)$ is the pressure given by the van der Waals equation of state

$$p(y) = \frac{\rho(y)T_0}{1-\frac{1}{3}\rho(y)} - \frac{9}{8}\rho(y)^2.$$

The constant temperature T_0 stands for the homogeneous initial temperature field.

12.2 Side-Heated Cavity

In this section, we consider the problem of the interaction between gravitational convection and the piston effect in a simple configuration. To this end, a 2-D square cavity 1 cm on a side is considered (see Fig. 12.1). The cavity contains CO_2 that is at critical density and is initially 1 K above its critical temperature. All the boundaries are thermally insulated except the wall at $x = 0$, which is considered to be a perfect heat conductor. At $t = 0$, its temperature is ramped by 10 mK in 1 s. Although the computations have been performed with dimensionless variables, it is more convenient to discuss the results in terms of dimensional quantities. Additional details are given in [6].

12.2.1 Piston-Effect Period

The temperature field at $t = 4.7$ s is given in Fig. 12.2b. We observe that temperature equilibration is almost achieved by the piston effect, while the effects of buoyancy are restricted to a low-density, narrow thermal plume visible at the hot wall and the top-left part of the cavity. This part of the flow field is described in detail later on.

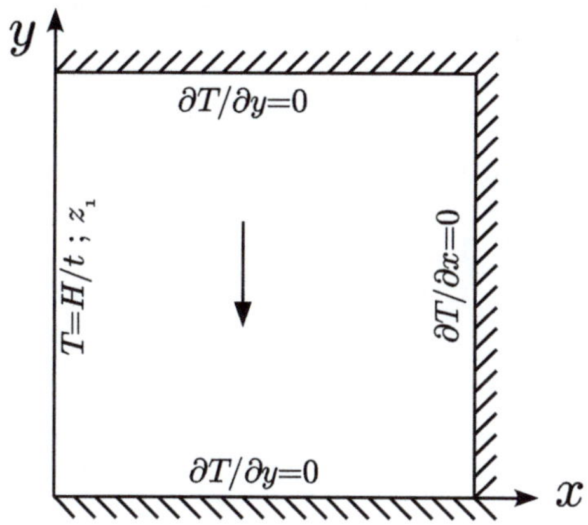

Fig. 12.1 The two-dimensional model of the cavity

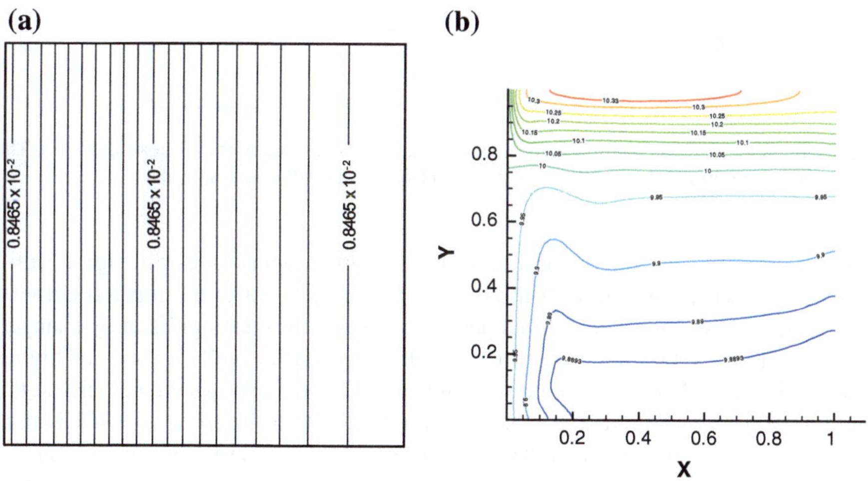

Fig. 12.2 Temperature differences from the wall temperature (isoclines labeled in K) at 4.5 s for **a** an ideal gas (isoclines // to the heating wall at $x = 0$) and **b** a supercritical fluid. Note the hot plume region in the *top left* part of (**b**), where the fluid temperature surpasses the wall temperature (see Sect. 12.2.3)

The thermal field of an ideal gas is given in Fig. 12.2a for the same time period. The straight isotherms parallel to the heating wall are characteristic of fast heat diffusion and weak fluid velocity. The density profiles would be homothetic to the temperature profiles. In contrast, in Fig. 12.2b, the homogeneous bulk temperature and density fields of a supercritical fluid occur at a time that is much shorter than the thermal diffusion time. They are the signature of the piston effect, and they show that there is no significant interaction between it and buoyant convection. Mass depletion does indeed occur in the thermal boundary layer, inducing a buoyant

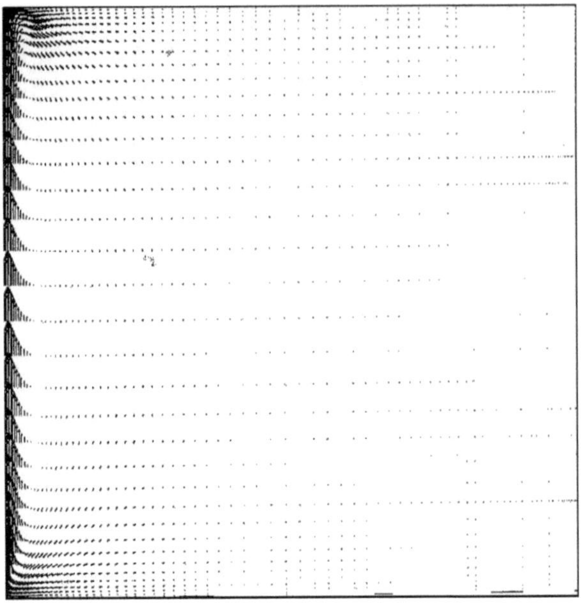

Fig. 12.3 Velocity field (vectors) in the supercritical fluid at the end of the 1 s heating period. The maximum vector plot is 7.2×10^2 μm s^{-1}(vertical vector along the heating wall)

vertical velocity component parallel to the isotherms, which does not significantly change the thermal structure of the boundary layer (except in the top-left corner). The net mass loss out of the thermal boundary layer is thus not significantly changed compared to the 1-D situation, although the buoyant vertical velocity is much higher than the horizontal one, as can be seen in Fig. 12.3, which shows the velocity at the end of the heating period ($t = 1$ s). The total mass transferred to the bulk is thus nearly convection independent. This is why the piston effect is responsible for the temperature equilibration of the bulk even in the presence of convection. However, it should be noted that the forced convection along the heated wall enhances the heat flux and thus the piston effect, since the velocity at the edge of the boundary layer is (as in the 1-D case) proportional to the heat flux at the wall [5].

The temperature difference (in mK) is plotted in Fig. 12.4a as a function of x (in cm) at $y = 0.5$ cm for several times (in s) during the temperature equilibration period for both the supercritical fluid and the ideal gas.

The supercritical fluid exhibits the characteristic profile of the piston effect (a very thin boundary layer followed by a homogeneous bulk phase). For the density field, some interesting features can be observed in Fig. 12.4b, which shows density profiles for both the supercritical fluid and the ideal gas near the heated wall at $y = 0.5$ cm for times on the piston-effect timescale. The first observation is that density inhomogeneities for the supercritical fluid are four orders of magnitude larger than those for the ideal gas. The second observation is that the boundary layer is much thinner for the supercritical fluid than for the ideal gas.

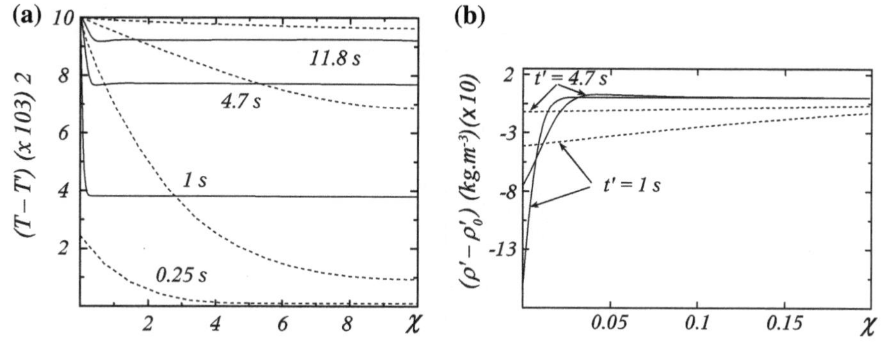

Fig. 12.4 **a** Evolution of the temperature increase (mK) as a function of x (cm) at $y = 0.5$ cm. *Continuous curve*: supercritical fluid; *dashed curve* ideal gas. **b** Evolution of the density decrease $\times 10$ ($\mathrm{kg\,m^{-3}}$) as a function of x (cm) at $y = 0.5$ cm. *Continuous curve* supercritical fluid; *dashed curve* ideal gas with time $\times\ 10^4$. Note that the density difference is plotted only near the wall

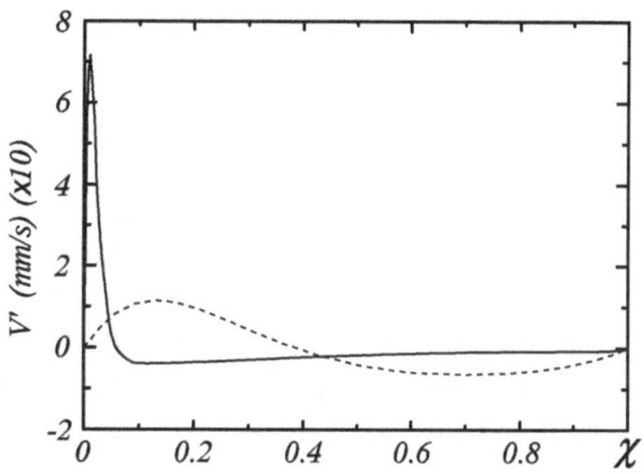

Fig. 12.5 Vertical velocity at $y = 0.5$ cm at the end of the heating period. The velocity in the ideal gas (*dashed*) is 10^4 times greater than the indicated velocity scale

Figure 12.5 shows the vertical velocity as a function of x (cm) at $y = 0.5$ cm for the supercritical fluid and the ideal gas. The result is a much larger vertical velocity occurring in a much thinner boundary layer for the supercritical fluid than for the ideal gas. While the driving force for convection (density gradient) fills the entire cavity for the ideal gas, thus producing a velocity field of homogeneous magnitude, the driving force for convection in the supercritical fluid is limited to a very thin boundary layer. Looking carefully at the bulk region located near the boundary layer, we can see that the fluid is drawn upward by viscous coupling (see [6]). This is confirmed in Fig. 12.6a, b, where the velocity vectors are shown for the two types of fluid at 4.7 s: while the ideal gas exhibits a classical circular one-roll pattern (Fig. 12.6a), convection in the supercritical fluid is located near the heated wall (Fig. 12.6b). It should also be noted that the intensity of convection for the supercritical fluid is much 300 times higher than that for the ideal gas. The results presented in Fig. 12.6a, b seem to contradict the fact that the thermal field is homogeneous. We address this paradox in the next section.

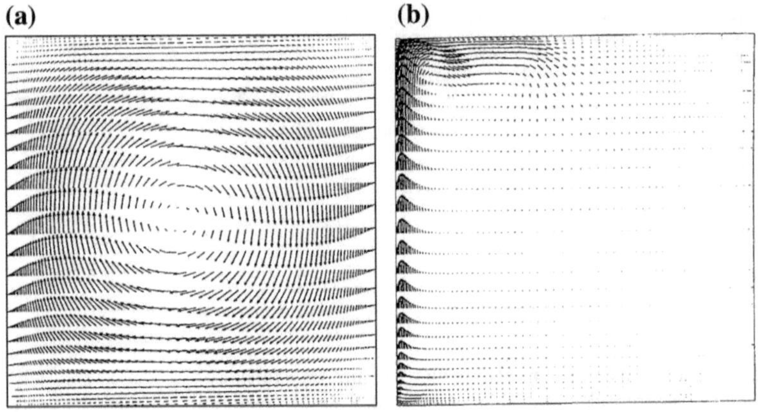

Fig. 12.6 Velocity fields for the ideal gas (**a**) and for the critical fluid (**b**) under ground conditions at 4.7 s. The maximum vector plot is 6.5 μm s^{-1} in (**a**) and 2.2×10^3 μm s^{-1} in (**b**)

12.2.2 Density Relaxation

The results of the above $1g$, 2-D calculations are in agreement with the piston-effect mechanism described in $0g$, 1-D approaches already presented in Chap. 4. Recall the two periods of the density relaxation: firstly, a short piston-effect time period during which the temperature relaxes to equilibrium (to first order) while the density relaxes to the order of magnitude it would have had if the fluid was an ideal gas; secondly, a divergent long heat-diffusion time period where density follows the relaxation of the second-order temperature inhomogeneities. Zero gravity experimental studies also report a very long density relaxation time [4]. In the $1g$ case, Boukari et al. [2] numerical solution for 1-D convectively stable flow equations has shown that the density profile can take hours to form, whereas temperature equilibrates very quickly through the piston effect. Similar $1g$ observations have been reported by Zhong and Meyer [8].

The presence of convection in a near-uniform thermal field can then be explained as follows. The second-order temperature inhomogeneities give rise to a first-order density inhomogeneity which relaxes very slowly, leaving time for gravity to generate significant convection in a near-isothermal fluid. The back flow also compensates for the mass depletion in the heated region, thus, in turn, speeding up density relaxation. This phenomenon has been called "isothermal critical convection" [6]. Under zero gravity, the equilibration of the density field takes longer, since only diffusion takes place. This can be seen in Fig. 12.7a, b, where the role of gravity is highlighted later on during the relaxation process ($t = 16.6$ s). Figure 12.7a, which shows temperature as a function of x at $y = 0.5$ cm, confirms that the piston effect is responsible for 85 % of the equilibration under zero gravity. Under $1g$, in the presence of convection, the equilibration is barely any faster (10 % quicker). It is also clear from Fig. 12.7b

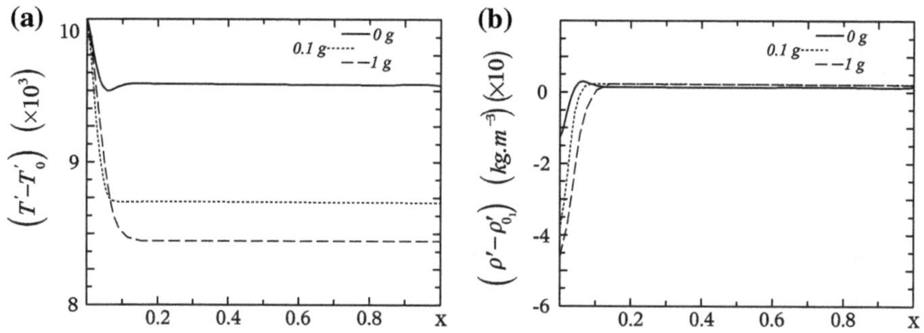

Fig. 12.7 Temperature difference (**a**) and density difference (**b**) fields for the supercritical field at $y = 0.5$ cm and $t = 16.6$ s as a function of x and at different levels of gravity. Temperature is mostly homogenized by the piston effect, while density is relaxed by diffusion and convection

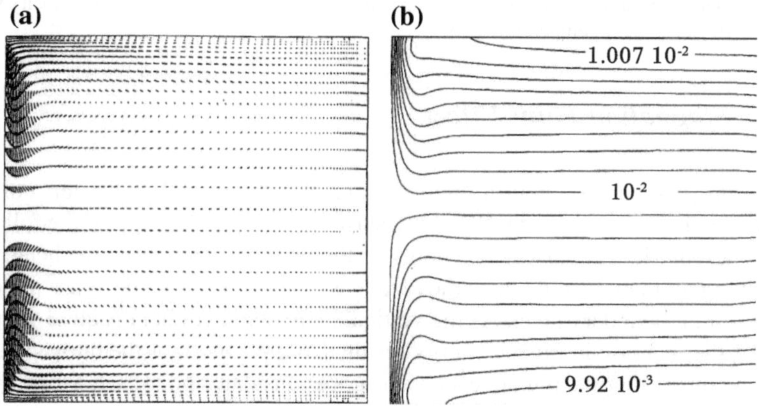

Fig. 12.8 Near-isothermal convection field at $t = 713$ s: **a** velocity, where the maximum vector plotted is $86\,\mu\text{m s}^{-1}$; **b** temperature

that the mass depletion in the boundary layer is reduced when gravity is increased. In near-critical fluids, convection clearly acts to homogenize density.

Examples of isothermal critical convection patterns at the latest computed time $t = 713$ s are given in Fig. 12.8a (velocity) and b (temperature). Unusual two-roll patterns are observed (see Fig. 12.8a), the origin of which is discussed in the next section. Note that the fluid velocity in the supercritical fluid more than 10 min after the heating was turned off is still larger than its magnitude in the ideal gas only a few seconds after the end of the heating period (see Fig. 12.6a). The convective pattern slowly disappears on the heat diffusion timescale without changing its two-roll configuration significantly. In addition, the convective pattern remains quasi-steady on the convective timescale owing to the long density homogenization time.

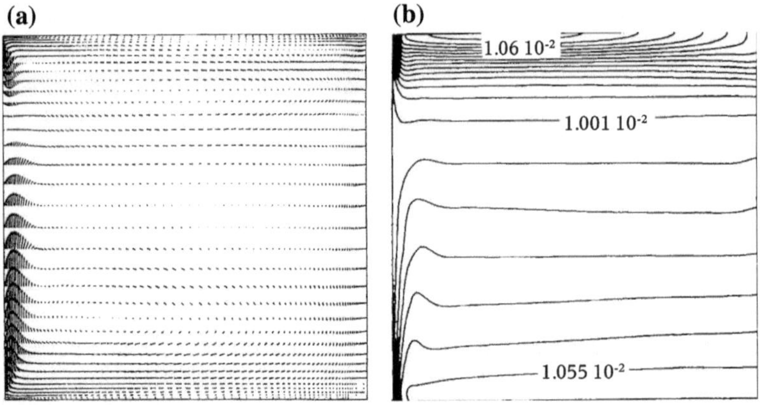

Fig. 12.9 The hydrodynamic field at $t = 35.5$ s. **a** Velocity, where the maximum velocity plotted is 2.24×10^2 μm s^{-1}. **b** Temperature (K)

12.2.3 Stagnation Point Effect

In Fig. 12.2, we can see that the temperature of the top wall in the left corner is higher than that of the heated wall. This a surprising result. We suggest the following interpretation. The gain in kinetic energy of the fluid particles along the heated wall is much higher than it would be in the ideal gas, due to the strong decrease in density in the thermal boundary layer. As the vertical hydrostatic gradient is quite small 1 K from T_c, the increase in kinetic energy results from the work of the buoyancy force alone. When the thermal plume reaches the upper, insulated boundary, a turning flow region forms that surrounds a stagnation point where the kinetic energy turns into internal energy and into work done by pressure forces to expand the fluid because of the high isothermal compressibility.

The existence of a spot on the upper wall that is hotter than the heated boundary is associated with the formation of the two-roll structure of the long-lasting quasi-isothermal near-critical convection, as shown in Fig. 12.9a, b, corresponding to $t = 35.5$ s. The hotter fluid at the insulated boundary in the upper left corner expands horizontally, but as it flows towards the left, cooler boundary, it becomes heavier and flows downward. A counterclockwise roll appears in the left corner. Then, as heat is slowly carried away by diffusion, the hot region widens and so does the corresponding roll (see Fig. 12.8a). At the same time, a back flow increases the density of the thermal boundary layer, and thus lowers the buoyancy forces and decreases the kinetic energy. The hot region tends to disappear over time through diffusion and decreasing kinetic energy.

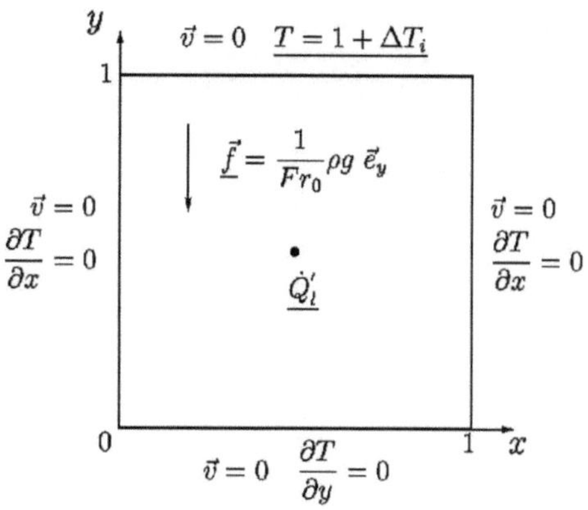

Fig. 12.10 Cavity configuration for the $1g$, 2-D model of the heating from an immersed point heat source. The gravitational force f is given in nondimensional form

12.3 Immersed Point Heat Source

We now consider the configuration of the experiment described in Chap. 9 where a point heat source (thermistor) is immersed in a fluid cavity with thermostated walls. Under $1g$, the buoyant thermal plume in near-critical CO_2 interacts with the upper, colder wall. This results in an enhanced cooling piston effect. The temperature equilibration comes from the balance between the two competing heating and cooling piston effects.

12.3.1 The Model

This section is devoted to the $1g$, 2-D numerical simulations [7] of the thermal behavior of near-critical CO_2 contained in a square cavity (10 mm side). The upper wall is maintained at the initial temperature while the other walls are adiabatic. The fluid is heated continuously by a point heat source located at the center (Fig. 12.10). The heating parameters were adjusted to fit with existing values of heat fluxes; e.g., a constant heat power injection of 0.74 $W\,m^{-1}$ (heat flux from a linear unit in 2-D modeling corresponding approximately to a heat flux of 5 mW dissipated by the thermistor). Initially, the fluid is set at 1 K above the critical temperature, and the very small density stratification due to gravity is accounted for. The model equations are similar to those of the previous section.

12.3.2 Flow and Thermal Field

Because the fluid's thermal expansion coefficient is large and its thermal diffusivity is low, the numerical simulation shows large density gradients (about 5×10^4 $kg\,m^{-4}$)

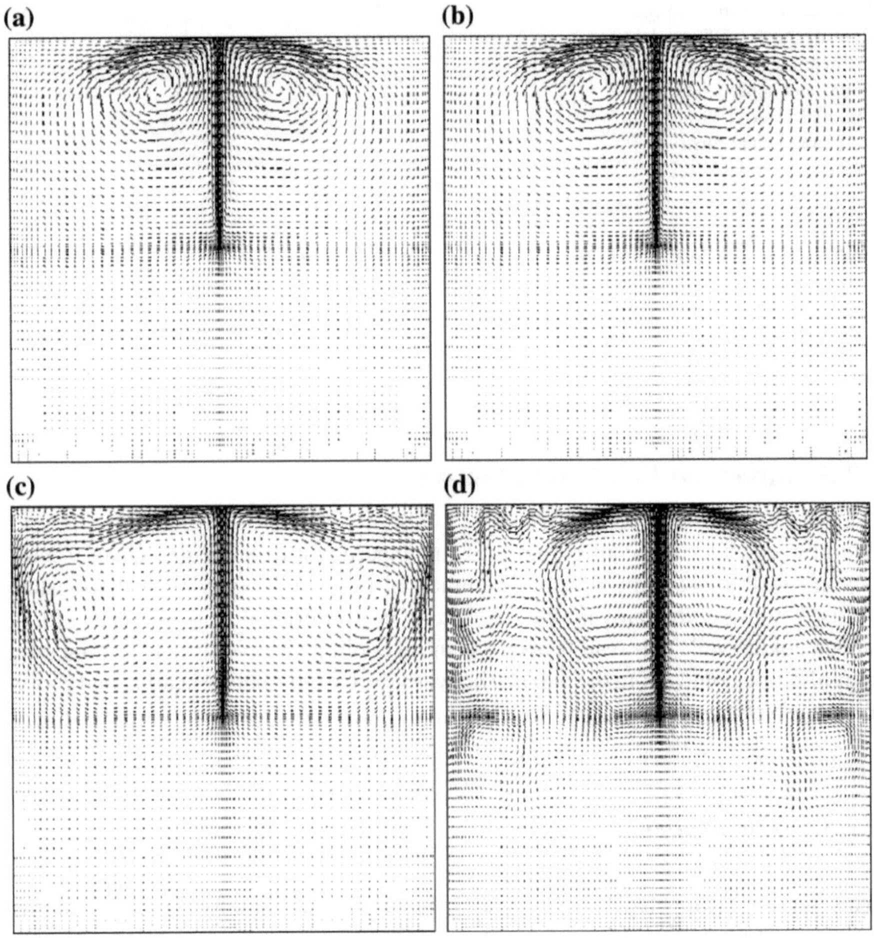

Fig. 12.11 Velocity field during the rise of the plume at **a** $t = 0.48$ s, **b** $t = 1.19$ s, **c** $t = 2.38$ s, **d** $t = 4.76$ s

close to the heat source. The temperature of the mesh point heat source increases up to approximately 2 K above the initial temperature. Consequently, the buoyancy forces generate a strong flow (the velocity reaches $15\ \mathrm{mm\,s^{-1}}$) and a thermal plume rises to the top wall, which has been observed experimentally [1, 3]. The main flow features can be seen in Fig. 12.11a–d, which represent the velocity field at different times during heating.

Two counter-rotating cells form beside the front of the rising flow by viscous coupling (Fig. 12.11a). Then, the plume strikes the upper boundary (Fig. 12.11b) and spreads along the thermostated wall (Fig. 12.11c). The flow becomes complex, with secondary vortices (Fig. 12.11c, d). In contrast, the lower part of the cavity is nearly unaffected by convection.

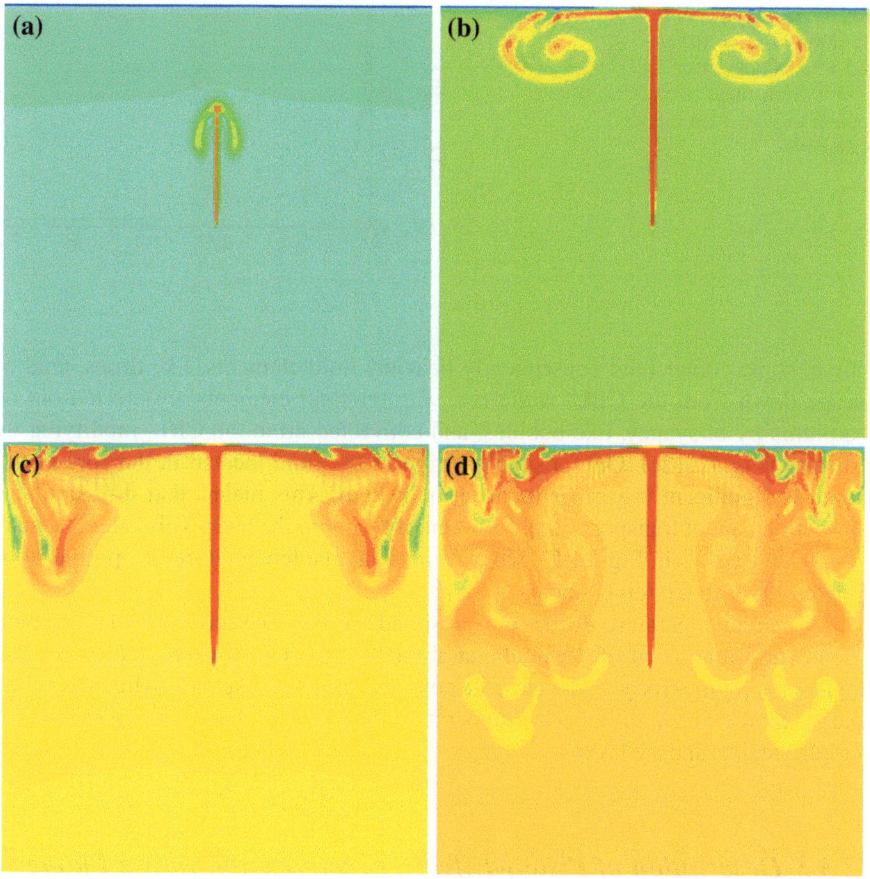

Fig. 12.12 Temperature field (*red* to *blue* depicts hot to cold) during the rise of the plume at **a** 0.48 s, **b** 1.44 s, **c** 2.38 s, **d** 4.76 s

The temperature distribution in the fluid during the same period is represented in Fig. 12.12a–d. The isothermal lines are plotted for temperature differences ranging from $\Delta T = 0$ to $\Delta T = 30$ mK. At this timescale, the heat penetration depth remains very small due to the low value of the thermal diffusivity. Figure 12.12a shows that the very local vortices extract some heat from the rising flow. The thermal field in the upper part of the cavity exhibits a complex structure as soon as the plume strikes the wall (Fig. 12.12b–d). Nevertheless, the presence of a thermal boundary layer along the top, thermostated wall shows that the bulk fluid temperature has risen. As neither convection nor diffusion from the heat source can bring heat into this area, this boundary layer is the hallmark of the piston effect, and it confirms the efficiency of the piston effect in the presence of convection.

The area of interaction between the cold boundary layer (CBL) and the hot plume progressively increases. In addition to the flow related to the spreading of the thermal

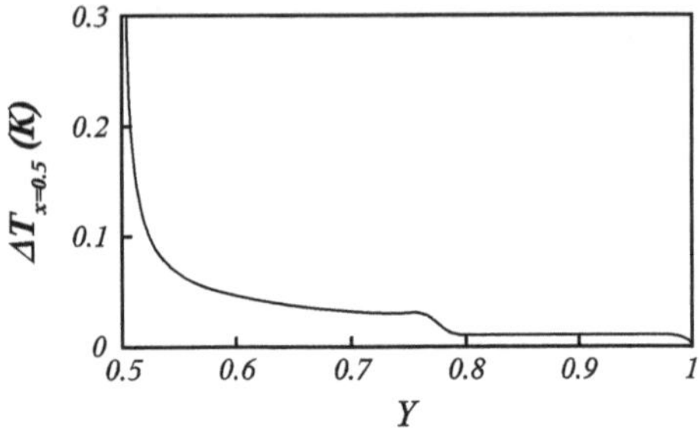

Fig. 12.13 Temperature elevation at $x = 0.5$ cm and $t = 0.48$ s from $y = 0.5$ cm (heat point source) to $y = 1$ cm (upper boundary)

plume, some cooler (and consequently heavier) fluid elements, like drops, tend to escape down from the CBL, while two warmer fluid elements are ejected out of the plume by buoyancy, like bubbles, and convected up to the CBL, modifying its structure (Fig. 12.12c). Only a small portion of the fluid located in the core of the plume is significantly warmer than the bulk fluid. This means that diffusion and convection cannot transport a large amount of heat into the bulk. This can be seen as early as $t = 0.48$ s in Fig. 12.13, which represents the temperature rise profile from the thermistor to the top boundary.

The steep temperature decrease corresponds to the convective transport, which extends up to the front of the plume at about $y \sim 0.76$ cm, where $\Delta T \sim 40$ mK. Beyond the plume (for $y \geq 0.8$ cm), the constant value corresponds to homogeneous heating by the piston effect. The temperature drop at $y \sim 1$ cm corresponds to the cold thermal boundary layer.

12.3.3 Description of Piston-Effect Thermalization with a Plume

The temperature of the bulk fluid results from competition between a heating piston effect (point heat source) and a cooling piston effect (upper, thermostated wall). Figure 12.14 shows the vertical velocity (a) and the temperature profiles (b) close to the upper wall at $x = 0.75$ cm and at 6 ms, before the plume's hot front reaches the edge of the cold boundary layer.

Solid lines correspond to the current configuration, while dashed lines represent, for the sake of comparison, the results of the same computations when the upper boundary is adiabatic. In the latter case, without CBL, the velocity goes linearly to zero at $y = 1$ (Fig. 12.14a). In contrast, when the wall is thermostated, a thermal CBL and a dynamic boundary layer appear simultaneously (Fig. 12.14a, b). When cooled in the CBL, the fluid contracts and induces an additional, homogeneous motion of the rest of the fluid toward the top wall. The bulk velocity is still linear but it is higher than in the adiabatic case, where the velocity derives only from the HBL expansion. Accordingly, the bulk density is lowered when compared to the adiabatic case, as are the pressure and bulk temperature. The decrease in temperature can be observed in

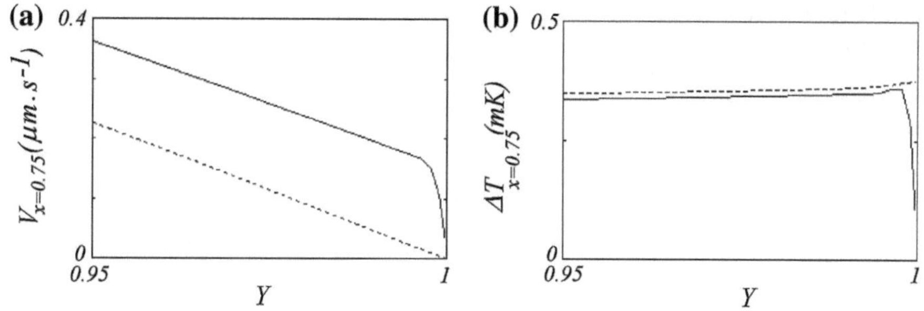

Fig. 12.14 Vertical velocity (**a**) and temperature (**b**) profiles versus y at $x = 0.75$ cm and $t = 6$ ms close to the top wall when the latter is thermostated (*solid*) or insulated (*dashed*)

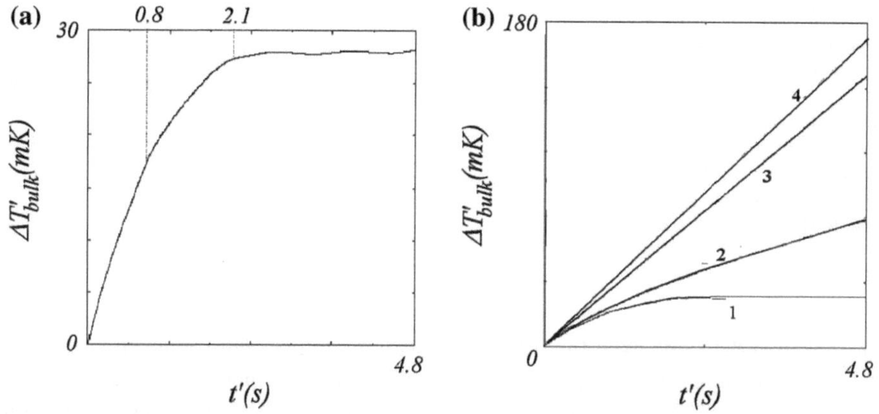

Fig. 12.15 Temperature elevation in the bulk: **a** $g = 1$ and thermostated top boundary. **b** thermostated top wall and $g = 1$ (*curve 1*) or $g = 0$ (*curve 2*), or insulated top wall and $g = 1$ (*curve 3*) or $g = 0$ (*curve 4*)

Fig. 12.14b: homogeneous cooling is added in the bulk to the heating caused by the heating piston effect.

12.3.3.1 Thermoacoustic Cooling Triggered by a Thermal Plume

The global piston-effect heating in the fluid is shown in Fig. 12.15a, which reports the bulk temperature variations with time. The temperature variations are measured in the bulk of the bottom part of the cavity, which receives no heat through diffusion or convection. Two main changes in the rate of increase in temperature can be observed: first around $t = 0.8$ s and then around $t = 2.1$ s, which finally leads to a nearly constant value (28 mK). These observations will be discussed in more detail later. This bulk temperature variation is different from that observed for $g = 0$ (curve 2 in Fig. 12.15b), where it exhibits a continuous increase. Note that curve 1 ($g = 0$)

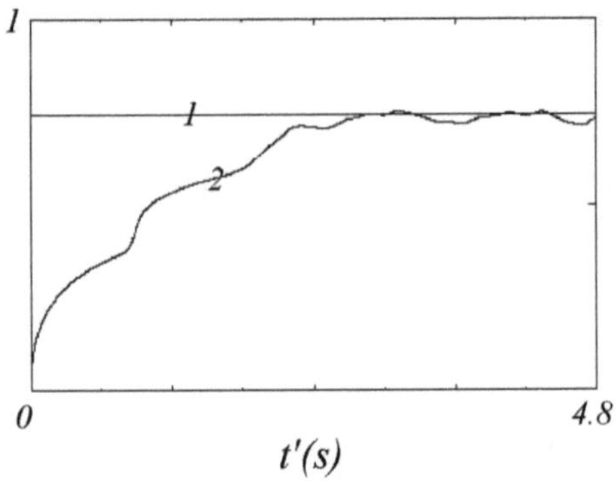

Fig. 12.16 Evolution of the heat flux injected by the point heat source (*curve 1*) and that of the heat flux extracted at the upper wall (*curve 2*)

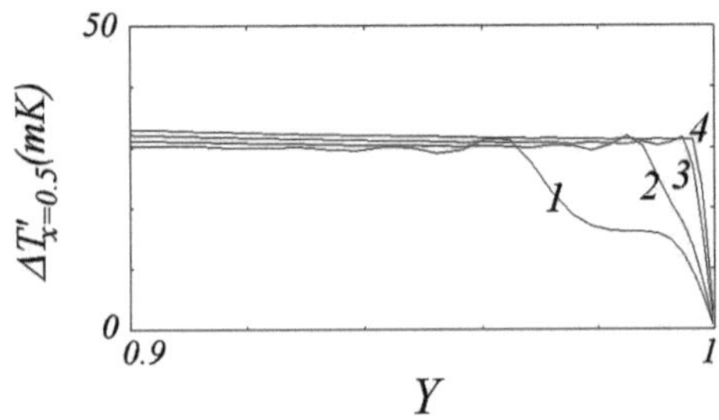

Fig. 12.17 Temperature profiles with respect to the vertical coordinate y around the cold boundary layer when the rising plume impinges the boundary at $x = 0.5$ for different times: $t = 0.76$ s (*curve 1*), $t = 0.81$ s (*curve 2*), $t = 0.86$ s (*curve 3*) and $t = 0.91$ s (*curve 4*)

in 12.15b is the same as in Fig. 12.15a but with the temperature scale changed. The temperature increase when the upper wall is insulated is depicted in curves 3 ($g = 1$) and 4 ($g = 0$). This observed difference is obviously linked to convection, but the exact mechanism is yet to be determined.

As shown in Sect. 12.2, the heating piston effect is not greatly affected by convection. The changes in the bulk fluid heating (curve 1 in Fig. 12.15b) when compared with the situation for weightlessness (curve 2 in Fig. 12.15b) must then be linked to cooling piston-effect modifications. This is also suggested by the plot in Fig. 12.16, which represents the evolution of the total heat flux at the boundaries: curve 1 for the constant heat source and curve 2 at the cold wall. The temperature evolution in Fig. 12.15a is well correlated with the heat flux extracted at the thermostated boundary (see Fig. 12.16). The bulk heating rate is directly proportional to the total heat flux and is thus governed by the changes in heat loss. These changes are connected to hydrodynamic events near the CBL, as shown in the following discussion.

In Fig. 12.15b, the small difference between curves 1 and 2 until $t \sim 0.8$ s is similar to that observed between curves 3 and 4. In contrast, the change in slope during the temperature evolution that occurs at $t \sim 0.8$ s is due to cooling piston-effect enhancement. At this time, the rising plume begins to interact with the middle of the upper wall and the structure of the CBL becomes locally modified. Figure 12.17 shows the temperature variations along the $x = 0.5$ vertical axis caused by this flow–wall interaction between $t = 0.76$ s and $t = 0.91$ s.

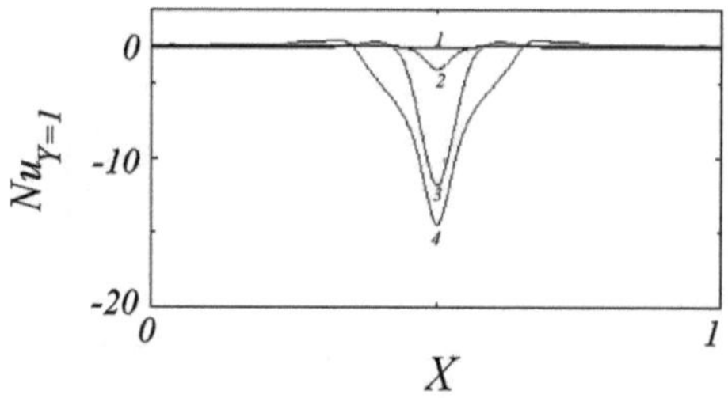

Fig. 12.18 Nusselt number at the upper wall at different times: $t = 0.71$ s (*1*), $t = 0.81$ s (*2*); $t = 0.91$ s (*3*); $t = 1.19$ s (*4*)

The outer boundary layer temperature becomes that of the flow, and the thermal diffusing layer suddenly becomes much thinner. This leads to much steeper temperature gradients in this region of the CBL and to a sudden increase in the heat loss through the thermostated wall, as shown by curve 2 in Fig. 12.16.

The fluid located in the CBL is then subjected to stronger cooling from the isothermal wall. Hence, it contracts more; the cooling piston effect is enhanced and the global heating piston effect decreases (Fig. 12.15a).

The Nusselt number is the ratio of the convective heat transfer coefficient to the conductive heat transfer coefficient. In the present situation, it is the ratio of the heat flux at the top wall to the conductive heat flux between the top and bottom walls.

The variation of the Nusselt number at the top wall with respect to x is plotted in Fig. 12.18 at different times. It illustrates the above mentioned sudden changes in the thermal structure of the boundary layer (see, for example, curves 1–3 at $t = 0.71,\ 0.81,\ 0.91$ s).

At $t = 1.19$ s (curve 4 in Fig. 12.18), the rate of change in heat loss decreases in the middle of the wall. This phenomenon corresponds to the fluid spreading toward the sides of the cavity (see also Figs. 12.11b and 12.12b).

Another crucial event must occur to explain the second significant change in the temperature rise at $t \sim 2.1$ s due to sudden, local modifications of the CBL structure.

Figure 12.19 shows the thermal field in the upper left side of the cavity (the flow is symmetric over the central vertical axis). This shows that some masses of warm fluid convected by the plume escape from the two vortices due to their buoyancy and strike the isothermal upper wall around $t = 2.1$ s. Heat losses are consequently increased (see Fig. 12.16), and the cooling piston effect is enhanced again (see Fig. 12.15a).

12.3.3.2 Long-Term Behavior

For $t \geq 2.4$ s, Figs. 12.15a and 12.16 show that bulk heating stops when the cooling piston effect balances the heating piston effect due to convection-forced heat diffusion at the cold wall. Consequently, a state of thermal quasi-equilibrium is reached in the bulk fluid where heat is transferred from the HBL to the CBL without bulk heating. All of the heat transferred to the bulk by the heating piston effect is instantaneously removed by the cooling piston effect. Under such conditions, the bulk fluid acts as

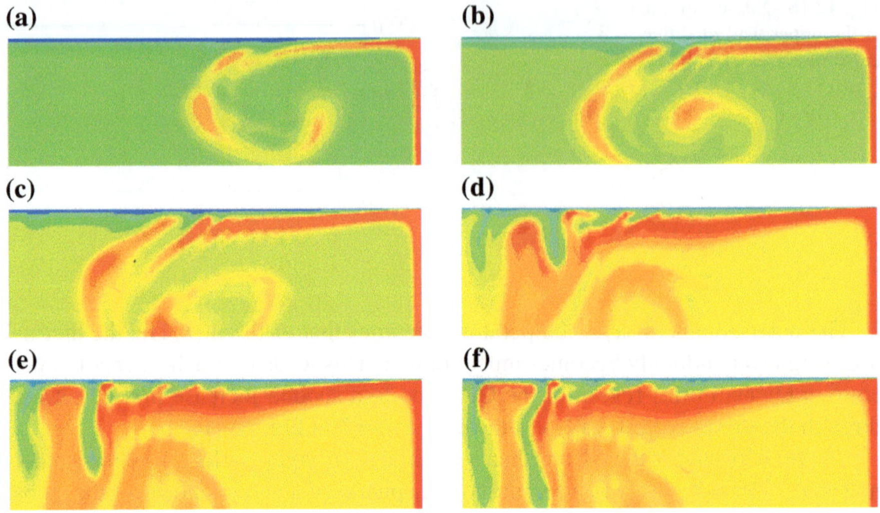

Fig. 12.19 Temperature field (*red to blue* is hot to cold) near the upper wall in the upper-left part (ordinate: $0.95 \le y \le 1$; abscissa: $0 \le x \le 0.5$) of the cavity when the plume impinges on the thermostated boundary: **a** $t = 1.19$ s, **b** $t = 1.42$ s, **c** $t = 1.67$ s, **d** $t = 1.84$ s, **e** $t = 2.1$ s, **f** $t = 2.2$ s

a thermal short circuit (we have already shown, in Sect. 5.2.1, that the bulk fluid acts as a thermal short circuit). However, in this latter case, when gravity-induced convective flows are absent, the stationary state is driven by pure conduction, while convection should prevent conduction from settling in the present case. In this case, a steady state could be reached. Nevertheless, Fig. 12.12c, d show that convection transports very little heat into the bulk between $t = 2.4$ s and $t = 4.8$ s. Heat brought by the thermal plume and the secondary flows around the CBL do not generate significant variations around the equilibrium temperature in the bulk fluid. The weak bulk temperature oscillations after $t = 2.4$ s (Fig. 12.15a) are related to the CBL instabilities (presumably of the Kelvin–Helmholtz type) that provoke heat loss variations (Fig. 12.16). It is likely that this behavior is maintained after $t = 4.8$ s. From Fig. 12.11c, d, it can be seen that, although the plume remains confined to the middle of the cavity, the lateral flows are very unsteady and can even be turbulent. Thus, strictly speaking, no steady state can be expected.

Farther from the critical point ($T_i = T_c + 15$ K, $\rho_i = \rho_c + 125$ kg m^{-3}, see Fig. 12.20a), equilibrium is no longer reached on a short timescale since the efficiency of the piston effect decreases while diffusion plays a more important role. The temperature map (Fig. 12.20b) shows that the instabilities in the CBL also disappear.

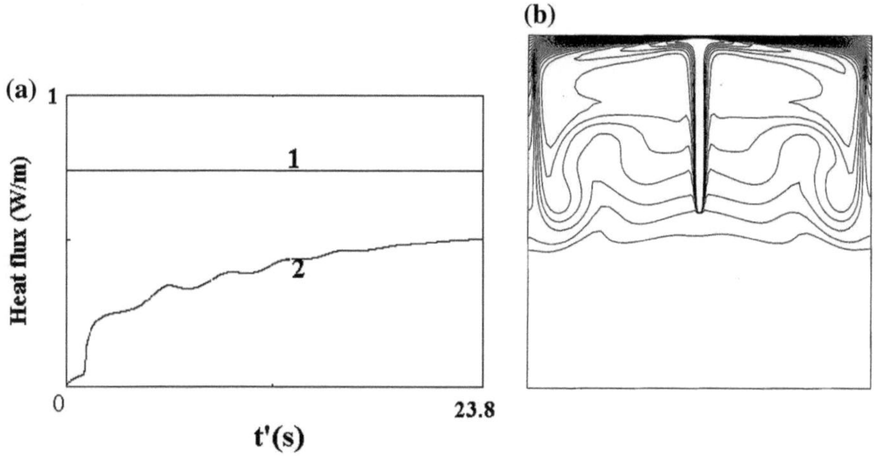

Fig. 12.20 Evolutions at $T_i = T_c + 15\,\text{K}$, $\rho_i = \rho_c + 125\,\text{kg}\,\text{m}^{-3}$: **a** heat flux brought to the fluid (*1*) and heat losses (*2*). **b** temperature pattern at $t = 19\,\text{s}$

References

1. Bonetti M, Perrot F, Beysens D, Garrabos Y (1994) Fast thermalization in supercritical fluids. Phys Rev E 49(6):R4779–R4782
2. Boukari H, Pego RL, Gammon RW (1995) Calculation of the dynamics of gravity-induced density profiles near a liquid–vapor critical point. Phys Rev E 52(2):1614–1626
3. Garrabos Y, Bonetti M, Beysens D, Perrot F, Fröhlich T, Carlès P, Zappoli B (1998) Relaxation of a supercritical fluid after a heat pulse in the absence of gravity effects: theory and experiments. Phys Rev E 57(5):5665–5681
4. Guenoun P, Khalil B, Beysens D, Garrabos Y, Kammoun F, Le Neindre B, Zappoli B (1993) Thermal cycle around the critical point of carbon dioxide under reduced gravity. Phys Rev E 47(3):1531–1540
5. Soboleva EB (2003) Adiabatic heating and convection caused by a fixed-heat-flux source in a near-critical fluid. Phys Rev E 68:042201
6. Zappoli B, Amiroudine S, Carlès P, Ouazzani J (1996) Thermoacoustic and buoyancy-driven transport in a square side-heated cavity filled with a near-critical pure fluid. J Fluid Mech 316:53–72
7. Zappoli B, Jounet A, Amiroudine S, Mojtabi A (1999) Thermoacoustic heating and cooling in near-critical fluids in the presence of a thermal plume. J Fluid Mech 388:389–409
8. Zhong F, Meyer H (1995) Density equilibration near the liquid–vapor critical point of a pure fluid: single phase $t > t_c$. Phys Rev E 51(4):3223–3241

Chapter 13
Rayleigh–Bénard and Rayleigh–Taylor Instabilities

Abstract The stability of a near-critical fluid is studied in the Rayleigh–Bénard configuration: the fluid is heated from below or presents itself as a layer of a heavier fluid is on top a lighter one. The Rayleigh–Bénard instability is first addressed showing a double Rayleigh–Bénard situation thanks to the presence of the piston effect. The stability criterium is then discussed and some similarities with geophysical flows described. In the case of a warm fluid layer topped by a cooler one, it is shown that the thermal diffusion layer behaves as an interface and gives rise to a Rayleigh–Taylor like instability. Thermo-acoustic oscillations at the Rayleigh–Bénard threshold are described. They come from the interaction between thermal plumes with the thermostated, horizontal upper wall.

13.1 Rayleigh–Bénard Instability

13.1.1 Rayleigh and Schwarzschild Criteria

When a square cavity containing a fluid is initially heated from the side, there is fluid motion for any value of the thermal gradient; the configuration is said to be unconditionally unstable since a vanishingly slow motion exists for a vanishingly small temperature gradient. When the cavity is initially heated from below, the configuration is said to be conditionally stable: for small and steady temperature differences between the bottom and top plates, the fluid initially remains at rest, and heat is transported by diffusion; for increasing values of the temperature gradient, the fluid suddenly starts to move at a given threshold value of the temperature gradient. In this case, internal motion starts within the fluid due to buoyancy, which attempts to even out the temperature throughout the sample. However, this motion is hindered by the density and pressure stratification of the fluid and dissipative processes. Usually, the effect of one of these two factors on the onset of convection is analyzed, and this leads to either the Rayleigh criterion or the Schwarzschild criterion, both of which

B. Zappoli et al., *Heat Transfers and Related Effects in Supercritical Fluids*,
Fluid Mechanics and Its Applications 108, DOI: 10.1007/978-94-017-9187-8_13,

are well known [19]. The resulting Boussinesq equation in which the density variations are neglected except in the term describing the buoyancy driving force is also applicable to a compressible fluid and an equivalent compressible Rayleigh number (which accounts for the two criteria listed below) can be derived from it [22].

In the case of near-critical fluids, the expansion coefficient diverges, which suggests that convective instabilities become more and more important. Gitterman and Steinberg [15] have used heuristic arguments to derive the convection threshold in the classical Rayleigh–Bénard problem for a fluid near its critical point. Experimental investigations of Rayleigh–Bénard convection in near-critical conditions can be found in [8, 17].

For a bidimensional layer of an incompressible fluid of infinite extension confined between two horizontal plates that are separated by a distance L, the following Rayleigh criterion applies:

$$\mathrm{Ra} = \frac{g_0 \rho c_p \beta_p e^3 \Delta T}{\nu \Lambda} = \frac{g_0 \rho c_p}{\Lambda \nu} \left(\frac{\partial \rho}{\partial T} \right)_p L^3 (T_L - T_U) \geq \mathrm{Ra}_c, \tag{13.1}$$

where Ra is the nondimensional Rayleigh number defined by Eq. (1.63). T_L and T_U are the temperatures of the lower and upper plates, respectively. Ra_c is the critical Rayleigh number, which is a threshold value above which the fluid motion starts. This critical value is equal to 657.5, 1707.8 or 1100.6 depending on whether the top and bottom boundaries are both stress-free, both solid and isothermal, or whether one boundary is solid and the other is a free boundary, respectively [12]. If the pressure-related density variations can be neglected; i.e., if

$$\left(\frac{\partial \rho}{\partial p} \right)_T \Delta p \ll - \left(\frac{\partial \rho}{\partial T} \right)_p \Delta T \text{ or } \frac{\Delta T}{\Delta p} \gg \frac{\kappa_T}{\beta_p}, \tag{13.2}$$

the Rayleigh criterion can be extended to compressible fluid cases.

When expansion is the leading phenomenon and viscosity and heat conductivity can be neglected (cloud physics), the instability starts for the Schwarzschild criterion

$$\frac{\partial T}{\partial y} \geq - \frac{g_0 \rho}{\beta_p} (\chi_T - \chi_S) = \frac{g_0 T \beta_p}{c_p}, \tag{13.3}$$

where χ_T and χ_S are, respectively, the isothermal and isentropic compressibilities, and β_p is the isobaric thermal expansion coefficient. The right-hand side of Eq. (13.3) represents the adiabatic gradient that is obtained by adiabatically lifting a fluid particle up a hydrostatic pressure gradient [22]. This notion is commonly used in atmospheric science.

The definition of the local Rayleigh number that accounts for these two effects for compressible flows can be written as a modified Rayleigh number [6]:

$$\mathrm{Ra} = \frac{g_0 L^4 \beta_p \rho^2 c_p}{\Lambda \mu} \left[\frac{T_L - T_U}{L} - \frac{g_0 T \beta_p}{c_p} \right]. \quad (13.4)$$

It is worth noting that the second term of this expression (the Schwarzschild term) is constant when one approaches the critical point, because β_p and c_p diverge with the same critical exponent. As the temperature difference $T_L - T_U$ is on the order of a few mK, the temperature in the Schwarzschild term is considered equal to the initial fluid temperature. For the same reason, β_p and c_p are calculated at this temperature.

Gitterman and Steinberg [15] developed an asymptotic model to determine the onset of convection in compressible and dissipative fluids and then extended it to near-critical conditions. They determined an approximate critical Rayleigh number that conforms to the following expression:

$$\mathrm{Ra}_c = \mathrm{Ra}_c^{in} \frac{1 + \left(\frac{L}{L_1}\right)^4 + \left(\frac{L}{L_1}\right)^2 \left(\frac{L_2}{L_1}\right)^2}{1 + \left(\frac{L_3}{L_1}\right)^4 \left(\frac{L_2}{L}\right)^2}, \quad (13.5)$$

where Ra_c^{in} is the critical Rayleigh number for an incompressible fluid, L is the length scale of the cavity, and $L_{1,2,3}$ are the characteristic lengths derived from Gitterman and Steinberg as

$$L_1 = \left[\frac{\mathrm{Ra}_c^{in} \nu D_T}{g_0 \left(\frac{\partial \rho}{\partial p}\right)_T \left(1 - \frac{c_V}{c_p}\right)} \right]^{\frac{1}{4}}$$

$$L_2 = \left[\frac{a_1 \nu D_T}{\left(1 - \frac{c_V}{c_p}\right)} \left(\frac{\partial \rho}{\partial p}\right)_T \right]^{\frac{1}{2}}$$

$$L_3 = \left[\frac{\mathrm{Ra}_c^{in} \nu D_T}{g_0^2 \left(\frac{\partial \rho}{\partial p}\right)_T} \right]^{\frac{1}{4}}$$

The magnitudes of the coefficients that appear in the above definitions depend on the boundary conditions. For two solid walls, $\mathrm{Ra}_c^{in} = 1707.8$. In the expression of L_2, the value of $a_1(\sim 70.5)$ is only known for the case where both surfaces are solid. We assume, following Zhong and Meyer [24], that this value also holds for the solid–free surface configuration. From the Gitterman and Steinberg analysis, the onset of Rayleigh–Bénard instability inside a near-critical fluid layer is characterized by the local Rayleigh number from Eq. (13.1), which has to be compared to the critical Rayleigh number derived from Eq. (13.5).

The analysis by Carles et al. in [11] discusses the approximations in Gitterman and Steinberg's theory and concludes that their developments lead to heuristic criteria through asymptotically formally incorrect equations. However, the deviation between both approaches does not exceed 0.1 % over a large range of temperature differences from the critical point. It also appears that Gitterman and Steinberg's model is more complex than needed for Eq. (13.5), and that the classical criteria from Spiegel [22] can be used to estimate the threshold of convection. If the Rayleigh number as defined by Eq. (13.4) exceeds the critical Rayleigh number for a Boussinesq fluid, the instability can start.

Different stable and unstable regimes are predicted depending on the proximity to the critical point and the height of the cell. Far from T_c ($T - T_c \geq 1\ K$) and for an infinite fluid layer, the onset of convection is governed by the classical Rayleigh number. However, near the critical point, the onset occurs for larger temperature differences and is ultimately governed by the Schwarzschild criterion. From Eq. (13.4), the temperature difference at the onset is

$$(\delta T)_{onset} = \mathrm{Ra}_c \frac{\lambda \mu}{g_0 L^3 \beta_p \rho^2 c_p} + \frac{g_0 T \beta_p L}{c_p}. \tag{13.6}$$

The second term on the right-hand side (corresponding to the adiabatic gradient) shows small variations regardless of the proximity to the critical point. The first term becomes asymptotically small close to the critical point, whereas it is dominant far from it. Convective stability will be then characterized by the Rayleigh criterion far from the critical point and by the Schwarzschild criterion close to the critical point. Similar reasoning shows that for ideal gases at large L, the second term is prominent: the Schwarzschild criterion drives the stability in the atmospheric sciences. For a comprehensive review of early works, the reader is advised to refer to the paper by Gitterman [14].

13.1.2 Model Configuration

The onset of instability is determined by numerical experiments that are based on the same method as that already used in previous sections. The side walls of a square cavity are thermally insulated while the top and bottom cavity boundaries are thermostated (see Fig. 13.1a). The fluid at critical density is initially at a temperature T_i that is slightly above the temperature of the liquid–gas critical point.

The onset of fluid motion is the result of an increase ΔT of the lower wall temperature, which is increased by a few mK in a few seconds. This temperature rise generates a thin thermal boundary layer that results in a steep density gradient owing to the very small D_T and large β_p. The layer diffuses upwards very slowly. Due to the piston effect, the bulk temperature T_{bulk} and density ρ_{bulk}increase homogeneously. As the top wall is thermostated at the initial fluid temperature, a cold boundary layer forms. Figure 13.1a sketches the bulk zone and the heat diffusion layers. The directions of expansion are indicated by arrows. The height of and the temperature difference in the hot bottom layer are h and δT, respectively.

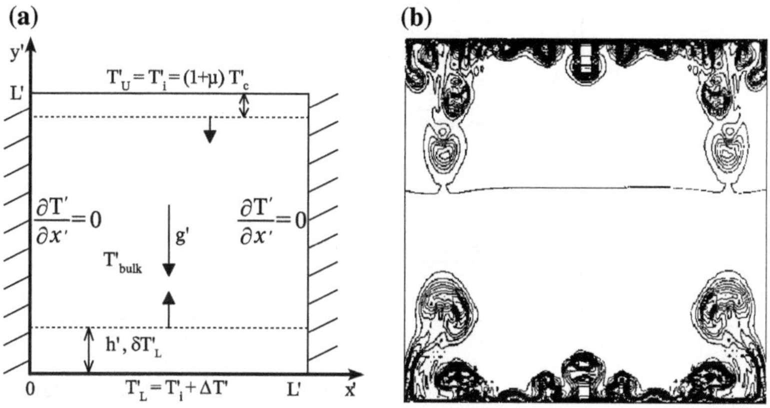

Fig. 13.1 **a** Geometry of the cavity of height $L = 10$ mm. Gravity is collinear with the y-axis. Velocity and temperature boundary conditions are indicated. The three characteristic zones are visualized: an isothermal bulk in the core; hot and cold layers near the bottom and top, and horizontal walls. The evolutions of the layers are shown by *arrows*. The height of and the temperature difference in the hot bottom layer are h and δT , respectively. **b** A typical flow configuration

The three regions and their main characteristics are as follows:

- *The boundary layer located above the rigid wall at the bottom.* Its upper boundary corresponds to the isotherm $T = T_{\text{bulk}}$ and to the isochore $\rho = \rho_{\text{bulk}}$ at the junction with the bulk. Its thickness h varies over time at the thermal diffusion speed. The upper boundary is open and the boundary conditions are only governed by continuity with the bulk. The temperature difference between the heated wall and the opened boundary is $\delta T_L = T_i - T_{\text{bulk}} + \Delta T$.
- *A bulk region* corresponding to uniform density and temperature fields due to the piston effect.
- *The upper boundary layer located under the rigid wall at the top*, and limited from below by the bulk region. The local temperature difference is $\delta T_U = T_{\text{bulk}} - T_0$.

Figure 13.1b gives the temperature pattern for $\Delta T = 10$ mK at $t = 8.3$ s (see also Fig. 13.12 and the related discussion below).

13.1.3 Stability Analysis

The only available model of analytical stability for unconfined layers considers a rigid lower wall with a stress-free upper boundary. It is assumed that the difference in the dynamic boundary conditions between the analytical model and the actual layer does not affect the threshold. The stability criterion where L is equated to h in Eq. (13.6) exhibits two opposite variations of δT_{onset}. Far from T_c, the Rayleigh

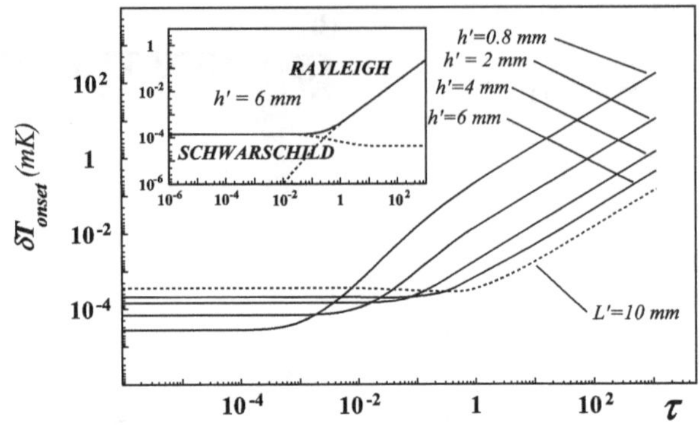

Fig. 13.2 Critical δT_{onset} (K) at the onset of convection, as derived from the Rayleigh and Schwarzschild criteria [Eq. (13.6)] versus the distance τ to the critical temperature. The characteristic length scale is the height h of the hot layer. The respective contributions of the Rayleigh and Schwarzschild criteria are identified and magnified in the *inset* for $h = 6$ mm

criterion gives a h^{-3} dependence, while the Schwarzschild limit close to T_c reveals a linear h dependence.

The stability of the upper layer has not yet been studied analytically. No results from a model involving a rigid upper wall and a stress-free boundary below are currently available. We shall see later that the upper layer is more stable than the lower one. Several solution scenarios can be identified: (i) both layers are stable; (ii) the lower layer is unstable and the upper one is stable; (iii) the upper layer is destabilized by the rising plumes issued from the lower layer after crossing the bulk region; (iv) both layers are unstable. Some of these situations are described and discussed below with respect to the stability analysis, which focuses mainly on the threshold.

Using Eq. (13.6), the temperature onset of convection can be found, and is shown versus the distance to the critical point τ in Fig. 13.2 and versus the thickness of the lower layer h in Fig. 13.3.

Figure 13.3 shows that the variation of δT_{onset} with respect to h is rather striking. For large values of h, the adiabatic gradient stabilizes the layer that would otherwise be destabilized by minute ΔT values. It also displays a transition from the adiabatic gradient condition (Schwarzschild branch with $\delta T_{onset} \sim h^{-3}$, close to T_c) to the Rayleigh condition (Rayleigh branch with $\delta T_{onset} \sim h$, far from T_c). The contribution of the two criteria to the actual stability of the system are shown in the top left corner of Fig. 13.2 for $h = 6$ mm. Figure 13.2 exhibits a narrow zone of $0.1 \leq \tau \leq 1$ where the actual criterion can differ by about 100 % from either the Rayleigh or Schwarzschild criterion if considered separately. In Fig. 13.2, the limits for δT_{onset} are displayed with respect to τ for several values of h ranging from 0.8 mm to 10 mm. These values, which vary as $\sqrt{D_T\, t}$, correspond to a few seconds of diffusion time after the start of heating. The corresponding values for the cavity height ($h = L = 10$ mm) are given as a limiting case. For this case, the δT_{onset} curve corresponds to the stability

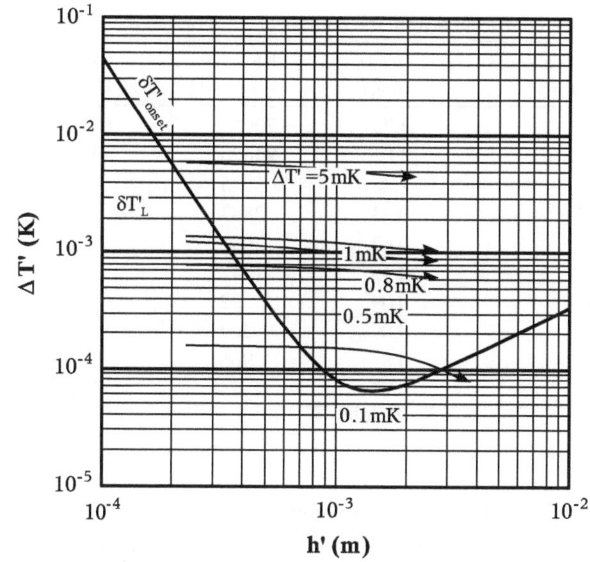

Fig. 13.3 Temperature onset δT_{onset} (K) as a function of height $h(m)$ of the hot layer as given by Eq. (13.6). The *curves with arrows* represent the evolution of the temperature difference δT_L between the edge of the hot bottom layer and the wall for $\Delta T = 0.1,\ 0.5,\ 0.8,\ 1$ and 5 mK. The *arrows* indicate the direction of variation of δT_L with h or time. The *crossing of the stability curve* gives an estimate of δT_L and h at the onset in the hot layer. The minimum is δT_{onset} = 0.065 mK

of the layer confined between two rigid walls. Looking at Fig. 13.3, the evolution of δT_L with respect to the layer height $h \sim \sqrt{D_T\, t}$ for values of ΔT larger than 0.1 mK shows that δT_L decreases due the bulk heating by the piston effect, and can even become smaller than δT_{onset} where the fluid restabilizes. Therefore, the possibility of convection damping exists, which will be explored further in Sect. 13.1.8.

13.1.4 Numerical Experiment

We now consider the two-dimensional square cavity of Fig. 13.1a, which is filled with CO_2 at critical density, initially at 1 K above its critical temperature. The numerical method used here is the same as that given in Sect. 12.1. We consider mesh-independent numerical solutions that use graded meshes of 101×81 to 121×121 in which about 50 % of the grid comprises points inside the thin layers to be studied. In particular, we concentrate about 30 points in the vertical direction within the layers, which are 0.5–1 mm thick. The mesh size near the walls is as small as 2×10^{-4} mm in the finer mesh. The time step is kept constant at 1.25×10^{-3} s. The accuracy of this numerical method has been studied by Amiroudine et al. [5, 7]. Control over the numerical accuracy is very important in the investigation of this time-dependent phenomenon.

Table 13.1 Parameters measured from the numerical solution

ΔT (mK)	B.L. thickness (h mm)	$(\delta T)_L$ (mK)	Ra Eq. (13.1)	Ra Eq. (13.4)	Ra_c Eq. (13.5)
10	0.8	5.73	70055	69719	1435
5	0.7	3.01	24626	24431	1296
2.5	0.6	1.28	6625	6519	1206
1	0.55	0.73	2883	2808	1175
0.8	0.52	0.62	2081	2021	1160
0.5	0.50	0.45	1342	1291	1151
0.1	0.35	0.09	92	80	1112
0.1	0.50	0.09	269	217	1151
0.1	0.75	0.85	906	646	1360
0.1	1	0.75	2148	1328	1920

The onset of convection is investigated for several linear temperature increases at the bottom wall, $\Delta T = 10, 5, 2.5, 1, 0.8, 0.5, 0.1$ mK, over time periods of 0.1–1 s. The temperature gap is initiated over the first two grid points near the bottom wall. In the following, we focus on the determination of the threshold value, and we give some insights into the flow structure. The different conditions that have been investigated are listed in Table 13.1. This gives the values of the different physical parameters measured from the numerical solution; i.e., the thickness of the thermal boundary, the local temperature gradient, and the corresponding local and critical Rayleigh numbers derived from Eqs. (13.1), (13.4) (including the Schwarzschild correction) and (13.5), respectively. Table 13.1 exhibits nearly the same values of the Rayleigh number as obtained from either Eq. (13.1) (pure Rayleigh term) or (13.4) (Rayleigh and Schwarzschild terms), which confirms that the Schwarzschild term is negligible.

13.1.5 Principle of Measurement and Comparison with Theory

The parameters needed to determine the instability threshold, such as the height h of the layer or the temperature drop δT_L, are measured from the output files. They are considered analogously to the idealized situations described in the previous section. The time at which the instability is triggered is determined by noting when the isotherms become disturbed by the flow field; in other words, when the deformations that first appear in the corners show an effect of vertical convective transport. The height h of the layer is deduced from its temperature profile. Taking h_1 to be the height of the bulk isotherm, we define the extension h_0 of the linearized temperature profile that reaches the bulk temperature and then $h = \frac{h_0+h_1}{2}$. The measurements of the height are given by the numerical code without any errors other than those arising from the numerical method and the model. It should also be noted that it is essential to account for the thermoacoustic effect when estimating the actual temperature drop in the layer, since the bulk temperature increases with time.

The use of the stability criterion for layers that are subjected to a steady gradient implies an assumption that the slowest growing mode still grows at a faster rate than the diffusion timescale, which governs the expansion of the unperturbed pure diffusion layer. The lower the heat diffusivity, the better the criterion, which is the case for a near-critical fluid. It is possible to check this hypothesis *a posteriori* by measuring the growth exponent as a function of the wavenumber of the perturbation. The simulation confirms that the instability does indeed grow much faster than the extension of the thermal boundary layer. The thermal boundary layer thickness grows as $h \sim \sqrt{D_T t}$. The amplification rate of the perturbation near the threshold is defined by $\frac{1}{v}\frac{dv}{dt}$, where v is the dimensional vertical component of the fluid velocity in the diffusion layer. At 1 K above T_c, the rate is found to be approximately ~0.3–0.7 s^{-1} for ΔT ~0.5–5 mK. As the expansion rate $\frac{1}{h}\frac{dh}{dt}$ of the thermal layer thickness decreases as $\frac{1}{t}$, comparison with the stability analysis leads us to the conclusion that the thermal layer is nearly frozen before the onset of convection.

13.1.6 Threshold at the Bottom Layer

The evolution of the local temperature difference between the bottom wall and the edge of the thermal boundary layer, $\delta T_L = T_i + \Delta T - T_{bulk}$, is represented in Fig. 13.3 for several values of ΔT. The isotherms inside the layer remain horizontal as long as the fluid velocity is not large enough to affect the temperature field after the threshold. The curves in the δT_L versus h diagram of Fig. 13.3 represent the evolution of the hot layer. They cross the Rayleigh stability curve. For $\Delta T \leq 0.1$ mK, no convection is observed. The corresponding curve in Fig. 13.3 is indeed below the stability extremum near the crossover point between the Rayleigh and Schwarzschild criteria. We note from Fig. 13.3 that for a sufficiently small h and time, δT_{onset} is always larger than ΔT (e.g., $\delta T_{onset} = 4.5 \times 10^{-2}$ K $> \Delta T = 1$ mK for $h =$ 0.1 mm). The fluid is thus always stable at small times.

The configuration becomes unstable when the curve ΔT = constant crosses the stability curve on the Rayleigh branch. For larger h and then later times, the curve can reach the Schwarzschild branch if $\Delta T \gtrsim 0.1$ mK, and the configuration can become stable again.

The computed Rayleigh numbers [using Eq. (13.4)] that correspond to the numerical experiment parameters listed in Table 13.1 are plotted against ΔT in Fig. 13.4 (log-log scale). For $\Delta T \succeq 0.5$ mK, convective instabilities are observed. For $\Delta T = 0.1$ mK, the fluid layer remains stable. The critical Rayleigh number calculated from Eq. (13.5) corresponds to the solid line. The Ra data appears to fit a unique curve. This is due to the fact that the detection of the destabilization of the thermal layer is observed for the smallest visible distortion of the isotherm. This occurs when a fluctuation has grown large enough to disturb the isotherm; that is,

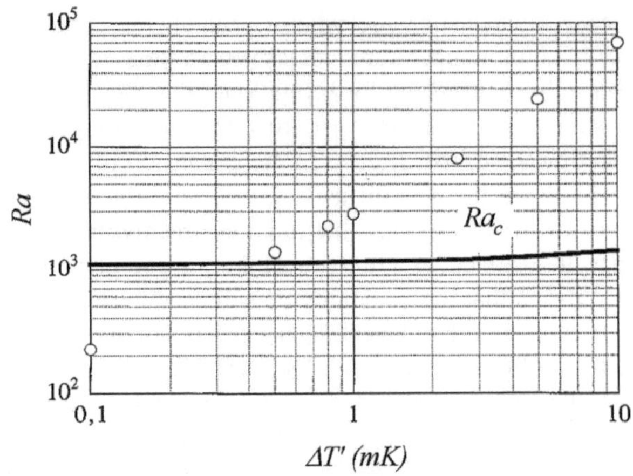

Fig. 13.4 *Solid line* critical Rayleigh number Ra_c calculated from Eq. (13.5) versus the imposed temperature difference ΔT (log-log scale). *Open circles* local Ra number calculated from Eq. (13.4) using the numerical experiment parameters given in Table 13.1. For $\Delta T \geq 0.5$ mK, convective instabilities are observed. For $\Delta T = 0.1$ mK, the fluid layer remains stable

after a time that depends on the distance $\mathrm{Ra} - \mathrm{Ra}_c$; this time decreases as $\mathrm{Ra} - \mathrm{Ra}_c$ decreases. This is confirmed by the corresponding ΔT_L versus h curve in Fig. 13.3.

The wavelength of the instability mode in all of the situations considered here is found to have an aspect ratio with respect to the height h of the thermal layer of about 1.6. It is nearly independent of ΔT in the range 0.5–10 mK. Another measure of the instability scale is the period of time that separates the rise of two successive thermal plumes in the numerical solution. This timescale decreases from 7 to 2 s as ΔT increases from 2.5 to 5 mK, and it becomes much larger (9–11 s) as the convection threshold is approached in the experiments at $\Delta T = 0.8$ mK and $\Delta T = 0.5$ mK. Additional, shorter, secondary oscillations subsequently appear after the onset of the preliminary oscillations at these two ΔT values.

13.1.7 Flow Field Description

Flow, density and temperature patterns are displayed during the significant stages of the convection process in Figs. 13.5, 13.6, 13.7, 13.8, 13.9, 13.10 and 13.11. Both hot and cold thermal layers remain diffusive for $\Delta T = 0.1$ mK. The vertical density profiles with respect to the initial value ($\rho_c = 467.8\,\mathrm{kg\,m^{-3}}$) show the evolution of the density gradient near the horizontal walls. The height of the hot layer grows by a factor of four during the process.

The variation of δT with h for different ΔT is given in Fig. 13.3. It is worth noting that for the case $\Delta T = 0.1$ mK, δT remains just below the stability limit. The h variations of the other computed cases, $\Delta T = 0.5$, 0.8, 1 and 5 mK, cross

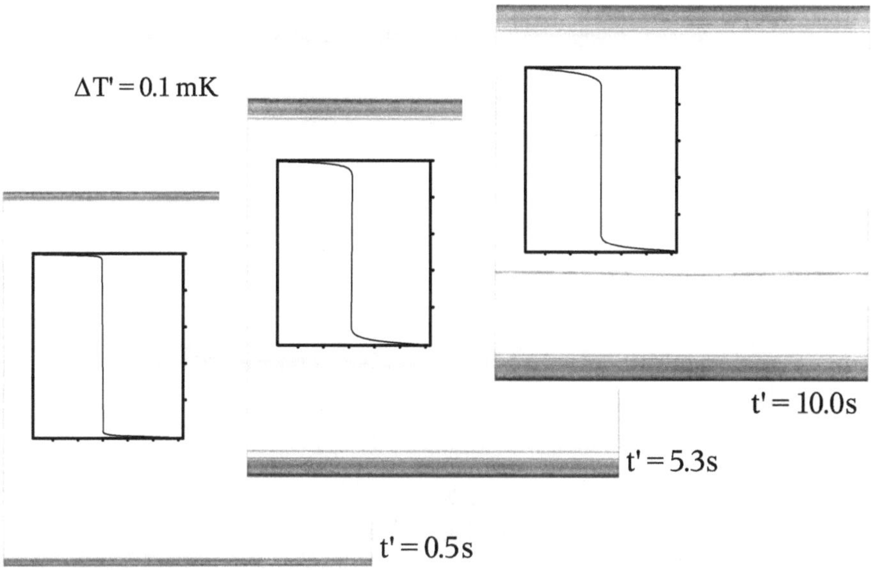

Fig. 13.5 Evolution of the density patterns and density profiles (*insert*) along the vertical axis during the heating process for $\Delta T = 0.1$ mK. The configuration is stable (no convection occurs). The density growth of the diffusive layer is shown at three differents timesteps. The density differences at the top and bottom walls and the density differences on the (*top* and *bottom*) iso-density lines at the indicated borders between the homogeneous bulk and the boundary layers are the following (with respect to the initial mean density $\rho_c = 467.8\,\mathrm{kg\,m^{-3}}$): at $t = 5.3\,s$: $-0.0674\,\mathrm{kg\,m^{-3}}$ (*top wall*); $-0.0002\,\mathrm{kg\,m^{-3}}$ (*top border*); $-0.0716\,\mathrm{kg\,m^{-3}}$ (*bottom wall*); $-0.0021\,\mathrm{kg\,m^{-3}}$ (*bottom border*): at $t = 10\,s$: $-0.0683\,\mathrm{kg\,m^{-3}}$ (*top wall*); $-0.00051\,\mathrm{kg\,m^{-3}}$ (*top border*); $-0.0711\,\mathrm{kg\,m^{-3}}$ (*bottom wall*); $-0.0046\,\mathrm{kg\,m^{-3}}$ (*bottom border*)

the stability curve, and this is consistent with the fact that they generate instabilities. When δT_L is sufficiently high and when the height h exceeds the threshold value, convection starts slowly and progressively invades the hot layer. During a time that can be on the order of a few seconds, convection remains weak and does not affect the density and temperature isoclines. The larger the magnitude of ΔT, the smaller the magnitude of the critical height h. Then, since the aspect ratio of the instability is scaled (with respect to h) so as to be constant and nearly 1.6 (as mentioned in the previous section), this suggests that the rate of convection is smaller near the onset.

We note that, in the temperature range $\Delta T \geqslant 2.5$ mK (see Figs. 13.10 and 13.12), there is a relation between the magnitude of the convection and the height h, since the time of the onset of disturbances at the edge of the hot layer is nearly constant and remains about 5 s. Later, the convection penetrates into the whole cavity and exhibits strong convective structures. We can also point out that, due to the larger critical h at larger ΔT values, the layer generates (for the same width L) a larger number of disturbances than it does for smaller ΔT values. This has a great impact on the thermal homogenization of the cell. These states are yet to be fully investigated,

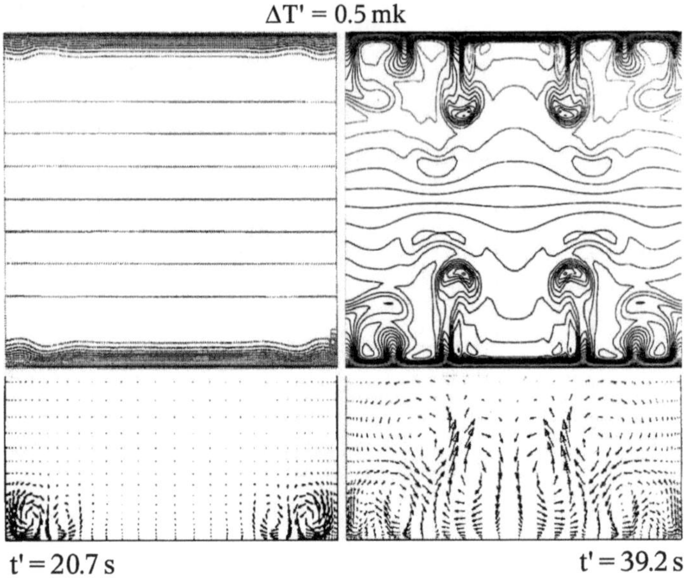

Fig. 13.6 Temperature patterns and relevant velocity fields at different times during the heating process for $\Delta T = 0.5\,\mathrm{mK}$. Note the onset of six thermal plumes. Density differences ($\mathrm{kg\,m^{-3}}$) with respect to the initial value are given at some locations of interest. The maximum velocity at 39.2 s is $0.61\,\mathrm{m\,s^{-1}}$, and the corresponding Reynolds number calculated with respect to the height of the layer is $\mathrm{Re} = 8.3$

but they would be worth exploring in order to determine the path to turbulence in hypercompressible media.

In addition to the upward hot plumes, there are downward cold ones from the upper, cold layer into the bulk of the cavity. Convection cells also transport cold eddies of gas to the heated wall that constrain the thermal layer by restricting its height. This can be observed by comparing temperature patterns before and after the initiation of convection in Figs. 13.6, 13.7, 13.8 and 13.9.

Velocity fields exhibit vortex motion, as shown in Figs. 13.6, 13.8 and 13.9, for $\Delta T = 0.5\,\mathrm{mK}$ (six plumes), 0.8 mK (six plumes) and 1 mK (five plumes). Structures that develop along the edge of the thermal boundary layer result from motion originating near the lateral, insulated walls.

The disturbances generate a similar number of cells in the cold layer. They are weaker and delayed with respect to those in the hot boundary layer because the upper temperature gradient is smaller (at fixed ΔT at the bottom wall). Successive cells progressively grow and fill the entire layer. They grow in magnitude and size and move downward in the bulk region. The local Reynolds number is estimated as $\mathrm{Re} = |V_{\max}|\frac{h}{\nu_0}$ from the maximum velocity in the vortices, $|V_{\max}|$, and from the boundary layer thickness, h, as the length scale. The number of cells remains small and ranges from about 1 to 13 for the three configurations during the simulation time under consideration. Figures 13.6 and 13.8 display about 20–40 s of the growth

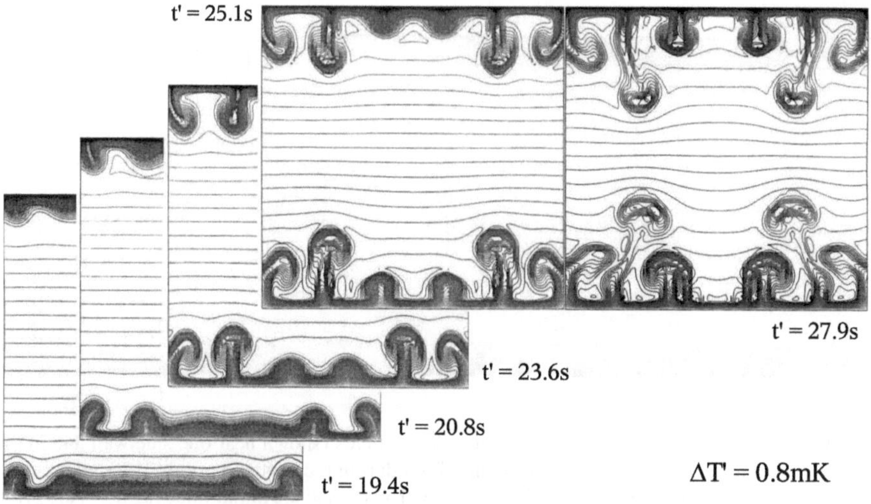

Fig. 13.7 Temperature patterns for $\Delta T = 0.8$ mK at several stages during the heating of the bottom wall. Note the onset of six thermal plumes. Some characteristic temperature differences with respect to the initial value are given at $t = 27.9$ s

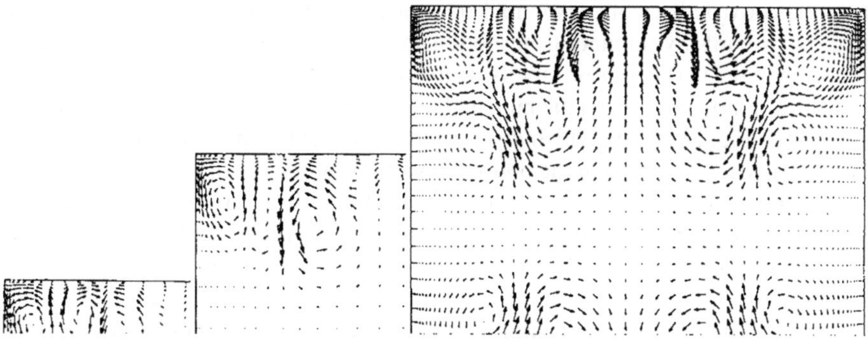

Fig. 13.8 Velocity field for $\Delta T = 0.8$ mK during the heating period. The maximum velocities and the corresponding Reynolds numbers at different times that are displayed are: (0.37 m s^{-1}, Re = 5), (0.85 m s^{-1}, Re = 11.5), (0.78 m s^{-1}, Re = 10.6) and (0.70 m. s^{-1}, Re = 9.5)

until plumes of hot gas break away from the hot layer and migrate up into the bulk. Secondary vortices are shown to arise below the primary corner vortices for 1 mK. Note that during this time period there is no real bulk interaction between the vortices originating from different layers.

It is also worth noting that in the temperature and density patterns at 5 and 10 mK (Figs. 13.10 and 13.12), the vortices grow and expand more quickly at larger ΔT. The 5 and 10 mK runs exhibit nine and seven plumes, respectively, for elapsed times as small as 9 and 10.5 s. Recall that 25–40 s are needed to generate 5–6 plumes for $\Delta T \leq 1$ mK. For about 10 s, the two families of hot and cold vortices remain within

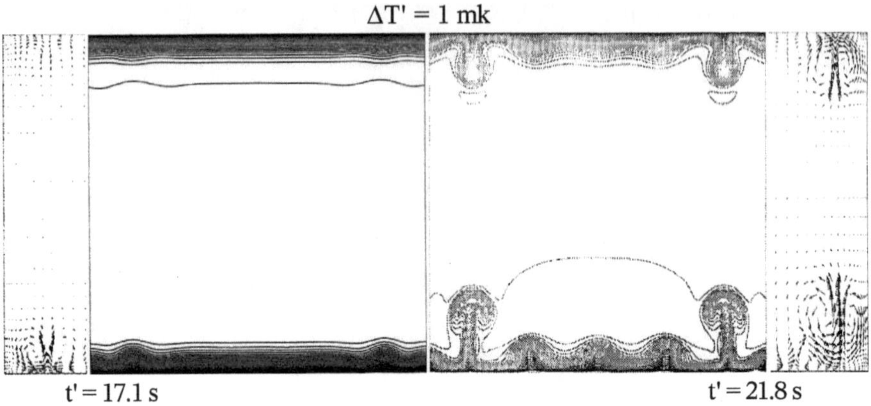

Fig. 13.9 Density patterns together with the velocity fields (shown near one side wall and in the *bottom layer*) for $\Delta T = 1$ mK at two different times during the heating period (21.8 s and 39.2 s). Onset of five thermal plumes. Density differences with respect to the initial value (kg m^{-3}) are given at some locations of interest. At 21.8 s, the maximum velocity is 0.94 mm s^{-1}, and the associated Reynolds number calculated with respect to the height of the hot layer is approximately Re = 12.9

their respective half-cavities. We tracked some particular convective structures in the case $\Delta T = 5$ mK by simulating the motion up to 45 s (175 timescales; see Fig. 13.11). The sizes of the vortices grow rapidly up to 60–70 % of the cavity height. Later, some of the lower structures cross the bulk and reach the upper layer. These structures are apparent in Fig. 13.11a, b, which display the density and temperature isoclines (that are close to each other) from the bottom to within about 45 % of their total variation. We observe three typical events. First, small plumes arise from the hot layer and move upward at about $t = 10.5$ s. Second, the central structure enlarges and reaches the cold wall at $t = 19.9$ s. Third, symmetry is broken following the impact of the structure on the upper wall. This is illustrated for two patterns at $t = 29.4$ s and $t = 41.2$ s. The symmetry breaking is related to the cold piston effect triggered by the interaction of the hot, rising plume with the cold upper wall. In Fig. 13.11a, we note that the structures are very thin and vertically elongated. They are very unstable but seem to remain confined to the lower half-cavity. Figure 13.11b concerns the upper layer and shows isoclines corresponding to about 42 % of the variation in density and temperature with respect to the cold wall. The pattern for $t = 19.9$ s reflects the interaction of the stronger structure with the upper boundary, and the two others, at $t = 29.4$ s and $t = 41.2$ s, represent the corresponding states after the symmetry is broken: the patterns exhibit similar features to those in the hot zone but are characterized by smaller structures and weaker disturbances.

For 10 mK, symmetry is broken earlier ($t = 12.54$ s). For 0.5 mK, the hot vortices remain confined inside the lower one-third of the cell after 70 s and the first interaction with the cold vortices occurs without breaking the symmetry after 89 s. At 0.8 mK, the confinement stage was found to last until up to 56 s of the simulation time.

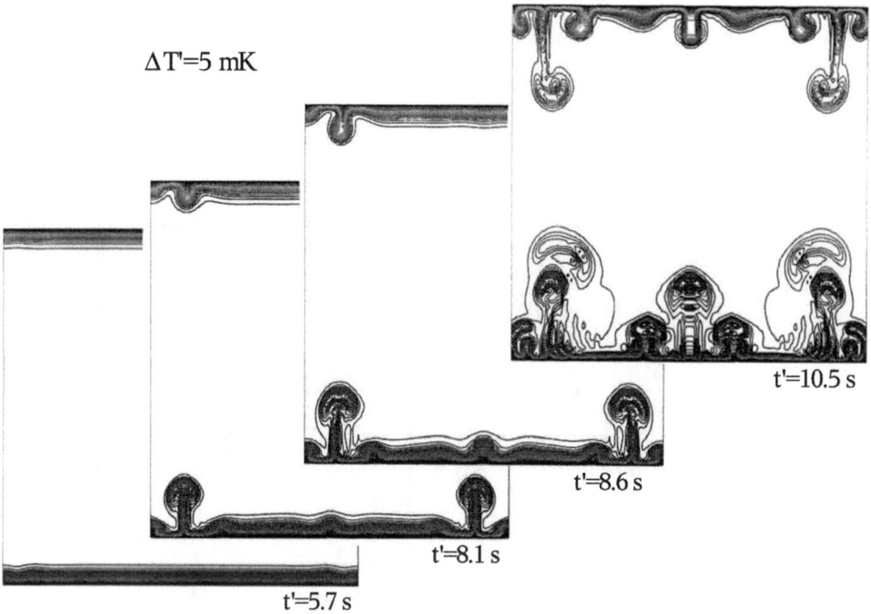

Fig. 13.10 Density patterns for $\Delta T = 5\,\text{mK}$ at several times during the heating period. Onset of seven thermal plumes. Density differences ($\text{kg}\,\text{m}^{-3}$) with respect to the initial value are given at some locations of interest

13.1.8 Reverse Transition to Stability Through the Schwartzschild Line

In compressible, highly expandable fluids, both Schwarzschild and Rayleigh criteria are relevant for comparable length scales, as shown in Fig. 13.3. Striking non-Boussinesq effects can be observed representing the reverse transition to stability through the Schwarzschild line in a layer a few millimeters thick [4]. The numerical method that has been used to perform the numerical experiment is accurate to second order in space and to third order in time. It was adapted from [20], with a vertically stratified initial density field being introduced [3]. It has been thoroughly validated [1], and the mesh size for a grid-independent solution depends on the heating of the bottom plate, with the finest being (100×140) nodes. Figure 13.3 shows that, after crossing the Rayleigh line, a return to stability can occur by crossing the Schwarzschild line if the boundary layer keeps growing in the unstable zone without any major deformation of the isotherms. To check this possible return to stability, we performed numerical simulations in a 2-D cavity that was 15 mm high and has periodic vertical boundary conditions to mimic a cavity of infinite width (Fig. 13.13).

The fluid is considered to be initially at rest and thermodynamic equilibrium at 1 K above its critical temperature. The bottom plate temperature is increased by a few mK in 1 s. Figure 13.14 displays (as does Fig. 13.3) the difference between the wall

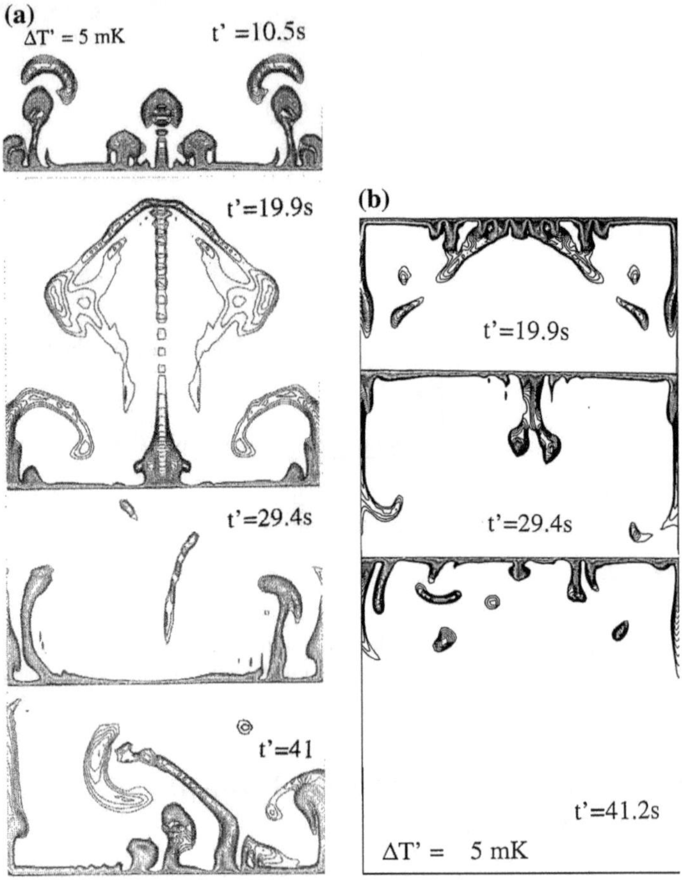

Fig. 13.11 Typical temperature pattern for $\Delta T = 5\,\text{mK}$ at several times during the heating. **a** hot layer *below*; small initial plumes arising from the hot layer at $t = 10.5$ s; large structure expanding up to the high of the cavity at $t = 19.9$ s; asymmetric small scale structures arising after the brake of symmetry near the bottom at $t = 19.4$ s; **b** associated structures obtained from the cold layer above. The displayed isovalues correspond to a range of about 40 % of the relative variation of the variable inside the cavity

temperature and the temperature at the edge of the boundary layer as a function of its height. Figure 13.15 displays the height of the bottom boundary layer as a function of time for different heating conditions.

In addition, a convenient measure of the fluid motion is given by the enstrophy field (noted Ens below), as defined by the following squared modulus of Curl **V** (also noted $\nabla \times V$):

$$\text{Ens} = \frac{1}{2}\left(\frac{\partial u}{\partial y} - \frac{\partial v}{\partial x}\right)^2, \tag{13.7}$$

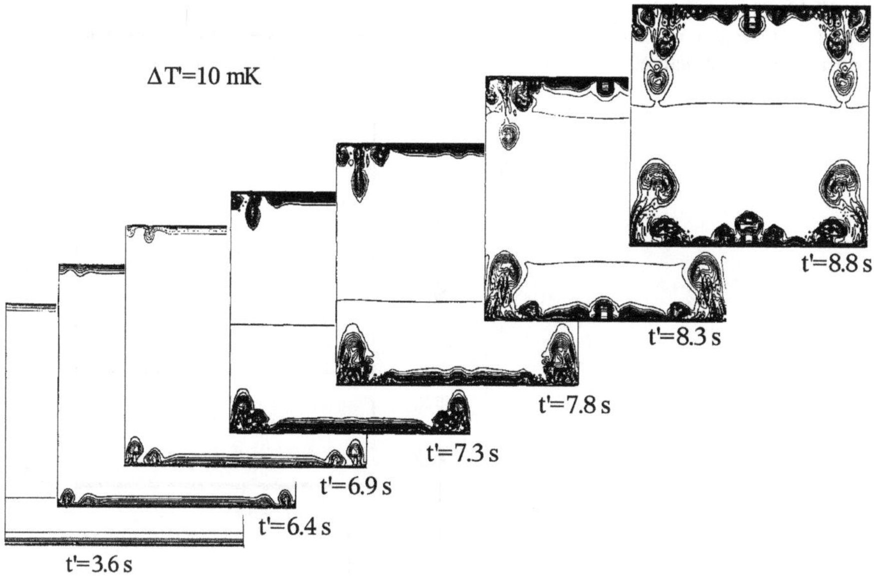

Fig. 13.12 Temperature patterns for $\Delta T = 10\,\mathrm{mK}$ at different times during the heating. Onset of nine thermal plumes beyond the threshold of convection. Some isotherm values defined with respect to the initial temperature (mK) are given at $t = 8.3\,\mathrm{s}$

where u and v are the velocity vector components. The enstrophy field is computed and the evolution of its mean value in the lower half of the cavity is used to detect the convection threshold accurately. The result is presented in Fig. 13.16.

A return to stability cannot be obtained for any heating. Under excessive heating, the intensity of the vortices produces enough convective motion to destroy the diffusive boundary layer. This is the case when the temperature is increased by 0.5 mK; here, the exponential increase of the enstrophy, as shown in Fig. 13.16, causes the boundary layer to collapse. The corresponding path is shown in Figs. 13.14 and 13.15: the boundary layer collapses about 70 s after the heating begins, which makes a return to stability impossible. The corresponding temperature field in Fig. 13.17 shows the onset of convection at 69.2 s.

In order to cross the Schwarzschild line, two scenarios are considered. (i) For weak heating, the convective transfer caused by the vortices might not deform the isotherms very much. The boundary layer can thus keep growing and stabilize again after crossing the Schwarzschild line. (ii) Starting with even weaker heating, the curve $\delta T(h)$ cannot enter the unstable zone through the Rayleigh line. However, when the curve drops under the Schwarzschild line, it becomes possible to cross the latter from below by increasing the bottom plate temperature again. These two scenarios are explored in further detail below.

(i) The return to stability through the Schwarzschild line is obtained when the heating is between 0.2 and 0.24 mK. Once the evolution curve has crossed the

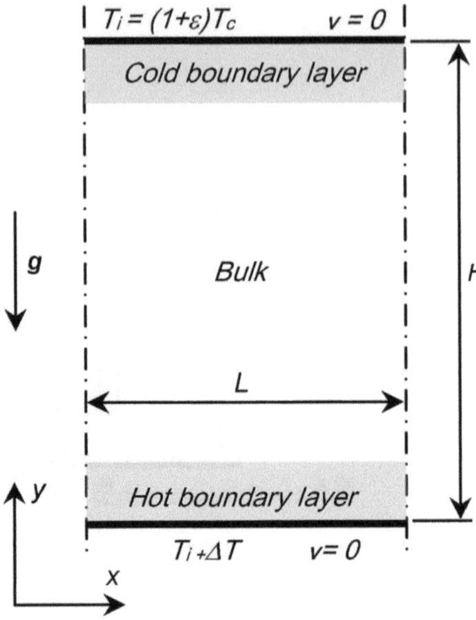

Fig. 13.13 The Rayleigh–Bénard configuration and the temperature field structured by the PE

unstable zone, convective cells start to form along the lower plate at the point denoted by an open circle in the RTS lines in Figs. 13.14, 13.15 and 13.16. The enstrophy rises exponentially over time, reaching a maximum, and then it starts to decrease (see the black circle in Fig. 13.16) once $\delta T(h)$ has crossed the Schwarzschild line towards the stable zone (see point B in Fig. 13.14). When its maximum is reached, the most intense deformation of the isotherms is obtained. However, this deformation is not large enough to induce the collapse of the boundary layer, which continues to grow, as shown in Fig. 13.15, and the curve $\delta T(h)$ remains in the stable zone. Once the convective motion is sufficiently damped out, the isotherm distortion in the boundary layer disappears, as shown in Fig. 13.17c, which corresponds to point C in Fig. 13.14.

(ii) The onset of convection according to the Schwarzschild criterion requires a two time-phase heating process that corresponds to the COS line in Fig. 13.14. In the first phase, a constant increase of 0.12 mK in the lower plate temperature is imposed. For this weak heating, the evolution curve $\delta T(h)$ remains in the stable zone. This first phase is maintained as long as necessary to bring $\delta T(h)$ under the Schwarzschild line ($h \sim 2$ mm, see Fig. 13.14). This phase duration is 58 s. The second heating phase consists of increasing the lower plate temperature to 0.3 mK in 87 s according to a cosine law. This new heating phase induces a new thermal boundary layer inside the previous one, which, through expansion, induces a new homogeneous bulk temperature increase due to the piston effect. The second heating phase is slow enough to prevent convection inside the new boundary layer according to the Rayleigh criterion. However, this second heating phase is fast enough for the temperature gradient inside the modified first boundary

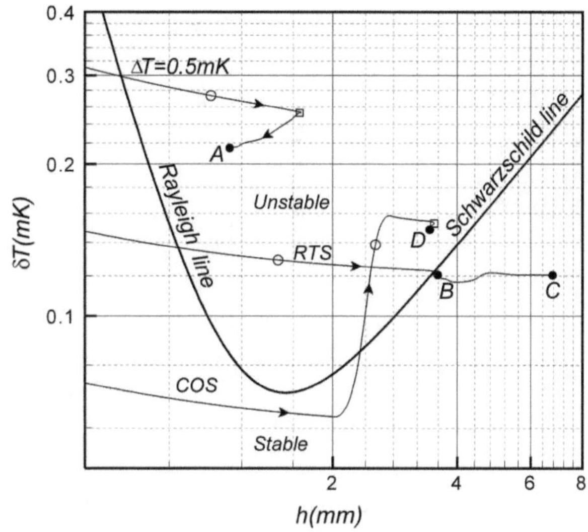

Fig. 13.14 Evolution of the temperature difference δT in the boundary layer as a function of its thickness h, according to the Rayleigh criterion (*Rayleigh line*) and the Schwarzschild criterion (*Schwarzschild line*); see Fig. 13.3. The return to stability corresponds to the *RTS line*, and the convection onset according to the Schwarzschild criterion corresponds to the *COS line* (see text). Each *open circle* represents a time where the intensity of the bottom plate vortices increases exponentially. Each *square* represents a time where the boundary layer convectively collapses. The *filled circles A, B, C and D* correspond to the snapshots of the temperature field given in Fig. 13.17

Fig. 13.15 Evolution of the boundary layer thickness. For the return to stability (*RTS line*) and the convection onset according to the Schwarzschild criterion (*COS line*), the boundary layer thickness increases continuously; however, for $\Delta T = 0.5\,\text{mK}$, it rapidly decreases ($\Delta T = 0.5\,\text{mK}$ *line*) when the thermal plumes emerge from the boundary layer, causing its collapse

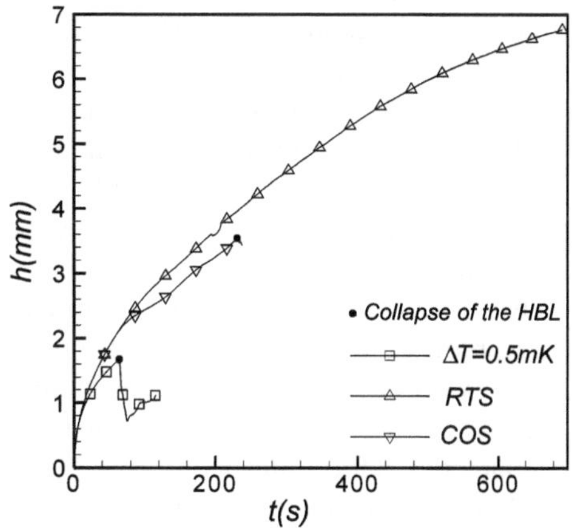

layer to exceed the adiabatic temperature gradient. Once the evolution curve $\delta T\ (h)$ has crossed the Schwarzschild line, convective cells appear in the first boundary layer (see Fig. 13.18), not in the second boundary layer. At the point noted with a circle in Figs. 13.14 and 13.16, the intensity of the vortices increases exponentially with time, deforming the isotherm with a number of convective

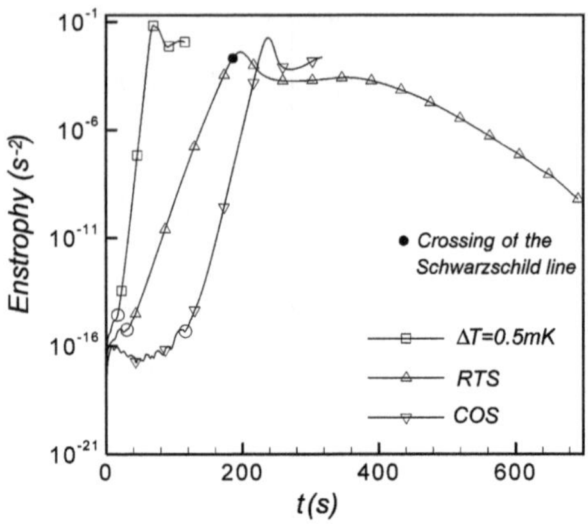

Fig. 13.16 Time evolution of the average enstrophy (squared vorticity vector modulus) field in the lower half of the cavity. The *circle* shows the beginning of the exponential increase that lasts until convection deforms the isotherms

plumes and causing the boundary layer to collapse. The temperature field after this collapse at $t = 238.6\,\mathrm{s}$ is shown in Fig. 13.17d, which corresponds to point D in Fig. 13.14.

13.1.9 A Reduced Model of Geophysical Flows

We saw above that convection can be triggered in a supercritical fluid layer a few millimeters thick according to the Schwarzschild criterion. This finding is of interest for simulating atmospheric-like phenomena at small length scales. The large isothermal compressibility of supercritical fluids is responsible for density stratification under the Earth's hydrostatic pressure gradient at constant temperature (see Sect. 1.8). Typically, in a cell that is 1 cm high and filled with CO_2 which is at critical density and maintained at 1 K above its critical temperature, the density gradient (in a nondimensional form) is about one thousand times that of the atmosphere. Therefore, convection-related mechanisms encountered in geophysical flows [13, 16] could be reproduced (at least qualitatively) in a supercritical fluid layer at the laboratory scale. Two-dimensional numerical simulations have demonstrated [2] the interaction between convection and stratification in a supercritical fluid in configurations usually encountered in the atmosphere. An additional component was used as a tracer in the fluid flow (see below for details, and [3] for the numerical method).

The first configuration addressed is the penetrative convection of a thermal plume, as induced by a local heating in a supercritical fluid. In [23], the fluid was initially assumed to be homogeneous (no density stratification); the results are discussed in Sect. 12.3. When an initial stratification is present, the evolution of the plume is slowed down, and it can be blocked due to its stratified environment (see Sect. 13.1.9.1 and [2]). The geophysical analogy is the buoyancy-induced rising of smoke, which rises up to a finite height in the presence of a stable stratification of the ambient atmosphere.

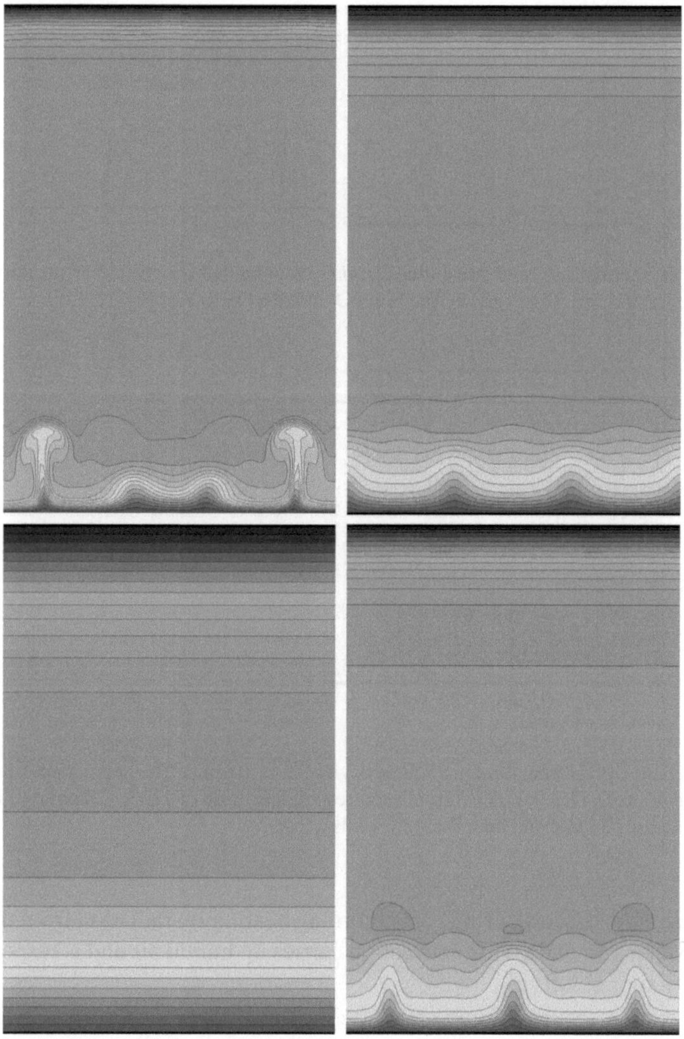

Fig. 13.17 Snapshots of the temperature field. **a** Convection at $t = 69.2$ s for $\triangle T = 0.5$ mK (point A in Fig. 13.14). **b** Beginning of the restabilization process at $t = 196.9$ s for $\triangle T = 0.24$ mK (point B on the RTS line in Fig. 13.14). **c** Stable fluid layer at $t = 692.3$ s for $\triangle T = 0.24$ mK (point C on the RTS line in Fig. 13.14). **d** Convection at $t = 238.6$ s for the second heating phase (see text and point D on the COS line in Fig. 13.14)

The second configuration was already addressed in the previous section, where convection in an infinite 2-D layer of supercritical fluid heated from below is triggered when crossing the Schwarzschild stability line from below. A similar destabilization scenario is studied below in Sect. 13.1.9.2, using a heat flux boundary condition at the bottom wall. Similar convective motions occur in the troposphere (the lowest

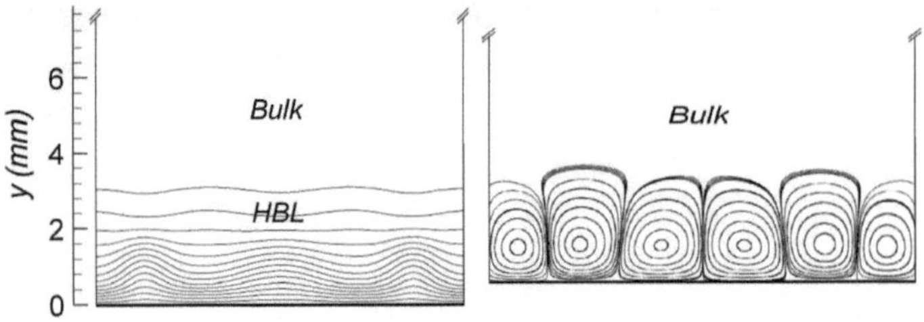

Fig. 13.18 Isotherms (*left*) and streamlines (*right*) showing that convection arises throughout the boundary layer at $t = 226.4$ s (just before point D on the COS line of Fig. 13.14)

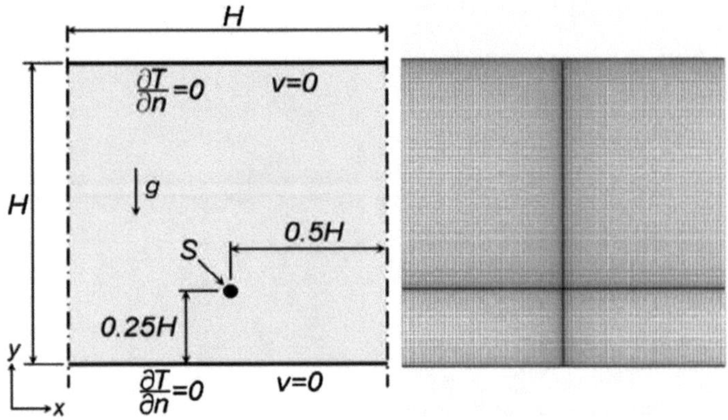

Fig. 13.19 *Left* supercritical fluid in a square cavity ($H = 10$ mm). The heating power $\mathcal{P} = 3$ mW is injected at point $(0.5H,\ 0.25H)$. *Right* the corresponding mesh of 120×120 points that is refined around the heating source and near the solid walls

layer of the atmosphere) during early morning heating by radiation from the ground (in the absence of wind). This convection carries up humid air and is responsible for the formation of clouds.

13.1.9.1 Stationary Plume in a Supercritical Fluid

The first configuration considers a supercritical fluid confined in a cavity of height $H = 10$ mm, which is subjected to the Earth's gravitational field (see the diagram on the left in Fig. 13.19). The horizontal walls of this cavity are solid and insulated, while periodic conditions are applied to the vertical walls. The fluid is initially at rest and in thermodynamic equilibrium on the critical isochore 1 K above its critical temperature. The simulation starts with a linear increase in the heating power for up to a constant value $\mathcal{P} = 3$ mW. Heat is considered to be supplied from a point at $(0.5H,\ 0.25H)$. As the heat flux becomes infinite at a point source, we consider the heat flux to originate from a circular fluid region centered at $(0.5H,\ 0.25H)$ of

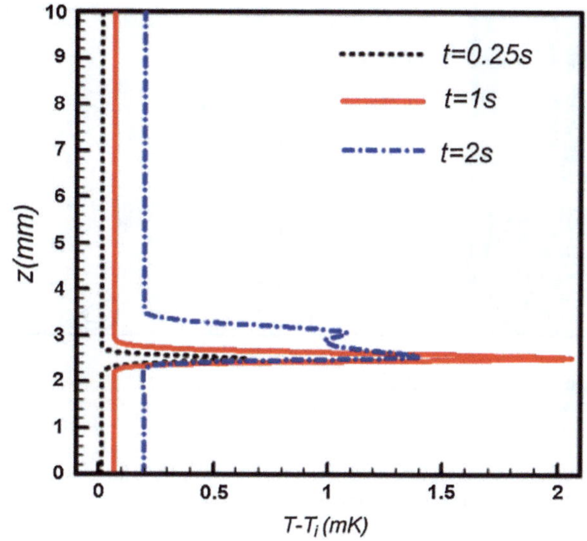

Fig. 13.20 Temperature profiles for $t = 0.25$, 1, 2 s at $x = 0.5H$ for $\mathcal{P} = 3$ mW showing the evolution of the fast homogeneous heating of the bulk fluid by the piston effect. The temperature profile is deformed by the thermal plume at $t = 2$ s (see text)

radius $e = 0.01H$. At a distance r from the center ($0 < r \leq e$), the heating intensity for $t \geq 1$ s follows

$$s(r) = \frac{\mathcal{P}}{\pi e} \frac{\pi^2}{\pi^2 - 4} \left(1 + \cos \frac{\pi r}{e}\right). \tag{13.8}$$

In addition to the Navier–Stokes equations, a transport equation of a passive scalar Q is implemented with a diffusion coefficient that is one thousand times less than the thermal diffusivity of the fluid. This passive scalar Q is thus the additional tracer that is simply advected by the flow, as mentioned above (see also [3]). The same source term as given by Eq. (13.8) is introduced in the transport equation of the passive scalar, for which boundary conditions $\frac{\partial Q}{\partial n}$ are imposed at all boundaries to avoid singularity problems.

Results were obtained for the case $\mathcal{P} = 3$ mW, which requires a grid of 120×120 points and a time step of 0.0125 s for a mesh and time step independent solution. Density gradients as large as $5\,\mathrm{kg\,m^{-3}/mm}$ are reached inside the circular heating zone. The expansion of the heated fluid leads to bulk homogeneous heating due to the piston effect, as shown in Fig. 13.20. On the other hand, these large density gradients result in a buoyancy-driven velocity that reaches $2\,\mathrm{mm\,s^{-1}}$, generating the thermal plume shown in Fig. 13.21.

As soon as the plume has risen, its motion is subjected to the damping effect of the fluid stratification. Figure 13.22 displays the evolution of the thermal plume by means of the passive Q field. It shows that it is practically stopped, and only evolves by diffusion: its propagation front progresses at a rate of about $0.1\,\mathrm{mm\,s^{-1}}$.

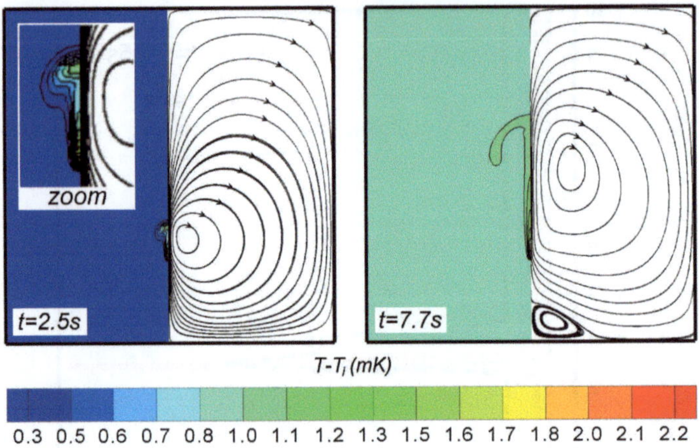

Fig. 13.21 Instantaneous temperature field (*colors*) and streamlines (*arrows*) for half of the domain; the solution is symmetric with respect to the line $x = 0.5H$

Such stationary behavior occurs because the density variations due to the stratification (about $0.11\,\mathrm{kg\,m^{-3}}$ across the fluid height) are comparable to those induced by the heat source: the largest density variation, which occurs at the heat source, is about $0.5\,\mathrm{kg\,m^{-3}}$ (see Fig. 13.22).

The same simulation performed using the low Mach number approximation was then repeated, with the initial density and pressure stratification suppressed in the model. The left half-domain of the right-hand figure in Fig. 13.23 show that the plume evolves much faster in this case. The dynamic pressure field, as shown in the right half-domain of the right-hand figure in Fig. 13.23, illustrates the impact of the plume on the top wall: an event that does not occur when the stratification is accounted for.

13.1.9.2 Atmospheric-Like Convection

Here, the supercritical fluid at critical density, set 1 K above its critical temperature and subjected to the Earth's gravity, is stratified between two horizontal, solid plates separated by $H = 15\,\mathrm{mm}$ (see Fig. 13.24a). A 10 mm-wide domain is considered with periodic boundary conditions. While maintaining the top plate at its initial temperature, the simulation starts when a spatially homogeneous, time-varying heat flux is applied to the bottom plate. Heating (see Fig. 13.24b) evolves according to a cosine law with a period of 800 s, simulating a single diurnal–nocturnal cycle. When the heating reaches its maximum value, $2.064\,\mathrm{mW\,m^{-2}}$, the temperature gradient induced by the bottom plate is six times larger than the adiabatic temperature gradient of $34.4\,\mathrm{mK\,m^{-2}}$. Here, $Q = 0$ at the top plate, while it is increased from 0 to 1 in one second at the bottom plate and then held constant. The vertical boundaries are periodic for the Q-transport equation. A wall-refined mesh of 90×200 points was used with a time step of 0.025 s.

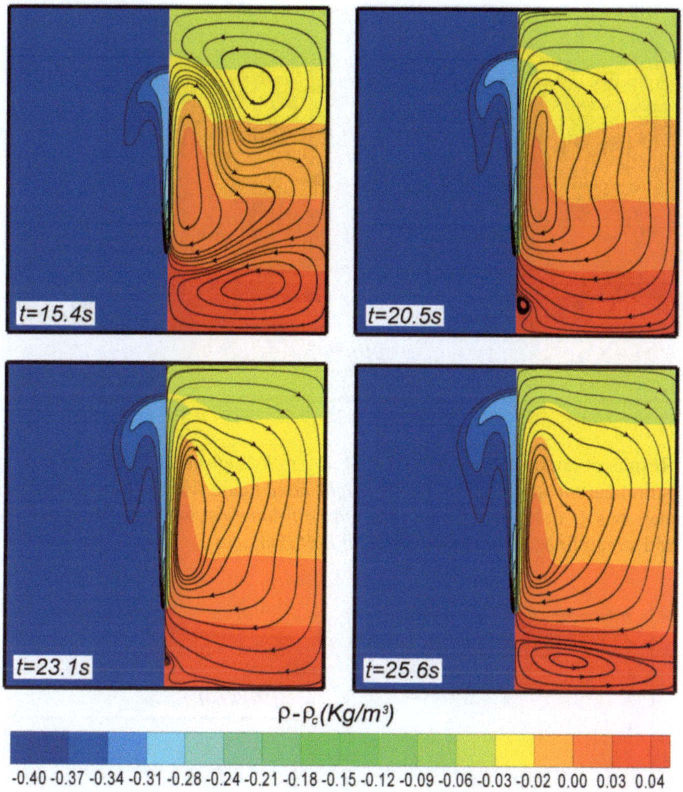

Fig. 13.22 *Left half-domains* instantaneous Q fields at $t = 15.4,\ 20.5,\ 23.1,\ 25.6$ s (the color scale is linearly distributed between $Q = 0$ at the wall and the maximal value at the heating source). *Right half-domains* density fields (*color bar*) and streamlines. The solution is symmetric with respect to the line $x = 0.5H$

As schematically shown in Fig. 13.24a, the temperature field consists of three distinct zones. The thickness h of each thermal boundary layer during heating grows as $\sqrt{D_T\, t}$. When the local Rayleigh number, based on h and on the temperature difference δT_L across the thermal boundary layer, exceeds the critical value of about 1100 given by Eq. (13.4), convection is triggered. For both HBL and CBL, δT_L is plotted in Fig. 13.25 as a function of h. (The thickness h of the boundary layer is defined as the distance from the wall to the place where the local temperature gradient is equal to $\frac{\Delta T}{H}$, where $\Delta T = T\,(y = 0) - T\,(y = H)$).

The intensity and evolution of the heating were adjusted (Fig. 13.24a) so that the evolution curve $\delta T_L\ (h)$ crosses the Schwarzschild line from the stable region to the unstable region at $t \simeq 250$ s. In order to confirm the destabilization of the layer, the enstrophy field [Eq. (13.7)] was computed. The evolutions of the mean values in the cold boundary layer (CBL; dotted blue line) and in the hot boundary layer (HBL; solid red line) are given in Fig. 13.26. Just after crossing the Schwarzschild

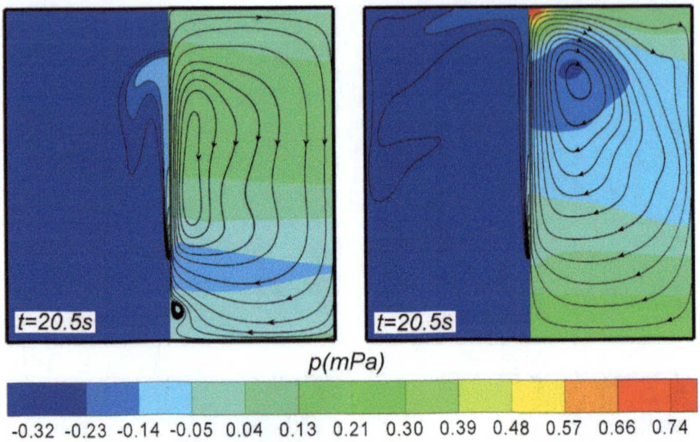

Fig. 13.23 Results obtained at $t = 20.5$ s for $\mathcal{P} = 3$ mW. *Left-hand figure* accounting for supercritical fluid stratification (see also Fig. 13.22). *Right-hand figure* without fluid stratification. In each figure, instantaneous Q fields are given in the *left half-domains*, while the dynamic pressure field (*color bar*) and streamlines are given in the *right half-domains*

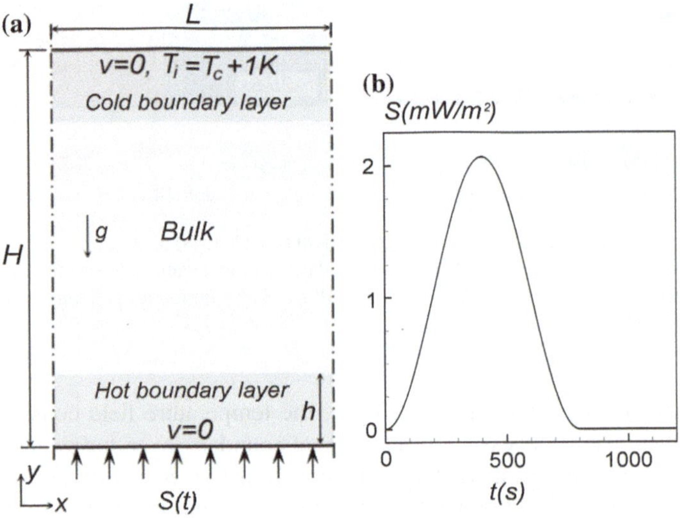

Fig. 13.24 **a** Supercritical fluid heated from below ($H = 15$ mm, $L = 10$ mm), and structured (schematic) by diffusion and the piston effect only. **b** Time evolution of the applied heating flux

line at $t \simeq 250$ s, perturbations of numerical origin are amplified. They give rise to convective cells, and the intensity of these cells increases exponentially with time until convective motion is produced. This motion is strong enough to deform the isotherms while remaining confined within the thermal boundary layers. This is illustrated in Fig. 13.27a (isotherms) and c (tracer) at $t = 420$ s, which correspond to point A in Fig. 13.26. Although the heating is maximum at $t \simeq 400$ s, the curve

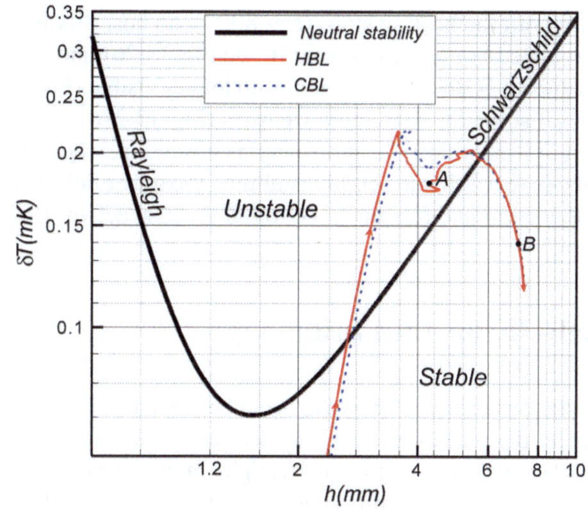

Fig. 13.25 Temperature difference δT_L in the bottom boundary layer versus its thickness h for the heating conditions given in Fig. 13.24. The stability criteria correspond to Schwarzschild's (see Fig. 13.11). The T and Q fields corresponding to points A ($t = 420$ s) and B ($t = 815$ s) are shown in Fig. 13.27

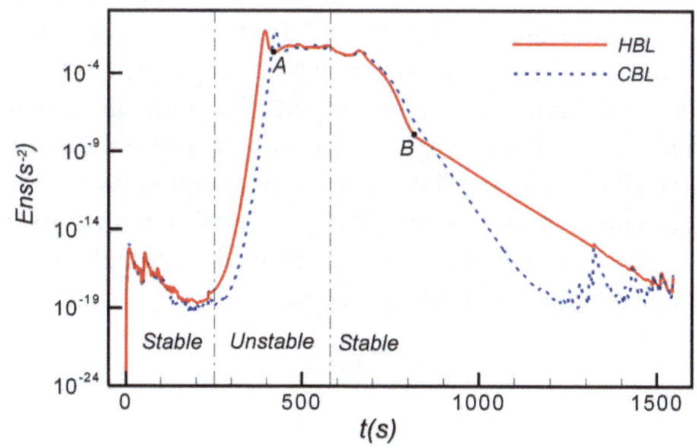

Fig. 13.26 Evolution of the mean enstrophy in the thermal boundary layers, reflecting the intensity of convection. The *vertical dash-dot lines* correspond to the times when the curve $\delta T(h)$ of the hot boundary layer crosses the Schwarzschild line (see Fig. 13.25). T and Q fields corresponding to points A and B are shown in Fig. 13.27

$\delta T(h)$ remains close to the Schwarzschild line. In Fig. 13.27a, the deformations of the isotherms occur in three spots with small extensions. In contrast, the corresponding tracer plots given in Fig. 13.27c show larger extensions of the three plumes.

As heating decreases, $\delta T_L(h)$ re-enters the stable region, again crossing the Schwarzschild line at $t \sim 580$ s, as illustrated by the second mixed line in Fig. 13.26. Convection is sustained until $t \sim 700$ s due to the inertia of the flow, and then it starts to decrease until the heating stops ($t \sim 800$ s): no heat flux is sent to the layer from the bottom plate. At $t = 815$ s (point B, see Fig. 13.26), the convective motion is sufficiently damped. Isotherm distortion is no longer visible (Fig. 13.27b), and the tracer plot shows that the fluid structure vanishes (Fig. 13.27d).

In the context of its similarity to atmospheric convection, where warm and humid air is convected due to a Schwarzschild instability, the tracer field Q could represent cloud droplets that form and remain at the top of the convective structures.

13.2 Rayleigh–Taylor-Like Instability

A Rayleigh–Taylor instability is known to occur between two immiscible fluid layers when the denser fluid is placed above the lighter fluid [21]. In the same configuration, two miscible fluid layers can also show a Rayleigh–Taylor-like instability. The surface tension is replaced by the mutual diffusion coefficient [9, 10]. In this section we aim to determine whether such an instability can take place between two near-critical fluid layers set initially to different temperatures.

We consider a diffusion layer that grows between two uniform near-critical fluid domains at the same pressure but at different temperatures, which are initially separated by a thermal discontinuity corresponding to a density discontinuity. The boundary and initial conditions are given in Fig. 13.28. For $t > 0$, the thermal constraint relaxes and a diffuse interface corresponding to a steep density gradient forms, similar to the gradient between two miscible liquids. Note that there is no piston effect here, as no parietal thermal boundary layer is present. The stability of the two layers is studied numerically by solving the Navier–Stokes equations with a van der Waals equation of state. The evolution of an initial perturbation with a given wavenumber k and its amplification growth rate ω are determined by calculating the exponential growth rate of the velocity. In the linear regime,

$$\omega = \frac{\ln(\frac{v_1}{v_2})}{t_1 - t_2},$$

where v_1 and v_2 are the velocity at a given location at times t_1 and t_2. The calculations are performed for CO_2 at 1 K above T_c. With R defined as

$$R = \frac{\rho_1 - \rho_2}{\rho_1 + \rho_2},$$

two different reduced density differences are considered, $R = 10^{-2}$ and $R = 10^{-3}$. These R values correspond to temperature differences between the top and the bottom layer of 15.2 mK and 1.49 mK, respectively. The values of $\omega(k)$ obtained are plotted in Fig. 13.29a.

The plot shows a cut-off wavenumber where the short wavelengths are damped by thermal diffusion. The cut-offs can be compared with the following expression from [18]:

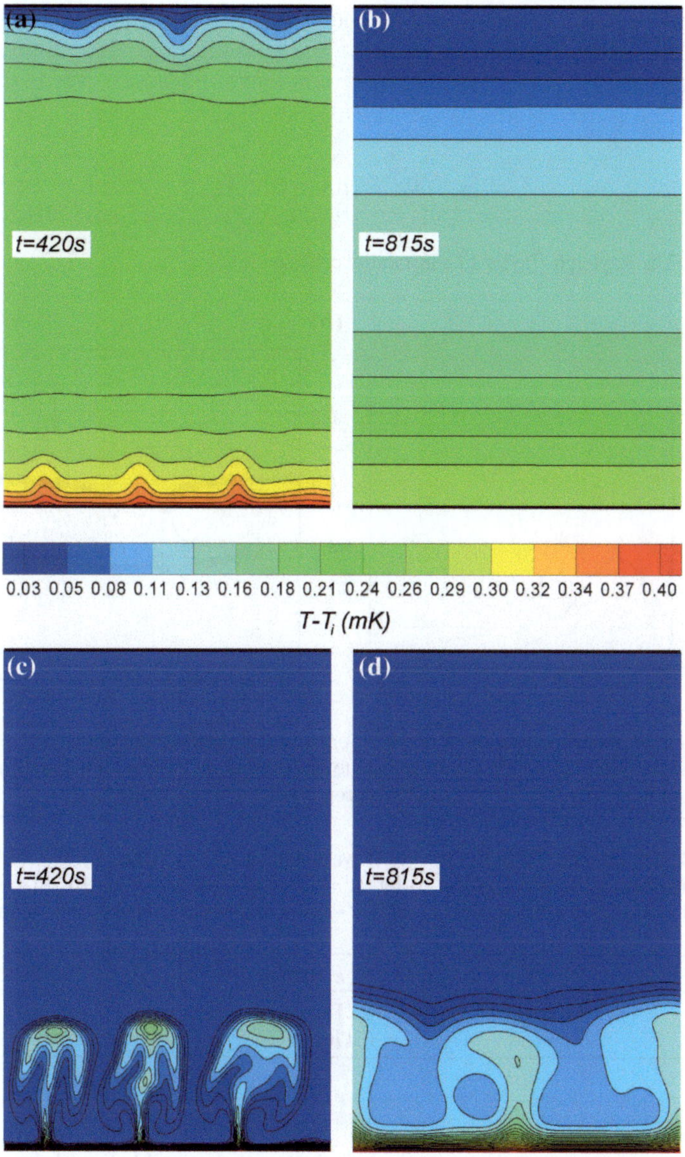

Fig. 13.27 **a** and **b** Instantaneous temperature (*color bar*). **c** and **d** Tracer Q field (the color levels are linearly distributed). **a** and **c** correspond to point A ($t = 420$ s) in the unstable region and **b** and **d** correspond to point B ($t = 815$ s) in the stable region (see Figs. 13.25 and 13.26)

$$\lambda_{c,K} = 2\pi \left(\frac{16\nu D_{12}}{3 g_0 R} \right)^{\frac{1}{3}}, \tag{13.9}$$

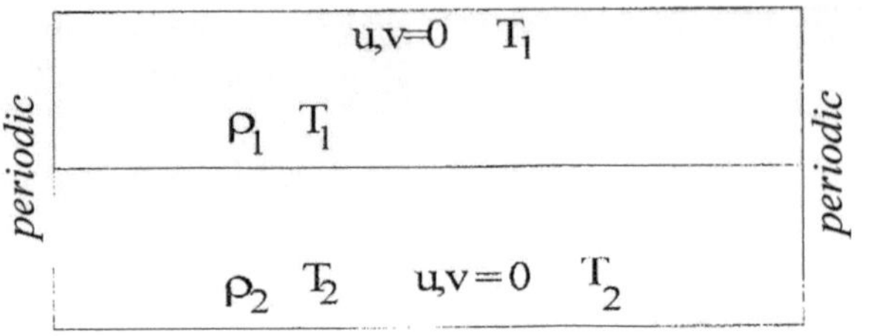

Fig. 13.28 The Rayleigh–Taylor-like instability configuration

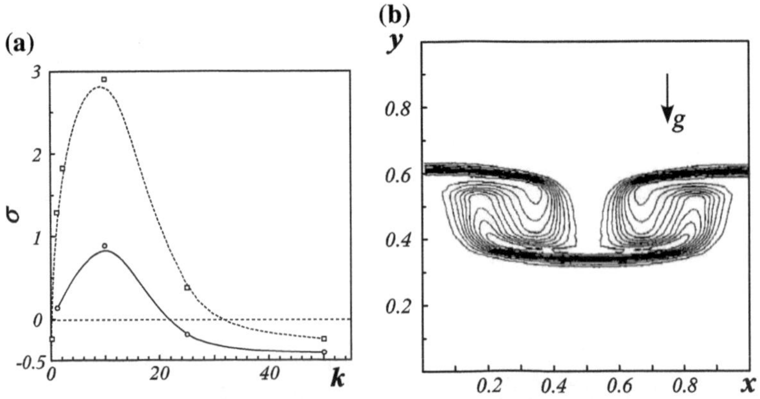

Fig. 13.29 **a** Dispersion curves for $R = 10^{-2}$ (*open squares and dotted line*) and $R = 10^{-3}$ (*full squares and full line*). The curves corresponds to smooth fits. **b** Typical temperature pattern in a convective structure for $R = 10^{-2}$. Density is proportional to temperature

Table 13.2 Comparison between the cut-off wavelengths $\lambda_{c,K}$ obtained by Eq. (13.9) (*column* 2) and numerical calculation (*column* 3) for $R = 10^{-2}$ (*row* 2) and $R = 10^{-3}$ (*row* 3), which correspond to the density differences $\Delta\rho = \frac{\rho_1 - \rho_2}{2}$ given in *column* 1, respectively (with $\rho_1 + \rho_2 = 2\rho_c$)

$\Delta\rho$ (kg m^{-3})	$\lambda_{c,K}$ (mm)	λ_c (mm)
4.6678	0.218	0.314
0.4668	0.470	0.452

where D_{12} corresponds to the mutual coefficient of two miscible liquids here replaced by the thermal diffusion coefficient D_T. The results are given in Table 13.2. The good agreement between the cut-off wavelength, even for the largest difference from the critical point, indicates that the thermal diffusion in nonisothermal supercritical fluids can behave as the mass diffusion in miscible liquids.

References

1. Accary G, Raspo I (2006) 3-d finite volume methods for the prediction of supercritical fluid buoyant flows in a differentially heated cavity. Comput Fluids 35:1316–1331
2. Accary G, Bontoux P, Zappoli B (2007) Convection in a supercritical fluid: a reduced model for geophysical flows. Phys Fluids 19:014104
3. Accary G, Raspo I, Bontoux P, Zappoli B (2005) An adaptation of the low mach number approximation for supercritical fluid buoyant flows. C R Mec 333(5):397–404
4. Accary G, Raspo I, Bontoux P, Zappoli B (2005) Reverse transition to hydrodynamic stability through the Schwarzschild line in a supercritical fluid layer. Phys Rev E 72(3):035301
5. Amiroudine S (1995) Modelisation Numerique des Phenomenes de Transport de Chaleur et de Masse dans les Fluides Supercritiques. PhD thesis, IRPHE (Institut de Recherche sur le Phenomènes Hors-Equilibre), Marseille
6. Amiroudine S, Larroudé P, Bontoux P, Zappoli B (1996) Numerical simulations of convective instabilities in near-critical fluids. In: 2nd European symposium on fluids in space, Naples, Italy, 22–26 Apr 1996
7. Amiroudine S, Ouazzani J, Carlès P, Zappoli B (1997) Numerical solution of 1-d unsteady near-critical fluid flows using finite volume methods. Eur J Mech B 16(5):665–680
8. Assenheimer M, Steinberg V (1993) Rayleigh-Bénard convection near the gas–liquid critical point. Phys Rev Lett 70(25):3888–3891
9. Authelin JR, Brochard F, de Gennes PG (1993) Instabilité gravitationnelle de deux fluides miscibles. C R Acad Sci II 317:1539
10. Brochard F, de Gennes PG (1994) Mélange de deux fluides avec changement de volume. C R Acad Sci Paris 318(II):27
11. Carles P, Ugurtas B (1999) The onset of free convection near the liquid–vapour critical point. Part 1: stationary initial state. Phys D 126(1–2):69–82
12. Chandrasekar S (1961) Hydrodynamic and hydromagnetic stability. Clarendon, Oxford
13. Getling AV (1998) Rayleigh-Bénard convection: structures and dynamics. World Scientific, Singapore
14. Gitterman M (1978) Hydrodynamics of near-critical fluids near a critical point. Rev Mod Phys 50:85
15. Gitterman M, Steinberg VA (1970) Criteria for the commencement of convection in a liquid close to the critical point. High Temp (USSR) 8(4):754
16. Jaluria Y (1980) Natural convection: heat and mass transfer. Pergamon, Oxford
17. Kogan AB, Murphy D, Meyer H (1999) Rayleigh-Bénard convection onset in a compressible fluid: ^{3}He near t_C. Phys Rev Lett 82(23):4635–4638
18. Kurowsky P, Misbah C, Tchourkine S (1995) Gravitational instability of a fictitious front during mixing of miscible fluids. Europhys Lett 29:309
19. Normand C, Pomeau Y, Vellarde MG (1977) Convective instabilities: a physicist's approach. Rev Mod Phys 49(3):581–624
20. Paolucci S (1982) On the filtering of sound from the Navier-stokes equations (sand 82-8257). Technical report, Sandia National Laboratories, Albuquerque
21. Rayleigh L, Hydrodynamic and hydromagnetic stability
22. Spiegel EA (1965) Convective instabilities in a compressible atmosphere. Astrophys J 141:1068
23. Zappoli B, Jounet A, Amiroudine S, Mojtabi A (1999) Thermoacoustic heating and cooling in near critical fluids in the presence of a thermal plume. J Fluid Mech 388:389–409
24. Zhong F, Meyer H (1995) Density equilibration near the liquid–vapor critical point of a pure fluid: single phase $t > t_c$. Phys Rev E 51(4):3223–3241

Chapter 14
Experiments in a Weak Acceleration Field and on Earth

Abstract Thermalization experiments performed in a weak acceleration field (as provided in SF_6 at liquid-like density by manoeuvres, rotations and positioning corrections of a Space Shuttle) and on Earth (CO_2 at gas-like density) are presented in a configuration close to most situations currently encountered in an industrial context. The sample fluid is filled in an interferometer cell with its walls maintained at constant temperature. When heat pulses are applied in the centre of the fluid by a thermistor, a long thermal transient is observed where the bulk fluid temperature reaches significantly below the initial temperature. This unconventional cooling originates from the fast decompression of the fluid, as induced by the rapid convectively disappearing hot boundary layer at the heat source. The effect is also more pronounced when the free fall acceleration is large. The result are analysed by using a simple one-dimensional model with ad hoc convective heat losses.

In this chapter, we describe and analyze thermalization experiments performed in a weak acceleration field and on Earth. These extend and illustrate the analysis given in Chaps. 12 and 13, where thermalization in the presence of the Earth's gravitational field was considered in two different configurations. The first configuration involved heating a vertical wall in a 2-D square sample that was thermally isolated at the other walls. The 2-D analysis of this configuration showed that, although the hot boundary layer develops as a plume along the wall and the top of the cell, thermalization was only marginally affected. The second configuration was that of Rayleigh–Bénard, where heating occurred from below and cooling from above, and the two vertical walls were thermally isolated. In this arrangement, convection developed in the boundary layers, which were much smaller than the sample's thickness, thus considerably increasing the stability of the heating process.

We now present a somewhat different configuration; one that is closer to most situations currently encountered in an industrial context. The fluid is confined in a temperature-controlled (cylindrical) optical interferometer sample, already described in Chap. 9 (see also Fig. 23.2). Two or three thermistors, depending on the setup, are used to measure the temperatures in the bulk fluid and in the HBL and CBL.

B. Zappoli et al., *Heat Transfers and Related Effects in Supercritical Fluids*, Fluid Mechanics and Its Applications 108, DOI: 10.1007/978-94-017-9187-8_14,

A thermistor (Th1) is used to send a heat pulse of 10–100 mW. The pulse duration can be varied. The density variations are visualized in an interferometer.

In general, the thermal behavior is governed by the balance of the heat flux between (i) the heat source and the diffuse HBL, which compresses the bulk fluid, and (ii) the thermostated walls and the diffuse CBL, which depresses the bulk fluid. In the presence of convection, the thickness of the thermal HBL is not or is only weakly affected by convection (see Sect. 12.3 and the experiments in Sect. 14.2 described below). In contrast, the CBL becomes considerably thinner under convection, which appreciably increases the heat losses and lowers the amplitude of the piston effect.

When heating stops, the compression caused by the HBL also stops. However, the decompression caused by the CBL is still present and vanishes only through a diffusion process. The pressure therefore drops, which can lead, as we will see later, to a bulk fluid temperature that is *below* the thermostat temperature. This phenomenon lasts for a significant period of time, corresponding to the diffusive vanishing of the CBL. In this sense, heating the fluid results in a (transient) cooling, which is in (apparent) contradiction to the second law of thermodynamics. This is a spectacular demonstration of the thermomechanical nature of the heat exchange by the piston effect. Another apparent violation of the second law in a two-phase, gas–liquid sample is analyzed in Sect. 10.2, where the gas becomes hotter than the heating wall [14, 15].

In order to describe the temperature evolution and this cooling effect in a simple way, in the following we develop a simplified, phenomenological 1-D model that only retains the energy equation. The heat losses with the thermostat that are increased by convection are described by an *ad hoc* cold source. This model explains the data obtained both on Earth and in the presence of small accelerations (as provided by maneuvers—rotations and positioning corrections—of the Space Shuttle) quite well. These accelerations (with dimensionless modulus noted γ and expressed in g_0 unit) range between $10^{-4}g_0$ and $10^{-2}g_0$ and act as "gravity" pulses with accurately determined start and end times. Their influence can then be analyzed in detail, which contrasts with the investigation on Earth, where convection dominates the process, even at very short times.

14.1 A Simplified 1-D Model

We consider the simplified energy-balance equation, which can be derived from Eq. (3.33), and has been already used by Boukari et al. [5]. We write this equation in spherical coordinates to represent the geometry of the problem. In the heating (or cooling) process, the pressure is assumed to be uniform, and the flow due to the dilatation (or contraction) of the boundary layers is neglected:

$$\frac{\partial T}{\partial t} = \frac{1}{\rho c_p}\frac{1}{r^n}\frac{\partial}{\partial r}\left(r^n \Lambda \frac{\partial T}{\partial r}\right) + \left(1 - \frac{c_V}{c_p}\right)\left(\frac{\partial T}{\partial p}\right)_\rho \frac{dp}{dt} + \frac{q}{\rho c_p}. \tag{14.1}$$

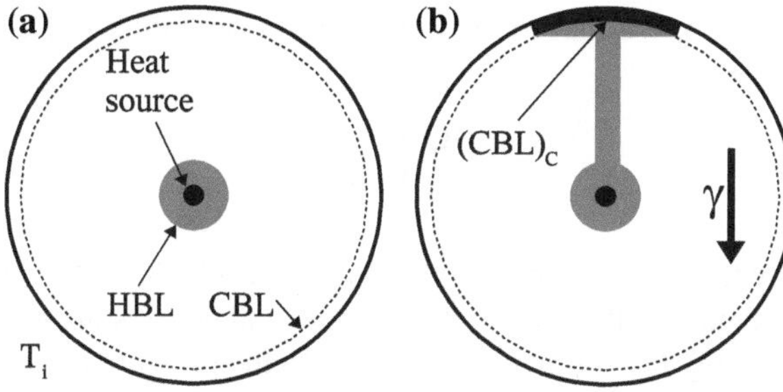

Fig. 14.1 Schematic of thermal behavior under an acceleration γ for a fluid confined in a sample at constant temperature T_i. **a** $\gamma = 0$. The temperature in the bulk is the result of the balance between a hot piston effect due to the expansion of the HBL and a cold piston effect arising from the contraction of the CBL. **b** $\gamma \neq 0$. The HBL is convected to the wall and the CBL thins out at the zone of contact [$(\mathrm{CBL})_c$], increasing the cooling effect

Here, q is a source term that is sufficient to model the thermal characteristics of the piston-effect dynamics (except the acoustic phenomena) in the absence of gravity and for a negligible hydrodynamic velocity. The parameter n represents the geometry of the problem: $n = 1, 2, 3$ for a linear geometry, a planar geometry, and a spherical geometry, respectively. We only consider in the following the spherical case ($n = 3$), and we simulate the convective behavior for the thermal evolution of the fluid by assuming the two following conditions (confirmed by experiments; see later and Fig. 14.1):

1. The HBL is only weakly affected by convection
2. The HBL is convected towards the cell wall, where a thin CBL forms, increasing the thermal exchange.

As the pressure drop due to the thin CBL persists longer than the pressure rise due to the HBL, which stops with heating, we model the influence of the thin CBL by including an additional cold source; i.e. by including a *negative* heat source in Eq. (14.1) that acts only in the HBL region:

$$q_{HBL}(r, t) = C_c\,[T(r, t) - T_b]\,. \tag{14.2}$$

This term—which we denote the "convective" term—is proportional to the difference between the local temperature $T(r, t)$ and the homogeneous bulk temperature T_b. The "convective" parameter C_c is phenomenological and accounts for the amplitude of the cooling effect. It must be determined from measurements. In order to adapt the above spherical model to the experimental cylinder configuration, we have to preserve the ratio of the heating (thermistor bead) to the cooling (cylindrical body plus plane windows) surfaces (see Sect. 8.2.2). This means that the mass is not conserved during the transformation. In order to account for this mismatch, we consider the following.

To a first approximation, for a given amount of heat δQ that is transferred to the bulk fluid, the amplitude δT_b of the bulk temperature change due to the piston effect is inversely proportional to the fluid mass m according to Eq. (4.7) (see also [11, 12]), leading to

$$\delta T_b = \frac{\delta Q}{m}\left(\frac{1}{c_V} - \frac{1}{c_p}\right). \tag{14.3}$$

We thus multiply the pressure term in the energy Eq. (14.1) by the mass ratio $\frac{m_{\exp}}{m_{\mathrm{mod}}}$, where $m_{\exp}$ is the mass of the fluid in the cylinder and m_{mod} is the mass in the model sphere. We then apply the following correction to the pressure term:

$$\left(\frac{\partial T}{\partial p}\right)_S = \frac{m_{\mathrm{mod}}}{m_{\exp}}\left[\left(\frac{\partial T}{\partial p}\right)_S\right]_{mod} = \frac{m_{\mathrm{mod}}}{m_{\exp}}\left[\left(1 - \frac{c_V}{c_p}\right)\left(\frac{\partial T}{\partial p}\right)_\rho\right]. \tag{14.4}$$

In this model, equal heating and cooling surface areas give $\frac{m_{\exp}}{m_{\mathrm{mod}}} \approx 1.2$. The direct application of this additional factor to the temperature profile during the calculation does not fulfill mass conservation. The thermalization process is then solved according to the energy Eq. (14.1) and using

$$\frac{dp}{dT} = \frac{\partial p}{\partial T} = -\frac{\int_V \left(\frac{\partial \rho}{\partial T}\right)_p \frac{\partial T}{\partial t} dV}{\int_V \left(\frac{\partial \rho}{\partial p}\right)_T dV}, \tag{14.5}$$

which stems directly from the mass conservation; i.e.,

$$\int_V \frac{d\rho}{dt} dV = 0, \tag{14.6}$$

and the expression for the density changes in terms of temperature and pressure variations:

$$d\rho = \left(\frac{\partial \rho}{\partial T}\right)_p dT + \left(\frac{\partial \rho}{\partial p}\right)_T dp. \tag{14.7}$$

In parallel, a second temperature profile is iterated using the above correction [Eq. (14.4)].

14.2 Behavior Under Acceleration

We now consider the experimental configuration where a heat source is located in a thermostated cylindrical sample that is 12 mm in diameter and 6.8 mm thick, as described in Chap. 9, and Fig. 14.2a.

The fluid is SF_6 at an off-critical density $\rho = 1.27\rho_c$. The hot source is a thermistor (Th1) that is 0.89 mm in diameter and 3 mm from the wall, and which also functions as a thermometer once the heating has stopped. The temperature of the bulk fluid is

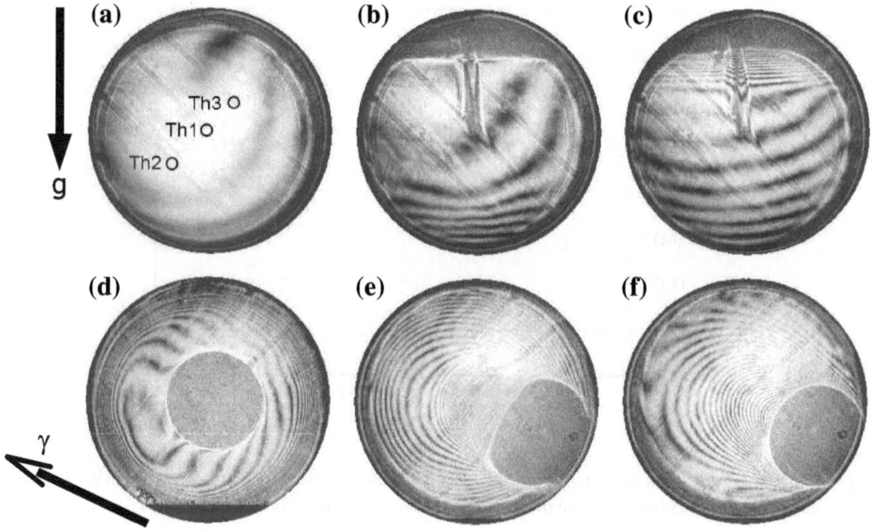

Fig. 14.2 A 100 mW, 15 s heat pulse is sent into off-critical SF_6 ($\rho = 1.27\,\rho_c$ $T = T_c + 0.95$ K). **a–c** Under Earth's gravity (1g). **d–f** Under a vector acceleration (indicated by the *arrow*) of dimensionless modulus $\gamma = 3 \times 10^{-2}$ as provided by the Space Shuttle *Columbia* (STS-55, IML-2, July 1994). **a** $t = 0$, homogeneous state. *Th1* heating and measuring thermistor, *Th2*, *Th3* measuring thermistors. **b** $t = 15$ s, end of the heat pulse. **c** $t \approx 120$ s, relaxation towards equilibrium. **d** $t = 5$ s. The HBL develops symmetrically around Th1. **e** $t = 20$ s, 5 s after the acceleration has started. The HBL is convected towards the cell wall, in the opposite direction to Th2. **f** $t = 42$ s during the relaxation period after the end of the heat pulse

measured by a thermistor (Th2) that is 3 mm from the wall in the opposite direction with respect to Th1. Again, this thermistor can be used for temperature measurements, but only after heating. The heating power is 100 mW. The positions of the thermistors are such that the hot convected fluid does not touch Th2. The evolution of the thermal boundary layers during a 15 s heat pulse at temperature $T_i = T_c + 0.95$ K is shown in Fig. 14.2. During the early stages, a thin HBL develops at the thermistor surface and becomes unstable (see Fig. 14.2b). The layer then remains at a constant thickness of 0.2–0.3 mm and hot, expanded fluid is convected (see Fig. 14.2c), leading to an accumulation of hot fluid in the upper part of the sample cell. In the part of the cell that is not perturbed by the convection, a uniform shift in the fringes shows that density and temperature homogeneously increase in the bulk, demonstrating the presence of the piston effect.

This behavior is qualitatively the same as that seen during much weaker acceleration ($\gamma = 3 \times 10^{-2}\, g_0$), as provided by the rotation of the Space Shuttle (Fig. 14.2d–f). During convection, the HBL remains nearly spherical. Temperature measurements after the heat pulse, as taken by Th1 and Th2, are shown in Fig. 14.3. A clear temperature decrease, on the order of 2 mK below the cell wall temperature, is observed 60 s after the end of the pulse. At the beginning of the relaxation period (Fig. 14.2d), the HBL is still convected, as it remains hotter than the bulk, explaining the cooling effect.

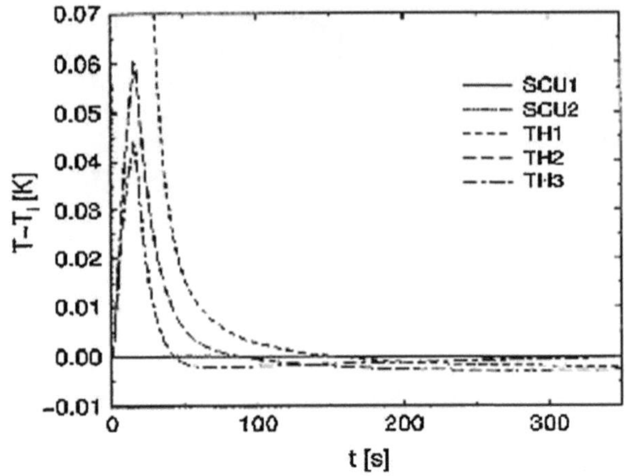

Fig. 14.3 Time responses of thermistors Th1 and Th2 during relaxation after the 100 mW, 15 s heat pulse of Fig. 14.2d–e. A temperature undershoot below the thermostat temperature $T_i = T_c + 0.95$ K is observed

The whole evolution is reproduced well by the 1-D model, as described in the previous section, if $C_c = 0.006\,\mathrm{W\,K^{-1}}$ (Fig. 14.3). The 1 g situation leads to a stronger effect (16 mK of cooling 12 s after the end of the pulse), and this can be reproduced by setting $C_c = 1\,\mathrm{W\,K^{-1}}$ (see Refs. [3, 4]).

14.3 Temperature Oscillations Near the Rayleigh–Bénard Threshold

Kogan and Meyer experimentally studied the heat transport in Rayleigh–Bénard cells close to the onset of convection on the Earth [8, 9]. They performed their experiments in ^{3}He close to critical conditions ($T_c = 3.317$ K, $\rho_c = 41.45\,\mathrm{kg\,m^{-3}}$, $p_c = 0.117$ MPa; see Table 1.1), and measured the temperature difference as a function of the imposed heat flux at the bottom plate of a Rayleigh–Bénard cell of height $h = 1.06$ mm and diameter 57 mm. The top plate was maintained at the initial temperature. Germanium thermometers with a resolution of 0.3 μK were used to control the temperature of the upper plate and to measure the temperature difference ΔT from the bottom plate where the heat flux is imposed. The top and bottom plates were made of OFHC copper and were separated by a thin sidewall made from stainless steel, which is a poor heat conductor. The copper surfaces in contact with the fluid were parallel to within 0.025 mm. The calculated adiabatic temperature difference from Eq. (13.3) is approximately $\Delta T_{ad} = \left(\frac{\partial T}{\partial y}\right) h = 3.6\,\mu\mathrm{K}$, and the estimated piston-effect time from for example Eq. (4.11) is on the order of 10^3 s at $\tau = 10^{-3}$ and 10 s at $\tau = 10^{-1}$.

The cell was filled with ^{3}He to a density of within 0.5 % of ρ_c. During the experiments, the temperature was controlled at a fixed value, with a short-term stability of

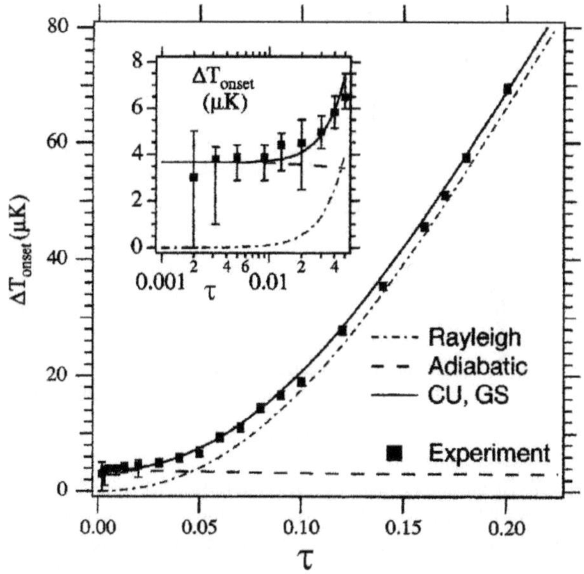

Fig. 14.4 Experimental ΔT_{onset} (*squares*) versus the distance τ to the critical point, and comparison with theory (*lines*). *Inset* semi-log plot in the region where the adiabatic gradient is dominant. (*Adiabatic* [13]; *CU* [6]; *GS* [7])

$\pm 0.8\,\mu$K and a drift of typically 3–5 μK over tens of hours. For a given measurement, the temperature difference across the fluid was measured as a function of time, and its steady-state value $\Delta T(q)$ was recorded for different values of the injected power q.

The transition from a mechanically stable to a convective regime is shown by a sudden change in slope of the curve $\Delta T(q)$. The resolution of $\Delta T(q)$ is ± 0.3 mK. The result given in Fig. 14.4 shows the onset value as a function of the distance τ to the critical point. This result is compared in the figure with the current theories of Refs. [6, 7, 10, 13], as described in Sect. 13.1.1.

Beyond the steady state, transient measurements of the temperature difference were also performed with the same setup. These highlighted the temperature oscillations shown in Fig. 14.5.

It is very difficult to obtain a spatially resolved temperature field experimentally, since the cell is only 1.06 mm high. In order to obtain the temperature, a numerical simulation of the instability process is performed (see [1]) using the same numerical method as in Chap. 22. We consider a 2-D rectangular cavity of height 1 mm (vertical coordinate: y) and length 2 mm (horizontal coordinate: x), which is heated from below, and has periodic lateral boundary conditions to simulate the large experimental aspect ratio of ~50, as in [8]. The upper wall is maintained at a fixed temperature, while a constant heat flux is imposed at the lower boundary. For ^{3}He at $\tau = 0.01$, we consider the following parameters:

- $D_T = 10^{-9}\,\mathrm{m^2\,s^{-1}}$
- $c_p = 5.18 \times 10^5\,\mathrm{J\,kg^{-1}\,K^{-1}}$

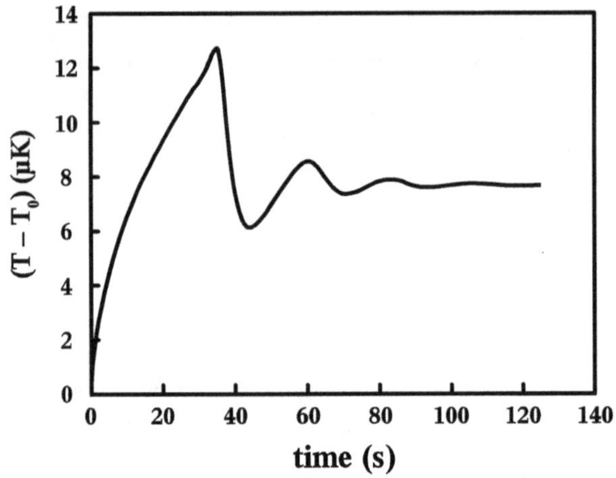

Fig. 14.5 Temperature of the upper plate of a Rayleigh–Bénard cell as measured by Kogan and Meyer [8] for a heat flux of 4.78×10^{-3} Wm^{-3} at the bottom plate. The initial reduced temperature is $\tau = 0.01$. Damped oscillations are observed, whereas critical damping or a simple overshoot can be seen for other values of the difference from the critical point and the heat flux

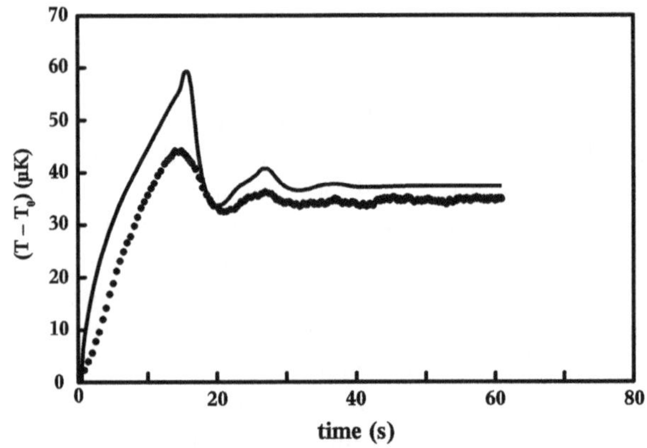

Fig. 14.6 Comparison of the experimental temperature at the bottom plate (*bold line*, see Fig. 14.5) with the numerical one [1] for a heat flux of $q = 6.65 \times 10^{-4}$ Wm^{-2} and a reduced temperature $\tau = 0.01$

- $\kappa_T = 4.3 \times 10^{-3}\,\mathrm{Pa}^{-1}$
- $\beta_p = 52\,\mathrm{K}^{-1}$

Validation of the numerical method is achieved by comparing the experimental and theoretical values of the steady-state temperature difference between the plates as a function of the heat flux. The numerical and experimental transient temperature differences are compared in Fig. 14.6 for a heat flux of $q = 6.65 \times 10^{-4}\,\mathrm{W\,m}^{-2}$ and for

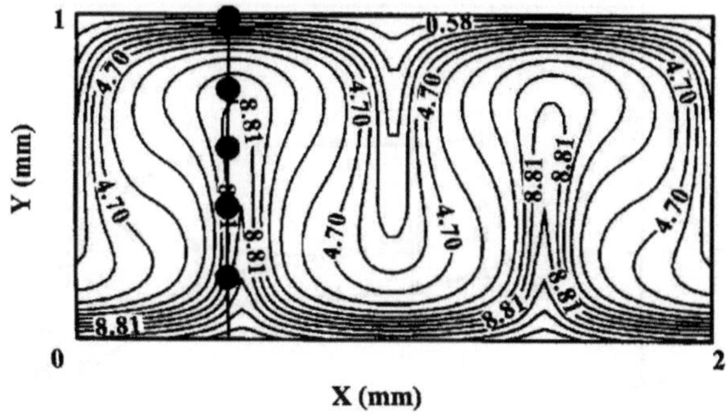

Fig. 14.7 Temperature plot ($T - T_0$) in μK at $t = 36$ s after initiating a heat flux step of $q = 6.65 \times 10^{-4}$ Wm^{-2}. The *dots* indicate the points for which temperature is plotted as a function of time in Fig. 14.8

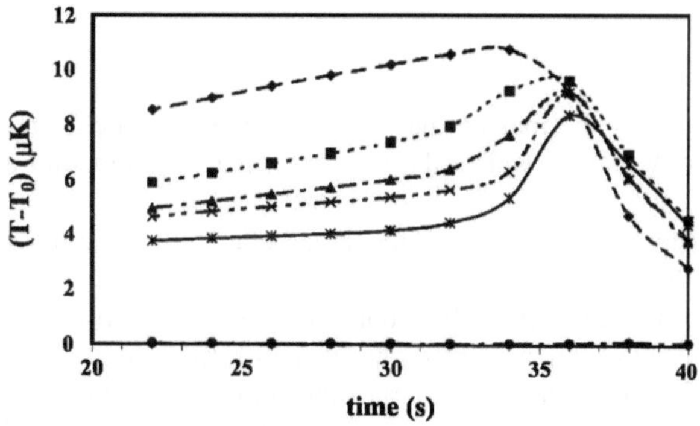

Fig. 14.8 Temperature differences as a function of time at different heights in the thermal plume for the heat flux 6.65×10^{-4} Wm^{-2} (*diamonds* $x = 0.25$ mm, $y = 0.5$ mm; *squares* $x = 0.25$ mm, $y = 0.2$ mm; *triangles* $x = 0.25$ mm, $y = 0.4$ mm; *crosses* $x = 0.25$ mm, $y = 0.6$ mm; *asterisks* $x = 0.25$ mm, $y = 0.8$ mm; *circlesx* $= 0.25$ mm, $y = 1$ mm). The quite simultaneous and homogeneous temperature drops in the curves for $t > 36$ s are a signature of the cooling piston effect triggered by the thermal plume. The drops at the point $x = 0.25$ mm, $y = 0.5$ mm, where there is no convection or diffusion, are particularly enlightening

$\tau = 0.01$. Although the maximun error is on the order of 22 %, good correspondence between of the experimental and numerical oscillations is obtained.

After the bottom wall starts to heat up, two diffusion layers form as a result of the piston effect. During this convection-free period, the bulk temperature evolves under the influence of the competition between the heating piston effect at the bottom plate and the cooling piston effect at the upper thermostated plate. When the diffusion penetration depth and the bulk temperature are such that the layers become unstable relative to the Rayleigh criterion, convection starts, which changes the temperature fields in both the top and bottom layers (see Chap. 13).

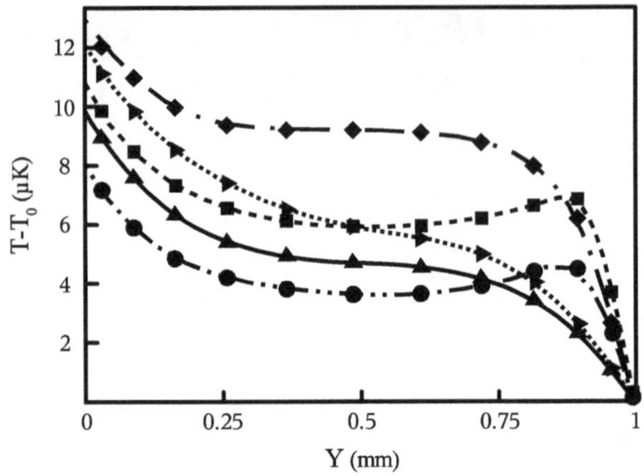

Fig. 14.9 Vertical temperature profiles at $x = 0.5$ mm and at different times (*up-pointing triangles* $t = 22$ s; *sideways-pointing triangles* $t = 32$ s; *diamonds* $t = 36$ s; *squares* $t = 38$ s; *circles* $t = 40$ s)

Figure 14.7 shows thermal plumes in the temperature field at $t = 36$ s and $q = 6.65 \times 10^{-4}$ W m^{-2}, as well as the points at which the temperature was measured numerically.

Figure 14.8 shows the evolution of the temperature at $x = 0.5$ mm and at different y values along the thermal plume (see Fig. 14.7). We first consider a position near the outer edge of the top thermal boundary layer ($y \simeq 0.8$ mm).

At about $t = 32$ s after the beginning of the heating, the temperature increases more rapidly, corresponding to the start of the convection. Fluid at the warmer bulk temperature is brought to the top boundary layer. Then, at about $t = 35$ s, there is an inflection point in the curve for $T(t) - T_0$.

The thermal plume reaches the top boundary region and brings with it warmer fluid from the bottom plate. This results in an increase in the temperature gradient, which strengthens the cooling piston effect. This causes a global and homogeneous temperature drop throughout the bulk fluid. $T(t) - T_0$ reaches a maximum and then decreases.

Figure 14.8 shows homogeneous drops that occur almost simultaneously at all locations, although hot fluid is convected to the top plate. As the normal temperature gradient at the bottom plate is constant, a decrease in the bottom plate temperature must accompany this bulk drop in temperature. The temperature drop occurs at different times depending on the location.

The reason for this is that the local rates of convection-induced heat transport temperature vary.

Figure 14.9 displays vertical temperature profiles along the plume at $x = 0.5$ mm. From $t = 22$ s to $t = 32$ s, they correspond to the convection-free piston effect (homogeneous bulk temperature and a diffusing boundary layer). Later on, convective profiles are observed. At $t = 36$ s, the bulk temperature reaches its maximum.

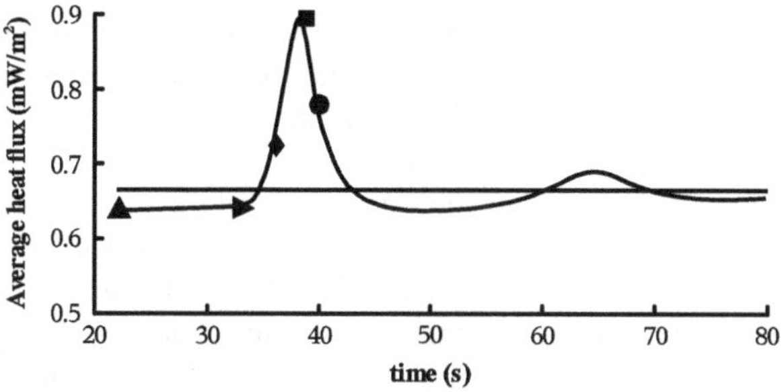

Fig. 14.10 Average heat flux at the upper (*curve*) and lower (*horizontal line*, $q = 6.65 \times 10^{-4}\,\text{Wm}^{-2}$) boundaries as a function of time. The *symbols on the curve* correspond to the symbols used in Fig. 14.9

At $t = 38$ s, convection continues to increase the temperature, but the piston effect decreases it in the bulk.

The temperature gradient at $y = 1$ becomes steeper and steeper and is maximum at $t = 38$ s. Between $t = 38$ s and $t = 40$ s, there is a fast, homogeneous drop in temperature in the bulk, and a corresponding drop in the bottom wall temperature to maintain a constant heat flux at the wall.

The driving force for convection is weakened considerably, and the velocity field exhibits only the remains of the thermal plumes. The thermal field is more homogeneous.

At this time, the system has been reset to its initial conditions by the piston effect somehow, except that the mean temperature is higher. The velocity field was calculated in [1]. It was found to be weak but nonzero.

The other thermal plume (see Fig. 14.7) also contributes to the increase in the cooling piston effect. It should be noted that the downward plume at the center of the cell does not generate any additional heating piston effect, since the heat flux (temperature gradient) is imposed at the bottom wall.

Another signature of the piston effect is the temperature evolution at a point where the velocity is close to zero at all times. For symmetry reasons, this point is located at $x = 0.25$ mm, $y = 0.5$ mm: the velocity is very small (3–4 mm s^{-1}), whereas it is on the order of some 500 mm s^{-1} in the thermal plume. Figure 14.8 gives the temperature evolution at this point ($x = 0.25$ mm, $y = 0.5$ mm). As heat diffusion is negligible at this location, the temperature drop that occurs between $t = 32$ s and $t = 40$ s, with a maximum at about $t = 38$ s, is the result of the competition between the heating and cooling piston effects.

The subsequent production of oscillations can be explained as follows. When the bulk temperature decreases, so does the temperature at the edge of the upper thermal boundary layer.

The temperature gradient and thus the cooling piston effect decreases, and the heating piston effect again becomes prominent, so a new heating phase begins as well as the scenario previously described.

After some damped oscillations, this process leads to an equilibrium steady state. This is the result, as mentioned in [2], of an equilibrium between the heating and the cooling piston effects driven by the thermal plumes.

Figure 14.10 shows the oscillations of the heat flux leaving the cell at the top wall. They are synchronized with the bulk temperature oscillations and with the interactions of the Rayleigh–Bénard plumes with the top wall. The cooling flux eventually reaches a steady state value equal to the incoming heat flux. We should emphasize that this state is not the final equilibrium steady state, since diffusion is still not equilibrated with convection because this timescale is rather short compared to the diffusion timescale.

References

1. Amiroudine S, Zappoli B (2003) Piston-effect-induced thermal oscillations at the Rayleigh-Bénard threshold in supercritical ^{3}He. Phys Rev Lett 90(10):105303
2. Amiroudine S, Kogan A, Meyer H, Zappoli B (2000) 20th ICTAM congress. In: ICTAM 2000, Chicago, IL, USA, 27 August–2 September 2000
3. Beysens D, Garrabos Y, Fröhlich T (2008) Heat can cool near critical fluids. In: Proceedings of 59th IAC congress, Glasgow, Scotland, 29 September–03 October, 2008, number paper IAC-08-A2.4.5
4. Beysens D, Fröhlich T, Garrabos Y (2011) Heat can cool near-critical fluids. Phys Rev E 84(5 pt 1):051201
5. Boukari H, Shaumeyer JN, Briggs ME, Gammon RW (1990) Critical speeding up in pure fluids. Phys Rev A 41(4):2260–2263
6. Carles P, Ugurtas B (1999) The onset of free convection near the liquid–vapour critical point. Part 1: stationary initial state. Phys D 126(1–2):69–82
7. Gitterman M, Steinberg VA (1970) Criteria for the commencement of convection in a liquid close to the critical point. High Temp (USSR) 8(4):754
8. Kogan AB, Meyer H (2001) Heat transfer and convection onset in a compressible fluid: ^{3}He near the critical point. Phys Rev E 63(5):056310
9. Kogan AB, Murphy D, Meyer H (1999) Rayleigh-Bénard convection onset in a compressible fluid: ^{3}He near t_C. Phys Rev Lett 82(23):4635–4638
10. Normand C, Pomeau Y, Vellarde MG (1977) Convective instabilities: a physicist's approach. Rev Mod Phys 49(3):581–624
11. Onuki A, Ferrell RA (1990) Adiabatic heating effect near the gas–liquid critical point. Phys A 164(2):245–264
12. Onuki A, Hong H, Ferrell RA (1990) Fast adiabatic equilibration in a single-component fluid near the liquid–vapor critical point. Phys Rev A 41(4):2256–2259
13. Spiegel EA (1965) Convective instabilities in a compressible atmosphere. Astrophys J 141:1068
14. Wunenburger R, Garrabos Y, Lecoutre C, Beysens D, Hegseth J, Zhong F, Barmatz M (2002) Thermal response of a two-phase near-critical fluid in low gravity: strong gas overheating as due to a particular phase distribution. Int J Thermophys 23:103–115
15. Wunenburger R, Garrabos Y, Lecoutre-Chabot C, Beysens D, Hegseth J (2000) Thermalization of a two-phase fluid in low gravity: heat transferred from cold to hot. Phys Rev Lett 84(18):4100–4103

Part IV
Influence of Time-Dependent Acceleration Fields

Chapter 15
Introduction to Influence of Time-Dependent Acceleration Fields

The main aim in this final part of the book (Part IV) is to introduce new theoretical tools for using acceleration fields to understand and control heat and mass transfer in supercritical and near-critical fluids. This is achieved by determining how such systems would be expected to respond when they are subjected to mechanical vibrations, and then validating these expected results using data from experiments performed under microgravity conditions. The effects of vibrations in liquids and ideal gases lead to a diverse collection of situations that have not yet been fully investigated. We should mention, however, the important contribution of the Russian school of Perm, as reviewed in the book *Thermal Vibrational Convection* by Gershuni and Lyubimov [3]. On Earth, vibrations couple with the Earth's gravitational field, which considerably complicates such studies. Thus, studying systems that are under the influence of microgravity can greatly simplify the mechanisms involved and/or reveal new phenomena that exist only during weightlessness. In the case of near-critical fluids, this subject is in its infancy, and research in this area is limited to some studies that have reported new behavior.

Vibrations act on density inhomogeneities in matter. Vibrations are thus able to induce mean flows that depend on the period of the vibration, as opposed to the inverse hydrodynamic time. The average effect of vibrations is to direct the inhomogeneity perpendicular to the direction of acceleration under the effect of a Bernoulli-like pressure $\Delta p_B \sim \Delta\left(\rho u^2\right)$. This is due to the differential velocities (Δu) that density inhomogeneities $(\Delta\rho)$ acquire through inertial effects. It is also a general effect that can be related to the case of a pendulum subjected to vertical vibrations that cause its mean center of gravity to be displaced from the vertical of the axis of rotation [4].

Part IV is therefore divided into two chapters that illustrate expected and observed new fluid motions which arise when it is assumed that each fluid element is subjected to a periodic acceleration with angular frequency ω. Chapter 16 presents the expected "solid-like" response of near-critical fluids to low-frequency vibrations. In this particular regime, which is never witnessed in ideal gases, the bulk part of the

B. Zappoli et al., *Heat Transfers and Related Effects in Supercritical Fluids*, Fluid Mechanics and Its Applications 108, DOI: 10.1007/978-94-017-9187-8_15,

fluid behaves almost like a solid body bouncing back and forth between two highly compressible thermal boundary layers [2]. Finally, Chap. 17 analyzes the thermal effects that occur in vibrated near-critical fluids under zero gravity (it is also worth noting that interesting features are observed during phase separation under vibration [1]). We show that the periodic acceleration caused by vibrations can have a marked influence on the behavior of such fluids.

References

1. Beysens D, Chatain D, Evesque P, Garrabos Y (2005) High-frequency driven capillary flows speed up the gas–liquid phase transition in zero-gravity conditions. Phys Rev Lett 95(3):034502
2. Carles P, Zappoli B (1995) The unexpected response of near-critical fluids to low-frequency vibrations. Phys Fluids 7:2905–2914
3. Gershuni GZ, Lyubimov DV (1998) Thermal vibrational convection. Wiley, Hoboken
4. Landau LD, Lifshitz EM (1973) Mecanique. Mir, Moscow

Chapter 16
Response to Low-Frequency Vibrations: Solid-Like Behavior

Abstract A one-dimensional analytical description of the response of a supercritical fluid to harmonic vibrations is given. The mechanical response of the bulk consists in a compression or in an expansion that induces a local heating or a cooling. In the case of thermostated boundaries, the thermal perturbation creates a thermal boundary layer. The fluid contained in this layer contracts or expands by thermo-mechanical coupling thus creating a non-zero velocity at the edge of the boundary layer. This non zero velocity replaces the zero velocity boundary condition for the bulk equations as written for a normally compressible fluid. For low frequency vibrations, the bulk behavior is changed and the fluid behaves almost as a solid body bouncing back and forth between two boundary layers. This is referred to as a "solid-like" response to low frequency vibrations.

16.1 Model and Governing Equations

We consider the 1-D model of a fluid cavity of length L as presented in Sect. 3.2.1. The 1-D slab-like cell is thermostated at $x = 0$ and at $x = 1$, respectively. The fluid is initially at rest and at thermodynamic equilibrium. This cell is submitted to a harmonic linear vibration that acts at $t > 0$. The vibration is directed perpendicular to the side walls, and the corresponding acceleration is given by $\gamma_v \sin \omega_v t_a t$, where $t_a = \frac{L}{c_s}$ is the dimensional acoustic time for the fluid cavity and c_s is the velocity of sound (with $c_s = c_{s0} = \sqrt{\gamma_{0,\mathrm{IG}} r T_c}$ sing a van der Waals fluid modeling, see Sect. (1.11.8)). The nondimensional variables and physical properties, as well as the corresponding governing equations, are identical to those defined in Sect. 3.2.1, except that the momentum equation involves a dragging force source term. The latter is assumed to be homogeneous. The nondimensional momentum equation can be written as

$$\rho\,(u_t + u u_x) = -\frac{1}{\gamma_{0,\mathrm{IG}}} p_x + \epsilon \frac{4}{3} u_{xx} + \rho \gamma \sin \omega t. \tag{16.1}$$

B. Zappoli et al., *Heat Transfers and Related Effects in Supercritical Fluids*,
Fluid Mechanics and Its Applications 108, DOI: 10.1007/978-94-017-9187-8_16,

The nondimensional acceleration γ is derived from the dimensional acceleration γ_v as

$$\gamma = \frac{L\gamma_v}{\gamma_{0,\mathrm{IG}} r T_c}.$$

The parameter γ is generally small. For example, if $\gamma_v = 10\,g_0$ is applied to a $L = 10\,\mathrm{mm}$ cell filled with CO_2 at critical density, we get $\gamma = 1.22 \times 10^{-5}$.

The nondimensional angular frequency ω is defined by

$$\omega = \omega_v t_a,$$

where ω_v is the dimensional pulsation of the vibration.

The presence of small parameters in the governing equations suggests that the problem can be solved by the asymptotic expansions technique. The vibration pulsation ω introduces a new timescale Θ_v such that $\Theta_v = \omega t$. For an ideal gas, the 1-D response to an imposed vibration depends on the value of Θ_v relative to the acoustic timescale only. Here, it can be anticipated that the impact of Θ_v on the fluid response depends on its value relative to the physical characteristic times in the nonvibrated van der Waals fluid, especially to the characteristic time of the piston effect, which corresponds to the piston-effect timescale Θ of Eq. (4.65) given in Sect. 4.2.3, which can be written as follows:

$$\Theta = \frac{\varepsilon}{\tau^{\frac{3}{2}}} t.$$

For a given distance to the critical point, $\frac{\varepsilon}{\tau^{\frac{3}{2}}}$ and t_a take fixed values, while ω depends on the imposed frequency. In other words, we have to define the low-frequency vibrations according to the characteristic acoustic time, *and then* low frequency means that

$$\omega \ll 1.$$

In the case of near-critical fluids, we can expect crossover behavior when the characteristic vibration time is equal to the characteristic piston-effect time; i.e., when

$$\omega = \frac{\varepsilon}{\tau^{\frac{3}{2}}}.$$

Also, as $\frac{\varepsilon}{\tau^{\frac{3}{2}}} \ll 1$, to overcome difficulties when the characteristic vibration time is on the same order of magnitude as the characteristic piston-effect time, we first consider the case where the piston-effect timescale is the shorter one, so that the following condition is satisfied:

$$\frac{\varepsilon}{\tau^{\frac{3}{2}}} \ll \omega \ll 1. \tag{16.2}$$

16.1.1 Bulk Response

The analysis presented in this section is formally similar to the analysis already performed in [2] for an ideal gas. As the volume dragging force acts on the bulk phase, this analysis is first performed for the nondimensional space variable $x = O(1)$. Boundary layers are further introduced near the boundaries to account for the thermal adaptation at a temperature difference ΔT above the critical temperature. The fluid variables are asymptotically expanded. The scaling functions are chosen so as to ensure that the continuity equation is not truncated and the pressure gradient is matched to the dragging force. This latter criterion derives from the fact that a volume force within a fluid initially generates a pressure gradient. The following expansions for the fluid variables are then obtained:

$$
\begin{aligned}
T &= 1 + \tau + \gamma T_b\left(x, \Theta_v\right) + o\left(\gamma\right) \\
p &= \tau + \frac{3}{8} + \gamma p_b\left(x, \Theta_v\right) + o\left(\gamma\right) \\
\rho &= 1 + \gamma \rho_b\left(x, \Theta_v\right) + o\left(\gamma\right) \\
u &= \omega\gamma u_b\left(x, \Theta_v\right) + o\left(\omega\gamma\right).
\end{aligned}
$$

Inserting these expansions into the governing equations (3.31) to (3.34) given in Sect. 3.2.1, where the momentum equation is now given by Eq. (16.1), we obtain

$$
\begin{aligned}
\rho_{b\,\Theta_v} &= -u_{b\,x} \\
p_{b\,x} &= \gamma_{0,\mathrm{IG}} \sin \Theta_v \\
T_{\Theta_v} &= -\frac{3}{2}\left(\gamma_{0,\mathrm{IG}} - 1\right) u_{b\,x} \\
p_b &= \frac{3}{2} T_b.
\end{aligned}
$$

Ignoring the transient period, the solution can be written as in [2]:

$$
\begin{aligned}
T_b\left(x, \Theta_v\right) &= \frac{3}{2}\gamma_{0,\mathrm{IG}}\left(x - \frac{1}{2}\right)\sin \Theta_v \\
p_b\left(x, \Theta_v\right) &= \gamma_{0,\mathrm{IG}}\left(x - \frac{1}{2}\right)\sin \Theta_v \\
\rho_b\left(x, \Theta_v\right) &= \frac{1}{c_{s0}^2}\left(x - \frac{1}{2}\right)\sin \Theta_v \\
u_b\left(x, \Theta_v\right) &= \frac{1}{c_{s0}^2}\frac{x}{2}\left(x - \frac{1}{2}\right)\cos \Theta_v.
\end{aligned}
\tag{16.3}
$$

The response of the bulk is thus a linear oscillation in pressure, density and temperature around the central node $x = \frac{1}{2}$ in the cell. It is clear that the above solution does not satisfy the boundary conditions. A boundary layer analysis must therefore be performed.

16.1.2 Boundary Layer Response

The boundary layer analysis is only performed in the left boundary layer near the origin, as the boundary layer located near $x = 1$ is symmetrical. The new boundary layer variable is

$$z = \frac{x}{\delta},$$

where δ is the boundary layer thickness.

The fluid variables are expanded for $\varepsilon \to 0$ and $\tau \to 0$, with $z = \frac{x}{\varepsilon} = O(1)$. The scaling functions are found in the same way as in the bulk, except that we impose matching between the diffusion and pressure force work terms in the energy equation. This matching leads to the boundary layer thickness $\delta = \sqrt{\frac{\varepsilon}{\omega}}\tau^{\frac{1}{4}}$. The variables are thus expanded as

$$\begin{aligned}
T &= 1 + \tau + \gamma \overline{T}_{bl}(z, \Theta_v) + o(\gamma) \\
p &= \tau + \frac{3}{8} + \gamma p_{bl}(z, \Theta_v) + o(\gamma) \\
\rho &= 1 + \frac{\gamma}{\tau}\rho_{bl}(z, \Theta_v) + o\left(\frac{\gamma}{\tau}\right) \\
u &= \omega\gamma\frac{\delta}{\tau}u_{bl}(z, \Theta_v) + o\left(\omega\gamma\frac{\delta}{\tau}\right),
\end{aligned} \tag{16.4}$$

which lead to the following boundary layer equations:

$$\begin{aligned}
\rho_{bl\,\Theta_v} &= -u_{bl\,z} \\
p_{bl\,z} &= 0 \\
D_c T_{bl\,zz} &= \frac{3}{2}u_{bl\,z} \\
p_{bl} &= \frac{3}{2}T_{bl} + \frac{9}{4}\rho_{bl},
\end{aligned}$$

with the boundary conditions:

$$\begin{aligned}
T_{bl}(z = 0) &= 0 \\
u_{bl}(z = 0) &= 0.
\end{aligned}$$

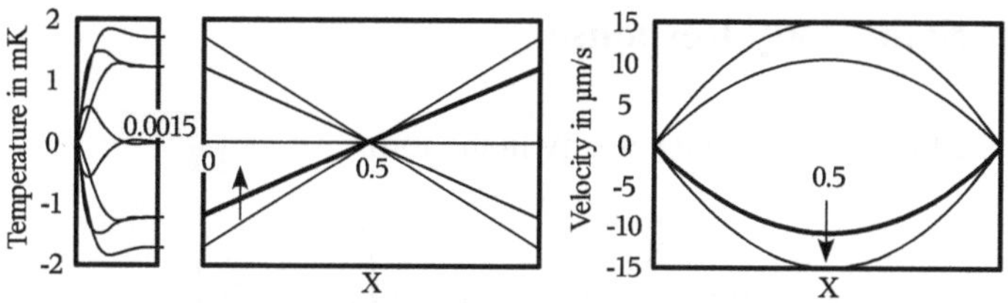

Fig. 16.1 Temperature and velocity profiles over one period of oscillation (for a frequency of 100 Hz). *Left* temperature in the boundary layer given every eighth of a period. *Center* and *right* temperature and velocity in the bulk, respectively. Note the matching between the bulk and boundary layer temperatures (only two of them match to zero)

The solutions are

$$T_{bl}\left(z,\Theta_v\right)=\frac{\gamma_{0,IG}}{3}\left[\exp\left(-\sqrt{\frac{2}{D_c}}z\right)\sin\left(\Theta_v-\sqrt{\frac{1}{2D_c}}z\right)-\sin\Theta_v\right] \quad (16.5)$$

$$p_{bl}\left(z,\Theta_v\right)=-\frac{\gamma_{0,IG}}{2}\sin\Theta_v$$

$$\rho_{bl}\left(z,\Theta_v\right)=\frac{2\gamma_{0,IG}}{9}\exp\left(-\sqrt{\frac{1}{2D_c}}z\right)\sin\left(\Theta_v-\sqrt{\frac{1}{2D_c}}z\right) \quad (16.6)$$

$$u_{bl}\left(z,\Theta_v\right)=\frac{2\gamma_{0,IG}}{9}\sqrt{D_c}\left[\sin\left(\Theta_v+\frac{\pi}{4}\right)-\exp\left(-\sqrt{\frac{1}{2D_c}}z\right)\right.$$
$$\left.\times\sin\left(\Theta_v+\frac{\pi}{4}-\sqrt{\frac{1}{2D_c}}z\right)\right]. \quad (16.7)$$

The bulk pressure is homogeneous and oscillates in phase with the vibration of the cell. The temperature exhibits the same type of oscillations, with the maximum amplitude occurring near the walls. However, since the walls are thermostated, temperature adaptation layers form.

The temperature and the velocity in the bulk and in the boundary layer are plotted in Fig. 16.1 for an imposed frequency of 100 Hz and an amplitude of 10 g_0. The container is $L = 10$ mm long and the initial temperature distance to the critical temperature is 0.2 mK. As pressure is homogeneous, high density gradients appear because of the piston effect. Unlike cases studied until now, temperature perturbations at the wall come from the bulk and not from outside the cell. The bulk behavior is similar to that of the ideal gas, but the density perturbations are $\frac{1}{\tau}$ larger inside the boundary layers.

16.2 The Solid Body Response

We can write Eq. (16.4) for the velocity in the form

$$u = \omega\gamma\sqrt{\frac{\frac{\varepsilon}{\tau^{\frac{3}{2}}}}{\omega}}u_{bl}\left(z, \Theta_v\right) + o\left(\omega\gamma\sqrt{\frac{\frac{\varepsilon}{\tau^{\frac{3}{2}}}}{\omega}}\right).$$

Note that, when $z \to \infty$, the first-order approximation for the velocity tends to a time-dependent homogeneous value, since $U_{bl \lim z\to\infty}\left(z, \Theta\right) = \frac{2\gamma_{0,\mathrm{IG}}}{9}\sqrt{D_c}\sin\left(\Theta_v + \frac{\pi}{4}\right)$. This indicates that the piston effect is driven at the second order in the bulk by the adaptation layer, as shown in [2]. The second-order perturbation in the bulk driven by this piston effect is on the order of $\mathcal{O}\left(\omega\gamma\frac{\frac{\varepsilon}{\tau^{\frac{3}{2}}}}{\omega}\right)$. Accordingly, this second-order expansion for the velocity in the bulk can be written as

$$u = \omega\gamma u_b\left(x, \Theta_v\right) + \omega\gamma\sqrt{\frac{\frac{\varepsilon}{\tau^{\frac{3}{2}}}}{\omega}}u_b^{(1)}\left(x, \Theta_v\right) + o\left(\omega\gamma\sqrt{\frac{\frac{\varepsilon}{\tau^{\frac{3}{2}}}}{\omega}}\right). \tag{16.8}$$

Under the conditions given by Eq. (16.2), the ratio $\frac{\frac{\varepsilon}{\tau^{\frac{3}{2}}}}{\omega}$ is a small parameter. The matching condition between the boundary layer and the bulk is fulfilled because $u_b\left(x, \Theta_v\right) \to 0$ when $x \to 0$. When $\frac{\frac{\varepsilon}{\tau^{\frac{3}{2}}}}{\omega}$ is no longer a small parameter (e.g., when the characteristic vibration time becomes of the same order of magnitude as the piston-effect timescale $\frac{\varepsilon}{\tau^{\frac{3}{2}}}$), the above matching condition is no longer satisfied. The description obtained for $\frac{\varepsilon}{\tau^{\frac{3}{2}}}$ tends to zero faster than ω does, and is not uniform when $\omega \to 0$. In other words, ω is a singular perturbation parameter when the ratio $\frac{\varepsilon}{\tau^{\frac{3}{2}}}$ is zero. The description obtained in the previous section is thus invalid when the frequency of the vibration becomes smaller and smaller. We expand the fluid properties for $\omega \to 0$ with a fixed value of an inner variable $\widetilde{\omega}$ defined by

$$\widetilde{\omega} = \frac{\omega}{\sigma(\varepsilon, \tau)},$$

where $\sigma(\varepsilon, \tau) \ll 1$ is to be determined. With the second term written in terms of the inner variable $\widetilde{\omega}$, Eq. (16.8) becomes:

$$u = \omega\gamma u_b\left(x, \Theta_v\right) + \omega\gamma\sqrt{\frac{\frac{\varepsilon}{\tau^{\frac{3}{2}}}}{\widetilde{\omega}\sigma(\varepsilon, \tau)}}u_b^{(1)}\left(x, \Theta_v\right) + o\left(\omega\gamma\sqrt{\frac{\frac{\varepsilon}{\tau^{\frac{3}{2}}}}{\widetilde{\omega}\sigma(\varepsilon, \tau)}}\right).$$

The above equation shows that the condition $\sigma(\varepsilon, \tau) = \frac{\varepsilon}{\tau^{\frac{3}{2}}}$ removes the singularity of the expansion, since the first and second terms are of the same order of magnitude.

16.2.1 Inner Description

16.2.1.1 Bulk Solution

Scaling functions of the inner description are obtained for $\omega \to 0$ and $\widetilde{\omega} = \mathcal{O}(1)$ fixed, using a similar approach to that described in Sect. 16.1.1 ($\omega \gg \frac{\varepsilon}{\tau^{\frac{3}{2}}}$). Introducing the superscript in, which stands for "inner", we obtain

$$
\begin{aligned}
T &= 1 + \tau + \gamma T_b^{in}\,(x, \Theta_v) + o(\gamma) \\
p &= \tau + \frac{3}{8} + \gamma p_b^{in}\,(x, \Theta_v) + o(\gamma) \\
\rho &= 1 + \gamma \rho_b^{in}\,(x, \Theta_v) + o(\gamma) \\
u &= \frac{\varepsilon}{\tau^{\frac{3}{2}}} \gamma u_b^{in}\,(x, \Theta_v) + o(\frac{\varepsilon}{\tau^{\frac{3}{2}}}\gamma),
\end{aligned}
$$

which leads to the following system:

$$
\begin{aligned}
\widetilde{\omega} \rho_{b\,\Theta_v}^{in} &= -u_{b\,x}^{in} \\
p_{b\,x}^{in} &= \gamma_{0,\mathrm{IG}} \sin \Theta_v \\
T_{b\,\Theta_v}^{in} &= -\frac{3}{2}\left(\gamma_{0,\mathrm{IG}} - 1\right) u_x^{in} \\
p_b &= \frac{3}{2} T_b^{in}.
\end{aligned}
$$

The above system yields

$$
\begin{aligned}
p_b^{in}\,(x, \Theta_v) &= \gamma_{0,\mathrm{IG}} \sin(\Theta_v)\, x + f(\Theta_v) \\
u_b^{in}\,(x, \Theta_v) &= -\frac{4}{9}\frac{1}{D_c}\widetilde{\omega} \cos(\Theta_v)\,\frac{x^2}{2} - \frac{4}{9}\frac{1}{D_c}\frac{d}{d\Theta_v} f(\Theta_v) + g(\Theta_v),
\end{aligned}
\tag{16.9}
$$

where the unknown uniform functions f and g are deduced through asymptotic matching with the boundary layer solutions.

16.2.1.2 Boundary Layer Solutions

The spatial variable is rescaled in the vicinity of $x = 0$ to account for the singularity introduced by the diffusion term in the energy equation. The rescaled space variable z is defined by

$$z = \frac{x}{\delta}.$$

The relevant expansions, performed with $\widetilde{\omega}$ fixed, $\omega \to 0$ with z fixed, are obtained by using a similar approach to that shown in Sect. 16.1.2:

$$\begin{aligned}
T &= 1 + \tau + \gamma T_{bl0}^{in}(z, \Theta_v) + o(\gamma) \\
p &= \tau + \frac{3}{8} + \gamma p_{bl0}^{in}(z, \Theta_v) + o(\gamma) \\
\rho &= 1 + \frac{\gamma}{\tau}\rho_{bl0}^{in}(z, \Theta_v) + o(\frac{\gamma}{\tau}) \\
u &= \frac{\varepsilon}{\tau^{\frac{3}{2}}}\gamma u_{bl0}^{in}(z, \Theta_v) + o(\frac{\varepsilon}{\tau^{\frac{3}{2}}}\gamma).
\end{aligned}$$

In [1], the same approach as the one developed in this book was used. Around $x = 0$, we obtain

$$\begin{aligned}
\widetilde{\omega}\rho_{bl0\,\Theta_v}^{in} &= -u_{bl0\,z}^{in} \\
p_{bl0\,z}^{in} &= 0 \\
D_c T_{bl0\,zz}^{in} &= \frac{3}{2}u_{bl0\,z}^{in} \\
p_{bl0}^{in} &= \frac{3}{2}T_{bl0}^{in} + \frac{9}{4}\rho_{bl0}^{in}.
\end{aligned}$$

For the homogeneous pressure in the boundary layers, the matching condition with the bulk gives

$$p_{bl0}^{in} = f(\Theta_v).$$

The previous system reduces to the diffusion equation

$$\left[T_{bl}^{in} - \frac{2}{3}f(\Theta_v)\right]_{\Theta_v} = \frac{D_c}{\widetilde{\omega}}\left[T_{bl}^{in} - \frac{2}{3}f(\Theta_v)\right]_{zz},$$

with the following boundary conditions:

$$\begin{aligned}
&T_{bl0}^{in}(z, \Theta_v) - \frac{2}{3}f(\Theta_v) \to 0 \quad \text{when } z \to \infty \\
&T_{bl0}^{in}(z, \Theta_v) - \frac{2}{3}f(\Theta_v) = 0 \quad \text{at } x = 0.
\end{aligned}$$

The Laplace transform technique gives the following solution in Laplace space:

$$LT_{bl0}^{in}(z, \Theta_v) = \frac{2}{3}F(s)\left[1 - \exp\left(-\sqrt{s\frac{\widetilde{\omega}}{D_c}}z\right)\right],$$

where $F(s)$ is the Laplace transform of the unknown function $f(\Theta_v)$ and s the Laplace variable corresponding to the time variable Θ_v. The solution for the velocity, written in Laplace space, is

$$Lu_{bl0}^{in}(z, \Theta_v) = -\frac{4}{9}\frac{1}{\sqrt{\widetilde{\omega}}\sqrt{D_c}}\sqrt{s}F(s)\left[1 - \exp\left(-\sqrt{s\frac{\widetilde{\omega}}{D_c}}z\right)\right].$$

From the above expression for the velocity in the boundary layer, we can deduce one equation for the matching condition with the bulk solution.

The other equation, that is needed to determine the second of the two unknown functions $f(\Theta_v)$ and $g(\Theta_v)$, is obtained by writing the matching condition between the bulk and the boundary layer located at $x = 1$. In the latter, the relevant spatial variable is

$$z^* = \frac{1-x}{\delta}.$$

For symmetry reasons, the physical properties are sought under asymptotic series expansions that are identical to those in the left boundary layer. The governing equations for the right boundary layer are

$$\begin{aligned}
\widetilde{\omega}\rho_{bl1\,\Theta_v}^{in} &= u_{bl1\,z^*}^{in} \\
P_{bl1\,z^*}^{in} &= 0 \\
D_c T_{bl1\,z^*z^*}^{in} &= -\frac{3}{2}u_{bl1\,z^*}^{in} \\
P_{bl1}^{in} &= \frac{3}{2}T_{bl1}^{in} + \frac{9}{4}\rho_{bl1}^{in}.
\end{aligned}$$

The above system results in the following diffusion equation:

$$\left[T_{bl1}^{in} - \frac{2}{3}\left(\gamma_{0,\mathrm{IG}}\sin(\Theta_v) + f(\Theta_v)\right)\right]_{\Theta_v} = \frac{D_c}{\widetilde{\omega}}\left[T_{bl1}^{in} - \frac{2}{3}\left(\gamma_{0,\mathrm{IG}}\sin(\Theta_v) + f(\Theta_v)\right)\right]_{z^*z^*},$$

which is associated with the boundary and matching conditions

$$T_{bl1}^{in}(z^*, \Theta_v) = 0 \text{ at } z^* = 0\ (x = 1)$$

$$T_{bl1}^{in}(z^*, \Theta_v) - \frac{2}{3}\left(\gamma_{0,\mathrm{IG}}\sin(\Theta_v) + f(\Theta_v)\right) = 0 \quad \text{when } z^* \to \infty.$$

The solution, expressed in Laplace space, is

$$LT_{bl1}^{in}(z^*,\Theta_v) = \frac{2}{3}F(s)\left(1-\exp\left(-\sqrt{s\frac{\widetilde{\omega}}{D_c}}z^\star\right)\right) - \frac{2}{3}\frac{\gamma_{0,\mathrm{IG}}}{1+s^2}\exp\left(-\sqrt{s\frac{\widetilde{\omega}}{D_c}}z^\star\right).$$

From this, it follows that

$$Lu_{bl1}^{in}(z^*,\Theta_v) = \frac{4}{9}\frac{1}{\sqrt{\widetilde{\omega}}\sqrt{D_c}}\sqrt{s}\left[F(s)+\frac{\gamma_{0,\mathrm{IG}}}{1+s^2}\right]\left[1-\exp\left(-\sqrt{s\frac{\widetilde{\omega}}{D_c}}z^\star\right)\right].$$

The two unknown functions f and g in Eq. (16.9) can be determined from the matching conditions for the bulk velocity. Therefore, we obtain the following solutions for the left boundary layer:

$$T_{bl}^{in}(z,\Theta_v) = \frac{\gamma_{0,\mathrm{IG}}}{3}\left[\exp\left(-\sqrt{\frac{2}{D_c}}z\right)\sin\left(\Theta_v-\sqrt{\frac{1}{2D_c}}z\right) - \sin(\Theta_v)\right] \tag{16.10}$$

$$P_{bl}^{in}(z,\Theta_v) = -\frac{\gamma_{0,\mathrm{IG}}}{2}\sin(\Theta_v)$$

$$\rho_{bl}^{in}(z,\Theta_v) = \frac{2\gamma_{0,\mathrm{IG}}}{9}\exp\left(-\sqrt{\frac{1}{2D_c}}z\right)\sin\left(\Theta_v-\sqrt{\frac{1}{2D_c}}z\right) \tag{16.11}$$

$$u_{bl}^{in}(z,\Theta_v) = -\frac{2\gamma_{0,\mathrm{IG}}}{9}\sqrt{D_c}\sqrt{\widetilde{\omega}}\left[\exp\left(-\sqrt{\frac{1}{2D_c}}z\right) \times\sin\left(\Theta_v+\frac{\pi}{4}-\sqrt{\frac{1}{2D_c}}z\right) - \sin(\Theta_v+\frac{\pi}{4})\right],$$

and for the bulk:

$$T_b^{in}(x,\Theta_v) = \frac{2}{3}\gamma_{0,\mathrm{IG}}(x-\frac{1}{2})\sin(\Theta_v) \tag{16.12}$$

$$P_b^{in}(x,\Theta_v) = \gamma_{0,\mathrm{IG}}(x-\frac{1}{2})\sin(\Theta_v)$$

$$\rho_b^{in}(x,\Theta_v) = \frac{4}{9}\frac{\gamma_{0,\mathrm{IG}}}{\gamma_{0,\mathrm{IG}}-1}(x-\frac{1}{2})\sin(\Theta_v)$$

$$u_b^{in}(x,\Theta_v) = \widetilde{\omega}\frac{4}{9}\frac{\gamma_{0,\mathrm{IG}}}{\gamma_{0,\mathrm{IG}}-1}\frac{x}{2}(1-x)\cos(\Theta_v) + \sqrt{\widetilde{\omega}}\frac{1}{9}\gamma_{0,\mathrm{IG}}\sqrt{D_c}\sin(\Theta_c+\frac{\pi}{4}). \tag{16.13}$$

The corresponding solution in the right boundary layer is deduced from the solution of Eq. (16.10) for symmetry reasons.

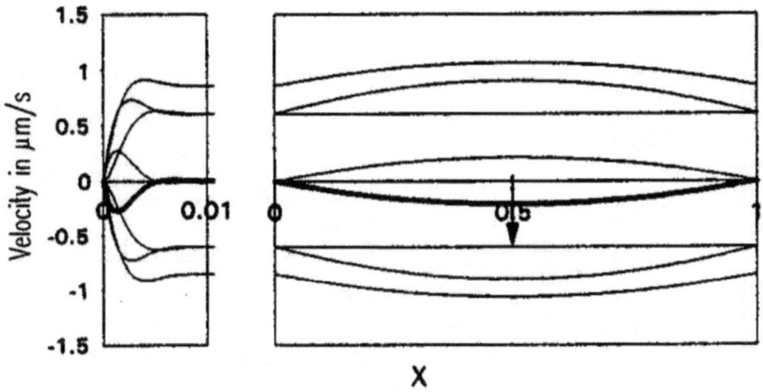

Fig. 16.2 Velocity profiles versus the spatial variable x for the quasi-solid regime over one period of oscillation, given every eighth of a period of the imposed frequency 2 Hz. *Right* $0 < x < 1$; *left* details of the left boundary layer $0 < x < 0.01$

16.2.2 The Solid Body Response of a Hypercompressible Fluid

The inner description again gives the solutions obtained for $\omega \ll \frac{\varepsilon}{\tau^{\frac{3}{2}}}$ and for $\omega \gg \frac{\varepsilon}{\tau^{\frac{3}{2}}}$ when the inner variable goes to zero or to infinity, respectively. The solution for $\omega \gg \frac{\varepsilon}{\tau^{\frac{3}{2}}}$ was obtained in Sect. 16.1.1. The second limit, $\omega \ll \frac{\varepsilon}{\tau^{\frac{3}{2}}}$, is precisely the solid body response. It cannot be reached continuously from Eqs. (16.3) and (16.6). As $\widetilde{\omega} \ll \sqrt{\widetilde{\omega}}$ when $\widetilde{\omega} \to 0$, the expansion of Eq. (16.13) shows that the velocity is homogeneous in the bulk. The fluid thus behaves as a solid body bouncing back and forth in-between the two very thin and hypercompressible boundary layers. In other words, the solid body regime is progressively reached as the imposed frequency is lowered below the inverse of the characteristic time of the piston effect. The parabolic term in Eq. (16.13) progressively decreases, giving prominence to the homogeneous term. Such a profile is given in Fig. 16.2 for $\widetilde{\omega} = 0.21$, which corresponds to a frequency of 2 Hz applied to a $L = 10$ mm container that is initially at 1 K from T_C.

The corresponding mechanism is as follows. When the bulk is drawn to the wall or off the wall by the dragging force, it is heated up or cooled down by the compression or expansion of the fluid in the boundary layer. As the pressure is homogeneous, the density increases or decreases. The adaptation layer provokes a fluid flow to the wall (into the boundary layer) or off the wall (out of the boundary layer), in the same direction as the bulk flow. When this velocity is much smaller than the bulk velocity, there is no modification of the boundary conditions. The parabolic profile is prominent. This situation corresponds to $\omega \gg \frac{\varepsilon}{\tau^{\frac{3}{2}}}$. When ω decreases to $\omega \lesssim \frac{\varepsilon}{\tau^{\frac{3}{2}}}$, the vibration leaves enough time for the velocity at the edge of the boundary layer to reach the same order of magnitude as the bulk velocity.

We now consider the maximum velocity in the bulk phase. Equation (16.13) shows that this maximum velocity is proportional to ω for $\omega \gg \frac{\varepsilon}{\tau^{\frac{3}{2}}}$ and proportional to $\sqrt{\omega}$ for $\omega \ll \frac{\varepsilon}{\tau^{\frac{3}{2}}}$, with a transition regime for $\omega \sim \frac{\varepsilon}{\tau^{\frac{3}{2}}}$.

The maximum velocity in the bulk is plotted in Fig. 16.3. The acoustic regime ($\omega \gg \frac{\varepsilon}{\tau^{\frac{3}{2}}}$) does not display any particular feature as compared to the ideal gas. However, when the initial

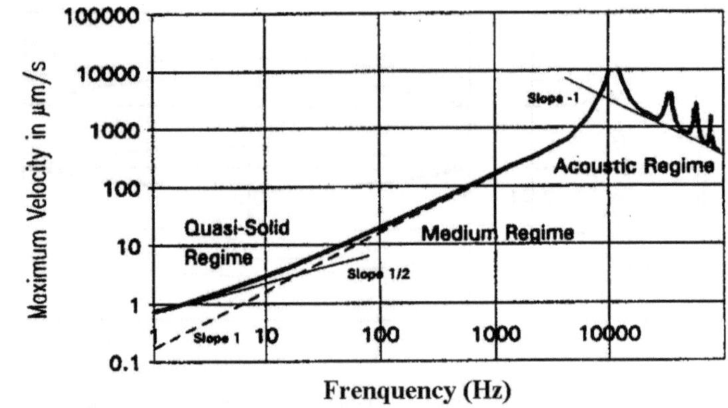

Fig. 16.3 Maximun bulk velocity versus frequency. The *solid lines* correspond to the supercritical fluid with three different regimes. The *dashed lines* correspond to the ideal gas with two regimes. The solid regime is characteristic of the supercritical fluid

temperature is very close to T_c in the saturated regime ($\tau \sim \varepsilon^{\frac{2}{3}}$), the acoustic regime is likely to be modified.

References

1. Jounet A (2002) Density relaxation of a near-critical fluid in response to local heating and low frequency vibration in microgravity. Phys Rev E 65(3):037301
2. Zappoli B (1990) Response of a solid–gas growth interface to a homogeneous time dependent acceleration field. Int J Heat Mass Transf 33(9):1829–1837

Chapter 17
Thermovibrational Effect

Abstract The effect of vibrations on thermal processes (thermo-vibrational effects) is investigated in fluids under a zero gravity field, far and close to their critical point. The effect of vibrations is made apparent in the Navier-Stokes equations when averaged in the limit of small vibration amplitudes (with respect to the sample size) and high frequencies (with respect to the inverse hydrodynamics times). Then the Rayleigh–Bénard configuration is studied, leading to the so-called vibrational Rayleigh–Bénard instability. Numerical simulations and experiments (during the MIR missions) with SF_6 are also reported where a point heat source triggers either convection parallel to the vibration direction or perpendicular to it, depending on the vibration amplitude and frequency parameters.

17.1 Basic Equations with Vibrations

Here, we will reproduce the main aspects of the theoretical development given in [8] that are needed to understand the vibrated fluid motion in the Rayleigh–Bénard-like experimental configuration, without an external gravitational acceleration field. Let us first consider an *incompressible* fluid at temperature T and pressure p vibrating with an amplitude of Ξ and an angular frequency of ω (frequency f, period ϑ), which is submitted to a thermal gradient $\nabla T \sim \frac{\Delta T}{e}$, where ΔT is a typical temperature difference across the typical sample length scale e. "Incompressible" means that we ignore the mechanical compression of the fluid, and the density variations due to temperature inhomogeneities (δT) remain relatively small compared to the average density; that is, $\beta_p \delta T \lll 1$. These density inhomogeneities will be taken into account only in the equation of motion, where they will produce buoyancy forces at the origin of the convection.

The maximum amplitude of the fluid vibration is a_v and the direction of vibration is characterized by the unit vector **n** such that

B. Zappoli et al., *Heat Transfers and Related Effects in Supercritical Fluids*, Fluid Mechanics and Its Applications 108, DOI: 10.1007/978-94-017-9187-8_17,

$$\Xi = a_v \cos(\omega t)\, \mathbf{n}. \quad (17.1)$$

The fluid is submitted to the periodic velocity

$$\mathbf{v} = -a_v \omega \sin(\omega t)\, \mathbf{n}, \quad (17.2)$$

and the periodic acceleration

$$\gamma_v = -a_v \omega^2 \cos(\omega t)\, \mathbf{n}. \quad (17.3)$$

Using Boussinesq's approximations, and replacing the Earth's acceleration by the vibrational acceleration of Eq. (17.3), we can write the hydrodynamic equations as

$$\begin{aligned} \frac{\partial \mathbf{v}}{\partial t} + (\mathbf{v} \cdot \nabla)\, \mathbf{v} &= -\frac{1}{\rho} \nabla p + \nu \Delta \mathbf{v} + \beta_p T a_v \omega^2 \cos(\omega t) \mathbf{n} \\ \frac{\partial T}{\partial t} + \mathbf{v} \cdot \nabla T &= D_T \Delta T \\ \nabla \cdot \mathbf{v} &= 0. \end{aligned} \quad (17.4)$$

According to these equations, an inhomogeneous fluid will move periodically over time. It is now useful to represent the velocity, temperature and pressure fields by superimposing the average and oscillating parts:

$$\mathbf{v} = \mathbf{v}_a + \mathbf{v}_\varpi; \quad T = T_a + T_\varpi; \quad p = p_a + p_\varpi. \quad (17.5)$$

Here, the subscripts a and ϖ stand for "averaged" and "pulsating" respectively.

A method of obtaining the equations corresponding to the average fields was proposed by Kapitza for a vibrating pendulum [10]. This assumes that (i) the vibrating times are smaller than the typical thermal and hydrodynamic times, and (ii) the amplitude of the vibration is small with respect to the typical length scale of the vibrated body. In this context, one notes that the average quantities in Eq. (17.5) vary slowly with time when compared to the vibration period.

Ignoring the small pulsating terms and integrating over the vibration period in Eq. (17.4), we then eventually obtain [8]

$$\begin{aligned} \frac{\partial \mathbf{v}_a}{\partial t} + (\mathbf{v}_a \cdot \nabla)\, v_a &= -\frac{1}{\rho} \nabla p_a + \nu \Delta \mathbf{v}_a + \Omega\, (\mathbf{o} \nabla)\, (T_a \mathbf{n} - \mathbf{o}) \\ \frac{\partial T_a}{\partial t} + \mathbf{v}_a \cdot \nabla T_a &= D_T \Delta T_a \\ \nabla \cdot \mathbf{v}_a = 0; &\quad \nabla \cdot \mathbf{o} = 0. \end{aligned} \quad (17.6)$$

Here, $\mathbf{o}$ is the rotational part of the vector field

$$T_a \mathbf{n} = \mathbf{o} + \nabla \phi$$
$$\nabla \cdot \mathbf{o} = 0, \tag{17.7}$$

where ϕ is a scalar. After substituting $\mathbf{o}$ into Eq. (17.6) and integrating $\frac{\partial \mathbf{v}_a}{\partial t}$, we get

$$\mathbf{v}_a = \left(\beta_p a_v \omega \sin \omega t\right) \mathbf{o}$$
$$T_a = \left(\mathbf{o} \cdot \nabla T_a\right) \beta_p a_v \cos \omega t. \tag{17.8}$$

Note that $\mathbf{o}$ corresponds to the slowly varying part of the velocity $\mathbf{v}_a$. The useful parameter that eventually measures the effect of high-frequency vibrations is defined as

$$\Omega = \frac{1}{2} \left(\beta_p a_v \omega\right)^2. \tag{17.9}$$

Here, amplitude and frequency contribute independently to the average flow.

17.2 Dimensionless Equations

The boundary conditions (denoted by the subscript bc) are as follows. Velocity is zero at the sample wall, and the temperature is fixed at the wall (as denoted by the subscript w):

$$\mathbf{v}_{bc} = 0$$
$$T_{bc} = T_w. \tag{17.10}$$

The viscous and thermal boundary layers are neglected, so the normal component o_n of $\mathbf{o}$ is zero (impermeability condition):

$$(o_n)_{bc} = 0. \tag{17.11}$$

The limitations on the amplitude and frequency of the vibration are mainly related to the following aspects:

(i) The Boussinesq conditions of an incompressible fluid. The acoustic wavelength $c_{s0}\vartheta$ should remain larger than any sample dimension e; that is,

$$\vartheta \gg \frac{e}{c_{s0}}. \tag{17.12}$$

(ii) The thermal time $t_D = \frac{e^2}{D_T}$ must remain larger than the vibration period, as must the viscous time

$$t_{\mu} = \frac{\rho e^2}{\mu} = \frac{e^2}{\nu}. \tag{17.13}$$

The limits of the above description are thus written as

$$\frac{e}{c_{s0}} \ll \vartheta \ll \min\left(\frac{e^2}{D_T}, \frac{e^2}{\nu}\right). \tag{17.14}$$

The above equations can be written in a dimensionless form. We consider the following characteristic scales:

$$\begin{aligned} \text{length scale} \quad & e \\ \text{time scale} \quad & \frac{e^2}{\nu} \\ \text{velocity scale} \quad & \frac{D_T}{e} \\ \text{temperature and } \mathbf{o} \text{ field scales} \quad & \Delta T \\ \text{pressure} \quad & \frac{\rho \nu D_T}{e^2} \end{aligned} \tag{17.15}$$

where ΔT is the reference temperature difference. After introducing the following dimensionless parameters:

$$\begin{aligned} \text{Prandt number Pr} &= \frac{\nu}{D_T} \\ \text{Vibrational Rayleigh number Ra}_v &= \frac{\Omega \Delta T^2 e^2}{\nu D_T} = \frac{(\beta_P a_v \omega \Delta T e)^2}{2\nu D_T}, \end{aligned} \tag{17.16}$$

we obtain

$$\begin{aligned} \frac{\partial \mathbf{v}_a}{\partial t} + \frac{1}{\text{Pr}} (\mathbf{v}_a \cdot \nabla)\, v_a &= -\nabla p_a + \Delta \mathbf{v}_a + \text{Ra}_v\, (\mathbf{o} \cdot \nabla)\, (T_a \mathbf{n} - \mathbf{o}) \\ \text{Pr} \frac{\partial T_a}{\partial t} + \mathbf{v}_a \cdot \nabla T_a &= \Delta T_a \\ \nabla \cdot \mathbf{v_a} &= 0 \\ \nabla \cdot \mathbf{o} = 0;\ \nabla \times \mathbf{o} &= 0 \\ \mathbf{v}_b = 0;\ T_b = T_w;\ (o_n)_b &= 0. \end{aligned} \tag{17.17}$$

The vibrational Rayleigh number will be very useful for defining the stable region for the flow. Note that it is not obtained from the regular Rayleigh number under $1g$ by naively substituting the vibrational acceleration $a_v \omega^2$ in place of the Earth's gravitational acceleration g in the definition of the Rayleigh number $\left(\text{Ra} = \frac{g_0 \beta_P \Delta T e^3}{\nu D_T}\right)$.

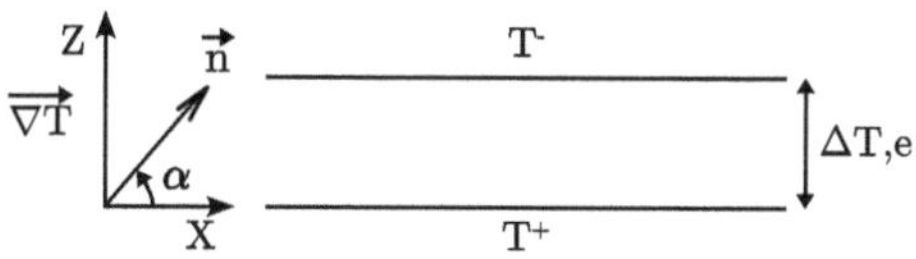

Fig. 17.1 The configuration of a vibrational Rayleigh–Bénard instability

The physical meaning of this vibrational Rayleigh number can be found by performing the following phenomenological analysis.

Let us first consider a fluid element of size e that is acted upon by Earth's gravity. This element starts to rise when the typical convective time across the fluid element is shorter than the diffusion time over the length scale e; that is, $\frac{e}{v_{Sg}} < \frac{e^2}{D_T}$. Here, v_{Sg} is the convective velocity as given by the Stokes equation

$$v_{Sg} = g_0 \frac{e^2 \Delta\rho}{18\rho_c \nu}. \tag{17.18}$$

The Rayleigh number gives a measure of the ratio of these two diffusive and convective times; i.e., $\mathrm{Ra} = g_0 \frac{\beta_p \Delta T e^3}{\nu D_T}$ (using the relation $\Delta\rho = \beta_p \Delta T$). Convection starts for Ra $\geq$1,700.

Let us now consider the same configuration that is not acted upon by gravity, but is subjected to vibrations. The criterion for buoyancy-driven convection can be replaced with a Bernoulli-like pressure difference arising from the velocity difference Δu between the fluid element and its surroundings, $\Delta p \sim \rho u^2 = \rho\,(\Delta\rho/\rho_c)^2\, a_v^2 \omega^2$. This results in a driving force $e^2 \Delta p$, oriented perpendicular to the direction of vibrations. Applying the same procedure that is used to define the Rayleigh number, another number is obtained that is precisely the vibrational Rayleigh number of Eq. (17.16). Note that simply replacing the Earth's gravitational acceleration g with the vibrational acceleration $a_v \omega^2$ in the Rayleigh number leads to a quite different formula from Eq. (17.16): $\mathrm{Ra}_{vRB} = a_v \omega^2 \frac{\beta_p \Delta T e^3}{\nu D_T}$. This other "vibrational" number corresponds to a Rayleigh–Bénard instability induced by the instantaneous acceleration of the vibration. In contrast, the present vibrational instability corresponds to the averaged effects of the vibration.

17.3 Rayleigh–Bénard Configuration Far From the Critical Point

We now consider the simple Rayleigh–Bénard configuration without an external gravitational field (see Fig. 17.1). The direction of vibration forms an angle α with the temperature gradient.

The 2-D and 3-D stability analyses from [8] enable the determination of a critical vibrational Rayleigh number Ra_{vc} above which the flow becomes unstable. According to the reasoning developed above, where the Bernouilli pressure drives

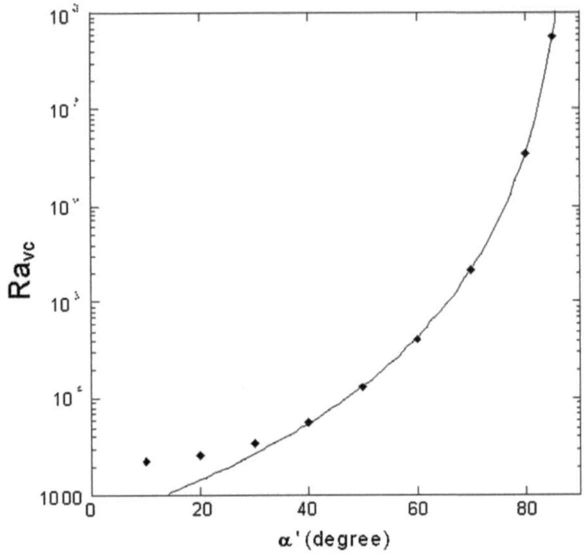

Fig. 17.2 Variation of the critical vibrational Rayleigh number Ra_{vc} with the angle α between the direction of vibration and that of the temperature gradient. The *solid line* corresponds to Eq. (17.20)

the convection perpendicular to the direction of vibration, the onset of instability will depend crucially on the angle α between the direction of vibration and that of the temperature gradient. When $\alpha = \frac{\pi}{2}$, the pressure difference drives any inhomogeneity to move perpendicular to the temperature gradient; the inhomogeneity stays at the same temperature. The fluid is thus stable, $\mathrm{Ra}_{vc} = \infty$.

More generally, this critical number depends on the angle α, as shown in Fig. 17.2. For $\alpha = 0$, the instability threshold occurs for the following value:

$$\mathrm{Ra}_{vc} \sim 2100. \tag{17.19}$$

Near $\alpha = 90°$, the fluid becomes more and more stable, as Ra_{vc} diverges:

$$\mathrm{Ra}_{vc} = 3.46 \times 10^{10} \, (90 - \alpha)^{-4} \, . \tag{17.20}$$

Here, α is in degrees.

17.4 Vibrational Rayleigh–Bénard Instability Near the Critical Point

Near the critical point, some of the approximations we needed to make in Chap. 4 to derive the above vibrational theory may become invalid due to the divergence of the thermodynamic and transport coefficients. Nevertheless, if we consider Eq. (17.14)

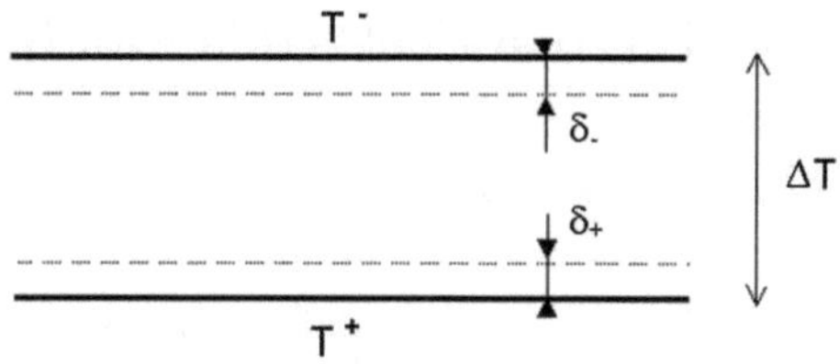

Fig. 17.3 Rayleigh–Bénard configuration in the presence of the piston effect: δ^+ (δ^-) are the thermal hot (cold) boundary layers where the thermal gradient is confined

for example, we can see that the small divergence exponent ($\frac{\alpha}{2} \approx 0.055$) of the velocity of sound (see Sect. 1.10.3) reinforces the validity condition, while the weak divergence exponent ($Y_\eta \approx 0.04$) of the viscosity (see Sect. 1.10.4) has very little effect. This gives us some confidence that the vibrational theory is valid. However, it neglects the compressible nature of the fluid and the role of the thermal boundary layers. For a sample at critical density, Eq. (17.16) yields the following temperature dependence:

$$\mathrm{Ra}_v = \frac{(a_v \omega \beta_p \Delta T e)^2}{2\nu D_T} \sim \tau^{-\nu-2\gamma} \approx \tau^{-3.11}. \tag{17.21}$$

As the vibrational Rayleigh number strongly diverges near the critical point, the fluid then becomes very unstable.

The presence of the thermal boundary layers (see Fig. 17.3) at the heart of the piston effect has not been accounted for in this (idealized) configuration. We conjecture that, analogously to the classical Rayleigh–Bénard configuration in near-critical fluids (see Chap. 13), instability will first arise in the (time-dependent) boundary layers. As a first approximation based on the thermodynamic analysis of the piston-effect timescale described by Eq. (4.11) , we thus replace the sample thickness e with $\delta = \frac{e}{\gamma_0 - 1}$, where $\gamma_0 = \frac{c_p}{c_V} \sim \tau^{-\gamma+\alpha}$ is the specific heat ratio. Equation (17.21) thus becomes

$$\mathrm{Ra}_v = \frac{\left(a_v \omega \beta_p \Delta T e\right)^2}{2\nu D_T \left(\gamma_0 - 1\right)^2} \sim \tau^{-\nu-2\alpha} \approx \tau^{-0.85}. \tag{17.22}$$

Therefore, the thermal layers do indeed become smaller near T_c, but with a smaller exponent than that obtained in Eq. 17.21. The fluid should still become more and more unstable as T tends to T_c.

Preliminary experiments in SF_6 and H_2 and simulation in CO_2 (not reported here, see Ref. [1]) indeed exhibits after a thermal quench under vibration a finger-like destabilization of the thermal boundary layer. These fingers are the hall-mark of convection rolls. They appear when the thermal boundary layer reaches a critical thickness e_c, or, equivalently, after a critical time t_c after the quench, t_c being the time needed for the boundary layer to reach by diffusion the length e_c.

17.5 Thermal Convection with Internal Heat Sources

17.5.1 Uniform Heating

The case of an internal, uniform heat source as provided by e.g. light absorption was treated in [8] for an incompressible fluid. We will just recall the main results here. First, we return to the general equations developed above in Sect. 17.1 and add a source of constant volumic heat power $\mathring{q}$. The thermal equation in Eq. (17.6) then becomes

$$\frac{\partial T_a}{\partial t} + \mathbf{v_a}\nabla T_a = D_T \Delta T_a + \frac{\mathring{q}}{\rho c_p}. \tag{17.23}$$

In order to make the equations dimensionless, we consider the reduction introduced in Eq. (17.15), but with a new reference temperature difference (temperature and **o** field scales):

$$\Delta T \Rightarrow \frac{qe^2}{\rho c_{P D_T}}. \tag{17.24}$$

The set of equation (17.6) can now be written

$$\frac{\partial \mathbf{v_a}}{\partial t} + \frac{1}{\Pr}(\mathbf{v}_a.\nabla)\mathbf{v}_a = -\nabla p_a + \Delta \mathbf{v}_a + \mathrm{Ra}_v(\mathbf{o}.\nabla)(T_a\mathbf{n} - \mathbf{o}) \tag{17.25}$$

$$\begin{aligned} \Pr\frac{\partial T_a}{\partial t} + \mathbf{v}_a.\nabla T_a &= \Delta T_a + 1 \\ \nabla.\mathbf{v}_a &= 0 \\ \nabla.\mathbf{o} = 0; &\qquad \nabla \otimes \mathbf{o} = 0 = \nabla T_a \otimes \mathbf{n} \\ \mathbf{v}_b = 0; &\qquad T_b = T_w; \qquad (o_n)_b = 0, \end{aligned} \tag{17.26}$$

with a new formula for the vibrational Rayleigh number:

$$\mathrm{Ra}_v = \frac{\Omega}{\nu D_T}\left(\frac{qe^3}{\rho c_{p D_T}}\right)^2 = \frac{1}{2\nu D_T}\left(\frac{\beta_p a_v \omega q e^3}{\rho c_{P D_T}}\right)^2.$$

It appears that this number corresponds to the formula in Eq. (17.16), where the temperature difference is expressed as in Eq. (17.24) [Ω being given by Eq. (17.9)]. An instability develops when $\mathrm{Ra}_v \geq \mathrm{Ra}_{vc}$. Different cases can be found depending on the boundary conditions (all isothermal; one isothermal wall and the other wall thermally isolated; one isothermal wall and one wall with heat exchanged; etc.). Some elements of the discussion related to this complex task can be found in Ref. [8]. Here, we will simply note that when the boundaries are kept isothermal, the stability analysis does not differ significantly from the vibrational Rayleigh–Bènard case.

17.5.2 *Point Heat Source and van der Waals Gas in an Isothermal Square Box*

We now discuss now the situation where there is a point heat source with heating power per unit volume q in a container of size $L \times L$ that is filled with near-critical CO_2. All of the walls of this container are kept isothermal. A 2-D numerical simulation was performed by Jounet [9] using the same approach as in Chap. 12; namely, the van der Waals equation of state and the mean-field behavior of the transport coefficients, especially the heat conductivity, were investigated. A vibration is introduced parallel to one axis (x) of the box. The fluid is at critical density, and initially both the fluid and the box are at rest close to but above T_c. From an experimental point of view, this situation is similar to vibrations in a cylindrical cell under zero gravity, as described in Sect. 17.5.3.

The dimensionless equations corresponding to mass and momentum are (the parameters a and b correspond to the specific coefficients of the van der Waals equation; see Table 1.7)

$$\nabla \cdot \mathbf{v} = \frac{(1-b\rho)\frac{dp}{dt} - (\frac{\gamma_{0,\mathrm{IG}}}{\mathrm{Pr}_0\mathrm{Re}_0})\nabla \cdot (\lambda \nabla T)}{-\gamma_{0,\mathrm{IG}}(p + a\rho^2) + 2a\rho^2(1-b\rho)} \tag{17.27}$$

$$\frac{\partial(\rho\mathbf{v})}{\partial t} + \nabla \cdot (\rho\mathbf{v} \otimes \mathbf{v}) = -\nabla \Pi + \frac{1}{\mathrm{Re}_0}[\nabla^2\mathbf{v} + \frac{1}{3}\nabla(\nabla \cdot \mathbf{v})] + \rho\dot{\gamma}\sin(2\pi t)\mathbf{e}_y. \tag{17.28}$$

In addition, the energy equation is

$$\frac{\partial(\rho T)}{\partial t} + \nabla \cdot (\rho\mathbf{v}T) = \frac{1}{\mathrm{Pr}_0\mathrm{Re}_0}\nabla \cdot (\Lambda \nabla T) + q + \frac{\gamma_{0,\mathrm{IG}} - 1}{\gamma_{0,\mathrm{IG}}}(1-b\rho)(\frac{dp}{dt} - 2a\rho^2\nabla\mathbf{v}). \tag{17.29}$$

Nondimensionalization has been performed as in Chap. 2, except for the following:

$$t \Rightarrow ft; \quad \mathbf{v} \Rightarrow \frac{\mathbf{v}}{U}; \quad q \Rightarrow \frac{q}{q_0}; \quad \dot{\gamma} \Rightarrow \frac{4\pi^2 a_v f^2}{U^2 L^{-1}}. \tag{17.30}$$

Here, $\dot{\gamma}$ is the reduced vibration acceleration and

$$q_0 = \frac{p_c f}{(\gamma_{0,\mathrm{IG}} - 1)}. \tag{17.31}$$

The velocity U corresponds to the vibration frequency f such that

$$U = fL. \tag{17.32}$$

Equation (17.28) contains the oscillating field $\rho\dot{\gamma}\sin(2\pi t)\mathbf{e}_y$. As introduced in Chap. 22 on acoustic filtering, Π represents the dynamic pressure and p, normalized by $r\rho_c T_c$ corresponds to the homogeneous part of the pressure.

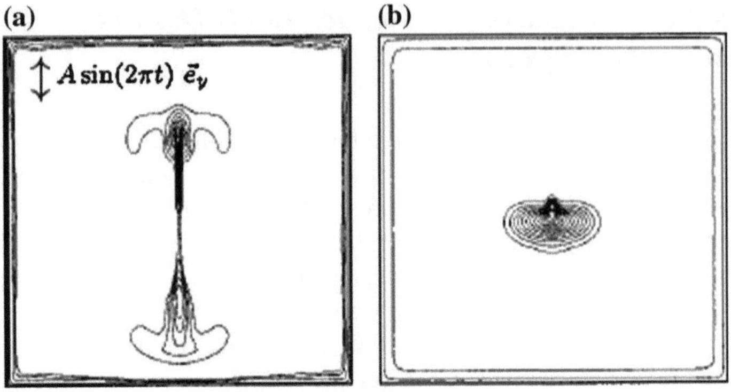

Fig. 17.4 Density fields for different types of vibration: **a** a V1-type vibration: $f = 0.8\,\mathrm{s}^{-1}$ and $a_v = 400\,\mathrm{mm}$ (acceleration $\gamma_v = 10\,\mathrm{ms}^{-2}$), after five periods; **b** a V2-type vibration: $a_v = 8\,\mathrm{mm}$, $f = 1.8\,\mathrm{s}^{-1}$ (acceleration $\gamma_v = 1.0\,\mathrm{ms}^{-2}$), after ten periods. Note that high densities (cold boundary layers) are observed on the thermostated cell walls (From [9])

In [9], the heating power was 0.82 mW, and two cases with different amplitudes and frequencies were investigated. The first case corresponds to a large amplitude with respect to the sample dimension and a moderate frequency, and is called a V1 type of vibration: $a_v = 400\,\mathrm{mm}$, $f = 0.8\,\mathrm{s}^{-1}(\gamma = 10\,\mathrm{m\,s}^{-2})$. The second is of smaller amplitude and higher frequency and is called a V2 type of vibration: $a_v = 8\,\mathrm{mm}$, $f = 1.8\,\mathrm{s}^{-1}(\gamma = 1.0\,\mathrm{m\,s}^{-2})$. The sample has $L = 10\,\mathrm{mm}$. It is filled with CO_2 at critical density and a temperature $T - T_c = 1\,\mathrm{K}$. It is easy to verify that the situation in both cases always corresponds to the "high-frequency case" (see Eq. (17.14)).

Figure 17.4a shows the results of the simulation. A hot boundary layer (HBL; low-density layer) forms around the hot spot and develops symmetrical plumes with vortices or rolls at each end due to viscous coupling with the bulk fluid. These vortices are similar to the plume vortices found under $1g$ conditions (see Sect. 12.3). Due to the low value of the thermal diffusivity, the density field moves mainly by convection. These symmetrical plumes stem from viscous coupling and buoyancy effects associated with the acceleration and inertial effects. When one plume stops because the acceleration has changed sign, the symmetrical plume develops again, and so on for every time period. A plume cannot be convected backwards because the density has changed, so buoyancy is no longer active. At longer times, the plume reaches the cell walls and spreads, forming a hot region with an "interface" *perpendicular* to the direction of vibration, which contrasts with the former extension *parallel* to the vibration.

Similar phenomena have been observed in equivalent numerical simulations [7] on near-critical SF_6 for vibrations with a frequency of 1.8 Hz and an amplitude of 0.38 mm. The same situation, when simulated in an ideal gas at $T = T_c$, shows that the convection rolls extend over a larger area than they do in the near-critical fluid,

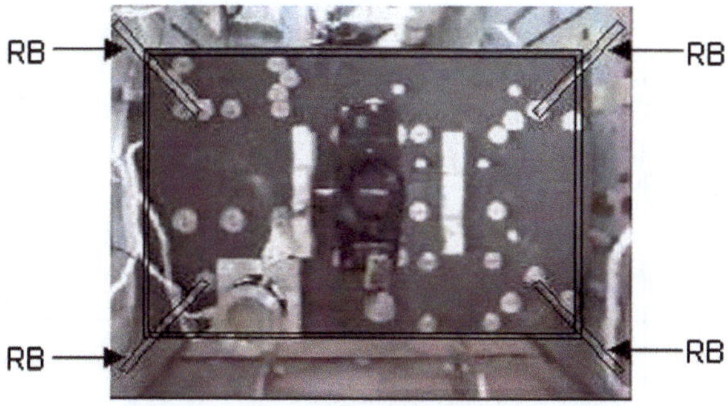

Fig. 17.5 The ALICE2 setup, which was attached to the MIR station by eight rubber bands (*RB*; only four bands are visible) in order to induce vibrations

where they are confined to the hot boundary layer due to the much lower thermal diffusivity.

As a result of mass conservation, the vortices eventually organize themselves into four counter-rotating convection rolls. Such rolls have been observed in all simulations of 2-D incompressible and inviscid fluids [7–9].

17.5.3 Experiments

17.5.3.1 Hardware

A number of experiments have been performed under the microgravity conditions that were present in the MIR station [7]. The fluid employed was SF_6 at critical density. The cell used was cylindrical (pancake cell; see Sect. 7.1.1), and it was equipped with three thermistors, one of which was used to send heat pulses. The cell was set into an interferometer.

The temperature was varied from a few mK to a few K above the critical point. In order to impart a sinusoidal oscillation to the fluid, the instrument was suspended by eight rubber bands inside a compartment of the *Priroda* module in MIR (see Fig. 17.5). One end of each belt was attached to a corner of the instrument, while the other end was secured to a wall of the compartment. The resonance was excited close to the center of mass of the instrument either manually or by an electronically controlled 4 kg mass shaker. Varying the tension of the belts changed the resonance frequency. Three-axis microaccelerometers measured the acceleration levels.

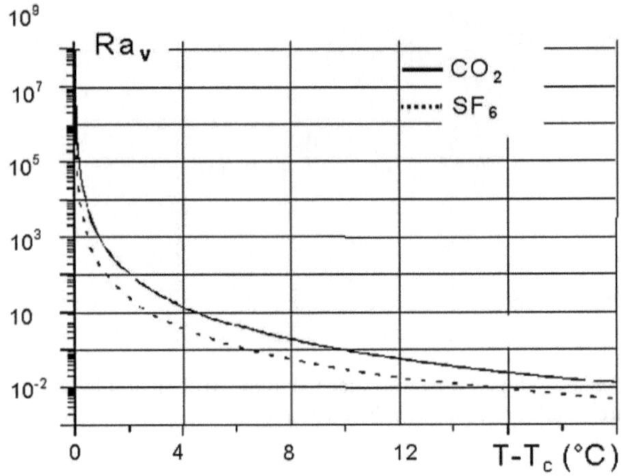

Fig. 17.6 Calculated behavior with respect to $(T - T_c)$ of the vibrational Rayleigh number Ra_v for CO_2 and SF_6. Convection is observed below $T - T_c \simeq 1$°C for $Ra_v \simeq 2000$

17.5.3.2 Instability Threshold

The measurements of flow velocities performed in the above experiment [2] qualitatively confirm the stability analysis of Eq. (17.22) (see Fig. 17.6). For an excitation with a frequency of 5 Hz and an amplitude of 0.1 mm (maximum acceleration $10^{-2}\,g_0$), as induced by the shaker, the heated zone undergoes a large amount of drift below $T - T_c = 1$ K, precisely where Ra_v is about a few 10^3.

17.5.3.3 Hot Boundary Layer

The above analysis also corresponds to what was observed in an experiment in the MIR space station [7] using the experimental arrangement described above in Sect. 17.5.3.1. Heat was generated by 0.828 mW of electrical power sent through a thermistor of 0.2 mm diameter in the center of the sample for a typical duration of either $\delta t = 3.6$ s (V1 case, denoted V′1) or $\delta t = 120$ s (denoted V1 and V2). The fluid employed was SF_6 at near-critical density. The temperature was varied from a few mK to a few K above the critical point.

The phenomena can be separated into two cases, as follows.

Large-Amplitude Case

At early times, the HBL is convected at each oscillation along the direction of vibration, and it spreads out in two symmetric plumes, as shown in Fig. 17.7. During the first complete half-period (1.25–3.75 s), a hot plume develops (due to buoyancy)

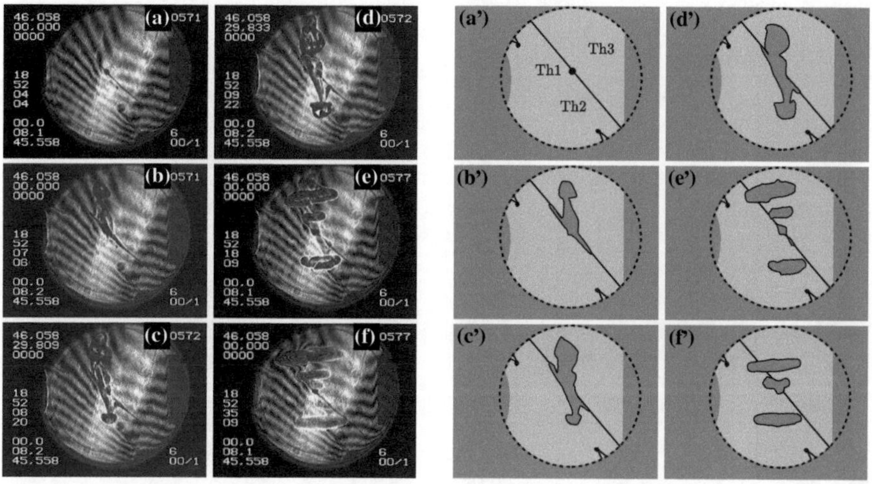

Fig. 17.7 Spread of the HBL at $T = T_c + 500\,\mathrm{mK}$ during a V'1 vibration (3.6 s heat pulse, 0.828 mW power). (a–f) Observations: **a** $t = 0$, **b** $t = 3.08$ s after starting the heat pulse. **c** $t = 4.64$ s, **d** $t = 5.72$ s, **e** $t = 14.2$ s, **f** $t = 31.2$ s. (a'–f') Schematic versions of (a–f). The *arrows* in (a') indicate the location of the three thermistors. The *large double arrow* in (b') indicates the direction of vibration

from the hot thermistor with a direction parallel to the vibration direction (vibration $V'1$, see Fig. 17.7b).

During the second complete half-period (3.75–6.25 s), the head of the plume does not return to its initial position; the head cannot move back by buoyancy into a plume region that exhibits nearly the same density. At the same time, a symmetric plume rises from the hot thermistor (see Fig. 17.7c) and develops just like the previous plume. This plume stops between the thermistor and the wall, at about 2.4 mm from the thermistor.

Figure 17.8 shows the same phenomenon under V1 vibrations. The heating period (120 s) is much longer than in the preceding case (3.6 s). The HBL shows a similar evolution, but it is noticeable that in this case the plume has enough time to reach the cell wall.

This phenomenon is then the symmetrical case to what is currently observed on Earth (such a convection plume is studied under $1g$ in [11]).

This difference in HBL propagation between the short ($V'1$) and long (V1) heat pulses can be understood as follows. Depending on the buoyancy velocity, which itself depends on the density difference $\Delta\rho$, the plumes are convected over a distance l, the value of which determines whether the plume reaches the cell walls (see Fig. 17.8, V1) or not (see Fig. 17.7, $V'1$). Assuming that the Stokes velocity v_S for the buoyancy of a hot fluid volume of diameter d_0 (the thermistor diameter) subjected to an acceleration $a_v\omega^2$ is

$$v_S = \frac{(d_0)^2\,\Delta\rho a_v\omega^2}{18\rho_c\nu}, \tag{17.33}$$

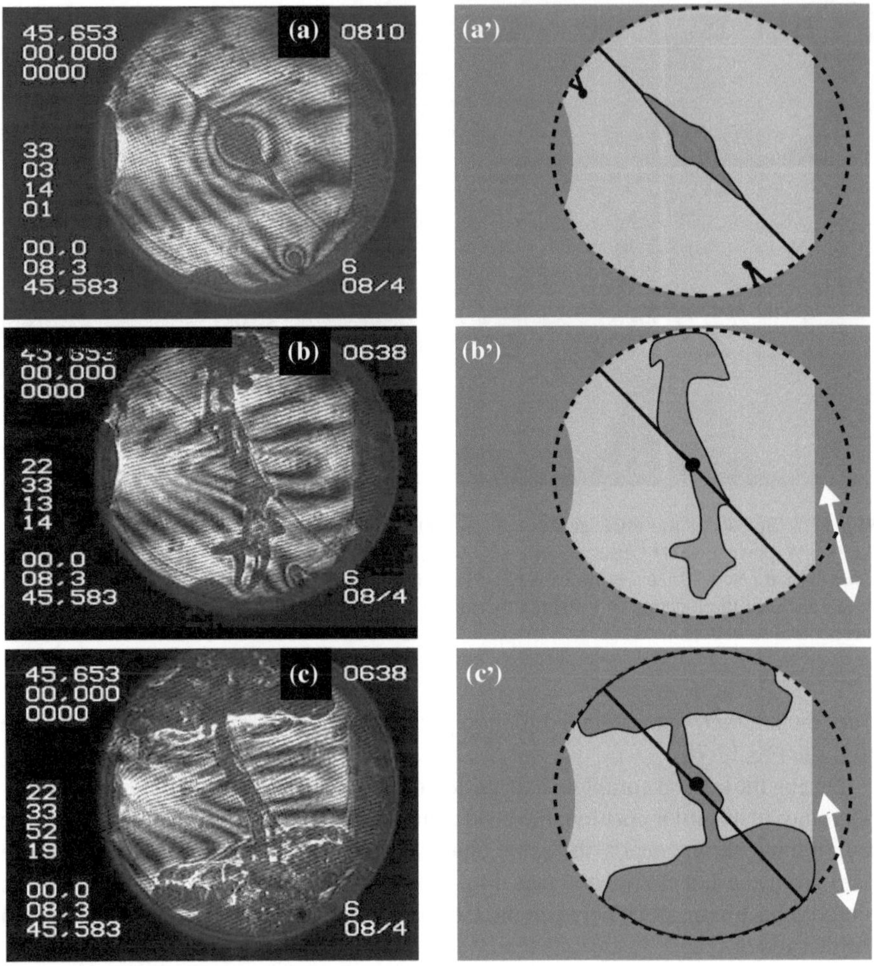

Fig. 17.8 Spread of the HBL during the heating by the thermistor (120 s time duration, 0.828 mW power). **a** Without vibration, $t = 10\,\text{s}$ after the start of the pulse. **b** Under a V1 vibration, $t = 9.4\,\text{s}$ after the start of the heat pulse. The HBL is convected as a double plume. **c** $t = 48.6\,\text{s}$ after the start of the heat pulse. (a'–c') Schematic versions of (a–c). The *double arrows* indicate the direction of vibration

we get

$$l = \frac{v_S}{2f} = \frac{\pi\,(d_0)^2\,\Delta\rho a_v \omega}{18\rho_c \nu}. \tag{17.34}$$

This distance depends only on the vibration velocity $a_v\omega$. The temperature change at Th3 can be estimated to be on the order of 0.1 K, corresponding to a density inhomogeneity of about $\frac{\Delta\rho}{\rho_c} \simeq 2.5\,\%$ (see Fig. 17.12, early times, V1, $T - T_c = 0.5\,\text{K}$).

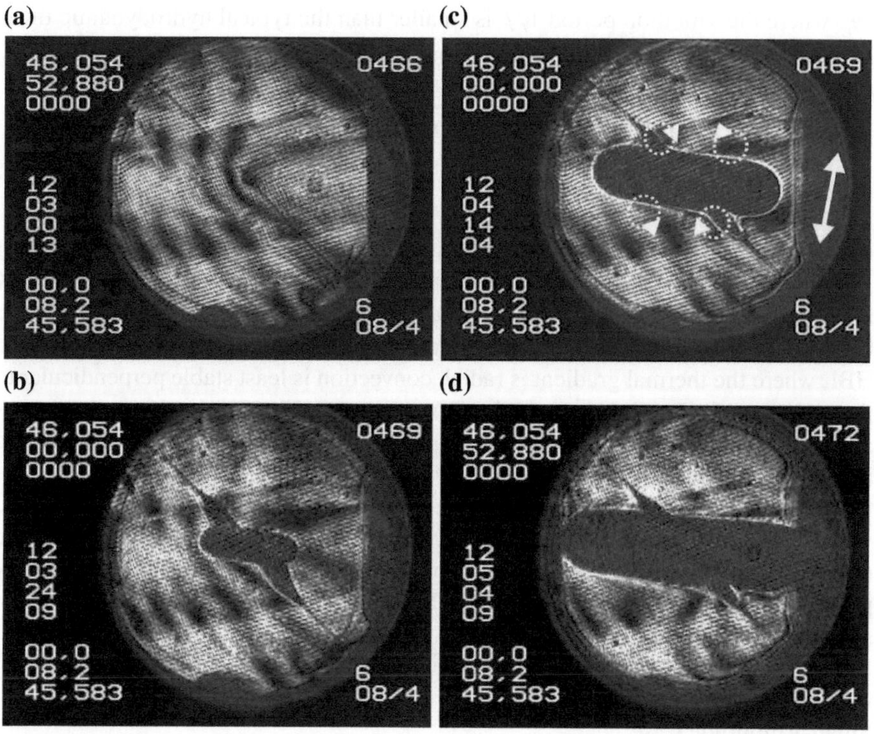

Fig. 17.9 Spread of the HBL at $T = T_c + 500\,\text{mK}$ during a V2 vibration (120 s heat pulse, 0.828 mW power). **a** Before activating the heat pulse. **b** $t = 20\,\text{s}$. **c** $t = 70\,\text{s}$. The *double arrow* indicates the direction of vibration. The *circles* that are drawn very schematically correspond to the presence of four counter-rotating convection rolls (see text). **d** $t = 120\,\text{s}$

Inverting Eq. (17.34) with $l = 2.4\,\text{mm}$ (V1 case) gives $\frac{\Delta\rho}{\rho_c} = 23\%$. Note that the density difference for V1 at 0.1 K from T_c, corresponding to $l \simeq 6\,\text{mm}$, would be larger than 60 %, which is unphysical. This latter estimate simply means that a linear analysis is not valid for such a high temperature rise so close to T_c.

At late times, since the hot fluid cannot be convected in the direction of vibration over a distance larger than l, it can only spread (i) by diffusion (an extremely slow process here) and (ii) by convection (due to the vortices resulting from the viscous coupling with the bulk fluid). The latter process makes the hot fluid spread perpendicular to the direction of vibration at the extremities of the plumes (see Fig. 17.7d–f), where the HBL develops as hot pancakes. This case is also reminiscent of what is observed on Earth when the interface between hot and cold fluid regions orders itself to eventually form a plane layer perpendicular to gravity.

The convective flow that extends the HBL perpendicular to the direction of vibration can also be understood as a vibrational thermal instability of Rayleigh–Bénard type in a cylindrical cell of radius R_0 that is subjected to the temperature difference ΔT. A vibrational Rayleigh number can be defined for high-frequency vibrations;

i.e., where the vibration period $1/f$ is smaller than the typical hydrodynamic times t_D (heat diffusion) and $t_\nu = \frac{L^2}{\nu}$. (This is clearly the situation here, as $1/f = 5$ s, $t_\mu \geq 40$ s, $t_D \geq 900$ s.)

From the $T - T_c$ dependence of β_p, ν, and D_T (with $R_0 = 6$ mm), we obtain

$$\mathrm{Ra}_{\nu V1} \approx 1.3 \times 10^6 \, (T - T_c)^{-1.77} \, \Delta T^2, \tag{17.35}$$

where the subscript V1 refers to V1 vibrations. In the present experiment, the value of ΔT corresponding to the threshold should be on the order of 160 mK at $T - T_c = 5$ K and 20 mK at $T - T_c = 0.5$ K. Because of the dependence of Ra_υ on the angle between the thermal gradient and the direction of vibration, for an initially spherical HBL where the thermal gradient is radial, convection is least stable perpendicular to the direction of vibration. As a result of mass conservation, convection rolls eventually form. Such rolls have also been observed in simulations of 2-D incompressible and inviscid fluids; see Sect. 17.5.2.

Note that the influence of the piston effect could induce a modification of the above vibrational Rayleigh–Bénard description. In particular, the Schwarzschild criterion (see Sect. 13.1.1) could be accounted for. This adjustment becomes increasingly pronounced as $T - T_c$ becomes smaller. How it couples with vibration is beyond the scope of the present book.

Small-Amplitude Case

A similar behavior to that seen for vibrations V1 and V′1 is observed for the vibration V2, although the amplitude (B) is much smaller. The extension l of the plume that is convected parallel to the direction of vibration, proportional to $B\omega$, is now reduced by a factor as large as 10. Thus, only the spread that occurs perpendicular to the vibration can be observed (see Fig. 17.9). We can estimate the vibrational Rayleigh number as

$$\mathrm{Ra}_{\upsilon V2} \approx 1.3 \times 10^4 (T - T_c)^{-1.77} \Delta T^2, \tag{17.36}$$

where the subscript V2 corresponds to the V2 vibration. At 0.5 K from T_c, the temperature gradient at the threshold corresponds to $\Delta T = 210$ mK. At 20 mK from T_c, it becomes as weak as $\Delta T = 12$ mK (see Fig. 17.12).

The dynamics of the propagation of the HBL can be characterized by the evolution of its shape. The two typical lengths are $\delta_\parallel$ and $\delta_\perp$, parallel and perpendicular to the direction of vibration, respectively:

$$\delta_\parallel = \frac{d - d_0}{2}$$

$$\delta_\perp = \frac{D - d_0}{2}.$$

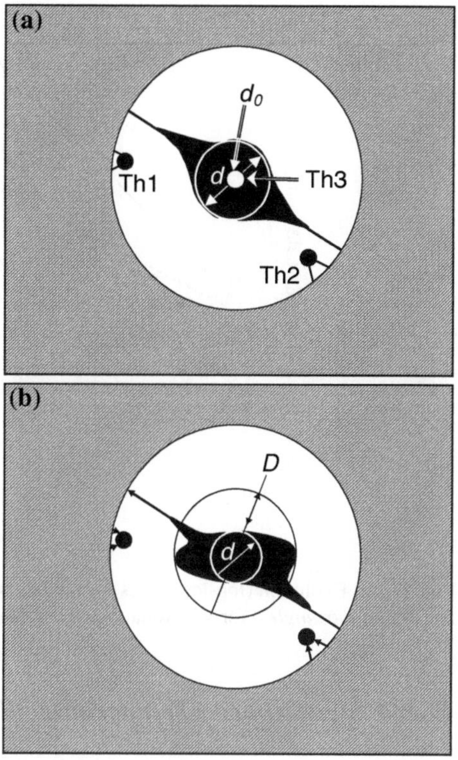

Fig. 17.10 Schematic representation of the hot boundary layer inside the cell. **a** Without vibration (NV), and **b** in the presence of vibration V2. *Th*1, *Th*2, and *Th*3 are thermistors. Th3 is also used as a heat source

Note that d and D are defined in Fig. 17.10. The evolutions of $\delta_{\parallel}$ and $\delta_{\perp}$ are presented in Fig. 17.11.

We also note the following:

(i) In the absence of vibration (NV), the HBL propagates symmetrically: $\delta_{\parallel} = \delta_{\perp}$. This length obeys a diffusion power law with an exponent that is not 0.5 but instead depends on the temperature range and reflects the nonlinear behavior of the thermophysical properties of the fluid near its critical point [3, 5]. The exponent here is 0.45. During this period, the uniform thermalization of the bulk fluid is ensured by the piston effect.

(ii) Under V2 vibrations, $\delta_{\parallel}$ increases after a small initial delay, as can be seen in Fig. 17.11. The length $\delta_{\perp}$ increases faster than it does without vibrations. The slope in the log-log plot of Fig. 17.11 is on the order of unity during most of the time that vibrations are occurring. This corresponds to a constant spreading velocity, and is indeed the hallmark of convective flow in the viscous Stokes law limit.

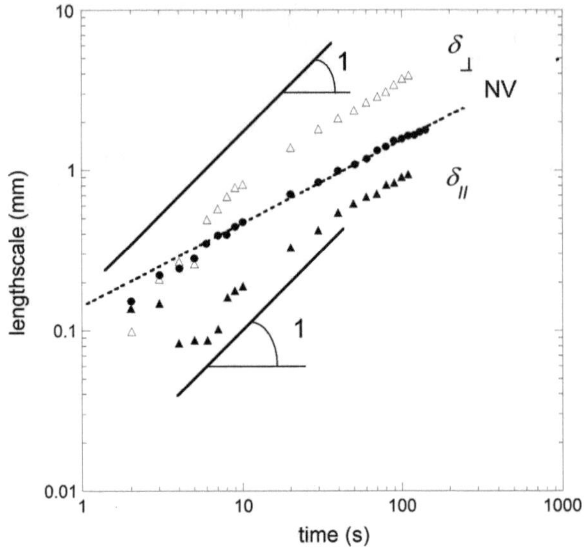

Fig. 17.11 Evolution (log-log scales) of the hot boundary layer length scales $\delta_{\perp}$ *(open triangles)* and $\delta_{\|}$ *(solid triangles)* for V2 excitations. No vibrations case *(NV)*: *black dots* ($T - T_c = 500$ mK.)

17.5.4 Heat Source Temperature Relaxation

We finally consider the temperature relaxation in the heater (Th3) after the heat pulse has stopped, when the HBL behavior crosses over from convective to diffusive spreading.

Figure 17.12 shows the typical temperature evolutions of thermistor Th3 [$\Delta T_3\,(t) = T_3\,(t) - T_3\,(\infty)$] without vibration (NV) and for vibrations V1 and V2. We can see in Fig. 17.12 that ΔT_3 exhibits different behaviors depending on the excitation. This fact simply implies that the heat exchange is affected by different processes: diffusion and the piston effect alone for NV; and diffusion, the piston effect and convection for V1 and V2.

In order to understand these different behaviors, we first consider the relative variations of the different timescales, $t_D = \frac{L^2}{D_T}$, with

$$L \approx \frac{1}{2}(e_0^{-1} + R_0^{-1}) \tag{17.37}$$

according to Eq. (8.2) and $t_{PE} = \frac{t_D}{\gamma_0(\gamma_0 - 1)}$. The longest timescale is the diffusion time, [t_D(5K) $=$ 900 s, and t_D(0.5K) $=$ 4200 s]. The shortest time is the piston effect time [t_{PE}(5K) $=$ 20 s and t_{PE}(0.5K) $=$ 0.4 s].

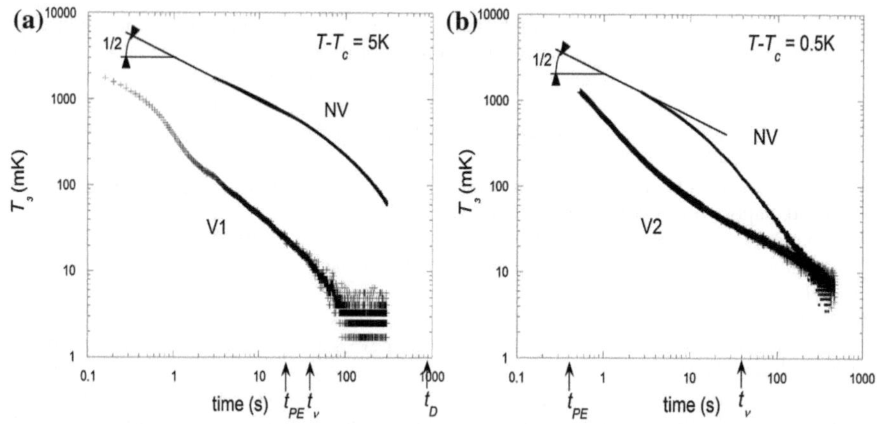

Fig. 17.12 Central thermistor (Th3) temperature relaxation ΔT_3 (log-log scale) after the heat pulse. Vibrations: *V1*, *V2*; no vibrations: *NV*. **a** V1 at $T - T_c = 5$ K, and **b** V2 at $T - T_c = 0.5$ K. The *vertical arrows* indicate the typical piston effect and viscous times. The *horizontal arrow* corresponds to the onset of the vibrational Rayleigh–Bénard instability (see text)

Without vibrations, equilibration occurs by the piston effect and is faster at temperatures closer to T_c. In Fig. 17.12, the early times ($t \leq t_{PE}$) correspond to relaxation slopes of -0.5, as expected [4, 6].

When dealing with V1 and V2 vibrations, we consider another time: the viscous time, corresponding to the viscous relaxation of the flow defined by Eq. (17.13), where e corresponds to the typical length L [as estimated from Eq. (17.37)]. This time does not vary much with temperature; $t_\nu \approx 40$ s. Looking at Fig. 17.12, we do indeed see that the temperature relaxation is nearly complete on this timescale. Convection appears to relax the temperature more efficiently than the piston effect does. Indeed, the fluid is still moving convectively until the temperature gradients vanish.

The temperature gradients at which convection stops can be estimated by calculating the temperature gradient for which Ra_v reaches its critical value $\mathrm{Ra}_{vc} \approx 2000$. This estimation, using Eqs. (17.35) and (17.36), gives $T = 160$ mK for V1 at 5 K from T_c (see Fig. 17.12a) and $T = 210$ mK for V2 at 0.5 K from T_c (see Fig. 17.12b). These temperatures are reached at precisely the moment that the temperature relaxation slows down.

References

1. Amiroudine S, Beysens D (2008) Thermo-vibrational instability in supercritical fluid under weightlessness. Phys Rev E 78:0362351
2. Avdeev SV, Ivanov AI, Kalmykov AV, Gorbunov AA, Nikitin SA, Polezhev VI, Putin GF, Zuzgin AV, Sazonov VV, Beysens D, Garrabos Y, Fròhlich T, Zappoli B (1997) Experiments in

the far and near-critical fluid aboard the mir station with the use of the "alice-1" instrument. In: Avduyevsky VS, Polezhaev VI (eds) Proceedings of the joint Xth European and VIth Russian symposium on physical science in microgravity, St., Petersburg (Russia), 15–21 June 1997. Institute for Problems in Mechanics, vol 1. RAS, Moscow, pp 333–340

3. Bonetti M, Perrot F, Beysens D, Garrabos Y (1994) Fast thermalization in supercritical fluids. Phys Rev E 49(6):R4779–R4782
4. Ferrell RA, Hao H (1993) Adiabatic temperature changes in a one-component fluid near the liquid–vapor critical point. Phys A 197(1–2):23–46
5. Frohlich T, Bouquet S, Bonetti M, Garrabos Y, Beysens D (1995) Self-similar solutions of the non-linear diffusion equation and application to near-critical fluids. Phys A 218(3–4): 419–436
6. Garrabos Y, Bonetti M, Beysens D, Perrot F, Fröhlich T, Carlès P, Zappoli B (1998) Relaxation of a supercritical fluid after a heat pulse in the absence of gravity effects: theory and experiments. Phys Rev E 57(5):5665–5681
7. Garrabos Y, Beysens D, Lecoutre C, Dejoan A, Polezhaev V (2007) Thermoconvectional phenomena induced by vibrations in supercritical SF_6 under weightlessness. Phys Rev E 75:056317
8. Gershuni GZ, Lyubimov DV (1998) Thermal vibrational convection. Wiley, Hoboken
9. Jounet A (2002) Density relaxation of a near-critical fluid in response to local heating and low frequency vibration in microgravity. Phys Rev E 65(3):037301
10. Landau LD, Lifshitz EM (1973) Mecanique. Mir, Moscow
11. Zappoli B, Jounet A, Amiroudine S, Mojtabi A (1999) Thermoacoustic heating and cooling in near-critical fluids in the presence of a thermal plume. J Fluid Mech 388:389–409

Part V
Appendices

Chapter 18
Scaling Laws, Universality, and Renormalization Group Theory

Abstract A short introduction of the scaling laws, the universality, and the renormalization group theory is provided.

18.1 Scaling Laws

In the immediate vicinity of a critical point, a number of thermophysical properties (Y = isothermal compressibility, specific heat at constant pressure, etc.) exhibit power law behavior. A property Y can diverge or tend to zero as a function of the reduced distance to the critical point

$$Y = Y_0 \lambda^{\psi}. \tag{18.1}$$

Here, ψ is the critical exponent. These power laws are observed only on particular thermodynamic paths that lead to the critical point: along the critical isotherm (constant temperature $T = T_c$), $\lambda = \Delta\rho = \frac{\rho - \rho_c}{\rho_c}$; along the critical isotherm (constant critical density $\rho = \rho_c$), $\lambda = \tau = \frac{T - T_c}{T_c}$; or along either the liquid or vapor side of the coexistence curve below the critical point ($\rho = \rho_L$ or $\rho = \rho_V$), $\lambda = -\tau = \frac{T_c - T}{T_c}$. The critical exponents depend on the thermophysical properties: α for the specific heat at constant volume; β for the coexistence curve; γ for the isothermal compressibility; δ for the pressure difference at critical pressure (see Table 1.2). Many inequalities have been found using thermodynamic arguments. Experiments and exact solutions from soluble models show that there are several relationships between exponents, and these have led to the homogeneous function approach or static scaling hypothesis. It is assumed in this hypothesis that any thermodynamic potential can be written as a generalized homogeneous function of its arguments [2]. All measurements of a given quantity for different fluids can be scaled such that they fall on a unique single curve.

B. Zappoli et al., *Heat Transfers and Related Effects in Supercritical Fluids*,
Fluid Mechanics and Its Applications 108, DOI: 10.1007/978-94-017-9187-8_18,

It also follows that the critical exponents above and below T_c are equal, and that exact relations between critical exponents—called scaling laws—are present. Examples of scaling laws include the Rushbrooke relation

$$\alpha + 2\beta + \gamma = 2 \tag{18.2}$$

and the hyperscaling relation

$$d\nu = 2 - \alpha, \tag{18.3}$$

where d is the space dimensionality and ν is the critical exponent for the correlation length on the critical isochore and the coexistence curve.

Another hyperscaling relation is

$$\delta = \frac{d + 2 - \eta}{d - 2 + \eta}, \tag{18.4}$$

where δ is the exponent for the variation of pressure with density at $T = T_c$ and η is the exponent that describes the power-law behavior of the two-point correlation function of density fluctuations at the critical point.

The relations between exponents correspond to similar relations between amplitudes. It follows that only two critical amplitudes are independent. This is the basis of two-scale-factor universality, where all of the critical exponents and amplitudes can be determined provided that at least two of them are known.

18.2 Universality

It has long been known that there are strong analogies between different phase transitions, such as gas–liquid and liquid–liquid critical points, the Ising model, magnetism, the liquid ^{4}He λ-point, and order–disorder transitions in binary alloys. The resulting universality concept asserts that all phase transitions can be divided into classes, where all transitions in a particular class show the same asymptotic critical behavior. These classes are defined by the space dimensionality d, the order parameter symmetry index n, and the range of the intermolecular potential. For example, the class $d = 3$, $n = 1$ (scalar order parameter) contains the uniaxial 3-D Ising model, liquid–gas critical points, and binary liquid consolute points. The $d = 3$, $n = 2$ (two-component order parameter) class comprises the lambda transition in ^{4}He, some ferromagnets, and smectic liquid crystals. However, distinctions must be made between isotropic systems where two-scale-factor universality is valid and anisotropic systems where nonuniversal amplitudes must be known in order to be able to characterize the critical behavior of the order parameter correlation function.

18.3 Renormalization Group Theory

In this context, the renormalization group (RG) approach (for which K. Wilson obtained the Nobel Prize in Physics in 1982) validates the observations and the conjectures concerning power laws, scaling, and universality. In the RG approach, an energy functional is defined that averages the microscopic details of the physical system. The most widely accepted model is the Landau–Ginzburg–Wilson functional, and the resulting field theory is the ϕ^4 (or RG) theory. The values of the critical exponents and universal amplitude ratios as well as the equation of state were initially determined by performing either a perturbation expansion at dimension $d = 3$ or an $\epsilon = 4 - d$ expansion. However, further perturbative extension of the series expansion for the free energy gave more accurate results. The RG approach has also been used to determine dynamic critical properties (transport properties), although the universality classes can be modified for dynamics.

Furthermore, the RG approach has been applied to determine the behavior in the crossover regime between asymptotic (close to the critical point) behavior and regular behavior (far from the critical point). In this regime, the correlation length for the critical fluctuations is on the order of the molecular scale of the system.

Further details and references can be found in the comprehensive review of Barmatz et al. [1], which inspired the above short presntation of the scaling laws, the universality, and the renormalization group theory.

References

1. Barmatz M, Hahn I, Lipa JA, Duncan RV (2007) Critical phenomena in microgravity: past, present, and future. Rev Mod Phys 79(1):1
2. Stanley HE (1971) Introduction to phase transitions and critical phenomena. Clarendon, Oxford

Chapter 19
The Ginzburg Criterion

Abstract A brief introduction to the Ginzburg criterion is given.

It is generally admitted to ignore fluctuations when formulating the free energy F in a classical thermodynamic approach. However, near the critical point, fluctuations become increasingly important. Therefore, Ginzburg [2] stated that, in order to be self-consistent, the fluctuations δM of the order parameter averaged over a suitable volume V should be small compared to the value of M (for more detail on this topic, see also [1]). Physically, fluctuations are important over linear dimensions on the order of the correlation length of fluctuations ξ, so the volume over which the contribution of fluctuations must be evaluated can be taken to be $V_\xi \sim \xi^d$ for a space of dimensionality d. Fluctuations are then negligible when the following inequality holds:

$$\frac{\langle \delta M^2 \rangle_V}{M^2} = \frac{\frac{1}{\xi^d}\int_{\xi^d} \delta M^2 dr}{M^2} = \frac{1}{\xi^d}\frac{k_B T \chi}{M^2} \ll 1. \tag{19.1}$$

Here, we have applied relation (1.33) between the susceptibility χ and the free energy F, noting that the fluctuations in M come from thermal noise with energy $k_B T$. Accounting for the temperature scaling laws of both M and χ, the inequality is fulfilled for all temperatures, even those close to T_c, provided the exponents and the space dimensionality obey the relation

$$d > d_c = \frac{\gamma + 2\beta}{\nu}. \tag{19.2}$$

In this inequality, the values of the exponents are assumed to be those of the mean field. It then follows that the dimension above which fluctuations can be ignored is the upper critical dimensionality

$$d_c = 4. \tag{19.3}$$

For dimensions that are less than the upper critical dimension, Landau theory is no longer correct because the free energy is dominated by the contribution from the

B. Zappoli et al., *Heat Transfers and Related Effects in Supercritical Fluids*, Fluid Mechanics and Its Applications 108, DOI: 10.1007/978-94-017-9187-8_19,

fluctuations. As the fluctuations dominate, the exponents ν and α are related by the so-called hyperscaling relation (see Eq. 18.3). This relation is obtained from Eq. 19.2 where $d\nu = \gamma + 2\beta$. Making use of the Rushbrooke relation (18.2) between α, β, and γ, we obtain $d\nu = 2 - \alpha$.

It is interesting to note that the exponents obtained from Landau theory are independent of d and therefore violate hyperscaling. They can only satisfy it at the upper critical dimension d_c, the dimension where the change from a mean-field- to a fluctuation-dominated regime occurs. For $d < d_c$, the exponents do depend on the dimension, and are different from those predicted by Landau theory.

References

1. Als-Nielsen J, Birgeneau RJ (1977) Mean field theory, the Ginzburg criterion, and marginal dimensionality of phase transitions. Am J Phys 45:554–560
2. Ginzburg VL (1961) Some remarks on phase transitions of the 2nd kind and the microscopic theory of ferroelectric materials. Sov Phys Solid State 2:1824–1834

Chapter 20
Conventional Theory of Nucleation and Spinodal Decomposition

Abstract A short summary of the conventional theory of nucleation and spinodal decomposition is provided.

A thorough review of phase transitions has been written by Onuki [5], which can be read for a more detailed approach of the conventional theory of nucleation and spinodal decomposition. Here it is only considered the case of a phase transition in a closed fluid sample in zero gravity with a constant mass (and thus mean density). This corresponds to the case in which the order parameter is conserved. The sample temperature is very rapidly changed ("quenched") from a state (at temperature T_i) where the fluid is homogeneous (a liquid, vapor, or supercritical state) to a state (at temperature T_f) below the coexistence curve where it is no longer thermodynamically stable. Vapor and liquid phases will eventually coexist at equilibrium. Changing the temperature has the advantage of not inducing hydrodynamics (flows) during the transition. A typical phase transition diagram for such a scenario is shown in Fig. 20.1.

A new liquid or vapor minority phase emerges at the expense of the initial majority phase when density fluctuations occur that lower the free energy of the system. However, the formation of a new phase necessitates the formation of a vapor–liquid surface tension at the phase border. The energy cost of this can be larger than the gain in free energy required to form the new phase, so the phase transition will not proceed and the system will remain in a metastable state. This process is characteristic of the *nucleation regime*. In another regime, called *spinodal decomposition*, the susceptibility of the system is negative and the fluctuations are unstable and grow, ultimately reaching the vapor and liquid compositions. Note that the vicinity of a critical point is characterized by large density fluctuations with a correlation length ξ, meaning that the above mean-field approach is inadequate in the critical point neighborhood [1, 2]. Only one regime is present: the so-called generalized nucleation regime.

B. Zappoli et al., *Heat Transfers and Related Effects in Supercritical Fluids*, Fluid Mechanics and Its Applications 108, DOI: 10.1007/978-94-017-9187-8_20,

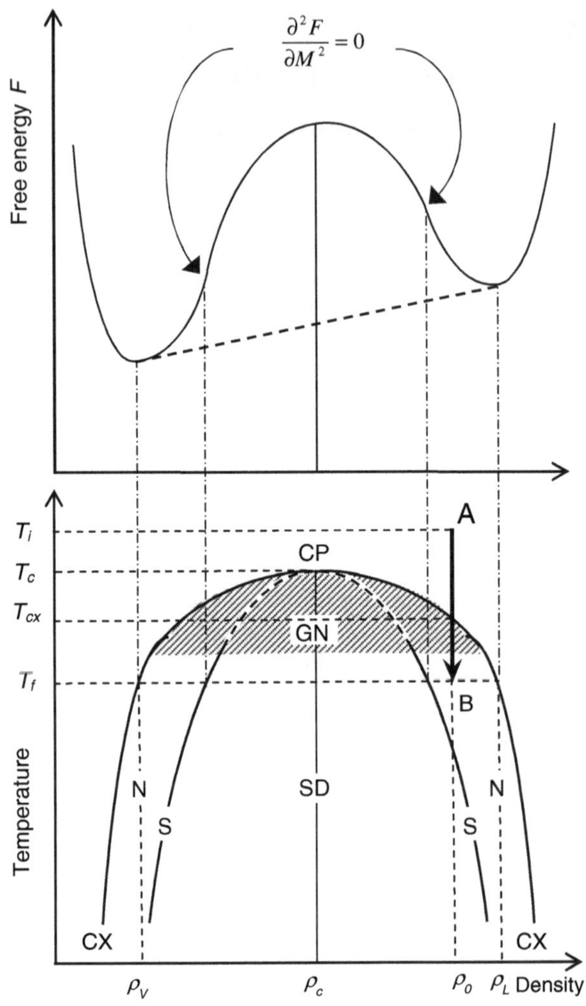

Fig. 20.1 Typical phase separation experiment performed at a constant mean density ρ_0, showing the coexistence curve (*CX*) and the shape of the free energy F at the final state. *CP* critical point, *A* initial state at temperature T_i, *B* final state at temperature T_f, T_{cx} temperature when crossing the coexistence curve, *N* nucleation (metastable) region, *SD* spinodal decomposition (unstable) region, *GN* (*hatched*) generalized nucleation region

20.1 Phase Diagram

The boundary of the unstable region is the coexistence curve (also called the binodal curve), which is determined by minimizing the free energy with respect to the order parameter M (in magnetic system notations the conjugated field is H). The term $-HM$ is also included when the initial order parameter is nonzero. The boundary is found

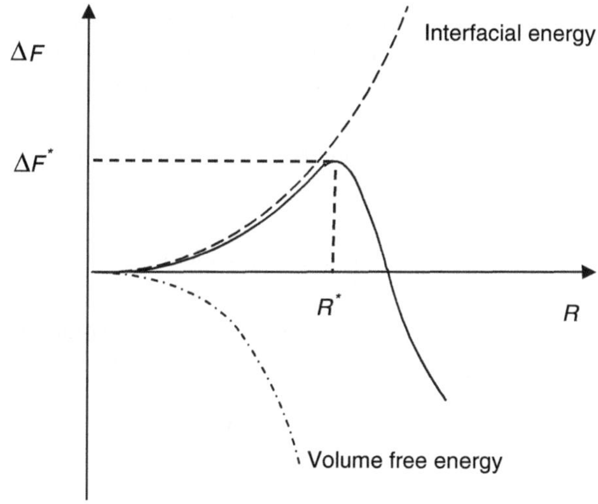

Fig. 20.2 Variation of free energy ΔF (the sum of the interfacial and volume energies) with respect to the cluster radius R

in Fig. 20.1 by constructing a tangent to the curve in the free-energy diagram. The stability of the fluctuations is indicated by the sign of the susceptibility $\chi = \left(\frac{\partial^2 F}{\partial M^2}\right)^{-1}$ (see Eq. 1.33). $\chi = 0$ corresponds to the spinodal curve, which separates a region of metastability ($\chi > 0$) where nucleation occurs from a region of instability ($\chi < 0$) where spinodal decomposition takes place. The coexistence and spinodal curves meet at the critical point, and they are homothetic since $\left(\frac{\partial^2 F}{\partial M^2}\right) = 0$ corresponds to Eq. 1.23 with an amplitude of $2/3$ instead of 2. There is no thermodynamic barrier to the growth of a new phase inside the spinodal. The spinodal therefore represents the limit of physical stability.

20.2 Nucleation

The creation of a nucleus implies the formation of an interface at the boundary of a new phase. When cooled below its coexistence curve, the fluid is supersaturated. The pressure in the newly formed phase is less than the vapor pressure, which brings about a change in free energy per unit volume, F_V, between the liquid and the newly created phase. This change in free energy is balanced by the energy gained upon creating a new volume and the energy cost of creating a new interface. When the overall change in free energy (ΔF) is negative, nucleation is favored (Fig. 20.2).

Some energy is required to form an interface between liquid and vapor phases. If a hypothetical nucleus is too small (i.e., an unstable nucleus or "embryo"), the energy that is released by volume formation is not enough to allow surface (interface) creation, so nucleation does not proceed. The critical size of a nucleus can be defined based on the radius R of the nucleus; nucleation occurs when $R = R^*$(the "critical radius"). For example, in the classic case of a spherical cluster that liberates $-F_V$ Joule per unit volume during formation (here, $-F_V$ is a negative quantity) but requires an energy σ Joule per unit surface to interface with its surroundings, the free energy needed to form a cluster of radius R is

$$\Delta F = \frac{4}{3}\pi R^3 F_V + 4\pi R^2 \sigma. \tag{20.1}$$

The first term represents the energy released due to the creation of a new volume, and the second term the energy lost due to surface tension at the new interface. It costs free energy to add molecules to this cluster (because $\frac{\partial F}{\partial R} > 0$) until the radius reaches

$$R^* = -\frac{2\sigma}{F_V}, \tag{20.2}$$

which corresponds to $\frac{\partial F}{\partial R} = 0$.

20.3 Critical Radius

The addition of new molecules to clusters larger than this critical radius releases available work. At that point, the growth of the cluster becomes limited by the supply of molecules (via diffusion).

The free energy needed to form this critical cluster is

$$\Delta F^* = \frac{16\pi\sigma^3}{3F_V^2}, \tag{20.3}$$

which occurs at the maximum ΔF^*, where $\frac{d\Delta F}{dR} = 0$. As the phase transformation becomes more and more favorable, nuclei can then grow until thermodynamic equilibrium is restored.

A greater degree of supercooling favors phase transformation, and ΔF can be related to the supercooling $\Delta T = T_{cx} - T_f$ by making use of the Clapeyron formula that relates F_V to ΔT and the release of enthalpy (latent heat) ΔH:

$$F_V = -\Delta H \frac{\Delta T}{T_{cx}}. \tag{20.4}$$

Then R^* and ΔF^* can be expressed as functions of ΔT:

$$R^* = \frac{2\sigma T_{cx}}{\Delta H}\frac{1}{\Delta T} \tag{20.5}$$

and

$$\Delta F^* = \frac{16\pi\sigma^3 T_{cx}^2}{3(\Delta H)^2}\frac{1}{(\Delta T)^2}. \tag{20.6}$$

The greater the supercooling, the smaller the critical radius, and the lower the amount of energy needed to form it.

20.4 Spinodal Decomposition

This mechanism differs from classical nucleation in that phase separation due to spinodal decomposition occurs uniformly throughout the material, not just at discrete nucleation sites. Since there is no thermodynamic barrier to the reaction inside the spinodal region, the decomposition is determined solely by diffusion. In the spinodal region of the phase diagram, the free energy can be lowered by allowing the components to separate, thus increasing the relative density of a component in a particular region of the fluid. The density will continue to increase until the component reaches the stable part of the phase diagram. The densities of very large volumes of the material will change only very slowly due to the amount of material that must be moved. On the other hand, very small regions will shrink away due to the energy cost of maintaining an interface between two dissimilar component materials. Cahn and Hilliard [3, 4] used the approximation for the mean-field Landau free-energy with the gradient term (Eq. 1.32, with $M = \frac{\rho - \rho_c}{\rho_c}$), obtaining

$$\tilde{f} = \tilde{f}_c + \int_V \left(\frac{a_0}{2}\tau M^2 + \frac{b_0}{4!}M^4 + \frac{d_0}{2}|\nabla M|^2\right) d^3r. \tag{20.7}$$

Diffusive motion dominates at the length scale of spinodal decomposition. The equation of motion for a diffusive system is

$$\frac{dM}{dt} = \nabla\left(m^*\nabla\mu^* + \varsigma(r)\right), \tag{20.8}$$

where m^* is the diffusive mobility, $\varsigma(r)$ is random noise such that $\langle\varsigma(r)\rangle = 0$, and the chemical potential μ^* is derived from the Landau free energy by a functional derivative:

$$\mu^* = \frac{\delta\tilde{f}}{\delta M} = a\tau M + \frac{b_0}{3!}M^3 - d_0\nabla^2 M. \tag{20.9}$$

We see that if $a_0\tau < 0$, small fluctuations around $M = 0$ have a negative effective diffusive mobility and will grow rather than shrink. To understand the growth dynamics, we disregard the fluctuating currents due to ς, linearize the equation of motion around $M_0 = \frac{\rho_0 - \rho_c}{\rho_c}$, and perform a Fourier transform into k-space. This leads to

$$\frac{dM_q\left(q,t\right)}{dt} = -m^*\left[\left(a\tau + \frac{b_0}{2}M_0^2\right)q^2 + d_0 q^4\right]M_q(q,t) = R(q)M_q(q,t), \tag{20.10}$$

which has an exponential growth solution:

$$M_q(q,t) = \exp\left[R(qt)\right]. \tag{20.11}$$

Since the growth rate $R(q)$ is exponential, the fastest-growing angular wavenumber

$$q_{sp} = \frac{2\pi}{\lambda_{sp}} = 2\pi\sqrt{-\frac{a\tau + \frac{b_0}{2}M_0^2}{2d_0}} \tag{20.12}$$

will quickly dominate the morphology. Spinodal decomposition thus results in domains with a characteristic length scale, called the *spinodal length*:

$$\lambda_{sp} = \frac{2\pi}{q_{sp}} = 2\pi\sqrt{-\frac{2d_0}{a\tau + \frac{b_0}{2}M_0^2}}. \tag{20.13}$$

The growth rate of the fastest-growing angular wavenumber is

$$\frac{1}{t_{sp}} = R\left(q_{sp}\right) = -m^*\left[\left(a_0\tau + \frac{b_0}{2}M_0^2\right)q_{sp}^2 + d_0 q_{sp}^4\right] = \frac{m^*\left[\left(a_0\tau + \frac{b_0}{2}M_0^2\right)^2\right]}{4d_0},$$

where t_{sp} is the *spinodal time*. The spinodal length and spinodal time can be used to nondimensionlize the equation of motion, resulting in universal scaling for spinodal decomposition.

References

1. Binder K (1984) Nucleation barriers, spinodals, and the Ginzburg criterion. Phys Rev A 29(1):341–349
2. Binder K (1991) Spinodal decomposition. In: Material science and technology: phase transitions in materials, Chapter 5. VCH Verlagsgesellschaft, Weinheim, pp 405–471 (and references therein)
3. Cahn JW, Hilliard JE (1958) Free energy of a nonuniform system. i. Interfacial energy. J Chem Phys 28:258–267
4. Cahn JW, Hilliard JE (1959) Free energy of a nonuniform system. iii. Nucleation in a two-component incompressible fluid. J Chem Phys 31:688–699
5. Onuki A (2002) Phase transition dynamics. Cambridge University Press, Cambridge

Chapter 21
Basic Equation of Fluid Mechanics

Abstract The basis of fluid mechanics is presented, with particular emphasis placed on its connection to the conservation laws of physics.

21.1 Notation

We will now provide a brief review of some elementary concepts regarding vectors and tensors. In the framework tensor theory, a scalar is a zeroth-order tensor, a vector is a first-order tensor, and a matrix is a second-order tensor. Higher-order tensors exist but will not be needed here.

21.1.1 Product of Vectors and Tensors

A vector is denoted by a bold lower-case letter. The velocity vector is written **v**, while its Cartesian components are (u, v, w), leading to

$$\mathbf{v} = u\mathbf{i} + v\mathbf{j} + w\mathbf{k}, \tag{21.1}$$

where **i**, **j**, **k** are the axis unit vectors. The tensor **T** and second-order operators, are denoted by bold upper-case letters, while its components are denoted T_{ij}, leading to

$$\mathbf{T} = \begin{vmatrix} T_{11} & T_{12} & T_{13} \\ T_{21} & T_{22} & T_{23} \\ T_{31} & T_{32} & T_{33} \end{vmatrix} = \left|T_{ij}\right|, \tag{21.2}$$

with $(i = 1, 3;\ j = 1, 3)$ where i is the line index and j the column index. The transposed tensor $\overline{\mathbf{T}} = \left|T_{ji}\right|$ to **T** is symmetrical with respect to the first diagonal and corresponds to

B. Zappoli et al., *Heat Transfers and Related Effects in Supercritical Fluids*,
Fluid Mechanics and Its Applications 108, DOI: 10.1007/978-94-017-9187-8_21,

$$\overline{\mathbf{T}} = \begin{vmatrix} T_{11} & T_{21} & T_{31} \\ T_{12} & T_{22} & T_{32} \\ T_{13} & T_{23} & T_{33} \end{vmatrix}. \tag{21.3}$$

The tensor (or direct) product of two vectors $\mathbf{v}$ and $\mathbf{v}'$ is a second-order tensor defined by

$$\begin{vmatrix} uu' & uv' & uw' \\ vu' & vv' & vw' \\ wu' & wv' & ww' \end{vmatrix} = \mathbf{u} \otimes \mathbf{v} = |u_i v_j| \tag{21.4}$$

The once-contracted (or scalar) product of two vectors is a scalar:

$$\mathbf{v} \cdot \mathbf{v}' = \sum_i v_i v_i' = uu' + vv' + ww'. \tag{21.5}$$

The once-contracted product of a second-order tensor with a vector is a vector:

$$\mathbf{v} \cdot \mathbf{T} = \sum_i v_i T_{ij} = \sum_j v_j T_{ji} = \overline{\mathbf{T}} \cdot \mathbf{v}. \tag{21.6}$$

The once-contracted product of two second-order tensors is a second-order tensor:

$$\mathbf{T} \cdot \mathbf{T}' = \sum_i T_{ij} T_{jk}'. \tag{21.7}$$

Finally, the twice-contracted product of two tensors is a scalar:

$$\mathbf{T} : \mathbf{T}' = \sum_{i,j} T_{ij} T_{ij}'. \tag{21.8}$$

21.1.2 Derivatives of Vectors and Tensors

The derivative with respect to the coordinates may be considered a vector (although it does not share all of the properties of vectors), denoted ∇, whose components are

$$\nabla = \left(\frac{\partial}{\partial x}, \frac{\partial}{\partial y}, \frac{\partial}{\partial z} \right). \tag{21.9}$$

The gradient of a scalar function ϕ is a vector, also denoted $\nabla\phi$, whose components are

$$\mathbf{grad}\,\phi = \nabla\phi = \left(\frac{\partial\phi}{\partial x}, \frac{\partial\phi}{\partial y}, \frac{\partial\phi}{\partial z}\right). \tag{21.10}$$

The gradient of a vector $\mathbf{v}$, also denoted $\nabla \otimes \mathbf{v}$, is the direct product of ∇ with $\mathbf{v}$, leading to

$$\mathbf{grad}\,\mathbf{v} = \nabla \otimes \mathbf{v} = \begin{vmatrix} \frac{\partial u}{\partial x} & \frac{\partial u}{\partial y} & \frac{\partial u}{\partial z} \\ \frac{\partial v}{\partial x} & \frac{\partial v}{\partial y} & \frac{\partial v}{\partial z} \\ \frac{\partial w}{\partial x} & \frac{\partial w}{\partial y} & \frac{\partial w}{\partial z} \end{vmatrix} = \frac{\partial v_i}{\partial x_j}. \tag{21.11}$$

The divergence of a vector $\mathbf{v}$ is the scalar product of ∇ with $\mathbf{v}$, leading to

$$div\,\mathbf{v} = \nabla \cdot \mathbf{v} = \sum_{ij} \frac{\partial v_i}{\partial x_i}. \tag{21.12}$$

The divergence of a second-order tensor is a vector defined by the once-contracted product of ∇ and $\mathbf{T}$, leading to

$$\mathbf{div}\,\mathbf{T} = \nabla \cdot \mathbf{T} = \sum_{i} \frac{\partial T_{ij}}{\partial x_i}, \tag{21.13}$$

where x_i $(i = 1, 3)$ denote the coordinates.

Straightaway, these lead us to some useful formulae:

- If $\mathbf{T}$ is a symmetric tensor, we have

$$\nabla \cdot (\mathbf{v} \cdot \mathbf{T}) = (\nabla \otimes \mathbf{v}) : \mathbf{T} + (\nabla \cdot \mathbf{T}) \cdot \mathbf{v} \tag{21.14}$$

and

$$\nabla \cdot (\mathbf{v} \cdot \mathbf{T}) = \sum_{j} \frac{\partial}{\partial x_j} \left(\sum_{i} v_i T_{ij}\right) = \sum_{ji} \frac{\partial v_i}{\partial x_j} T_{ij} + \sum_{ji} v_i \frac{\partial T_{ij}}{\partial x_j}. \tag{21.15}$$

Therefore, by definition, we have

$$\mathbf{T} : (\nabla \otimes \mathbf{v}) = \sum_{ij} T_{ij} \frac{\partial v_i}{\partial x_j} \tag{21.16}$$

and

$$(\nabla \cdot \mathbf{T}) \cdot \mathbf{v} = \sum_{j} v_j \left(\sum_{i} \frac{\partial T_{ij}}{\partial x_i}\right). \tag{21.17}$$

- Furthermore, if $\mathbf{T}$ is such that $\mathbf{T} = t\mathbf{I} + \boldsymbol{\Upsilon}$, with $\mathbf{I}$ being a unit tensor such that

$$\mathbf{I} = \begin{vmatrix} 1\,0\,0 \\ 0\,1\,0 \\ 0\,0\,1 \end{vmatrix} = \delta_{ij}, \tag{21.18}$$

where $\delta_{ii} = 1$ and $\delta_{ij} = 0$ if $i \neq j$, then we have

$$\mathbf{T} : (\nabla \otimes \mathbf{v}) = t\nabla \cdot \mathbf{v} + \boldsymbol{\Upsilon} : (\nabla \otimes \mathbf{v}) \,. \tag{21.19}$$

21.2 Fluid State as a Continuous Medium

The macroscopic theory of fluid motion is based on the hypothesis of a continuous fluid at the macroscopic scale. However, a fluid is not a continuous medium at the microscopic level. We can define a macroscopic hydrodynamic value at a given point as the mean value of the corresponding microscopic property across an arbitrarily small domain surrounding the geometric point. When there are enough particles in that arbitrarily small domain, the corresponding macroscopic property is a continuous function of space and time and the fluid is a continuous medium. Moreover, when the mean free path is small compared to the characteristic dimension of the fluid sample, the macroscopic properties (i.e., density, temperature and velocity for a simple gas) obey the Navier–Stokes equations. When the mean free path is much larger than the characteristic dimension, other theories must be used, such as the kinetic theory of ideal gases [1].

When the fluid is continuous, all of the physical properties that characterize fluid motion are continuous functions of time and of the location at which the property is measured. This property can be either a scalar field, such as the density $\rho(\mathbf{x}, t)$ and the temperature $T(\mathbf{x}, t)$, or a vector field, such as the fluid velocity $\mathbf{v}(\mathbf{x}, t)$, or even a tensor field, such as the fluid velocity gradient $\mathbf{grad}\,\mathbf{v}\,(\mathbf{x}, t)$. Here, $\mathbf{x}$ denotes the location and t the time.

A fluid particle is an elementary mass dm of fluid that is present in an elementary (arbitrarily small) volume dV which surrounds the point M located at $\mathbf{x}$ at time t. dm and dV must be vanishingly small at the macroscopic level. In order to be able to postulate that the fluid particle can be identified with the geometric point, the elementary volume must be infinitely large compared to the volume of the molecules and infinitely small compared to the experimental cell's dimensions. Under these conditions, the density of the fluid particle containing a mass dm of fluid in the elementary volume $\delta\mathcal{V}$ that surrounds the geometrical point M located at $\mathbf{x}$ at time t is defined by

$$\rho(\mathbf{x}, t) = \frac{dm}{dV}.$$

The fluid particle can then also be characterized by its temperature $T(\mathbf{x}, t)$ and pressure $p(\mathbf{x}, t)$, and it can move with a velocity $\mathbf{v}(\mathbf{x}, t)$, which has the components $u(\mathbf{x}, t)$, $v(\mathbf{x}, t)$, and $w(\mathbf{x}, t)$ [sometimes also denoted $v_i(\mathbf{x}, t)$]. These variables are suitable for describing a one-component, continuous and compressible fluid in a one-phase domain (i.e., without surface tension effects). However, we need to review some of the mechanical properties of any continuous medium before we introduce the specific nature of this fluid system.

21.2.1 Forces Acting on a Continuous Medium

21.2.1.1 Volume Forces Acting on Any Continuous System

For any macroscopic continuous system, we can define a closed domain $\mathcal{D}$ of volume V, which is limited by a concave external boundary $\delta\mathcal{D}$ of surface area Σ. The volume forces act on a particle of the system surrounding any point P inside the volume V (see Fig. 21.1 for a schematic illustration). If we now introduce the volumic density of forces $\mathbf{f}$ (P)—i.e., the force per unit volume, the elementary force $d\mathbf{f}$ acting on any elementary volume dV can be then written as

$$d\mathbf{f} = \mathbf{f}\,(\mathrm{P})\,dV.$$

$\mathbf{f}$ (P) is a volumic distribution of forces that represents the external forces due to the external systems acting on $\mathcal{D}$. For example, in the case of gravitational forces, $\mathbf{f}\,(\mathrm{P}) = \rho(\mathbf{x}, t)\mathbf{g}$.

21.2.1.2 Surface Forces Surrounding a Point M on $\delta\mathcal{D}$

Now, if we consider an elementary surface $d\Sigma$ surrounding a point M located on the external boundary $\delta\mathcal{D}$, we can also define the external unit vector $\mathbf{n}$ normal to the boundary $\delta\mathcal{D}$ and a surface density of forces $\mathbf{f}_s$ (M, $\mathbf{n}$)—i.e., a force per unit area. The elementary force $d\mathbf{f}_s$ acting on any elementary surface $d\Sigma$ can be then written as

$$d\mathbf{f}_s = \mathbf{f}_s\,(\mathrm{M}, \mathbf{n})\,d\Sigma.$$

$\mathbf{f}$ (M, $\mathbf{n}$) is the surface distribution of forces that relates to the forces acting on an elementary surface normal to n and surrounding a point M on $\delta\mathcal{D}$; see Fig. 21.1 for a schematic illustration. $\mathbf{f}$ (M, $\mathbf{n}$) is also classically defined as the stress vector $\mathbf{t}$(M, $\mathbf{n}$) during mechanical contact between bodies, thus introducing the forces due to the other domains of the continuous medium that are in the vicinity of the unit vector $\mathbf{n}$.

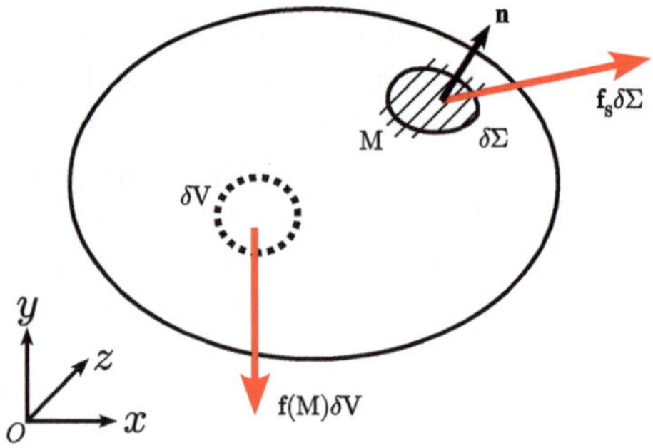

Fig. 21.1 Surface and volume forces that act on a closed fluid domain

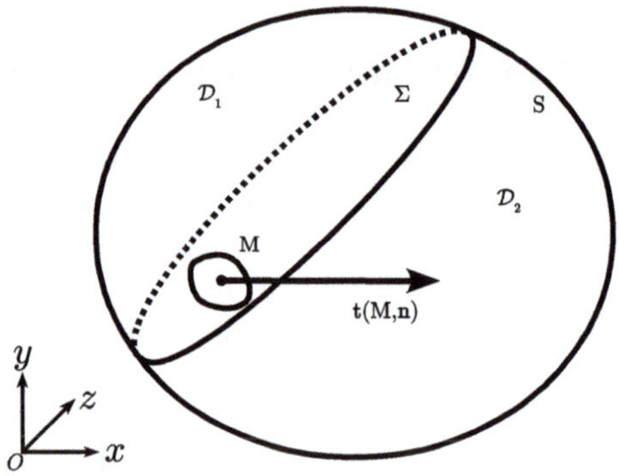

Fig. 21.2 Stresses that act on a surface element

21.2.2 Particle Stress

21.2.2.1 Stress on a Fluid Particle

Let us consider a volume $\mathcal{D}$, limited by a closed surface S, that is separated into two closed volumes $\mathcal{D}_1$ and $\mathcal{D}_2$ by a surface Σ, as shown in Fig. 21.2. Let **n** be the vector normal to Σ at a point M, directed from $\mathcal{D}_1$ to $\mathcal{D}_2$. This normal vector **n** is directed towards the interior of the body that acts, or toward the exterior of an isolated body. The mechanical actions of $\mathcal{D}_1$ on $\mathcal{D}_2$ are by definition contact actions.

The contact actions are represented by a force field defined on the surface Σ. Let $d\Sigma$ be a surface element that contains point M and belongs to Σ. The force field applied to $d\Sigma$ is reducible to a unique force $d\mathbf{f}$ that is applied to M and to a torque $d\mathbf{c}$; the latter will be considered to be a vanishingly small second-order quantity. Under these conditions, when $d\Sigma$ tends to zero in the vicinity of point M, the vector $\frac{d\mathbf{f}(\mathrm{M})}{d\Sigma}$ tends to a finite-limit vector such that

$$\lim_{d\Sigma \to 0} \frac{d\mathbf{f}(\mathrm{M})}{d\Sigma} = \mathbf{t}(\mathrm{M}, \mathbf{n})$$

Fig. 21.3 Normal and shear stresses

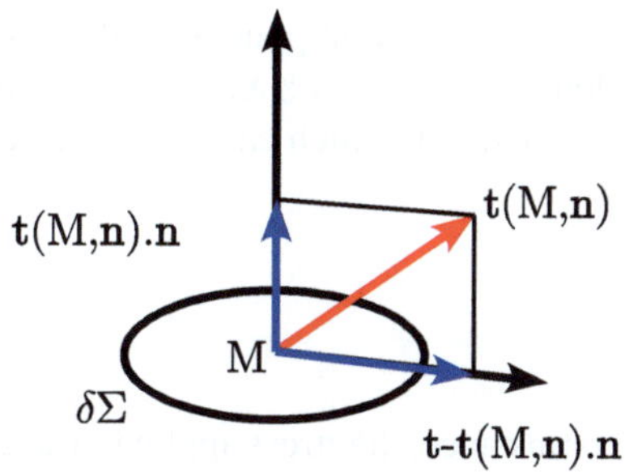

where $\mathbf{t}(\mathrm{M}, \mathbf{n})$ is the stress vector at point M on $d\Sigma$. The vector $\mathbf{t}$ represents the surface density of the forces acting at M when the vector unit normal to the surface at point M is $\mathbf{n} \cdot \mathbf{t}$ is a vector function that depends on five variables: the three components of the location of point M and the two angular directions of the normal vector $\mathbf{n}$ to $d\Sigma$ at M. Newton's third law implies that the following relation should hold:

$$\mathbf{t}(\mathrm{M}, \mathbf{n}) = -\mathbf{t}(\mathrm{M}, -\mathbf{n}).$$

The projection of $\mathbf{t}$ onto the normal vector $\mathbf{n}$ is the normal stress at point M on $d\Sigma$. This normal stress is unchanged when $\mathbf{n}$ is changed to $-\mathbf{n}$:

$$\mathbf{t}(\mathrm{M}, -\mathbf{n}) \cdot -\mathbf{n} \equiv \mathbf{t}(\mathrm{M}, \mathbf{n}) \cdot \mathbf{n}\,.$$

The vector defined by

$$\mathbf{t}(\mathrm{M}, \mathbf{n}) - [\mathbf{t}(\mathrm{M}, \mathbf{n}) \cdot \mathbf{n}] \cdot \mathbf{n}$$

is the tangential stress at point M, called the shearing stress, which is thus the projection of vector $\mathbf{t}$ at point M onto the surface element $\delta\Sigma$ in the plane that contains the vector normal to the surface. See Fig. 21.3 for an illustration of these normal and shear stress vectors.

21.2.2.2 Stress Tensor

Knowledge of the stress vector at point M depends on the knowledge of the stress tensor $\mathbf{T}$ at point M, which is such that the stress vector at point M is given by

$$\mathbf{t}(\mathrm{M}, \mathbf{n}) = \mathbf{T} \cdot \mathbf{n}, \tag{21.20}$$

where

$$\begin{aligned}\mathbf{T} \cdot \mathbf{n} = (\sigma_{xx} n_x + \sigma_{xy} n_y + \sigma_{xz} n_z)\mathbf{i} + (\sigma_{yx} n_x + \sigma_{yy} n_y + \sigma_{yz} n_z)\mathbf{j} \\ + (\sigma_{zx} n_x + \sigma_{zy} n_y + \sigma_{zz} n_z)\mathbf{k}.\end{aligned}$$

The n_i are the components of the normal vector **n**. Here **i**, **j** and **k** are the unit vectors along the x, y and z axes. The σ_{ij} are the components of the stress tensor **T** that is symmetrical, which can therefore be written

$$\begin{vmatrix} \sigma_{xx} & \sigma_{xy} & \sigma_{xz} \\ \sigma_{yx} & \sigma_{yy} & \sigma_{yz} \\ \sigma_{zx} & \sigma_{zy} & \sigma_{zz} \end{vmatrix}.$$

For instance, the stress applied to an elementary surface perpendicular to the direction of the axis x is given by

$$\mathbf{t}(M, \mathbf{i}) = \sigma_{xx}\mathbf{i} + \sigma_{xy}\mathbf{j} + \sigma_{xz}\mathbf{k}.$$

It should be emphasized that if

$$\sigma_{nn} = \mathbf{t}(M, \mathbf{n}) \cdot \mathbf{n} \geq 0,$$

when the normal component of the stress is oriented in the same direction as that of the normal vector, (e.g., oriented towards the body that acts), the action is a traction of the volume element following the normal vector **n**. In the opposite case, the action is a compression of the volume element.

The description for the case where the system is a one-component fluid is given in the Sect. 21.3.

21.3 Fluid Properties

In classical mechanics of solids, when a stress is applied to an elastic solid, leading to deformation, the solid continues to deform further until the applied force is kept constant. When the deformation of the solid remains after the force is removed, the deformation is said to be plastic. In the case of fluids, the situation is different, since the fluid will continue to deform even when the increase in force is set to be arbitrarily small. This is what we mean when we say that a fluid "flows". With this in mind, we need to review some basic definitions of the mechanical properties of a fluid.

21.3.1 Mechanical Definition of a Fluid

21.3.1.1 Definition

A fluid is defined as an isotropic medium for which the stress tensor depends only on the deformation rate tensor **D**, defined by

$$\mathbf{D} = \frac{1}{2}\left(\nabla \otimes \mathbf{v} + \overline{\nabla \otimes \mathbf{v}}\right),$$

leading to a functional dependence

$$\mathbf{T} \equiv \mathbf{T}(\mathbf{D}),$$

which can be made explicit when some basic hypotheses are put forward to characterize the mechanical nature of the fluid, as in the following case of a Newtonian fluid.

21.3.1.2 Newtonian Fluid

The stress tensor of a Newtonian particle fluid flowing with a velocity $\mathbf{v}$ (sometimes called a Navier–Stokes fluid) is, by definition,

$$\mathbf{T} = -\mathbf{P} = -p\mathbf{I} + [\mu_L \nabla \cdot \mathbf{v}\mathbf{I} + 2\mu\mathbf{D}], \qquad (21.21)$$

where p is the thermodynamic pressure p and μ_L and μ are the Landé and shear viscosity coefficients, respectively (see also Eq. (21.23)). The stress tensor $\mathbf{T}$ characterizes the effect of the outer fluid on the fluid particle under consideration. The tensor $\mathbf{P}$ is the pressure tensor, which corresponds to the tensor opposite to $\mathbf{T}$. By separating out the contribution from the thermodynamic pressure, we can define the viscous stress tensor $\boldsymbol{\tau}$ by

$$\boldsymbol{\tau} = \mu_L \nabla \cdot \mathbf{v}\mathbf{I} + 2\mu\mathbf{D},$$

and Eq. (21.21) can thus be written

$$\mathbf{P} = p\mathbf{I} - \boldsymbol{\tau}. \qquad (21.22)$$

A Newtonian fluid has no "memory", because the state of the stresses at a given time depends only on the state variables at this same given time. When an isolated fluid is at rest ($\mathbf{v} = 0$), the stress tensor reduces to the normal pressure stress $p\mathbf{I}$ only because the thermodynamic pressure p is related, through an equation of state, to the other thermodynamic variables such as the density or temperature of the one-component fluid. The two viscosity coefficients μ_L and μ are generally not independent. For example, the quantity

$$\mu_B = \frac{1}{2}(3\mu_L + 2\mu) \qquad (21.23)$$

is the volume viscosity coefficient.

Following the Stokes hypothesis, μ_B is generally zero for most fluids when they are considered to be ideal gases, but the issue of the divergence of the volume viscosity upon approaching the liquid–vapor critical point is still subject to debate

(See Sect. 1.10. Under the condition that Eq. (21.23) applies, the stress tensor can be written

$$\mathbf{P} = \left(p + \frac{2}{3}\mu\nabla \cdot \mathbf{v}\right)\mathbf{I} - 2\mu\mathbf{D}. \tag{21.24}$$

The kinematic viscosity coefficient is given by

$$\nu = \frac{\mu}{\rho}. \tag{21.25}$$

It is also called the vorticity diffusion coefficient, with dimensions of $\left[L^2T^{-1}\right]$, and is thus denoted D_μ as the unsteady penetration depth of a velocity perturbation in the fluid at time t is proportional to $\sqrt{D_\mu t}$. This mechanical (tensor) description is then similar to the thermal (vector) one, where the isotropic fluid is subjected to a temperature gradient. The heat flux vector is thus given by Fourier's law:

$$\mathbf{j} = -\Lambda\nabla T. \tag{21.26}$$

In Eq. (21.26), Λ is the heat conduction coefficient. The heat diffusion coefficient D_T, with dimensions of $\left[L^2T^{-1}\right]$, is given by (see Eq. (1.44))

$$D_T = \frac{\Lambda}{\rho c_p}, \tag{21.27}$$

and the heat diffusion penetration length is proportional to $\sqrt{D_T t}$. Recall that the viscosity and heat conduction coefficients depend only on temperature in the kinetic theory of ideal gases.

21.3.1.3 Viscosity Stress Vector

Let us consider an incompressible fluid flow (such that $\nabla \cdot \mathbf{v} = 0$) in a two-dimensional channel (x, y) that extends infinitely in the z direction perpendicular to the plane (x, y) of the page. The flow is assumed to be steady. The velocity at the free surface located at $y = y_0$ is assumed to be constant.

$$\mathbf{v} = u(y)\mathbf{i} + 0 \times \mathbf{j}$$

has only one horizontal component, which depends only on the vertical coordinate y ($\mathbf{i}$ and $\mathbf{j}$ are, respectively, the unit vectors in the x and y directions).

We can calculate the effort which is exerted by the fluid on an elementary surface $d\Sigma$ that is parallel to the plane xOz. The part of the fluid located above the elementary surface is denoted 2, while the part that is below the surface is denoted 1 (see Fig. 21.4). The normal to the surface is directed towards the medium that acts; let us say that this is fluid part 2. Based on its definition, the stress applied by 2 to 1

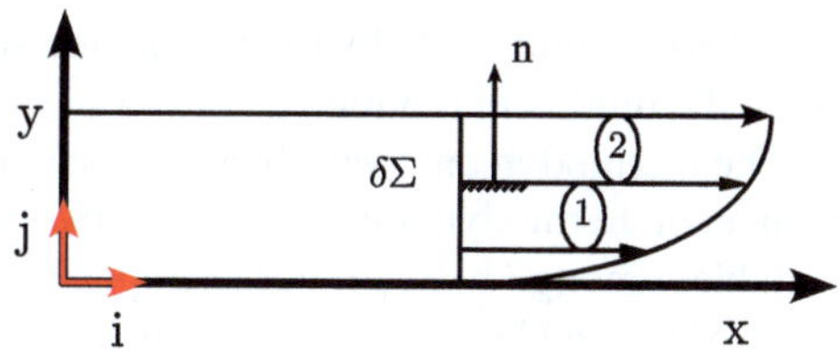

Fig. 21.4 Viscosity stress in a shear flow. The upper boundary is a free surface

through $d\Sigma$ is

$$\mathbf{t}(\mathrm{M}, \mathbf{n}) = \mathbf{T} \cdot \mathbf{n} = (-p\mathbf{I} + 2\mu\mathbf{D}) \cdot \mathbf{j} = -p\mathbf{j} + 2\mu\mathbf{D} \cdot \mathbf{j},$$

and according to the definition of $\mathbf{D}$ in this particular configuration,

$$\mathbf{D} = \frac{1}{2}(\nabla \otimes \mathbf{v} + \overline{\nabla \otimes \mathbf{v}}) = \begin{vmatrix} \frac{\partial u}{\partial x} & \frac{1}{2}(\frac{\partial u}{\partial y} + \frac{\partial v}{\partial y}) & \frac{1}{2}(\frac{\partial u}{\partial z} + \frac{\partial w}{\partial x}) \\ 0 & \frac{\partial v}{\partial y} & 0 \\ 0 & \frac{1}{2}(\frac{\partial v}{\partial z} + \frac{\partial w}{\partial y}) & \frac{\partial w}{\partial z} \end{vmatrix} = \begin{vmatrix} 0 & \frac{1}{2}\frac{\partial u}{\partial y} & 0 \\ 0 & 0 & 0 \\ 0 & 0 & 0 \end{vmatrix},$$

so

$$\mathbf{t} = -p\mathbf{i} + \mu\frac{\partial u}{\partial y}\mathbf{i} = -p\mathbf{i} + \boldsymbol{\tau}.$$

The vector

$$\boldsymbol{\tau} = \mu\frac{\partial u}{\partial y}\mathbf{i}$$

is called the viscosity stress vector at point M on $d\Sigma$. This vector is the tangential stress applied by region 2 to region 1, which is proportional to the viscosity coefficient μ and to the shear flow rate $\frac{\partial u}{\partial y}$. We deduce that the elementary force acting on the surface immersed within the fluid is

$$d\mathbf{f}(\mathrm{M}, \mathbf{j}) = \mathbf{t}\, d\Sigma = -p\mathbf{j}\, dx\, dz + \mu\frac{\partial u}{\partial y}\mathbf{i}\, dx\, dz.$$

The first term is the effort exerted by the static pressure (thermodynamic), while the second is the shear stress from the viscosity. The viscosity thus tends to accelerate the slower fluid particles and to slow the faster ones.

21.3.2 Thermodynamic Definition of a Fluid

The state of a pure single-phase fluid at rest and in thermodynamic equilibrium is defined by three thermodynamic variables: the pressure p, the temperature T, and the specific mass ρ (i.e., the mass per unit volume). Sometimes this state is termed the "thermostatic" state by hydrodynamicists in order to distinguish it from the case when the fluid is in motion.

The thermodynamic variables are related by an equation of state that reduces the number of thermodynamic variables to two; in other words, to any pair of independent variables among (p, T, ρ). For example, for ideal gases at equilibrium, this equation is the Mariotte Gay-Lussac equation.

In the case of a moving fluid, the thermodynamic relations can still be applied locally: the local variables for the moving fluid are related by the same variables as if the local state was an equilibrium state.

This fundamental hypothesis is called the *hypothesis of local equilibrium*. It becomes invalid at two opposite limits. The first of these is reached when the characteristic dimension of the system reaches the same order of magnitude as the mean free path of the particles; in other words, when the Knudsen number is no longer small [1]. The second limit occurs at high pressures, close to the critical conditions. The size of the critical density fluctuations, which are on the order of the diverging correlation length (defined from Eq. (1.39) and Table 1.2), may become as large as the smaller hydrodynamic scales, such as the thermal boundary layer thickness on the acoustic timescale. In this situation, the local thermodynamic equilibrium cannot be assumed.

21.3.2.1 Some Frequently Encountered Thermodynamic Relations

When the hypothesis of local equilibrium holds, thermodynamic variables are utilized in their specific forms (i.e., per unit mass of the selected one-component fluid). These forms are denoted by lower-case letters. Temperature and pressure are defined by

$$T = \left(\frac{\partial e}{\partial s}\right)_v \quad p = -\left(\frac{\partial e}{\partial v}\right)_s,$$

where p is the pressure in the corresponding thermostatic state, and s and e are the specific entropy and internal energy, respectively.

The first law of thermodynamics for closed systems undergoing equilibrium transformations can thus be used locally, and can be written as follows:

$$de = Tds - pdv = Tds + \frac{p}{\rho^2}d\rho, \tag{21.28}$$

where ρ is the specific mass (or density) of the fluid, defined as $\rho = \frac{1}{v}$.

The specific internal energy is thus a function of the specific entropy and the specific mass: $e \equiv e(s, \rho)$.

The specific enthalpy is defined as

$$h = e + \frac{p}{\rho}.$$

Since this relation defines the Legendre transform of e with respect to the specific mass, h is a function of s and p:

$$dh = Tds + \frac{1}{\rho}dp.$$

Therefore,

$$\frac{p}{\rho^2} = \left(\frac{\partial e}{\partial \rho}\right)_s \tag{21.29}$$

$$\frac{1}{\rho} = \left(\frac{\partial h}{\partial p}\right)_s. \tag{21.30}$$

The variations of e and h with temperature are related to the specific heats at constant volume and constant pressure:

$$c_V = T\left(\frac{\partial s}{\partial T}\right)_\rho = \left(\frac{\partial e}{\partial T}\right)_\rho \text{ and } c_p = T\left(\frac{\partial s}{\partial T}\right)_p = \left(\frac{\partial h}{\partial T}\right)_p. \tag{21.31}$$

The isothermal changes of e and h with volume or pressure can be expressed only in terms of pressure, density, temperature and their mutual derivatives, as for example in Eqs. (21.29) and (21.30) for their isentropic changes. However, more work is needed to introduce two supplementary calorimetric coefficients ℓ_ρ and ℓ_p, which appear from the use of e and h, or equivalently from the use of s. ℓ_ρ and ℓ_p are then defined by the following two equations:

$$\ell_\rho = -T\rho^2\left(\frac{\partial s}{\partial \rho}\right)_T = p - \rho^2\left(\frac{\partial e}{\partial \rho}\right)_T \tag{21.32}$$

$$\ell_p = T\left(\frac{\partial s}{\partial p}\right)_T = -\rho^{-1} + \left(\frac{\partial h}{\partial p}\right)_T, \tag{21.33}$$

where the derivatives of the three thermodynamic functions are the most useful for the present study. In addition to this, similar equations are obtained from the associated Maxwell rules for the entropy. Thus,

$$\left(\frac{\partial s}{\partial \rho}\right)_T = -\frac{1}{\rho^2}\left(\frac{\partial p}{\partial T}\right)_\rho \tag{21.34}$$

$$\left(\frac{\partial s}{\partial p}\right)_T = +\frac{1}{\rho^2}\left(\frac{\partial \rho}{\partial T}\right)_p \tag{21.35}$$

$$\left(\frac{\partial s}{\partial \rho}\right)_p = -\frac{1}{\rho^2}\left(\frac{\partial p}{\partial T}\right)_s \tag{21.36}$$

$$\left(\frac{\partial s}{\partial p}\right)_\rho = +\frac{1}{\rho^2}\left(\frac{\partial \rho}{\partial T}\right)_s . \tag{21.37}$$

Similar equations for e and h are then obtained by differentiating the Gibbs–Helmholtz equations, which relate the Helmholtz free energy f to e ($f = e - Ts$) and the Gibbs free energy g to enthalpy h ($g = h - Ts$):

$$-\rho^2\left(\frac{\partial e}{\partial \rho}\right)_T = -p + T\left(\frac{\partial p}{\partial T}\right)_\rho = -p + \ell_\rho \tag{21.38}$$

$$\rho^2\left(\frac{\partial h}{\partial p}\right)_T = \rho + T\left(\frac{\partial \rho}{\partial T}\right)_p = \rho^2\left(\rho^{-1} + \ell_p\right). \tag{21.39}$$

Equations (21.38) and (21.39) are known as the two general thermodynamic equations of state, which are valid for *any given system in equilibrium*. The first equation of state (Eq. (21.38)) is the equation that was initially used above to gain a good physical understanding of the microscopic interaction energy in a real fluid. In both Eqs. (21.38) and (21.39), the four calorimetric coefficients c_V, c_p, ℓ_ρ, and ℓ_p are interrelated by the so-called Clapeyron rules obtained by equating the two expressions for the differentiation of the entropy $ds = \frac{c_V}{T}\,dT - \frac{\ell_\rho}{T\rho^2}\,d\rho = \frac{c_p}{T}\,dT + \frac{\ell_p}{T}\,dp$, which gives

$$c_p - c_V = -\frac{\ell_\rho}{\rho^2}\left(\frac{\partial \rho}{\partial T}\right)_p \tag{21.40}$$

$$\ell_p = \frac{\ell_\rho}{\rho^2}\left(\frac{\partial \rho}{\partial p}\right)_T \tag{21.41}$$

$$\ell_\rho = T\left(\frac{\partial p}{\partial T}\right)_\rho = pT\beta_\rho, \tag{21.42}$$

where $\beta_\rho = \frac{1}{p}\left(\frac{\partial p}{\partial T}\right)_\rho$ is the isochoric thermal coefficient that measures the pressure increase when the temperature of a closed constant volume of the fluid increases. In addition to the microscopic nature of the terms $T\left(\frac{\partial p}{\partial T}\right)_\rho$ and $\rho^2\left(\frac{\partial e}{\partial \rho}\right)_T$ of the general thermodynamic Eqs. (21.38), (21.40), (21.41), and (21.42) provide an immediate understanding of the calorimetric coefficients or their ratios. ℓ_ρ represents the contribution of the repulsive forces, which leads to the natural tendency of the fluid to expand when the temperature increases. ℓ_ρ is related to the thermal pressure coefficient β_p of the fluid. $\beta_p = -\frac{1}{\rho}\left(\frac{\partial \rho}{\partial T}\right)_p = -\frac{\rho(c_p - c_V)}{l_\rho}$ is the isobaric thermal expansion coefficient (see Eq. (1.62)). This is the macroscopic result of the competition between the attractive forces and the repulsive forces associated with the natural tendency of a fluid to expand when the temperature increases; it *measures* the thermal bulk dilatability of the fluid. A fluid is said to be *compressible* when the density can vary with pressure at constant entropy. When the density can vary with pressure at constant temperature, the fluid is said to be *expandable*. The change in κ_T with temperature can be associated with the change in β_p with pressure. $\kappa_T = \frac{1}{\rho}\left(\frac{\partial \rho}{\partial p}\right)_T = \frac{\rho \ell_p}{l_\rho}$ is the isothermal compressibility (see Eq. (1.43)), which represents the competition between the attractive forces and the repulsive forces associated with the natural tendency of the fluid to contract when pressure increases, and it *measures* the isothermal bulk compressibility of the fluid. The isothermal compressibility is different from the isentropic compressibility defined by

$$\kappa_s = \frac{1}{\rho}\left(\frac{\partial \rho}{\partial p}\right)_s,$$

which appears in the thermodynamic definition of the sound velocity (see Eq. (1.84)). As the critical point is approached, κ_T and β_p diverge while $\beta_\rho \propto \gamma_V = \left(\frac{\partial p}{\partial T}\right)_\rho$ (see Sect. 1.4), and ℓ_ρ correspondingly has a finite value at the critical point. Consequently, ℓ_p and $(c_p - c_V)$ have similar critical divergences to β_p and κ_T, respectively.

We note that ℓ_ρ, ℓ_p, and $c_p - c_V$, can be calculated using the knowledge of the equation of state only, while determining one heat capacity among c_p or c_V requires more information: $e(T, \rho)$ or $h(T, p)$ for example.

21.3.2.2 The Helmholtz Equation

The Helmholtz equation is an important equation in scenarios when isobaric and adiabatic processes occur simultaneously. Moreover, the Helmholtz equation can be used to calculate the difference between c_p and c_V. This allows a better understanding of the physical meaning of this difference which complements our previous analysis of the contribution of the repulsive forces. The Helmholtz equation follows from the properties of the second cross-derivatives of entropy obtained by inverting Eq. (21.78):

$$ds = \frac{1}{T}de - \frac{p}{T\rho^2}d\rho. \tag{21.43}$$

Since e can be expressed as a function of ρ and T, we have $de = \left(\frac{\partial e}{\partial \rho}\right)_T d\rho + \left(\frac{\partial e}{\partial T}\right)_\rho dT$. Using this expression in Eq. (21.43), we obtain

$$ds = \left[\frac{1}{T}\left(\frac{\partial e}{\partial \rho}\right)_T - \frac{p}{T\rho^2}\right]d\rho + \frac{1}{T}\left(\frac{\partial e}{\partial T}\right)_\rho dT. \tag{21.44}$$

The coefficients of $d\rho$ and dT can now be identified as the derivatives $\left(\frac{\partial s}{\partial \rho}\right)_T$ and $\left(\frac{\partial s}{\partial T}\right)_\rho$, respectively. As expressed by the appropriate Maxwell rule, since the second *cross-derivatives* of $s(\rho, T)$ must be equal, we have

$$\left(\frac{\partial}{\partial T}\left[\frac{1}{T}\left(\frac{\partial e}{\partial \rho}\right)_T - \frac{p}{T\rho^2}\right]\right)_\rho = \left(\frac{\partial}{\partial \rho}\left[\frac{1}{T}\left(\frac{\partial e}{\partial T}\right)_\rho\right]\right)_T . \tag{21.45}$$

It is a simple calculation to show that Eq. (21.45) leads to the Helmholtz equation

$$\rho^2\left(\frac{\partial e}{\partial \rho}\right)_T = T^2\left(\frac{\partial}{\partial T}\left[\frac{p}{T}\right]\right)_\rho . \tag{21.46}$$

This equation enables us to calculate the variation in the energy with density if the equation of state is known. In particular, it can be concluded that, for an ideal gas, the equation of state implies that the specific energy e at constant T is independent of density. In addition to the first (usual) equation of state $p = \rho r T$ (where $r = \frac{R}{m}$), there is the second independent equation of state of an ideal gas $e = c_V T$. For real gases, due to molecular interactions, energy is no longer only a function of the temperature. Because the interaction energy of the molecules depends on the distance between them, a change in volume at a fixed temperature causes a change in energy, so the term $\left(\frac{\partial e}{\partial \rho}\right)_T$ does not vanish for a real gas. For example, the Helmholtz equation for a van der Waals gas is $\left(\frac{\partial e}{\partial \rho}\right)_T = -a$. More generally, given the equation of state of a real gas, the Helmholtz equation can be integrated to obtain an explicit expression for $c_{V,\text{real}}$. For any equation of state in which p is a linear function of T, such as the van der Waals equation, we have $c_{V,\text{real}} = c_{V,\text{ideal}}$. Because the intermolecular distances are unchanged at constant volume, the value of c_V is unaffected by the molecular forces. c_V mainly reflects the change in the kinetic energy of the molecules per unit change in temperature.

These remarks are also important when the Helmholtz equation is used to calculate the difference between c_p and c_V. Consider, for example, the change of enthalpy with temperature at constant pressure, which reads

$$\begin{aligned}\left(\frac{\partial h}{\partial T}\right)_p &= \left(\frac{\partial\left(e + \frac{p}{\rho}\right)}{\partial T}\right)_p \\ &= \left(\frac{\partial e}{\partial T}\right)_\rho + \frac{1}{\rho^2}\left[\rho^2\left(\frac{\partial e}{\partial \rho}\right)_T - p\right]\left(\frac{\partial \rho}{\partial T}\right)_p .\end{aligned} \tag{21.47}$$

Hence, using Eq. (21.46), the first Clapeyron rule is obtained, and this can be rewritten in the following useful form:

$$c_p - c_V = \frac{T}{\rho^2}\left(\frac{\partial p}{\partial T}\right)_\rho\left(\frac{\partial \rho}{\partial T}\right)_p . \tag{21.48}$$

The right-hand side of Eq. (21.48) is equal to the additional amount of heat required in a constant-pressure process to compensate for the energy expended in increasing the volume. The heat capacity difference is equal to the work done during expansion only if $\left(\frac{\partial e}{\partial \rho}\right)_T$ is zero, as in the case of a perfect gas. However, for liquids, the second term on the r.h.s. of Eq. (21.48) is far from being negligible, and is generally larger than the third term. Moreover, this difference exhibits a critical divergence for critical fluids, and goes to infinity at $T = T_c$.

21.3.2.3 The Particular Case of Adiabatic Transformations

Since the thermodynamic relations presented above describe local equilibrium processes, the local entropy production is second order, and thus adiabatic transformations for which the local heat transfer is null do not involve any entropy exchange—they are isentropic. Taking into account Eq. (21.44) and the definition for the specific heat at constant volume given by Eq. (21.31), the following relation holds for adiabatic, local equilibrium transformations:

$$\left[\left(\frac{\partial e}{\partial \rho}\right)_T - \frac{p}{\rho^2}\right] d\rho = c_V \, dT. \tag{21.49}$$

By evaluating the derivative on the right-hand side of the Helmholtz equation (21.46), it is easy to check that $\left[\left(\frac{\partial e}{\partial \rho}\right)_T - \frac{p}{\rho^2}\right] = \frac{T}{\rho^2}\left(\frac{\partial p}{\partial T}\right)_\rho = \frac{\ell_\rho}{\rho^2}$. Hence, using Eq. (21.49), we obtain

$$\ell_\rho dv = -c_V \, dT, \tag{21.50}$$

where we have made the substitution $v = \frac{1}{\rho}$ for the specific volume. For an adiabatic process, the work involved in expansion corresponds to an instantaneous decrease in the energy, and for a fixed value of the external pressure, the efficiency of this process is given by the ratio $\frac{c_V}{\ell_\rho}$. For a polyatomic van der Waals gas where it is assumed that $c_{V,\,\mathrm{vdW}} = c_{V,\,\mathrm{IG}} = \frac{5}{2} r$, the dimensionless form $T_c \rho_c \frac{c_V}{\ell_\rho}$ of this ratio is given by $T_c \rho_c \frac{c_V}{\ell_\rho} = \frac{1}{Z_c} \frac{c_V}{r} \left[\frac{T}{p_c}\left(\frac{\partial p}{\partial T}\right)_{\rho_c}\right]^{-1} = \frac{5}{3}(1 + \tau)$ along the critical isochore. Furthermore, it is possible to write Eq. (21.50) in terms of the ratio $\gamma_0 = \frac{c_P}{c_V}$, using the general Eqs. (21.40), (21.42), and (21.41), finally obtaining the following simple expression:

$$(\gamma_0 - 1) \frac{dv}{v} = \beta_p \, dT. \tag{21.51}$$

21.4 Kinematics of Fluids

A fluid, a continuous material system in which infinitely small fluid particles can be isolated around any geometrical points in the 3-D physical space, can be described in two ways:

- The Lagrangian description, which involves following each fluid particle considered individually during its motion along its trajectory.

- The Eulerian description, which involves studying the variations in the fluid properties over time at each point in the geometrical space.

The analogy that is sometimes given for the Eulerian description is that of a fisherman who only watches the water around the tip of his fishing rod.

These two descriptions obviously yield the same results, but depending on our ultimate goal, one description may be more appropriate or useful than the other.

21.4.1 Lagrangian Description of Fluid Motion

The motion of a fluid particle is determined by three coordinates in the physical space referred to a Cartesian 3-D reference frame. These coordinates are functions of time, $x_i(t),\ i = \{1, 3\}$. The trajectory of M is the curve in space that is constructed from the different positions occupied by point M at different times. In the case of a continuous fluid medium, there are an infinite number of trajectories corresponding each to a given fluid particle.

Let us consider a finite fluid domain $\mathcal{D}$ bounded at time t_0 by a closed surface Σ that is moving with respect to a given reference frame. We say that we follow the motion of $\mathcal{D}$ if, at each time t, we can define the domain $\mathcal{D}_t$ that is the image of $\mathcal{D}$ at time t_0. If M_0 is the position of any fluid particle belonging to domain $\mathcal{D}$ at time t_0, then its position at any other time t will be given by a vector function $\mathbf{q}(\mathrm{M}_0, t)$, which is a one-to-one function of the initial position of M, say $\xi_i,\ i = \{1, 3\}$, and of time. The four independent variables (ξ_i, t) are called the Lagrange variables, and the vector function of four independent variables $\mathbf{q}$ is the Lagrangian description of the motion of the fluid domain $\mathcal{D}$. We will not focus on the properties that must fulfill the function $\mathbf{q}$ here; we will just say that a closed fluid domain is always transformed over time into another closed fluid domain, and that a fluid domain is never transformed into a point. A given fluid domain that we follow along the streamlines thus maintains a constant total mass; this is a particularly important condition that is needed to achieve a thermodynamic description of the fluid.

The trajectories of the fluid particles are given as parametric representations with time as a parameter:

$$x_i = g_i(\xi_1, \xi_2, \xi_3, t).$$

The trajectory of a fluid particle can be visualized by injecting particles that have a density equal to that of the fluid into the fluid at a given geometric point.

The velocity and the acceleration of a fluid particle M at time t is given by the time derivative of the function $\mathbf{q}(\mathrm{M}_0, t)$:

$$\mathbf{v}_L(\mathrm{M}_0, t) = \frac{\partial}{\partial t}\mathbf{q}(\mathrm{M}_0, t)$$
$$\mathbf{a}_L(\mathrm{M}_0, t) = \frac{\partial^2}{\partial t^2}\mathbf{q}(\mathrm{M}_0, t).$$

21.4.2 Eulerian Description of Fluid Motion

In this description, we focus on what happens at a given geometrical point M at different times. We thus aim to elucidate the different fluid properties at the given point. The Euler variables are four independent variables (x_i, t) where x_i are the coordinates of point M. The Euler unknowns are the three components $\left[u(x_{i,}t),\ v(x_i,t),\ w(x_i,t)\right]$ of the velocity $\mathbf{v}_L(\mathrm{M}_0, t)$ of the fluid particle that passes through point M at time t and was at point M_0 at time $t = 0$. We can change from the Lagrange to Euler variables and vice-versa without changing the description of fluid motion, as described below.

- From the Lagrange to Euler variables:

The motion of a fluid particle M is given by the function $\mathbf{x} = \mathbf{q}(\mathrm{M}_{0,}t)$. Provided the function $\mathbf{q}$ is one-to-one and continuous, the preceding relation can be inverted to give the positions (ξ_i) at time $t = 0$ of the fluid particle that is at a given location (x_i) at a given time; e.g., to give a function $\mathbf{h}$ such that its components are:

$$\xi_i = h_i(x_1, x_2, x_3, t),$$

and thus find the expression of the velocity $\mathbf{v}(x_i, t)$ in the Eulerian description from

$$\mathbf{v}(x_i, t) = \mathbf{v}_L\left[h_i(x_1, x_2, x_3), t\right] = \mathbf{v}\left(\xi_i, t\right).$$

- From the Euler to Lagrange variables:
  Let us suppose that the Euler variables (u, v, w), which are the components of the velocity, i.e.,

$$u = \frac{dx_1}{dt},\quad v = \frac{dx_2}{dt},\quad w = \frac{dx_3}{dt},$$

are known. This ordinary system of differential equations with the initial conditions

$$x_i(t = 0) = \xi_i$$

can be solved to give

$$x_i(\xi_i, t),$$

which is the Lagrangian description of motion.

21.4.3 Total Derivative

21.4.3.1 Definition

When we follow the motion of a fluid particle, the evolution in time of a scalar quantity ϕ that is characteristic of a fluid particle depends only on time and the associated derivative with respect to time. This is called the material derivative of ϕ, the *total derivative*, or the derivative when following the motion of the fluid particle.

21.4.3.2 Lagrangian Description

In the Lagrangian description, we follow a moving particle that is at point M_0 at time t_0 and at point M at time t. Let ϕ be a characteristic quantity associated with the fluid particle M. ϕ is a function of M_0 and of time t:

$$\phi \equiv \phi_L(M_0, t).$$

In this case, the material derivative reduces to the partial derivative of ϕ_L

$$\frac{D\phi_L}{Dt} = \frac{\partial}{\partial t'}\phi_L(M_0, t) = \frac{\partial}{\partial t}\phi_L(\xi_i, t),$$

where ξ_i is fixed because we are following one particular particle (so M_0 is fixed).

21.4.3.3 Eulerian Description

In the Eulerian description, the independent variables are the three coordinates x_i of point M, the point at which we wish to measure the fluid properties, and time t; accordingly,

$$\phi \equiv \phi(x_i, t) = \phi\left[x_i(\xi_1, \xi_2, \xi_3, t), t\right] = \phi_L(\xi_i, t).$$

The material derivative becomes

$$\frac{D\phi(x_i, t)}{Dt} = \frac{D\phi\left[g_i(\xi_1, \xi_2, \xi_3, t), t\right]}{Dt} = \sum_i \frac{\partial \phi}{\partial g_i}\frac{\partial g_i}{\partial t} + \frac{\partial \phi}{\partial t}.$$

As

$$\left.\frac{\partial g_i}{\partial t}\right|_{\xi_i} = \frac{dx_i}{dt},$$

we obtain the following expression for the material derivative in the Eulerian description of the fluid motion:

$$\frac{D\phi(x_i, t)}{Dt} = \sum_i \frac{\partial \phi}{\partial x_i}\frac{dx_i}{dt} + \frac{\partial \phi}{\partial t}.$$

Taking into account the fact that $\frac{dx_i}{dt}$ are the components of velocity at the point M, the previous relation can be written as follows:

$$\frac{D\phi(x_i, t)}{Dt} = \frac{\partial \phi}{\partial t} + \mathbf{v} \cdot \nabla\phi(x_i, t),$$

where $\mathbf{v} \cdot \nabla$ is a scalar operator such that

$$\mathbf{v} \cdot \nabla = \sum_i v_i \frac{\partial}{\partial x_i}.$$

The first of the two terms above corresponds to the possible nonsteady variation of ϕ; the second corresponds to the motion of the fluid particle at M, which is commonly called the convection term.

For the density ρ, the material derivative can be written as

$$\frac{D\rho}{Dt} = \frac{\partial \rho}{\partial t} + \mathbf{v} \cdot \nabla \rho.$$

For a vector function ϕ, the material derivative is the set of vector components that represent the material derivatives of the components of ϕ:

$$\left.\frac{D\phi}{Dt}\right|_i = \frac{\partial \phi_i}{\partial t} + \sum_j v_j \frac{\partial \phi_i}{\partial x_j} = \frac{\partial \phi_i}{\partial t} + (\mathbf{v} \cdot \nabla)\,\mathbf{v}.$$

Making use of the tensor notation

$$(\mathbf{v} \cdot \nabla)\,\mathbf{v} = (\nabla \otimes \mathbf{v})\,\mathbf{v} = \mathbf{v} \cdot \overline{\nabla \otimes \mathbf{v}},$$

we will employ

$$(\nabla \otimes \mathbf{v}) = \nabla \mathbf{v} = \left|\frac{\partial v_i}{\partial x_j}\right|$$

so that

$$\left.\frac{D\phi}{Dt}\right|_i = \frac{\partial \phi_i}{\partial t} + \sum_j v_j \frac{\partial \phi_i}{\partial x_j} = \frac{\partial \phi_i}{\partial t} + \nabla \mathbf{v} \cdot \mathbf{v}.$$

21.5 General Aspects of Conservation Equations

In what follows, we derive the general forms of conservation equations that are needed to obtain the fluid properties at any point and at any time within the domain under consideration. We will thus apply the results in the next sections in order to derive the local equations for the mass (Sect. 21.6), the momentum (Sect. 21.7), and the energy (Sect. 21.8) balances for a given mass of moving fluid that we follow along its trajectory.

21.5.1 General Balance Equations for the Fluid Property F

Let us follow, during its motion along the trajectories, a given deformable domain $\mathcal{D}$ within which the fluid variables are continuous. The domain $\mathcal{D}$ is a closed thermodynamic system that may exchange momentum and energy with the external fluid. It exchanges energy in the form of the work done by pressure and viscous forces acting on its bounding surface, as its envelope Σ is deformable. It also exchanges heat by diffusion through Σ. We thus need to express balances of extensive quantities such as mass, momentum and energy for domain $\mathcal{D}$ in order to obtain the local balance equations for the corresponding specific variables; i.e., the corresponding quantities per unit mass.

Let f be the specific quantity (quantity per unit mass) of any mathematical form (scalar, tensor, etc.) that corresponds to the amount of fluid property F present in domain $\mathcal{D}$. If we follow domain D during its motion, the total variation in quantity F present in $\mathcal{D}$ is equal to the sum of the flux of f through Σ and the production of f within $\mathcal{D}$:

$$\frac{D}{Dt}\int_{\mathcal{D}} \rho f dV \int_{\Sigma} \mathbf{J}_F \cdot \mathbf{n} \, d\Sigma + \int_{\mathcal{D}} \mathring{\omega}_F dV, \tag{21.52}$$

where $\mathbf{n}$ is the vector normal to Σ that is directed towards the exterior of $\mathcal{D}$, $\mathbf{J}_F$ is the vector flux of F through Σ, and $\mathring{\omega}_F$ is the volume rate of production of F within $\mathcal{D}$. The function's properties are assumed to be such that we can derive under the summation sign; in other words, we can write Eq. (21.52) as

$$\frac{D}{Dt}\int_{\mathcal{D}} \rho f dV = \int_{\mathcal{D}} \frac{D}{Dt}(\rho f dV) = \int_{\mathcal{D}} \frac{D}{Dt}(\rho f) dV + \int_{\mathcal{D}} \rho f \frac{D}{Dt}(dV), \tag{21.53}$$

where $\frac{D}{Dt}(dV)$ is the material derivative of the volume element dV.

A quantity for which $\mathring{\omega}_F = 0$ is called a conservative quantity, since its variations as we follow the motion of the fluid domain are only due to the corresponding flux through the boundary. Any quantity that has both its flux and production terms equal to zero remains constant along the streamlines.

21.5.1.1 Material Derivative of a Vanishingly Small Vector and Volume Elements

Let us consider an arbitrarily small vector joining point A (vector $\overrightarrow{OA}$ denoted $\mathbf{a}$) and point B (vector $\overrightarrow{OB}$ denoted $\mathbf{b}$). Both points are "convected" by the fluid, so they move at velocities $\mathbf{v}(\mathbf{a})$ and $\mathbf{v}(\mathbf{b})$, respectively. The material derivative of $d\mathbf{a}$ is given by

$$\frac{D}{Dt}(d\mathbf{a}) = \frac{d\mathbf{a}(t+dt) - d\mathbf{a}(t)}{dt}$$

for dt going to zero; in other words,

$$\frac{D}{Dt}(d\mathbf{a}) = \frac{\mathbf{b}(t+dt) - \mathbf{b}(t) - [\mathbf{a}(t+dt) - \mathbf{a}(t)]}{dt}.$$

Taking into account the fact that $\mathbf{a}$ moves with the fluid,

$$\mathbf{a}(t+dt) = \mathbf{a}(t) + \mathbf{v}(\mathbf{a})dt$$

$$\mathbf{b}(t+dt) = \mathbf{b}(t) + \mathbf{v}(\mathbf{b})dt,$$

so that

$$\frac{D}{Dt}(d\mathbf{a}) = \mathbf{v}(\mathbf{b}) - \mathbf{v}(\mathbf{a}),$$

and noting that

$$\mathbf{b} = \mathbf{a} + d\mathbf{a},$$

we can deduce, for example, that

$$\left.\frac{D}{Dt}(d\mathbf{a})\right|_i = \sum_j \frac{\partial v_i(\mathbf{a})}{\partial x_j} dx_j = \nabla \otimes v \cdot d\mathbf{a},$$

where $\nabla \otimes v$ is the tensor gradient of $\mathbf{v}$.

In the same way, we can show that

$$\frac{D}{Dt}(dV) = \nabla \cdot \mathbf{v} dV \tag{21.54}$$

We can thus write the material derivative of the total amount of any fluid property f in $\mathcal{D}$ in the following form:

$$\frac{D}{Dt}\int_{\mathcal{D}} \rho f dV = \int_{\mathcal{D}} \left[\frac{D}{Dt}(\rho f) + \rho f \nabla \cdot \mathbf{v}\right] dV. \tag{21.55}$$

21.5.1.2 General Expression for the Material Derivative of Any Fluid Property F (of Density f)

Taking into account Eq. (21.55), the balance equation (21.52) of any property F can be written as

$$\int_{\mathcal{D}} \left[\frac{D}{Dt}(\rho f) + \rho f \nabla \cdot \mathbf{v} + \nabla \cdot \mathbf{J}_F - \overset{\circ}{\omega}_F\right] dV = 0,$$

which implies, for an arbitrarily chosen closed domain $\mathcal{D}$, the following general local equation that expresses the balance of any property F:

$$\frac{D}{Dt}(\rho f) + \rho f \nabla \cdot \mathbf{v} + \nabla \cdot \mathbf{J}_F = \overset{\circ}{\omega}_F \tag{21.56}$$

In this equation, $\mathbf{J}_F$ is the flux that passes through a surface element (which follows the motion of the fluid), and $\overset{\circ}{\omega}$ is the volume production term.

21.6 Mass Balance

The mass balance is obtained by setting $f = 1$ and $\mathbf{J}_F = 0$ in Eq. (21.56) (since the total local mass diffusion flux should be zero for a pure fluid):

$$\frac{\partial \rho}{\partial t} + \nabla \cdot (\rho \mathbf{v}) = 0. \tag{21.57}$$

Anticipating the modeling approaches developed in Chaps. 3, 4, and 5, we note that the one-dimensional form of the mass balance equation reads as follows:

$$\frac{\partial \rho}{\partial t} + \frac{\partial}{\partial x}(\rho u) = 0. \tag{21.58}$$

21.7 Momentum Balance

The momentum balance corresponding to $f = \mathbf{v}$ is given by Newton's law written in the laboratory reference frame, considered to be Galilean. When we follow the domain $\mathcal{D}$ during its motion along the streamlines, the associated momentum change is the result of the work done by all external forces acting on the fluid domain, including the surface and volume forces, leading to

$$\frac{D}{Dt}\int_{\mathcal{D}} \rho \mathbf{v} dV = \int_{\Sigma} \mathbf{t}(M, \mathbf{n}) d\Sigma + \int_{\mathcal{D}} \rho \mathbf{f}\, dV, \tag{21.59}$$

where $\mathbf{t}(M, \mathbf{n})$ is the stress at point M on the surface element $d\Sigma$ and $\mathbf{f}$ is the volume force that acts at each point of the volume within $\mathcal{D}$. Introducing Eq. (21.20), we have

$$\mathbf{t}(M, \mathbf{n}) = \mathbf{T} \cdot \mathbf{n} = -\mathbf{P} \cdot \mathbf{n},$$

where $\mathbf{P}$ is the pressure stress tensor. Equation (21.59) can be written as

$$\frac{D}{Dt}\int_{\mathcal{D}} (\rho \mathbf{v})\, dV = \int_{\Sigma} -\mathbf{P} \cdot \mathbf{n}\, d\Sigma + \int_{\mathcal{D}} \rho \mathbf{f}\, dV.$$

Taking into account Eq. (21.55) (in which $f = \mathbf{v}$) and applying the divergence theorem to the tensor surface integral, the momentum balance can be written in the following two alternate forms:

$$\frac{D}{Dt}(\rho \mathbf{v}) + \rho \mathbf{v} \nabla \cdot \mathbf{v} + \nabla \mathbf{P} = \rho \mathbf{f}$$

and

$$\frac{\partial}{\partial t}(\rho \mathbf{v}) + \nabla \cdot \rho \mathbf{v} \otimes \mathbf{v} + \nabla \mathbf{P} = \rho \mathbf{f}.$$

If we now consider the mass balance equation, the two equivalent forms are

$$\rho \frac{D\mathbf{v}}{Dt} + \nabla \mathbf{P} = \rho \mathbf{f} \tag{21.60}$$

$$\rho \frac{\partial \mathbf{v}}{\partial t} + \rho \nabla \otimes \mathbf{v} \cdot \mathbf{v} + \nabla \mathbf{P} = \rho \frac{\partial \mathbf{v}}{\partial t} + \rho \left(\mathbf{v} \cdot \nabla \right) \mathbf{v} + \nabla \mathbf{P} = \rho \mathbf{f}. \tag{21.61}$$

It thus appears that the pressure tensor can be identified with the momentum vector flux through the surface Σ that bounds the fluid domain $\mathcal{D}$, while the volume force $\mathbf{f}$ appears to be the local momentum production per unit volume.

Just as in the previous case of the 1-D mass balance of Eq. (21.58), when the fluid model is not subjected to an external field that is able to create additional surface and volume forces, we note that the one-dimensional form of the momentum balance equation reads as follows:

$$\rho \frac{\partial u}{\partial t} + \rho u \frac{\partial u}{\partial x} = -\frac{\partial p}{\partial x} + \frac{4}{3} \eta \frac{\partial^2 u}{\partial x^2}, \tag{21.62}$$

where Eq. (21.24) was used to derive the right-hand side of the above equation, which accounts for the respective contributions of the pressure forces and the viscous forces.

21.8 Energy Balance

21.8.1 Thermodynamic Aspects

We need to establish the local equation that expresses the energy balance of any closed fluid domain while following its motion along its trajectory. The variation in the total energy of the fluid domain is given, in these circumstances, by extending the first law of thermodynamics to macroscopically moving closed thermodynamic systems. To this end, let us consider K to be the total kinetic energy of the fluid domain under consideration. The mechanical energy theorem indicates that the variation in the mechanical energy of the fluid domain over time is equal to the power it receives from all nonconservative internal and external forces:

$$\frac{D}{Dt} \left[K + E_{p,ex} + E_{p,in} \right] = \mathcal{P}_{in}^{nc} + \mathcal{P}_{ex}^{nc}, \tag{21.63}$$

where $E_{p,ex}$ and $E_{p,in}$ are, respectively, the potential energies of the conservative external and internal forces. $\mathcal{P}_{in}^{nc}$ and $\mathcal{P}_{ex}^{nc}$ are, respectively, the powers of the internal and external nonconservative forces. If we do not separate the conservative and nonconservative powers, we can write

$$\frac{D}{Dt} K = \mathcal{P}_{in} + \mathcal{P}_{ex}. \tag{21.64}$$

On the other hand, the internal energy E is such that the sum of it, the macroscopic kinetic energy and the potential energy of conservative external forces is constant when the fluid domain exchanges neither heat nor nonconservative work with the exterior. We thus obtain the following expression for the first law of thermodynamics of a moving system:

$$\frac{D}{Dt}\left[E + E_{p,ex} + K\right] = \mathcal{P}_{ex}^{nc} + \overset{\circ}{Q}. \tag{21.65}$$

$\overset{\circ}{Q}$ is the heat power received by the fluid domain:

$$\overset{\circ}{Q} = \int_{\mathcal{D}} \overset{\circ}{q} dV - \int_{\Sigma} \mathbf{j} \cdot \mathbf{n} d\Sigma, \tag{21.66}$$

where $\overset{\circ}{q}$ is the volume heat source power within the fluid domain and $\mathbf{j}$ is the heat flux vector at its bounding surface.

If we do not separate conservative from nonconservative external powers, the first law of thermodynamics can be written as

$$\frac{D}{Dt}\left[E + K\right] = \mathcal{P}_{ex} + \overset{\circ}{Q}. \tag{21.67}$$

Subtracting Eq. (21.63) from Eq. (21.65) gives

$$\frac{D}{Dt}(E - E_{p,in}) = \overset{\circ}{Q} - \mathcal{P}_{in}^{nc}. \tag{21.68}$$

Equation (21.68) indicates in particular that for purely mechanical systems (planets for example), i.e., for those systems in which both $\overset{\circ}{Q}$ and $\mathcal{P}_{in}^{nc}$ are zero, the internal energy can be considered to be the potential energy of the internal forces.

Equation (21.68) can be also be written in an equivalent form:

$$\frac{DE}{Dt} = \overset{\circ}{Q} - \mathcal{P}_{in}, \tag{21.69}$$

where $\mathcal{P}_{in}$ is the power of the inner forces.

21.8.2 Kinetic Energy Balance

Taking into account the equation for the conservation of mass, Eq. (21.57) can be written as

$$\frac{DF}{Dt} = \frac{D}{Dt}\int_{\mathcal{D}} \rho f dV = \int_{\mathcal{D}} \left[\frac{D}{Dt}(\rho f) + \rho f \nabla \cdot \mathbf{v}\right] dV = \int_{\mathcal{D}} \rho \frac{Df}{Dt} dV.$$

When applied to the kinetic energy K of the fluid domain, we obtain (since the kinetic energy per unit mass is $\frac{\mathbf{v}^2}{2}$)

$$\frac{DK}{Dt} = \int_{\mathcal{D}} \rho \frac{D\frac{v^2}{2}}{Dt} dV.$$

Multiplying both sides of the momentum equation (21.61) by $\mathbf{v}$ and integrating over the fluid domain $\mathcal{D}$, we obtain the kinetic energy balance

$$\int_{\mathcal{D}} \rho \frac{D\frac{v^2}{2}}{Dt} dV = \frac{DK}{Dt} = -\int_{\mathcal{D}} \mathbf{v} \cdot \boldsymbol{\nabla} \cdot \mathbf{P} dV + \int_{\mathcal{D}} \rho \mathbf{v} \cdot \mathbf{f} dV. \tag{21.70}$$

Following Eq. (21.64), the total power provided by the internal and external surface and volume forces is thus

$$\mathcal{P}_{in} + \mathcal{P}_{ex} = -\int_{\mathcal{D}} \mathbf{v} \cdot \boldsymbol{\nabla} \cdot \mathbf{P} dV + \int_{\mathcal{D}} \rho \mathbf{v} \cdot \mathbf{f} dV. \tag{21.71}$$

This term, equal to the product of a displacement rate by a volume force, can be identified as the power provided by all outer and inner stresses during the motion of the fluid domain.

21.8.3 Power of External and Internal Stresses

Considering that the power of the external surface forces is equal to the outer stress multiplied by the displacement per unit time integrated over the whole surface of the domain, i.e.,

$$\begin{aligned}
\mathcal{P}_{ex} &= \int_{\Sigma} \mathbf{v} \cdot (\mathbf{T} \cdot \mathbf{n}) d\Sigma + \int_{\mathcal{D}} \rho \mathbf{v} \cdot \mathbf{f}\, dV \\
&= -\int_{\Sigma} (\mathbf{v} \cdot \mathbf{P}) \cdot \mathbf{n}\, d\Sigma + \int_{\mathcal{D}} \rho \mathbf{v} \cdot \mathbf{f}\, dV \\
&= -\int_{\mathcal{D}} \boldsymbol{\nabla} \cdot (\mathbf{v} \cdot \mathbf{P}) dV + \int_{\mathcal{D}} \rho \mathbf{v} \cdot \mathbf{f} dV,
\end{aligned} \tag{21.72}$$

and taking into account the fact that

$$\int_{\mathcal{D}} \boldsymbol{\nabla} \cdot (\mathbf{v} \cdot \mathbf{P}) dV = \int_{\mathcal{D}} \mathbf{v} \cdot \boldsymbol{\nabla} \cdot \mathbf{P}\, dV + \int_{\mathcal{D}} \mathbf{P} : \boldsymbol{\nabla} \otimes \mathbf{v}\, dV,$$

where the notation $\otimes$ is given by Eq. (21.8), we find (with the aid of Eq. (21.71)) that the power provided by the inner surface stresses is

$$\mathcal{P}_{in} = \int_{\mathcal{D}} \mathbf{P} : \nabla \otimes \mathbf{v} \, dV. \tag{21.73}$$

This term, proportional to the product of a stress by a rate of deformation, represents the power provided by the internal surface stresses during the deformation of the domain.

Noting that the pressure tensor can be written as

$$\mathbf{P} = p\mathbf{I} + \frac{2}{3}\mu \nabla \cdot \mathbf{v}\mathbf{I} - 2\mu\mathbf{D} = p\mathbf{I} - \boldsymbol{\tau},$$

it is possible to separate the internal power of the pressure stresses due to the deformation of the fluid domain, i.e.,

$$\mathcal{P}_{in}^{pr} = \int_{\mathcal{D}} p\nabla \cdot \mathbf{v} \, dV,$$

from that of the viscous stresses due to the same deformation, i.e.,

$$\mathcal{P}_{in}^{vis} = -\int_{\mathcal{D}} \boldsymbol{\tau} : \nabla \otimes \mathbf{v} \, dV.$$

If we define the tensor **K** as

$$\mathbf{K} = \nabla \otimes \mathbf{v},$$

it can be split into a symmetric term

$$\mathbf{D} = \frac{1}{2}(\mathbf{K} + \overline{\mathbf{K}})$$

and an antisymmetric term

$$\boldsymbol{\Omega} = \frac{1}{2}(\mathbf{K} - \overline{\mathbf{K}})$$

such that $\nabla \otimes \mathbf{v}$ can be written as

$$\nabla \otimes \mathbf{v} = \mathbf{D} + \boldsymbol{\Omega},$$

which leads to the alternate classical form of $\mathcal{P}_{in}^{vis}$, called the dissipation function ϕ_D:

$$\mathcal{P}_{in}^{vis} = \phi_D = -\left(\frac{2}{3}\mu\,(\nabla \cdot \mathbf{v})\,\mathbf{I} - 2\mu\mathbf{D}\right) : \nabla \otimes \mathbf{v} \tag{21.74}$$

$$= -\frac{2}{3}\mu\,(\nabla\cdot\mathbf{v})^2 + \frac{1}{2}\mu\left(\nabla\otimes\mathbf{v} : \overline{\nabla\otimes\mathbf{v}}\right)^2. \tag{21.75}$$

This function, which is always positive, is part of the entropy production due to inner friction.

In the same way, we can separate the power of the outer pressure (surface) forces and volume forces, i.e.,

$$\mathcal{P}_{ex} = -\int_{\mathcal{D}} \nabla\cdot(\mathbf{v}\cdot p)dV + \int_{\mathcal{D}} \rho\mathbf{v}\cdot\mathbf{f}\,dV,$$

from that of the outer viscous forces, i.e.,

$$\mathcal{P}_{ex}^{vis} = \int_{\mathcal{D}} \nabla\cdot(\mathbf{v}\cdot\boldsymbol{\tau})dV.$$

Finally, we can separate the power that all the inner and outer pressure and volume stresses impart to the fluid element during its displacement,

$$\mathcal{P}_{in}^{pr} + \mathcal{P}_{ex} = -\int_{\mathcal{D}} \mathbf{v}\cdot(\nabla\cdot p)\,dV + \int_{\mathcal{D}} \rho\mathbf{v}\cdot\mathbf{f}\,dV,$$

from that imparted by all the inner and outer viscous forces during the same displacement of the domain:

$$\mathcal{P}_{in}^{vis} + \mathcal{P}_{ex}^{vis} = \int_{\mathcal{D}} \mathbf{v}\cdot(\nabla\cdot\boldsymbol{\tau})\,dV.$$

Note that the power of the outer surface stresses is due to both the deformation and the displacement.

21.8.4 Total Energy Balance

From Eq. (21.67), into which we insert Eq. (21.72), we obtain

$$\frac{D}{Dt}[E+K] = -\int_{\mathcal{D}} \nabla\cdot(\mathbf{v}\cdot\mathbf{P})\,dV \tag{21.76}$$
$$+\int_{\mathcal{D}} \rho\mathbf{v}\cdot\mathbf{f}dV + \int_{\mathcal{D}} \mathring{q}dV - \int_{\mathcal{D}} \nabla\cdot\mathbf{j}\,dV,$$

so that, in the absence of external volume force and volume heat source terms, the total energy balance can be written as

$$\frac{D}{Dt}[E+K]+\int_{\mathcal{D}} \nabla\cdot(\mathbf{v}\cdot\mathbf{P}+\mathbf{j})dV=0.$$

Upon comparing with Eq. (21.52), the term

$$\mathbf{v}\cdot\mathbf{P}+\mathbf{j}$$

is thus the local flux of energy through the bounding surface of the fluid domain. In other words, the total energy is conservative. As a matter of fact, the variation in the total energy present in the domain is due only to the flux through the boundary, just as it is for the total mass. We may also note that when the kinetic energy is constant and there are no external volume forces acting in the fluid element, then, from Eq. (21.71), $\mathcal{P}_{in}=-\mathcal{P}_{ex}$; in other words, the total power brought to the fluid by the inner and outer stresses during the displacement is balanced by viscous dissipation.

21.8.5 Internal Energy Balance

By subtracting Eq. (21.70) from (21.76), we obtain the internal energy balance

$$\begin{aligned}\frac{DE}{Dt}&=\int_{\mathcal{D}}\rho\frac{De}{Dt}dV\\&=\int_{\mathcal{D}}\overset{\circ}{q}\,dV-\int_{\Sigma}\mathbf{j}\cdot\mathbf{n}\,d\Sigma-\int_{\mathcal{D}}p\nabla\cdot\mathbf{v}\,dV+\int_{\mathcal{D}}\phi_D\,dV,\end{aligned}\qquad(21.77)$$

or, under its global formulation,

$$\frac{DE}{Dt}=\overset{\circ}{Q}-\mathcal{P}_{in}.$$

Note that the internal energy is not conservative, since its evolution is partly due to the heat flux term. One source term is the power of the nondissipative part of the inner pressure stress ($-\int_{\mathcal{D}}p\nabla\cdot\mathbf{v}dV$). This term will be shown to diverge near the critical point, whereas it is null for incompressible fluids. In other words, the internal energy of a nondissipative incompressible fluid is constant during its motion provided there are no heat sources within the bulk fluid.

The standard local form of Eq. (21.77) can be written as

$$\rho\frac{De}{Dt}=-p\nabla\cdot\mathbf{v}+\phi_D-\nabla\cdot\mathbf{j}+\overset{\circ}{q}.\qquad(21.78)$$

In order to obtain a closed form of the former equation, we must express the internal energy as a function of the relevant local independent thermodynamic variables. Such a working method is possible provided we can apply the hypothesis of local equilibrium; in other words, if the internal energy of any small fluid particle within the fluid domain can be written in the same way as if thermodynamic equilibrium prevails in any small fluid element within $\mathcal{D}$. The result can be written (see Eq. (21.38)) as follows:

$$de(T, v) = c_V\, dT + \left(\ell_\rho - p\right) dv,$$

where $v = \frac{1}{\rho}$ is the local specific volume, or the local volume per mass unit. The material derivative, or the time derivative when following the motion of a given fluid domain, can be thus written as

$$\frac{De}{Dt} = c_V \frac{DT}{Dt} + \left(\ell_\rho - p\right) \frac{D}{Dt}\left(\frac{1}{\rho}\right),$$

where ℓ_ρ (see Eq. (21.42)) is determined from the equation of state using its definition

$$\ell_\rho = T \left.\frac{\partial p}{\partial T}\right|_\rho = pT\beta_\rho.$$

Finally the equation that expresses the local internal energy balance can be written as

$$\rho c_V \frac{DT}{Dt} = -\rho\left(\ell_\rho - p\right) \frac{D}{Dt}\left(\frac{1}{\rho}\right) - p\nabla\cdot\mathbf{v} + \phi_D - \nabla\cdot\mathbf{j} + \overset{\circ}{q}. \qquad (21.79)$$

As the internal energy balance will be used in the 1-D model developed in Chaps. 3, 4, and 5, the corresponding 1-D form of the above equation can be also written as

$$\rho c_V \frac{\partial T}{\partial t} + \rho c_V u \frac{\partial T}{\partial x} = -T\left(\frac{\partial p}{\partial T}\right)_\rho \frac{\partial u}{\partial x} + \Lambda \frac{\partial^2 T}{\partial x^2} + \frac{4}{3}\mu\left(\frac{\partial u}{\partial x}\right)^2. \qquad (21.80)$$

21.8.6 Enthalpy Balance

The enthalpy balance for the fluid domain is another form of the energy balance. From the definition of the specific enthalpy

$$h = e + \frac{p}{\rho},$$

its material derivative can be written as

$$\frac{Dh}{Dt} = \frac{De}{Dt} + \frac{1}{\rho}\frac{Dp}{Dt} - \frac{p}{\rho^2}\frac{D\rho}{Dt},$$

which gives (using the mass balance equation)

$$\rho \frac{Dh}{Dt} = \rho \frac{De}{Dt} + \frac{Dp}{Dt} + p\nabla \cdot \mathbf{v}.$$

Replacing $\frac{De}{Dt}$ by its expression from the internal energy balance leads to

$$\rho \frac{Dh}{Dt} = \frac{Dp}{Dt} + \phi_D - \nabla \cdot \mathbf{j} + \overset{\circ}{q}.$$

As for the internal energy, the preceding equation can be closed by expressing the specific enthalpy using the hypothesis of local thermodynamic equilibrium (the relations that are valid for the thermodynamic equilibrium are still valid locally in nonequilibrium flows):

$$\frac{Dh}{Dt} = c_p \frac{DT}{Dt} + \left(\ell_p + \frac{1}{\rho}\right) \frac{Dp}{Dt}, \tag{21.81}$$

where ℓ_p (see Eq. (21.41)) is determined from the equation of state using its definition

$$\ell_p = -T \left.\frac{\partial v}{\partial T}\right|_p = \frac{\ell_\rho}{\rho^2}\left(\frac{\partial \rho}{\partial p}\right)_T = -\frac{T}{\rho}\beta_p,$$

so that the equation that expresses the local enthalpy balance can be written as

$$\rho c_p \frac{DT}{Dt} = -\rho \ell_p \frac{Dp}{Dt} + \phi_D - \nabla \cdot \mathbf{j} + \overset{\circ}{q}$$

or in the alternate form

$$\rho c_p \frac{DT}{Dt} = T\beta_p \frac{Dp}{Dt} + \phi_D - \nabla \cdot \mathbf{j} + \overset{\circ}{q}.$$

It is interesting to note here that for very low fluid velocities such as those produced by local expansion under weak heating, the dissipation function as well as the convective terms in the temperature and pressure material derivatives can be neglected. This approximation leads to a purely thermodynamic equation, such as that written in Chap. 3.

21.8.7 Entropy Balance

As Eq. (21.43) applies to material derivatives, the entropy balance can be written as

$$\frac{Ds}{Dt} = \frac{1}{T}\frac{De}{Dt} - \frac{p}{T\rho^2}\frac{D\rho}{Dt}$$

Taking into account the expression derived in the previous section for $\frac{De}{Dt}$ and for $\frac{D\rho}{Dt}$, the entropy balance equation becomes

$$\rho\frac{Ds}{Dt} = \frac{1}{T}\left(\phi_D - \nabla \cdot \mathbf{j}\right)$$

Considering the Fourier law for the heat flux See Eq. (21.26) the above equation becomes

$$\rho\frac{Ds}{Dt} = \frac{1}{T}\left(\phi_D + \nabla \cdot \boldsymbol{\Lambda}\nabla T\right) \tag{21.82}$$

Rearranging Eq. (21.82) gives

$$\rho\frac{Ds}{Dt} = \frac{\phi_D}{T} + \Lambda\frac{(\nabla T)^2}{T^2} - \nabla \cdot \left(\frac{\mathbf{j}}{T}\right)$$

The first two terms are the entropy production that is obviously positive according to the second law of thermodynamics; the third term is the entropy flux that has no assigned direction.

Reference

1. Kogan MN (1969) Rarefied gas dynamics. Plenum, New York

Chapter 22
Numerical Method

Abstract The main aspects of the numerical methods used to explore the hydrodynamics of supercritical fluids are discussed. The numerical solution of a system of partial differential equations like the Navier–Stokes equations first requires us to discretize the equations (i.e., replace the continuous variables with discrete ones). The initial system of equations is replaced with linear equations that are solved by appropriate methods according to an algorithm of resolution. The treatment of nonlinear terms in the equations and the simultaneity of the solutions of the different equations are addressed in the algorithm of resolution. These two operations are not independent, since a good choice of discretization scheme often leads more simply and quickly to the solution. Its precision depends on the discretization scheme used, while the efficiency of the resolution in terms of computation time depends on both the discretization scheme and the algorithm.

22.1 Discretization in Space and Time

All of the governing equations can be written in the following convection–diffusion form, where ϕ is either density, velocity, or temperature:

$$(\rho\phi)_t + \nabla \cdot (\rho \mathbf{v}\phi) = \nabla \cdot \big(\Gamma_\phi (\nabla\phi)\big) + S_\phi. \tag{22.1}$$

Γ_ϕ is a diffusion coefficient, and S_ϕ is a volume source term. For the continuity equation, ϕ is 1. It is the velocity $\mathbf{v}$ for the momentum balance equation and the temperature T for the energy balance equation. Equation (22.1) is discretized using the finite volume scheme introduced by Patankar and Spalding (see [11]), which is now presented.

B. Zappoli et al., *Heat Transfers and Related Effects in Supercritical Fluids*, Fluid Mechanics and Its Applications 108, DOI: 10.1007/978-94-017-9187-8_22,

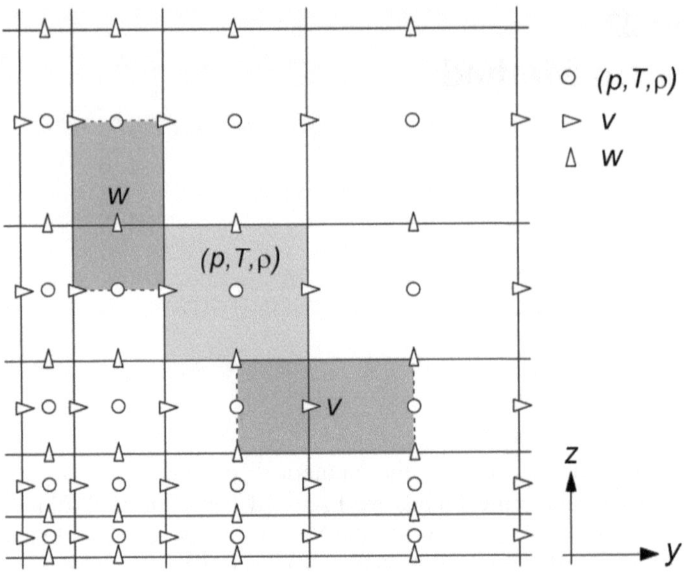

Fig. 22.1 Structure of the mesh and localization of each variable for the calculated points in the (y, z) plane. The *gray zones* indicate slices of the control volume; the variables are calculated at the center of this volume. Further details are shown in Fig. 22.2

22.1.1 Numerical Mesh and Localization of the Variables

We consider a physical parallelepipedic domain that contains $N \times M \times K$ cells, where N, M, and K are even numbers. The mesh is refined in the boundary layers located along the wall in order to properly compute the gradients and achieve a precise and reliable solution in these regions. For instance, if N is the (even) number of cells in the x direction, which is perpendicular to a boundary located in the (y, z) plane, the interfaces of the cells are located at positions given by

$$x(i) = \frac{1}{2}\left(2 \times \frac{i-1}{N}\right)^q, \qquad i = 1, \ldots, \frac{N}{2} + 1$$

$$x(i) = 1 - x(N - i + 2), \quad i = \frac{N}{2} + 2, \ldots, N + 1.$$

The interfaces thus follow a q-power law ($q \geq 1$). For $q = 1$ the mesh is uniform. The variables are staggered: scalar variables (P, ρ, T) are defined at the centers of the cells; velocity components are defined at the centers of the interfaces, which are perpendicular to the component under consideration. For each variable, Eq. (22.1) is integrated over a control volume. By doing so, convective terms can be calculated at the center of the control volume without interpolation. This procedure decreases the risk of obtaining nonphysical solutions [10], and is well adapted to simple geometries. Figure 22.1 shows the mesh and the localization of the variables in the (y, z) plane;

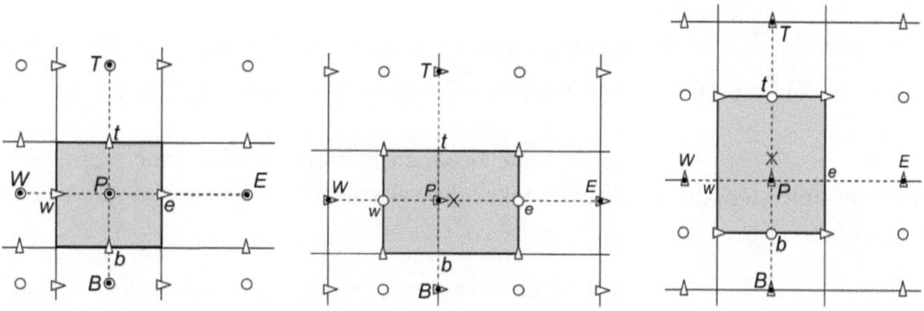

Fig. 22.2 Slices in the (y, z) plane of the control volume for scalar variables and for the three components of the velocity, as noted in Fig. 22.1

the structure is the same in the (x, y) and (x, z) planes. Figure 22.2 shows slices of the control volume in the (y, z) plane for each variable. The same notation as that used in [10] is employed, as follows. Point P is the position of the variable in the control volume under consideration. Points S (south), N (north), W (west), E (east), B (bottom), and T (top) represent its immediate neighbors in the three directions in space. The letters s, n, w, e, b, and t correspond to the faces of the control volume under consideration.

22.1.2 Time Scheme

At time t^n, where the superscript denotes the number of this time step, Eq. (22.1) can be written

$$(\rho\phi)_t^n + \nabla \cdot (\rho \mathbf{v} \phi)^n = \nabla \cdot \left(\Gamma_\phi (\nabla\phi)\right)^n + S_\phi^n, \qquad (22.2)$$

where the superscript n represents the value of the derivative at the time step n. Γ denotes the diffusion coefficient here. This scheme is implicit (all the terms are calculated at the same time); it is thus unconditionally stable with respect to the time step. The unsteady term is approached using a third-order Euler scheme that involves four time steps. The time step can be varied during the course of the calculation without downgrading the precision in time. The discretization in time of Eq. (22.1) can be written as

$$\frac{\partial}{\partial t}(\rho\phi)^n = C_0\,(\rho\phi)^n + C_1\,(\rho\phi)^{n-1} + C_2\,(\rho\phi)^{n-2} + C_3\,(\rho\phi)^{n-3}\,, \qquad (22.3)$$

where

$$C_3 = \frac{\delta t_1\,(\delta t_1 + \delta t_2)}{\delta t_3\,(\delta t_2 + \delta t_3)\,(\delta t_1 + \delta t_2 + \delta t_3)} \times \frac{\delta t_1\,(\delta t_1 + \delta t_2 + \delta t_3)}{\delta t_2 \delta t_3\,(\delta t_1 + \delta t_2)}$$

$$C_1 = \frac{(\delta t_1 + \delta t_2)(\delta t_1 + \delta t_2 + \delta t_3)}{\delta t_1 \delta t_2 (\delta t_1 + \delta t_2)}$$
$$C_0 = -(C_1 + C_2 + C_3),$$

and $\delta t_1 = t^{n-i+1} - t^{n-1}$ is the time between two consecutive time steps. For a constant time step, the variable C_i takes the values $C_1 = -\frac{1}{3}$; $C_2 = \frac{3}{2}$; $C_1 = 3$; so $C_0 = \frac{11}{6}$.

22.1.3 Integral Formulation and Discretization in Space

Equation (22.1) is integrated for each discrete variable over its own control volume. This integral formulation is based on the divergence theorem

$$\int_V \nabla \cdot \mathbf{v} dV = \int_{\partial\pm} \mathbf{v} \cdot \mathbf{n} dS, \tag{22.4}$$

where $\partial\pm$ is the boundary of the domain and $\mathbf{n}$ is the outer normal to the boundary. When applied to the convection and diffusion terms, this calculation requires the velocity and its gradient to be known on the faces of each control volume. When discretizing the integral on the right-hand side of Eq. (22.4), the term $\mathbf{v} \cdot \mathbf{n}$ is considered to be constant on the faces of each control volume. This limits the precision to second order in space. Some criteria must be fulfilled in order to reach this precision when integrating Eq. (22.2) with a nonuniform mesh; these are described below:

- Integration of the unsteady and source terms over a control volume V_p is written as

$$\int_{V_p} \phi dV = \bar{\phi}_p \overline{V}_p. \tag{22.5}$$

Second order is reached provided that $\bar{\phi}_p$ is the value of $\bar{\phi}$ at the center of the control volume. For scalar variables, $\bar{\phi}_p$ is $\bar{\phi}$ (see Fig. 22.2a). For the y and z components of the velocity, the calculation point P is not the center of the control volume (see Fig. 22.2a, b). The density, defined at the centers of the faces of the control volume, must be interpolated at that point (using a linear interpolation). For example, referring to Fig. 22.3a, the value of $\bar{\phi}_p$ considered in Eq. (22.5) can be written as

$$\bar{\phi}_p = \phi_p + \frac{1}{2}\left(1 - \frac{\delta_1^v}{2\,\delta_2^v}\right)(\phi_E - \phi_p). \tag{22.6}$$

In the case of a uniform mesh, we can easily check that $\phi_p = \bar{\phi}_p$.

- Taking into account Eq. (22.5), the discretization of the integration of the diffusive term can be written as

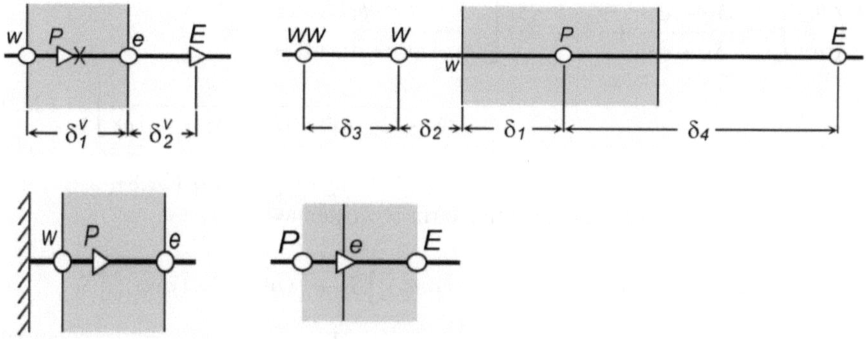

Fig. 22.3 Structure of the mesh and notation in the y direction, for the places where the variables are calculated: **a** for the component v of the velocity; **b** for any variable ϕ; **c** for the component v in the vicinity of the wall; **d** for the velocity–pressure coupling. The notation is similar for the other directions. The *gray zones* indicate the control volumes

$$\int_{V_P} \nabla \cdot (\Gamma\,\nabla\phi)\cdot dV = \left[\Gamma_n\left(\frac{\partial\phi}{\partial x}\right)_n - \Gamma_s\left(\frac{\partial\phi}{\partial x}\right)_s\right] S_x + \left[\Gamma_e\left(\frac{\partial\phi}{\partial y}\right)_e - \Gamma_w\left(\frac{\partial\phi}{\partial y}\right)_w\right] S_y + \left[\Gamma_t\left(\frac{\partial\phi}{\partial z}\right)_t - \Gamma_b\left(\frac{\partial\phi}{\partial z}\right)_b\right] S_z, \tag{22.7}$$

where S_x, S_y, and S_z are the surfaces of the control volume V_P that are perpendicular to the directions x, y, and z, respectively.

The diffusion coefficient Γ is constant for the momentum equation. For the energy equation, Γ is defined at the nodes of the scalar variables. Its value on each face of the control volume is the harmonic mean of the values taken at the nodes located on each side of the face under consideration. For example, the value of Γ_w on face w is written as

$$\Gamma_w = \left[\frac{\delta_2}{\delta_1+\delta_2}\frac{1}{\Gamma_W} + \frac{\delta_1}{\delta_1+\delta_2}\frac{1}{\Gamma_P}\right]^{-1}.$$

In Eq. (22.7), the partial derivative of a velocity component v, for example in the direction y on face e in Fig. 22.3, is written as

$$\left(\frac{\partial v}{\partial y}\right)_e = \frac{v_E - v_P}{2\delta_2^v}. \tag{22.8}$$

This scheme has second-order precision, since point e is the middle of the PE segment. In contrast, the scheme is not of order 2 for $\frac{\partial v}{\partial x}$ or $\frac{\partial v}{\partial z}$, or for the scalar variables in the three directions (see Fig. 22.2). In order to maintain the second-order precision of the scheme, such derivatives are calculated using the following interpolation, which is written as follows for the face w in Fig. 22.3 (for example):

$$\left(\frac{\partial \phi}{\partial y}\right)_w = \frac{\phi_E - \phi_P}{\delta_1 + \delta_2} + (\delta_1 - \delta_2)\left[\frac{\phi_W}{(\delta_1 + \delta_2 + \delta_4)(\delta_1 + \delta_2)} - \frac{\phi_P}{(\delta_1 + \delta_2)\,\delta_4} + \frac{\phi_E}{(\delta_1 + \delta_2 + \delta_4)\,\delta_4}\right]. \quad (22.9)$$

For a regularly spaced mesh ($\delta_1 = \delta_2$), Eq. (22.9) again gives Eq. (22.8).

The discretization of the convective term is written as

$$\int_{V_P} \nabla \cdot (\rho v \phi)\, dV = \left[(\rho u \phi)_n - (\rho u \phi)_s\right] S_x + \left[(\rho v \phi)_e - (\rho v \phi)_w\right] S_y + \left[(\rho w \phi)_t - (\rho w \phi)_b\right] S_z. \quad (22.10)$$

A hybrid scheme of order 2 (SHYBRID) is considered in order to calculate the convective terms (see [8]). The scheme uses a four-point formulation to interpolate the value of the variable ϕ on the faces of its control volume. It combines the QUICK scheme, the second-order UPWIND scheme, and a centered scheme (see [10]). In this reference, the respective weights in the formulation depend on the local value of the Péclet number (the ratio of the convective flux to the diffusive flux). In reference to Fig. 22.3b, the value of ϕ on face w is derived as follows:

$$\text{If } v_w \geq 0, \quad \text{then } \phi_w = (1 - \alpha)\,\phi_W + \alpha \phi_P - q^- \left(\phi_P - \left(1 - \beta^-\right)\phi_W + \beta^- \phi_{WW}\right) \quad (22.11)$$

$$\text{If } v_w \leq 0, \quad \text{then } \phi = (1 - \alpha)\,\phi_W + \alpha \phi_P - q^+ \left(\phi_P - \left(1 - \beta^+\right)\phi_W + \beta^+ \phi_{WW}\right), \quad (22.12)$$

where $\alpha = \frac{\delta_2}{\delta_1 + \delta_2}$, $\beta^+ = \frac{\delta_1 + \delta_2}{\delta_3}$, and $\beta^- = \frac{\delta_1 + \delta_2}{\delta_4}$.

Depending on the values of q^- and q^+, the derivatives in Eqs. (22.11) and (22.12) are equivalent to a centered scheme ($q^- = q^+ = 0$), or to a second-order UPWIND scheme ($q^- = \alpha$ and $q^+ = 1 - \alpha$), or to the QUICK scheme [$q^- = \alpha\delta_1/(\delta_1 + \delta_2 + \delta_3)$ and $q^+ = (1 - \alpha)\,\delta_2/(\delta_1 + \delta_2 + \delta_4)$]. The values of q^- and q^+ are automatically adjusted during the simulation depending on the Péclet number value Pe_w on face w in order to reduce the oscillatory potential of the scheme. This is done by minimizing the contributions of the distant nodes and maintaining positive coefficients between close neighbors. The result of this minimization is

$$Pe_w = 0 \Rightarrow q^- = q^+ = 0$$

$$Pe_w \neq 0 \Rightarrow \begin{cases} q^- = max\left(0,\, (1 - \alpha) - \frac{1}{|\mathrm{Pe}_w|}\right) \\ q^+ = max\left(0,\, \alpha - \frac{1}{|\mathrm{Pe}_w|}\right). \end{cases}$$

When the transport is mainly diffusive (Pe_w is very small), the SHYBRID scheme is quite centered. When the transport is convective (Pe_w is very large or infinite), the scheme tends to an UPWING scheme of order 2. For intermediate values of Pe_w, it can become a QUICK scheme for a certain value of Pe_w. As a result, the SHYBRID scheme is of order 2 for a large range of Péclet numbers (see [10]). In order to comply with the discretization scheme, the continuity equation is integrated over the control volume of the variable ϕ and then multiplied by $\bar{\phi}_P$ [see Eq. (22.6)] and subtracted from the discretized general transport equation [see Eq. (22.2)]. Quantities F (convective flux) and J (total flux) are introduced that are defined on the faces of the control volumes. For face e and the y direction, these fluxes can be written as

$$F_e = (\rho v \phi)_e$$
$$J_e = F_e - \Gamma_e \left(\frac{\partial \phi}{\partial y}\right)_e .$$

The transport equation for the discrete variable ϕ_P can therefore be written as

$$\begin{aligned} C_{lhs} \times V_P \bar{\phi}_P = & \left[\left(J_s - F_s \bar{\phi}_P\right) - \left(J_n - F_n \bar{\phi}_P\right)\right] S_x \\ & + \left[\left(J_w - F_w \bar{\phi}_P\right) - \left(J_e - F_e \bar{\phi}_P\right)\right] S_y \\ & + \left[\left(J_b - F_b \bar{\phi}_P\right) - \left(J_t - F_t \bar{\phi}_P\right)\right] S_z \\ & + \left(\bar{S}_P + C_{rhs}\right) V_P \end{aligned} \tag{22.13}$$

with

$$C_{lhs} = \left(-C_1 \bar{\rho}_P^{(1)} - C_2 \bar{\rho}_P^{(2)} - C_3 \bar{\rho}_P^{(3)}\right)$$

and

$$C_{rhs} = \left(-C_1 \bar{\rho}_P^{(1)} \bar{\phi}_P^{(1)} - C_2 \bar{\rho}_P^{(2)} \bar{\phi}_P^{(2)} - C_3 \bar{\rho}_P^{(3)} \bar{\phi}_P^{(3)}\right).$$

$\overline{\phi}_P$ represents the value of ϕ at the center of the control volume of the discrete variable ϕ. Equation (22.13) is written at time t^n [the exponent n in Eqs. (22.2) and (22.3) has been dropped and $\bar{\phi}_P^{(i)} = \bar{\phi}_P^{(n-i)}$].

Taking into account the discretization of the convection terms as given by Eq. (22.10) and the discretization of the diffusion terms [see Eqs. (22.8) and (22.9)], as well as Eq. (22.6) for $\bar{\phi}_P$, Eq. (22.13) can be written as follows:

$$\begin{aligned} a_P \phi_P = & a_S \phi_S + a_N \phi_N + a_W \phi_W + a_E \phi_E + a_B \phi_B + a_T \phi_T + a_{SS} \phi_{SS} \\ & + a_{NN} \phi_{NN} + a_{WW} \phi_{WW} + a_{EE} \phi_{EE} + a_{BB} \phi_{BB} + a_{TT} \phi_{TT} + b. \end{aligned} \tag{22.14}$$

Determining the discrete field of variable ϕ_P thus consists of solving a sparse 13-diagonal system. However, the nonimmediate neighbors ($\phi_{WW}, \phi_{EE}, \ldots$) of Eq. (22.14) from the discretization of order 2 make a small contribution, so Eq. (22.14) can be written as

$$a_P \phi_P = a_S \phi_S + a_N \phi_N + a_W \phi_W + a_E \phi_E + a_B \phi_B + a_T \phi_T + b. \tag{22.15}$$

22.1.4 Boundary Conditions

Various types of boundary conditions can be implemented in the numerical method:

- Fixed values of the variables (Dirichlet type)
- Zero boundary flux (von Neumann type) on either face
- Periodic or symmetric ones in the x and y directions, etc.

The spatial precision is maintained. In the case of von Neumann type conditions, the normal gradient [expressed via Eq. (22.9)] is staggered towards the interior of the domain. In the case of a velocity component (see Fig. 22.3), and for the nearest control volume from the wall, a second-order centered scheme is used to compute the convected flux orthogonal to the wall: $v_w = \frac{v_P}{2}$. Boundary conditions are thus treated as in [10].

22.2 Pressure–Velocity Coupling

Two versions of the Semi-Implicit Method for Pressure-Linked Equations (SIMPLE) algorithm developed by Patankar and Spalding in 1972 [11] have been implemented for the velocity–pressure coupling [7]. The basic principle of these algorithms is to transform the continuity equation into an equation for pressure, and to determine (via an iterative process) the pressure field for which the associated velocity field satisfies the continuity equation. We present the main features of the algorithm here, taking the calculation of the component v of the pressure field as an example; the equations for the other components are formally identical. The v component is obtained by discretizing the integrated y-direction momentum equation on the control volume of Fig. 22.3b to give

$$a_a v_e = \sum a_{nb} v_{nb} + A_e (p_P - p_E) + b_e^{'}, \tag{22.16}$$

where the coefficients a_e, a_{nb}, A_e, and $b_e^{'}$ are calculated together with the solution at the iteration $(k-1)$. The calculated velocity field u, v, w does not satisfy the continuity equation at this stage since the pressure field is not known at iteration k. A temporary velocity field $\hat{v}$ is defined as

$$\hat{v} = \frac{\sum a_{nb} v_{nb} + b_e^{'}}{a_e}. \tag{22.17}$$

Equation (22.16) can be written as

$$v_e = \hat{v}_e + \frac{A_e}{a_e} (p_P - p_E). \tag{22.18}$$

In the discretized integral form of the continuity equation over the control volume for density, Eq. (22.18) and its counterparts for u and w are reported to give the Poisson equation

$$d_P^p p_P = d_S^p p_S + d_N^p p_N + d_E^p p_E + d_B^p p_B + d_T^p p_T + b^n, \tag{22.19}$$

where

$$\begin{aligned} b^n = {} & \left[(\rho\hat{u})_s - (\rho\hat{u})_n\right] S_x + \left[(\rho\hat{v})_w - (\rho\hat{v})_e\right] S_y + \left[(\rho\hat{w})_b - (\rho\hat{w})_t\right] S_z \\ & - \left(C_0 \rho_P + C_1 \rho_P^{(1)} + C_2 \rho_P^{(2)} + C_3 \rho_P^{(3)}\right) V_P. \end{aligned} \tag{22.20}$$

This equation gives the pressure field associated with the velocity field u, v, w [calculated from Eq. (22.18)] that satisfies the continuity equation.

In the SIMPLER (where added R stands for "Revised") algorithm (for more details, see [10]), a correction p' to the pressure field is introduced and the algorithm can be described by the following sequence until convergence:

1. Compute $\hat{v}$ from the velocity at the previous time step using Eq. (22.17)
2. Predict a pressure field by solving Eq. (22.19)
3. Use this predicted pressure field to compute a velocity field v^* by explicitly solving Eq. (22.16)
4. Compute the correction p' to the pressure field using Eq. (22.19), where the velocity v^* replaces $\hat{v}$ via Eq. (22.2)
5. Correct the velocity field using Eq. (22.18), where v^* replaces $\hat{v}$ and p' replaces p.

The Pressure Implicit with Splitting of Operators (PISO) algorithm proposed by Issa [6] is another possibility for performing the velocity–pressure calculations. The sequence for this is as follows:

1. Use the pressure field at the previous time step to predict the velocity field v by implicitly solving Eq. (22.16)
2. Compute $\hat{v}$ using Eq. (22.17) and predict the pressure field p from Eq. (22.19)
3. Correct the velocity field using Eq. (22.18).

The first step is computed only once, whereas steps 2 onwards (updating the pressure) are repeated until convergence. In most cases, the convergence of the dynamic field occurs rather quickly compared to that of the van der Waals equation, which limits the overall convergence process.

22.3 Specific Aspects of the Numerical Method

22.3.1 Calculating the Density and the Thermodynamic Pressure

As the fluid is confined in all of the cases studied here, the thermodynamic pressure Π is calculated at any time using the conservation of the total mass, as expressed by

$$\int_D \rho(t)\, dV = \int_D \rho(t=0)\, dV = M_0.$$

Determining Π requires knowledge of the equation of state; e.g., $\rho = \rho(\Pi, T)$. The use of a linearized form of the equation of state speeds up the computation.

22.3.2 Overall Iterative Process

Figure 22.4 shows the two overall algorithms for the PISO and SIMPLER methods that are used to compute the velocity pressure. The global iterative process couples the energy, momentum, and state equations by the Newton method under the constraints of the local and global mass conservation statements. Each time step k begins with the calculation of the temperature T^k, which is achieved by solving the discretized energy equation. After that, the thermodynamic pressure field Π is updated via the mass conservation equation. The latter is used to compute the density field ρ^k using the linearization process $\rho^k = f\left(\rho^{k-1}, T^k, \Pi^k\right)$. The dynamic field is then determined via the PISO or SIMPLER algorithm. In the case of the PISO algorithm, there is only one predictor step for the velocity, whereas the corrector steps for the pressure and velocity field are repeated until convergence.

22.3.3 Preconditioning of the Poisson Equation

Linear systems such as (22.15) are solved by iterative methods, and the convergence speed of these methods is very sensitive to the dominant character of the diagonal terms. The diagonal term is said to be dominant if it is larger than the sum of the corresponding line and column. Dominant character is ensured by the unsteady terms and by boundary conditions of the Dirichlet type. It can be strengthened by the under-relaxation method described in [10]. This method basically consists of expressing the solution at iteration step k in the form of a linear combination of the solutions at iteration steps k and $k-1$, with the under-relaxation coefficient being that of iteration k.

In contrast, no boundary conditions are imposed on the Poisson equation (22.19) for pressure. The dynamic pressure p is defined within a constant (it only participates

Fig. 22.4 Overall algorithms for PISO (**a**) and SIMPLER (**b**)

through its gradient in the momentum equation). The coefficient a_P^n in Eq. (22.19) is the sum of the coefficients of its immediate neighbors. The linear system corresponding to the equation for pressure has a zero determinant. In addition, relaxation techniques cannot be used, as the velocity field calculation during the course of such a relaxation process would not satisfy the continuity equation. In [10], Patankar suggests fixing the value of the pressure at one point in the domain, but in the case of highly expandable fluids, this procedure leads to strong perturbations of the dynamic field at the location where the pressure is fixed. Another technique that can be used is the artificial compressibility technique proposed by Chorin [5].

Further details and references to the numerical methods applied to the hydrodynamics of supercritical fluids are given in [1–4, 9].

References

1. Accary G, Raspo I (2006) 3-d finite volume methods for the prediction of supercritical fluid buoyant flows in a differentially heated cavity. Comput Fluids 35:1316–1331
2. Accary G, Raspo I, Bontoux P, Zappoli B (2005) Reverse transition to hydrodynamic stability through the Schwarzschild line in a supercritical fluid layer. Phys Rev E 72(3):035301
3. Amiroudine S (1995) Modelisation Numerique des Phenomenes de Transport de Chaleur et de Masse dans les Fluides Supercritiques. PhD thesis, IRPHE (Institut de Recherche sur le Phenomènes Hors-Equilibre), Marseille
4. Amiroudine S, Ouazzani J, Carlès P, Zappoli B (1997) Numerical solution of 1-d unsteady near-critical fluid flows using finite volume methods. Eur J Mech B 16(5):665–680
5. Chorin AJ (1967) A numerical method for solving the incompressible and low speed compressible equations. J Comput Phys 72:277
6. Issa RI (1982) Solutions of the implicit discretized fluid flow equations by operator splitting (Mech. Eng. Rep. FS/82/15). Technical report, Imperial College, London
7. Jang DS, Jetii R, Acharya S (1986) Comparison of the piso, simpler, and simplec algorithms for the treatment of pressure velocity coupling in steady flow problems. Numer Heat Transf 10:209
8. Li Y, Rudman M (1995) Assessment of higher order upwind schemes incorporating FCT for convection-dominated problems. Numer Heat Transf B 27:1
9. Ouazzani J, Garrabos Y (2013) A novel numerical approach for low mach number application to supercritical fluids. In: Proceedings of ASME 2013 summer heat transfer conference : HT2013, 14-19 July 2013, Minneapolis, Minnesota, USA. edited by the American Society of Mechanical Engineers. New York (N. Y.): ASME, 2013. p HT2013-17732 (19 p)
10. Patankar SV (1980) Numerical heat transfers and fluid flows. Hemisphere, Washington
11. Patankar SV, Spalding DP (1972) A calculation procedure for heat, mass and momentum transfer in three-dimensional parabolic flows. Int J Heat Mass Transf 15:1787–1806 (Une methode de calcul du transfert de chaleur, de masse et de quantite de mouvement dans les ecoulements paraboliques tridimensionnels.)

Chapter 23
The Hardware

Abstract A brief presentation of the hardware of two main facilities (ALICE in MIR space station and CPF onboard the Space Shuttle) dedicated to the experiments analyzed in the book. The availability of long-duration experimental runs (typically greater than one week long) and the high quality of the microgravity environment (residual accelerations typically in the range 10^{-4} to $10^{-3}\, g_0$) have allowed fast temperature equilibration and slow density relaxation processes to be investigated in detail using these facilities. In addition, the principle and the limitations of a H_2 magnetic levitation setup used to create microgravity conditions on Earth are also presented. Although it does not compensate for gravity in a uniform manner, this setup provides much longer experimental times and is very useful for calibrating instrumentation and performing exploratory research, in particular to investigate the effects of vibrations.

23.1 The Facilities Used in Microgravity Conditions

The ALICE facility (Analyse des LIquides Critiques dans l'Espace) [19, 22] developed by the French Space Agency (Centre National d'Etudes Spatiales: CNES) and the CPF (the Critical Point Facility) [10] developed by the European Space Agency (ESA) were used from 1992 to 2000 to investigate heat and mass transfer and the piston-effect mechanism under microgravity conditions. ALICE was flown on the Russian MIR space station during seven CNES commercial or cooperative missions (Antares 1992; Altaïr 1993; Cassiopeia 1996; German-MIR'97 1998; Pegasus 1998; and Perseus and post Perseus 1999–2000, as well as NASA-GMSF 1999). CPF was used onboard the US Space Shuttle (during the IML1-1992 and IML2-1994 missions, where IML stands for International Microgravity Laboratory). The designs of these facilities are presented briefly in the next section. The availability of long-duration experimental runs (typically greater than one week long) and the high quality of the microgravity environment (residual accelerations typically

B. Zappoli et al., *Heat Transfers and Related Effects in Supercritical Fluids*, Fluid Mechanics and Its Applications 108, DOI: 10.1007/978-94-017-9187-8_23,

in the range 10^{-4} to 10^{-3} g_0, where $g_0 = 9.81\,\mathrm{m\,s^{-2}}$ is the Earth's gravitational acceleration constant) have allowed fast temperature equilibration and slow density relaxation processes to be investigated in detail using these facilities.

During the same decade, several experiments were also performed onboard Space Shuttles and the EURECA platform using other facilities (not presented here) in order to investigate specific critical behaviors. These included the HPT-HYDRA (high-precision thermostat using a critical SF_6 calorimeter) experiment during the German Spacelab missions (D-1 in 1985 and D-2 in 1993); the HPT critical adsorption experiment on the EURECA platform in 1992–1993; the ZENO (critical fluid light scattering in xenon) experiment during the USMP-2 (1994) and USMP-3 (1996) missions (USMP: United States Microgravity Platform); and the critical viscosity of xenon (CVX) experiment during STS-85 (1997) and STS-105 (2003). More details can be found using the references given in the dedicated reviews of Beysens and Garrabos [5] and Barmatz et al. [2].

Not presented here are other experimental modules that did not differ significantly from ALICE or CPF in their main thermal and optical characteristics. These other modules were developed in the same timeframe and shipped onboard the TEXUS (Technologische Experimente unter Schwerelosigkeit) sounding rocket in an ESA-funded program. The sounding rockets provided free-fall times of 360–420 s, and the residual accelerations were lower than 10^{-4} g_0. Such conditions were appropriate to investigate the initial stages of processes with fast dynamics (such as liquid–vapor phase separation), but not long relaxation processes. The few seconds of microgravity that can be provided by Earth-based testing facilities, such as drop tubes, drop towers, or aircraft parabolic flights, is not nearly long enough to achieve the necessary initial mechanical equilibrium. A detailed and complete review of the important results obtained from these space experiments is given by Barmatz et al. in [2].

23.1.1 The ALICE and CPF Instruments

The ALICE (see Fig. 23.1a) and CPF (see Fig. 23.1b) instruments were designed and developed to perform scientific investigations on the thermal behavior of transparent fluids such as CO_2 and SF_6 near their critical points. At the core of either ALICE or CPF is a chamber where an exchangeable thermostat unit (ThU), as shown in Fig. 23.1a, can be inserted. To illustrate a sounding rocket investigation, where the thermostat has a comparable size and design to a ThU, Fig. 23.1c displays a picture of the TEM 06–10 module (TEXUS program).

ALICE and CPF provide very precisely defined and stable thermal and mechanical conditions for a long period of time. The facilities are also equipped with optical and thermal diagnostic instrumentation and several specific stimuli. These facilities are designed to run experiments automatically according to a predefined timeline. Moreover, CPF has communication capabilities, making it possible to modify the initial timeline from the ground and to monitor the experiments in real time.

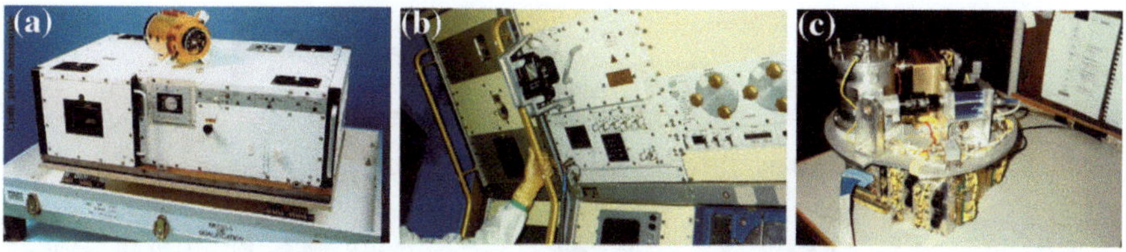

Fig. 23.1 Three main pieces of space hardware: **a** Analyse des LIquides Critiques dans l'Espace (ALICE), as developed by CNES and flown on the Russian MIR space station; **b** Critical Point Facility (CPF), as developed by ESA for the International Mission Laboratory 1 and 2 onboard the US Space Shuttles Discovery (1992) and Columbia (1984); **c** high-precision thermostat that was used for investigations of critical-point phenomena, and was accommodated on the TEM 06–10 module (with a video observation setup) for flight on the TEXUS 25 sounding rocket (ESA)

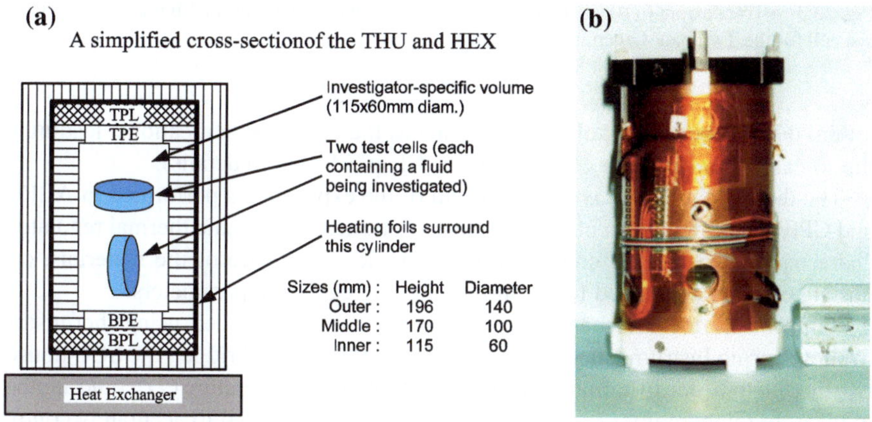

Fig. 23.2 **a** Schematic cross-section of the thermostat unit (ThU) and of a Peltier-driven heat exchanger (HEX) [11]; **b** picture of a sample cell unit (SCU) located in the investigator-specific volume within the ThU

23.1.2 Thermal Control Subsystems and Related Measurements

ALICE and CPF are able to accommodate experiments through the use of exchangeable ThU units. The latter provide high-quality temperature control of a dedicated sample cell unit (SCU). Figure 23.2a shows a schematic cross-section of the ThU, and Fig. 23.2b is a picture of an SCU. More detailed descriptions can be found in [11, 28].

The ThU is placed in thermal contact with a cold plate in order to control the rate of heat exchange with the external environment. This contact plays an essential role in SCU heating or cooling. The cold plate included a Peltier heat exchanger (HEX) that controls the input/output heat from ambient to ThU. Each SCU hosts two test cells. These cells have different optical observation modes: (i) for direct transmission observations, and (ii) for interferometric observations (Fig. 23.3). Each test cell hosts

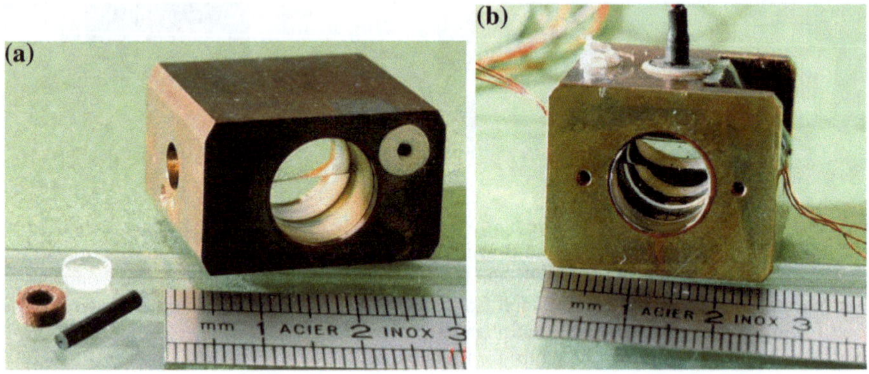

Fig. 23.3 **a** Transparent (TR) test cell used in transmission observations, and **b** interferometry (IF) test cell for the Twyman–Green interferometer

a thin, disk-shaped fluid volume that matches the optical specifications, leading to the so-called *pancake-like* cell design (see Chap. 8 for more details).

The thermostat design is similar for all of the experiments performed in ALICE and CPF. All of the different thermostats have basically the same thermal regulation characteristics. Small differences in the SCU design (nature of the materials, cell design, etc.) are accounted for by modifying the regulation parameters.

Temperature measurements for thermal control are obtained using thermistors (Yellow Spring Instruments, YSI) mounted in a Wheatstone bridge. The SCU houses two thermistors, one to monitor the thermostat setpoint temperature (T_{set}) and another to measure temperature (T_{mes}). A fine operating mode is used to accurately control both the SCU temperature and temperature gradients. This regulation mode authorizes heat power dissipation within the SCU due to sensors, optical diagnoses, stimuli, etc. Practically speaking, the typical maximum value (q_{SCU}) for the dissipated heat power inside a SCU made with electrolytic copper or pure aluminum of high thermal efficiency is on the order of 5 mW.

This fine regulation mode provides the SCU with excellent temperature stability, better than 10 μK/day, with spatial temperature gradients of $<0.5\,\mu\mathrm{K\,mm}^{-1}$. This high-level accuracy for the controlled thermal stability enables T_c to be determined within 25 μK. The set temperature range of ALICE or CPF allows CO_2 and SF_6 to be studied across a temperature range that is typically larger than four decades ($100\,\mu\mathrm{K} \leq T - T_c < 20\,\mathrm{K}$). The SCU temperature can be changed to either a thermal quench (*quenching mode*) or a thermal ramp (*ramping mode*). A thermal quench is a very fast and well-defined temperature step down (or up). The quench time duration corresponds to the time needed to remove or inject the heat corresponding to the temperature change from or into the SCU. In the ramping mode, T_{set} is continuously changed in very small steps so as to make the SCU temperature change linearly over time.

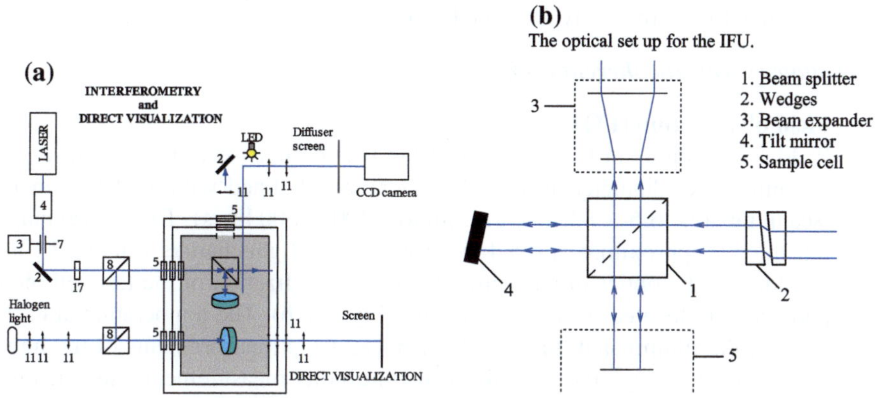

Fig. 23.4 The two main observation channels of the CPF instrument or ALICE (schematics from Ref. [11]): **a** optical setup for transmission observations of the transparent (TR) cell; **b** optical setup for the Twyman–Green interferometer (IF) cell

The minimum time response of the thermal control subsystem is on the order of 2 s, which corresponds to the T_{set} thermistor time response. Considering the SCU and ThU designs and the heat power needed to control the SCU setpoint temperature in the fine regulation mode, a typical value of $\approx$3–10 s can be inferred for the time response of the whole thermostat. These time responses can be correlated with the typical values of the power q_{ThU} dissipated in the whole ThU (20 mW $\leq q_{\mathrm{ThU}} \leq$ 5 W). It follows that the upper order of magnitude for the temperature variations of the SCU in the quenching and ramping modes is $\sim 10\,\mathrm{mK\,s^{-1}}$.

23.1.3 Optical Diagnosis Subsystems and Related Measurements in Space Experiments

The ALICE and CPF facilities have similar optical diagnostic systems. As illustrated in Fig. 23.4a, b, two sources of illumination are used: incoherent light from light emitting diodes (LEDs) and coherent light from a 1 mW He-Ne laser beam. The optical setup is sketched for direct observations (in transmission, TR) and for the Twyman–Green interferometer (IF). In the latter case, the cell is located in one arm of the interferometer. Note that the light transmission and interferometry diagnoses give measurements integrated across the cell thickness. The combination of different optical components yields four diagnoses when working with incoherent light (numbered from 1 to 4 in Fig. 23.4a) and three diagnoses when working with coherent light (numbered from 5 to 7 in Fig. 23.4a).

These diagnoses are briefly described below:

- ***Diagnoses with Incoherent Light***

  1. **Field observation (FO)**
     Field observations in transmission of the whole cross-section of the test cell (12 mm object diameter at the cell center) are obtained with a CCD camera (scanning area: $6.6 \times 8.8\,\text{mm}^2$, resolution: 400×600 lines). The optical magnification is approximately 0.5. FO corresponds to light channel 1 in Fig. 23.4a. In addition, the optics of CPF provide direct visualization of the test cell on a photocamera located outside the facility (Fig. 23.1b). The temperature at each point in the illuminated fluid can be deduced from measurements of the light transmission through the cell (also called turbidity measurements; see the discussion of diagnoses with coherent light below). The average temperature and local temperature gradients can also be computed independently.
  2. **Microscopy observation (MO)**
     Microscopy observations in transmission of the same test cell are able to magnify a 1 mm field object to a 6.6 mm field image onto the same CCD video camera (magnification 6.6). MO corresponds to light channel number 1 in Fig. 23.4a.
  3. **Grid observation (GO)**
     The grid observation or grid shadow technique consists of inserting a grid lattice into the optical path of either full-field or microscopic observations. The distortion of the shadow of the grid allows refractive index gradients and then density gradients to be determined, as detailed in Ref. [16]. Due to its noninvasive nature, this technique is mainly used to measure local density variations in the TR cell.

- ***Diagnoses with Coherent Light***

  The laser beam is split into a narrow beam (≈ 0.5 mm in diameter) that is used for transmission and diffusion diagnoses and a wide beam (12 mm in diameter) for interferometry. The noninvasive interferometry technique is used, just as it is for the grid shadow device, to measure the local density behavior.

  1. **Light transmission (LT) and light scattering (LS)** Direct light transmission and light scattering signals collected on fiber-optic guides are complementary to the previously described incoherent diagnoses of the TR cell [14, 15]. The phase-transition temperature is easily determined using these transmission and diffusion diagnoses. In particular, a program for automatically determining T_c based on turbidity measurements of the light transmission is available for the ALICE facility (see, for example, Ref. [20] and references therein).
  2. **Interferometry (IF)** When the Twyman–Green [17, 27] interferometer is used, the widened laser beam illuminates the interferometry (IF) cell, which facilitates the recording of an image of the fluid on a CCD video camera. Figure 23.4b displays a sketch of the optical setup for the interferometric observation channel.

The IF cell is placed on one arm of the Twyman–Green interferometer located in the SCU. The interference pattern provides quantitative measurements of the fluid's refractive index and thus its density. At each equilibrium temperature, the reference interferogram of the sample cell exhibits parallel and equidistant straight fringes. The number and the direction of the fringes (using an arbitrary axis as reference) can be controlled by adjusting the tilt of one of the mirrors of the interferometer.

Changes in the fluid density along the beam path through the fluid correspond to averaged variations in the refractive index, denoted δn, which consequently change the optical path difference in the interference pattern. The corresponding variation in the interference order δf is related to the laser light wavelength λ, the path length of the light in the sample ($L = 2e$, where e is the fluid thickness), and the refractive index variation δn, through

$$\delta f = \frac{2e}{\lambda}\delta n. \tag{23.1}$$

Density changes in the fluid thus result in a shift in or distortion of the fringe pattern. In order to determine the connection density or interference order, an accurate relation between n and ρ is required. It is generally assumed that this relation is given by the Lorentz–Lorenz relationship

$$\left(\frac{n^2 - 1}{n^2 + 1}\right)\frac{1}{\rho} = Q_{LL}. \tag{23.2}$$

Here, Q_{LL} is the Lorentz–Lorenz constant, which only depends on the fluid system. The linear expansion of the above equation around the critical density leads to

$$n - n_c = C\left(\frac{\rho - \rho_c}{\rho_c}\right), \tag{23.3}$$

where n_c is the critical refractive index and

$$C = \frac{\left(n_c^2 - 1\right)\left(n_c^2 + 2\right)}{6n_c}. \tag{23.4}$$

For the density range $0.95\rho_c < \rho < 1.05\rho_c$, the precision of the linear approximation of the Lorentz–Lorenz expression is better than 0.1 %. In this approximation, the relation between density and refractive index around the liquid–gas critical point is completely determined by the value of C; i.e., by the literature values of n_c and ρ_c (see Table 1.6). Typically, $C \sim 0.1$ for the one-component fluids of interest here.

Combining of Eqs. (23.1) and (23.3) shows that there is a simple linear relation between changes in the interference order of the interferogram and the local density difference $\delta\rho = \rho - \rho_c$ in the critical region, such that

$$\delta\rho \approx \rho_c \frac{\lambda}{2eC}\delta f \approx \frac{\rho_c}{n_c - 1}\frac{\lambda}{2e}\delta f. \tag{23.5}$$

In the above equation, we have used the value $n_c \approx 1.1$ as a typical critical value for the refractive index (see Table 1.6) of a SF_6 or CO_2 sample filled to critical density. Taking $e = 1$ mm (cell thickness) and $\lambda = 633$ nm (laser wavelength), the corresponding deviation in the relative density is

$$\frac{\delta\rho}{\rho_c} \simeq 3.2 \times 10^{-3}\,\delta f. \tag{23.6}$$

The fringe shift δf is the number of fringes that cross a reference line parallel to the initial fringe orientation. When the tilt of the mirror is adjusted to produce 1 or 2 fringes per mm across the whole diameter (~12 mm) of the test cell, the resulting complete interferogram corresponds to a good compromise between the resolution of the fringe distortions and the position of the fringe displacement for a convenient range of density gradient values. When the resulting interferogram assumes the maximum size on a typical picture of 500 × 500 pixels, a resolution of at least 24 μm/pixel is reached. The noise in this optical measurement is independent of temperature and corresponds to approximately $\frac{1}{20}$ of the fringe. Thus, a dynamic range for the density variation data of two or three orders of magnitude is easily obtained, except very close to T_c, where the interferograms are affected by turbidity, especially for large sample cell thicknesses ($e \geq 5$ mm). In practice, for $e = 1$ mm, the currently covered range for density variation measurements is $0.06 \gtrsim \frac{\delta\rho}{\rho_c} \gtrsim 1.5 \times 10^{-4}$ in the temperature range $T - T_c \gtrsim 1$ mK, while for $e \simeq 5$ mm, these density variations can only be measured for $T - T_c \gtrsim 10$ mK. In addition, the sign of the density variations can be inferred from the direction of the pattern shift.

23.1.4 Timeline Programs, Dedicated Equipment and Measurements

The ALICE and CPF facilities run automatically according to a predefined timeline program called the experiment parameter table (EPT). Each EPT consists of a header followed by a dedicated number of sequential steps called "action points". The header is a standard table that contains specific data such as the thermostat and experiment identification numbers, the predetermined value of the critical temperature of the test cell, the maximum duration of the experiment, the calibration parameters for regulation and thermal exchange control, the checksum of the thermal cycle, etc.

Each action point defines up to four commands and the time at which they are executed. These commands control the stimuli and diagnoses, and one command is always reserved for temperature control purposes. Therefore, the action points are primarily used to monitor the temperature difference between the SCU setpoint temperature and the critical temperature of the fluid. The action points are also used to monitor the appropriate optical diagnostics.

In addition to these basic sets of thermal and optical stimuli and diagnoses, the ALICE and CPF contain a number of other subsystems that can operate dedicated equipment on request for specific action point commands on the EPT. This dedicated equipment is briefly presented in the following.

23.1.4.1 The Three Optional Thermistors

One branch of the extra Wheatstone bridges implemented on ALICE or CPF can be equilibrated using a thermistor of the same type as the SCU's YSI thermistor located in the fluid sample. Up to three thermistors can be individually switched (i) to the high-sensitivity mode, where the power dissipation is larger than 10 μW; (ii) to the low-sensitivity mode, where the power dissipation is lower than 1 μW; or (iii) off. The exact power dissipation of a switched-on thermistor depends on its resistance at a given T_{set} value. In the following comparisons of different local heating stimuli of the fluid sample, we adopt $q_{Thermistor} \approx 1\,\mu W$ as the typical order of magnitude of the power dissipation of any thermistor in any measurement mode. As justified below from our discussion of the bottleneck effect (see Chap. 8) close to the critical point, the surface temperature of the thermistor should follow the temperature of the fluid more and more closely as T approaches T_c. Therefore, fluid temperature measurements obtained using theseimmersed thermistors provide *local* temperature data. Typically, within the critical fluid sample, a resolution of a few μK was achieved, and temperature evolution rates as low as $1\,\mu K\,s^{-1}$ were measured using these optional thermistors (see, for example, the fluid temperature evolution during a heat pulse reported in Sect. 7.1.3).

Moreover, reducing the size of the thermistor bead should also increase the spatiotemporal resolution and the *local* nature of the temperature measurements. This is a significant advantage over interferometry, shadowgraphy or turbidity techniques, which integrate the density variations along the optical path and are increasingly affected by scattered light as T approaches T_c.

23.1.4.2 The Two Optional Pressure Sensors

The pressure range 0–10 MPa full scale results in measured voltages in the range 0–250 mV full scale. The slow rate of measurement corresponds to 1 sample s^{-1}, with a resolution of 70 μV between 0 and 250 mV full scale and a resolution of 24 μV between 0 and 100 mV full scale. This is achieved with a 12-bit converter.

The fast rate of measurement corresponds to 150 sample s^{-1}. In this fast mode, the resolution of the measurements remains at 24 μV, but the measurement range is reduced to 6.25 mV full scale by transmitting only the last 8 bits (see, for example, the pressure sensor measurements performed by Fröhlich and coworkers [13]).

23.1.4.3 The Heating (Thermistor) Source Subsystem

One of the additional thermistors located inside the fluid sample can be switched to the *pulse heating mode*. In this mode, the thermistor is used in a self-heating regime and delivers a calibrated power to the fluid. In the last updated version of this equipment, the power delivered to the sample could be varied between 0 and 1 W, in steps of $\frac{1}{64}$ mW. The maximum power value is in practice limited by the power dissipation of the thermistor, which is typically a few 10 mW, depending on the diameter of the thermistor bead. As an example, the typical power delivered by a thermistor bead about 0.254 mm in diameter (Thermometrics B10, time constant of 10 ms in steady water) is in the range of 0.015 mW–10 mW. The heat flux can then reach $5 \times 10^4\,\mathrm{W\,m^{-2}}$ within 640 steps of $77\,\mathrm{W\,m^{-2}}$. Recall that the order of magnitude of the maximum power monitored by the fine regulation mode of the thermal control subsystem is $q_{\mathrm{SCU}} \approx 5\,\mathrm{mW}$. Accordingly, monitoring pulse heating with a thermistor bead delivering up to 10 mW must be done with care in order to avoid significant disturbances to the SCU temperature.

23.1.4.4 The Electric (Current or Voltage) Source Subsystems

The current source subsystem (CSS) supplies a well-controlled amount of power to the sample by delivering a calibrated current to a resistor (resistance heater) inside the test cell. The constant current ranges from 0 to 1 A, in steps of 20 μA.

The typical useful power delivered to the fluid by this stimulus (*power heating mode*) is in the range of 0.02 mW to 2 mW, and the energy pulse duration can last up to few minutes. This subsystem was used by Michels and coworkers [11, 12, 23, 29] to operate internal heating pulses from a gold plate heater deposited on a quartz substrate (details are given in Sect. 7.1.3). In this experiment, the power delivered by the plate heater was varied between 0.04 mW and 0.18 mW, while the pulse duration was varied between 60 s to 300 s. Due to the large plate heater area ($210\,\mathrm{mm}^2$), the total heat flux ranged between approximately $0.2\,\mathrm{W\,m^{-2}}$ and $0.9\,\mathrm{W\,m^{-2}}$. However, as noted in [11], this power is split between the fluid and the quartz substrate. Due to the decreasing thermal impedance of the fluid, the energy dissipated in the substrate decreases as the critical temperature is neared.

This CSS was also used by Wilkinson and coworkers [33] to create density inhomogeneities within their sample in order to study density relaxation. In the specific configuration of their test cell, the electrical power is applied to an external foil heater cemented onto a part of the cell. This results in asymmetric heating of the fluid sample. The pulse energy was varied between 0.1 J and 2.7 J, while its duration was varied between 5 s and 2346 s. Their test cell is described in depth in Sect. 7.1.2.

A voltage source subsystem (VCS) delivers a calibrated voltage of up to 500 VDC to the SCU in order to maintain an electric field in the sample fluid. This subsystem was used by Zimmerli and coworkers [35, 36] to create a local density variation in acritical fluid that was kept isothermal. An electrostriction effect resulted from the electrostrictive pressure of the electric field. This isothermal process in an ideally dielectric fluid is not strictly within the scope of heat-transfer studies. Turning on the electric field resulted in an electrocaloric effect, generating a simultaneous adiabatic density change in the cell. The density inhomogeneities then relax while the fluid returns to its initial temperature, which requires an increasingly long time as T_c is approached. This thermal process, which is responsible for small adiabatic density changes, affects the measurement of isothermal density changes upon applying the electric field.

23.2 Heating Conditions and Stimuli

In order to complement the above presentation of test cells that exhibit large differences in design and materials, particular attention is now given to the heating (or cooling) conditions. When external heating/cooling is produced at the external boundary of the fluid volume, the fluid sample acts as a heat accumulator. When heating/cooling is produced by an internal heater with the external boundary temperature maintained constant, the fluid sample acts as a heat exchanger. The following section presents these two main ways of operating and discusses additional aspects that must be considered regarding control over temperature changes (thermostat monitoring, specific heater devices, local temperature measurements, etc.).

23.2.1 External Heating (EH)

External heating (EH) is a thermal process that corresponds to the quenching mode (a very fast temperature change, see Fig. 23.5) of ThU operation in the ALICE or CPF facilities. The temperature of the fluid is changed by imposing a temperature variation $\Delta T(t)$ on the whole outer surface A_F of the fluid volume. This transient heating corresponds to the 1-D model of adiabatic cell heating (or cooling) analyzed in Chap. 4. However, the process of changing the temperature of the cell inner wall in contact with the fluid requires a finite time Δt_W that depends on the thermal behavior of the cell body, the SCU, the ThU, and their interrelated transient responses. The minimum value of Δt_W depends on the largest available instantaneous heat flux. For a given amplitude of the temperature step, the quenching mode is then the fastest mode to change the inner cell wall temperature. When the duration of the temperature step is shorter than the characteristic time of the piston effect, the quenching mode can explore the adiabatic temperature equilibration of the fluid. Considering the expected critical speeding up of the piston effect when the temperature goes to the critical temperature of the fluid, it becomes difficult to fulfill this condition in a cell that is centimeters in size, as the minimum Δt_W values remain on the order of 10–30 s, as discussed below.

The ramping mode corresponds to the application of a time-independent power across the whole area A_F, which leads to a linear temperature rise $\Delta T \simeq kt$. Here, k

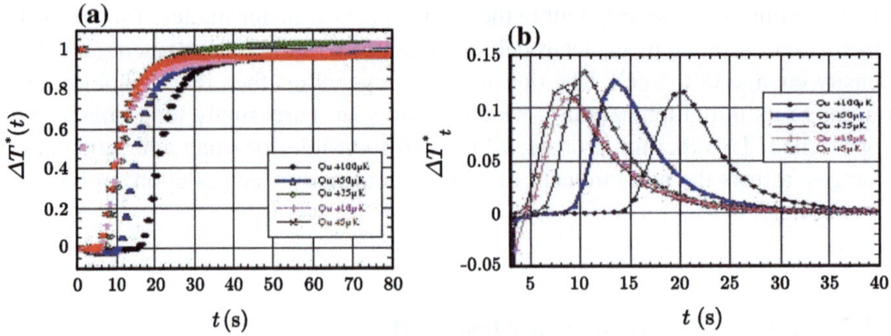

Fig. 23.5 **a** Normalized SCU temperature–time profile during an external temperature heating mode (EH). **b** Time derivative of the temperature evolution: this gives the order of magnitude of the duration of the heat flux. The diffusive time response of the thermostat is around ~10–20 s, while SCU temperature equilibration is achieved before ~100 s for the maximum temperature change of 100 mK

must be kept constant, which is a difficult condition to achieve when only considering the conduction mechanisms. It is the appropriate heating mode in the case of an adiabatic cell (see Sect. 7.1.4).

During EH, the SCU temperature is monitored using a thermostat whose power varies in the range 5 mW $\lesssim q_{ThU} \lesssim$ 20 W, with a characteristic time ranging from 2 s (YSI response time) to 100 s (ThU response time × 10). A typical value for the heat capacity of the SCU is about 10^3 J K^{-1} (i.e., 2.5 kg of copper or 1.0 kg of aluminum) if the thermal diffusivity of the SCU material is about 10^{-5} m^2 s^{-1}. The typical length of the SCU is on the order of 50 mm, which is generally ten times larger than the characteristic length of the sample fluid. A simple 1-D model of the heat conduction process in such a SCU indicates a temperature change of between 20 mK s^{-1} (fastest quenching mode) and 5 μK s^{-1} (slowest ramping mode).

SCU temperature profiles, as measured during quenches from 100 mK to 5 mK (see Fig. 23.5a), confirm the above estimations. The effective quench durations Δt_W are on the order of 10–25 s (see Fig. 23.5b). A 10–30 % overshoot in temperature can also be observed in a time period ranging between 10 s (100 mK quench) and 100 s (0.1 mK quench). These typical times make the fast dynamics of the piston effect very difficult to estimate in this EH mode. In addition, the fluid acts as an adiabatic heat accumulator, as its external boundary is in thermal contact with the thermally active wall (except for the heat losses due to the connecting wires of specific devices immersed in the fluid). As a consequence, the late time evolution of the temperature inhomogeneities is governed only by the boundary layer diffusion.

23.2.2 Internal Heating (IH)

In the internal heating (IH) mode, a heater is immersed in the fluid. The fluid acts as a heat exchanger, as it is in thermal contact with both the heater device and the cell wall (which is thermally connected with the thermostat). The heater must be supported by a substrate that can be entirely or partially immersed in the fluid. For example, in the test cells described above, the heater can be a thermistor bead, a thermistor bead integrated onto a solid substrate, a resistive film layer deposited onto a wall substrate, or a resistive wire that crosses the fluid volume.

A temperature change in the fluid is induced by a heat flux j_H per unit area generated at the heater–fluid wall (area A_H). Heat is generally produced by the Joule effect: a current $i(t)$ is delivered to a heater with a resistance $R(t)$, leading to the heat power $q_H(t) = [R(t)]^2 i(t)$. In the general case, the temperature response of the fluid during the IH period depends simultaneously on the heater's thermal response and on the thermal boundary conditions at the cell wall. A typical example was given in Chap. 9 which considers the self-heating mode of a thermistor bead. However, using an appropriate control system, heat can also be generated from a time-independent power q_H, starting at time $t = 0$ and lasting for an adjustable time Δt_H, which leads to the total heating energy $Q_H = q\Delta t_H$. Let us consider the operating conditions of the thermal subsystems of ALICE (or CPF) thermostats. The power ranges from a minimum (the power dissipation of a thermistor operating in the low-sensitivity measurement mode, $q_{\text{Thermistor}} \approx 1\,\mu\text{W}$) to a maximum (the power dissipation within the SCU, $q_{SCU} \approx 5\,\text{mW}$). Thus, q_H typically covers four orders of magnitude.

The IH at constant heat power can then be related to the 1-D model developed in Chap. 5, where a constant heat power is injected at one limiting wall of area A_{HD}, while the opposite wall (which has the same area, $A_W = A_{HD}$) is maintained at a constant temperature. However, at least two important characteristic parameters of the real heater device and cell design must be accounted for:

- The ratio $\frac{A_{HD}}{A_W}$ of the area of the heater to that of the surrounding container. For example, the role of this parameter in the heat transfer will certainly be significant when a small thermistor bead immersed in the fluid is used as the internal heater.
- The finite minimum duration of initial heating $\Delta t_{H,min}$. This minimum period is restricted *a priori* to a heating time that is slightly greater than the time response of the heater device itself. The typical time response of the heater itself can be then relatively low (e.g., 10 ms when a small thermistor bead is used as the internal heater). However, in the ALICE and CPF cases, the minimum time that separates two successive action points is 40 ms. The minimum heating time is then defined as $\Delta t_{H,min} = 120\,\text{ms}$ in order to guarantee that at least two successive video frames (25 fps) are recorded during the shortest IH period. Such a limiting time must be then accounted for in the modeling.

The comparison between the two minimum periods, 120 ms for IH and 20 s for EH, shows that the IH mode is more appropriate for studying a fast thermal response

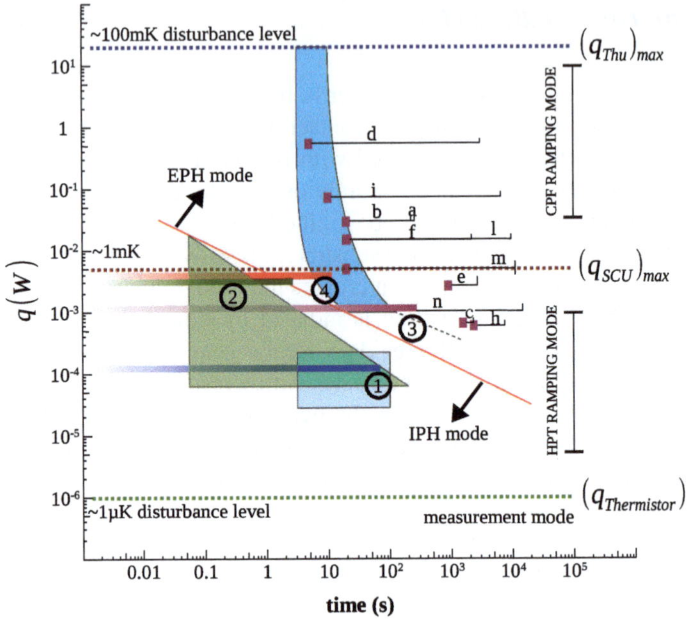

Fig. 23.6 Heating power q as a function of heating time t

of the fluid. Practically speaking, the available times range from a few hundreds of milliseconds (heating period) to a few hundreds of seconds (relaxation period).

We can now summarize the experimental investigations performed in ALICE, CPF and HPT-HYDRA under either external or internal heating. We have used a q–t diagram drawn on a log–log scale, as given in Fig. 23.6, where a heating power q was supplied for a time t. The power is separated into regions by three horizontal (dotted) lines that correspond to the following characteristic levels:

(i) The maximum heat flux $(q_{\text{Themistor}})_{max} \simeq 1\,\mu$W (lower dotted line) when operating a thermistor in the measurement mode. A heat flux that is lower than this limiting value is needed to ascertain fluid temperature measurements within $1\,\mu$K.
(ii) The maximum heat flux $(q_{SCU})_{max} \simeq 5$ mW (median dotted line) that can be produced inside the SCU to limit its temperature perturbation to less than 1 mK.
(iii) The maximum heat flux $(q_{ThU})_{max} \simeq 20$ W (upper dotted line) when operating a thermostat unit in ramping or quenching mode. This heat flux is able to produce a maximal temperature step of about 100 mK in a typical time of about 10 s.

The SCU time response is on the order of a few seconds. In order to distinguish the IH and EH modes, a conduction equation, $q = \frac{5}{\sqrt{t}}$, is used (q in mW , t in s; see the red line in Fig. 23.6). Such an equation corresponds to a constant (1 mK) temperature perturbation in SCU due to a constant heat flux $j_H = \frac{q}{A_H}$ as delivered for a time t through an ideal cell-heater contact of area A_H. In this diagram, the EH modes lie above the red line. The two vertical segments address either the CPF or HPT-HYDRA ramping modes. The upper and lower blue areas indicate heating times of the same order of magnitude as the typical diffusive time response of the SCU or the ThU. Close to T_c (typically $T - T_c \lesssim 10$–$20\,\text{mK}$), or for low-temperature disturbances (typically $\lesssim 1\,\text{mK}$), the diffusion in the fluid is the limiting factor, as shown by the dashed line of slope $-\frac{1}{2}$. The history and the pulse energies of the density disturbance events observed by Wilkinson *et al* in CPF are represented by the corresponding horizontal segments labeled a to q (see Table III in [34]). The red square indicates the end of the heat pulse, while the length of the segment indicates the total duration of the analysis of the density relaxation phenomena. The experimental results obtained at $T - T_c = 8.5\,\text{mK}$, as previously illustrated in Fig. 7.3, correspond to the event "n" (see label 3). In this latter case, the total energy delivered into the SCU was 0.28 J. Similarly, the experiments performed by Straub *et al* in HPT-HYDRA and reported in Sect. 7.1.4 correspond to label 2.

The IH modes lie below the red line. The experiments performed by Michels *et al* (using CPF) and Garrabos *et al* (using ALICE and CPF) are in this latter mode. The green area corresponds to the ALICE experiments with the immersed thermistor. Labels 1 and 4 indicate the heating pulse conditions for the experimental results analyzed in Sect. 7.1.3 and Chap. 9, respectively.

Obviously, only the IH mode can be used when the objective is to characterize the fast equilibration temperature of the bulk fluid. Nevertheless, the analysis of experiments performed very close to the critical point should account for the 3-D cell geometry, for the real thermal properties of the materials, and for the diverging fluid properties. In the next chapter, we present a procedure to account for these factors in a simplified manner, using the so-called pancake test cell.

23.3 Magnetic Compensation of Earth's Gravitational Forces in Hydrogen Near its Critical Point

23.3.1 *The Principle*

The magnetic levitation of matter involves compensating for the weight of the matter using magnetic forces. Most substances exhibit only weak diamagnetic susceptibilities, so strong magnetic fields are needed for magnetic levitation. Such field intensities have been attained only recently, through the use of superconducting magnets. Ever since the first demonstration of magnetic levitation of diamagnetic substances [3, 9], this technique has been applied to the containerless handling of liquids, solids or living bodies [9, 32], the simulation of low-gravity or weightlessness conditions for life science [26], and fluid physics (the noncoalescence of liquid drops in contact [31], the deformation modes of drops [32], and investigations of the lambda transition in helium [18]).

In some previous works using magnetic compensation, precise measurements of the magnetic field [18] as well as detailed analytical and numerical calculations of the force field applied to the sample were performed [4, 9, 18, 30, 31]. Measurements of the residual gravity were also performed [21]. In the latter case, the method takes advantage of the very low gas–liquid surface tension and capillary length near the critical point, leading to the precise analyses of the shape of the gas–liquid interface and the behavior of the capillary length of H_2 near its critical point.

Containerless handling applications only require global compensation of the total weight of the body and stable levitation, whereas compensation for gravity requires that more conditions are fulfilled. Indeed, when a diamagnetic sample is placed in a magnetic field **B**, each of its molecules is subjected to a force that is proportional to its magnetic susceptibility χ_B, which can be positive (paramagnetic substances) or negative (diamagnetic substances), and to the local gradient of $\mathbf{B}^2$. The resulting magnetic acceleration vector $\mathbf{a}_m$ induced by the magnetic field **B** is

$$\mathbf{a}_m = \frac{\chi_B}{\mu_0 \rho} \nabla \left(\frac{1}{2} \mathbf{B}^2 \right), \tag{23.7}$$

where ρ is the sample density and $\mu_0 = 4\pi \times 10^{-7}\,\mathrm{H\,m^{-1}}$ is the magnetic permeability of the vacuum. The total acceleration vector applied to the sample on the Earth is

$$\mathbf{g} = \mathbf{g}_0 + \mathbf{a}_m, \tag{23.8}$$

where $\mathbf{g}_0$ is the Earth's gravitational acceleration vector. Let g^* be the reduced acceleration modulus (effective gravity):

$$g^* = \frac{\|\mathbf{g}\|}{g_0}. \tag{23.9}$$

g^* can be zero, larger or smaller than 1 due to the vectorized nature of Eq. (23.8). The acceleration field **g** is homogeneous at the length scale of the sample provided that the gradient of $\mathbf{B}^2$ and the chemical composition of the sample are homogeneous at the sample length scale. Under these conditions, the Earth's gravitational acceleration can be increased, partially compensated for, or canceled out exactly. This permits the realization of *tunable artificial gravity*, ranging from large to low values of g^* and ultimately to weightlessness ($g^* = 0$).

However, a uniform force field cannot be produced, as it would violate the laws of electromagnetism [25]. Thus, the gravity can only be tuned to a certain precision. It is therefore important to evaluate the maximum deviation of g^* from its mean value $\langle g^* \rangle$ within the sample:

$$\Delta g^* = Max\rfloor_{\text{sample}} \left\{ g^* - \langle g^* \rangle \right\}. \tag{23.10}$$

When dealing with the global compensation of weight (levitation $g^* = 0$), Δg^* is called the "residual gravity". This is a quantitative evaluation of the quality of the weightlessness, and this parameter allows the various low-gravity facilities (magnetic levitation, parabolic flights, sounding rockets, orbital platforms, ...) to be compared. Note that the use of saturated ferromagnetic inserts [25] leads to great improvements in force field homogeneity.

23.3.2 The Setup

The magnetic levitation setup described below utilizes hydrogen, which exhibits a high magnetic susceptibility high magnetic susceptibility ($\frac{\chi_B}{\rho} = -2.51 \times 10^{-8}$ m^3kg^{-1}). Although it does not compensate for gravity in a uniform manner, this setup provides much longer experimental times and is very useful for calibrating instrumentation and performing exploratory research. It has been used in particular to investigate the effects of vibrations, as discussed in Chaps. 16 and 17 of this book (see also [6–8] for example).

The H_2 critical temperature ($T_c = 33$ K) is low, which makes it easier to use a superconducting coil. This apparatus can also be regarded more generally as a variable effective gravity facility that allows the effect of gravity to be studied in fluid physics. Variable acceleration can be obtained in the range $g^* = 0$ (levitation) to $g^* = 2.2$. In particular, the gravities of the Moon and Mars ($g^* = \frac{1}{6}$ and $g^* = \frac{1}{3}$, respectively) can be reproduced.

The apparatus consists of two parts: the cryostat containing the superconducting coil and the vacuum chamber containing the sample and the optics. This is shown schematically in Fig. 23.7a. The cryostat has two stages:

- A Dewar containing liquid N_2 surrounds another Dewar containing liquid He maintained at 2.17 K with the help of a heat exchanger.
- A superconductive coil with a total height of 200 mm is immersed inside the helium bath. This supports a current of up to 65 A and produces a magnetic field of up to 10 T along its axis, which is parallel to the direction of Earth's gravity.

The vacuum chamber is a cylindrical container that is partially immersed inside the helium bath, and stands along the axis of the coil. Vacuum (10^{-4} Pa) is maintained in this chamber for thermal isolation purposes. The axis of the cell containing the H_2 sample coincides with the axis of the coil. The position of the center of the cell with respect to the coil corresponds to the maximum intensity of the upward magnetic force; i.e., at the upper end of the coil, 85 mm above its median plane.

On the coil axis, $\mathbf{g}$ is vertical, and Eq. (23.8) reduces to $g = g_0 - a_m$ at the upper end (and $g = g_0 + a_m$ at the lower end). Levitation is attained when the electric current in the coil I is equal to 63.1 A, corresponding to a magnetic field of magnitude 6.65 T in the cell. According to the linear relation between the electric current in the coil and the magnetic field, g^* is computed using the relation

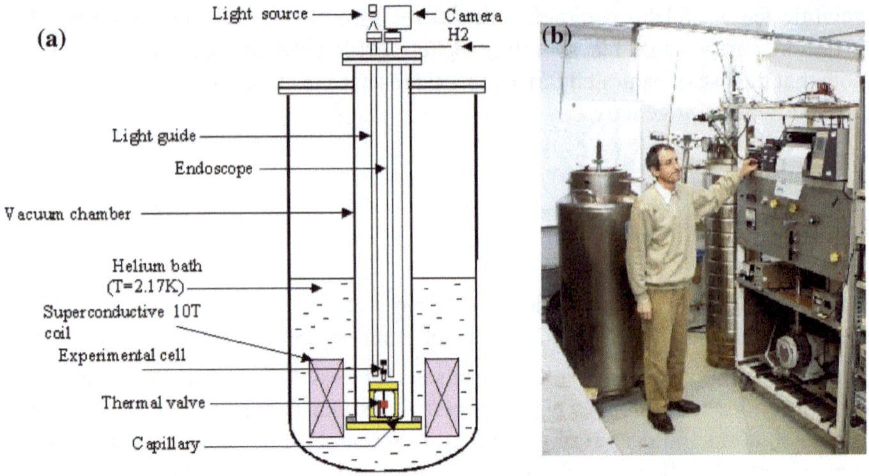

Fig. 23.7 Experimental setup: **a** schematic, **b** photo

$$g^*(I) = g_0 \left[1 - \left(\frac{I}{63.1} \right)^2 \right]. \tag{23.11}$$

The cell body can be made of electrolytic copper or sapphire (sapphire is mandatory when the cell is vibrated, see Sect. 23.3.3). The cavity containing the fluid is cylindrical (3 mm in diameter) and its axis is horizontal, perpendicular to the coil axis. There are two parallel sapphire windows 3 mm apart at each end that are sealed by indium rings.

The gas–liquid system is observed in light transmission. Spatially coherent (parallel) light is issued from a pinpoint light source at the focus of a lens. The light beam is made to cross the sample parallel to the cell axis using mirrors. The sample is imaged on a CCD camera by a lens. The light source, the camera and the lenses are located outside the cryostat.

The cell is in thermal contact with the helium bath. Its temperature is measured with a Cernox thermometer and controlled using an electrical resistance heater connected to a closed-loop temperature control device. The temperature control device is accurate to within 0.3 K at 33 K, and the working temperature range is 15–40 K. A power of a few mW is needed to maintain the cell temperature at 33 K. The cell is filled with purified, pressurized H_2 at room temperature through a capillary. This capillary is closed with an H_2 ice floe, which is allowed to form locally by positioning a thermal conductor in contact with the helium bath (H_2 solidification temperature: 14 K). The cell is filled to critical density, which is verified by checking the position of the meniscus in the middle of the vessel at the critical temperature [24].

When H_2 is cooled in the cell, the ortho–para hydrogen equilibrium is shifted and the percentage of para-H_2 increases from 25 % at room temperature to 96 % at 30 K

[1]. The corresponding critical coordinates can be found in Table 1.6. In particular, the critical temperatures are different: T_c(n-H_2)= 33.24 K and T_c(p-H_2)= 32.976 K.

The time course for the slow conversion of n-H_2 to p-H_2 is followed by determining the temperature at which the meniscus disappears (the critical temperature of the ortho–para mixture) at regular time intervals. We find that the critical temperature decays exponentially with a time constant of around 50 h. When the critical temperature reaches a constant value (within 1 mK), equilibrium is assumed to have been attained.

23.3.3 Vibration Setup

In order to vibrate a fluid sample, an electric stepper motor is equipped with an adjustable-eccentricity cam. The motor is fixed to the top of the cryostat. The alternating motion is transmitted to the cell by means of a thin tube with an inner diameter of 4.5 mm and an outer diameter of 5 mm, which induces an angular vibration with the axis of rotation $D_0 = 20$ mm below the center of the sample (see Fig. 23.8a). The latter is thus subjected to a near-linear vibration (x-axis) directed perpendicular to the axis of the coil such that $x = a \sin \omega t$. The frequency range is 0–60 Hz and the amplitude range is 0–500 μm. As the vibrational amplitude remains small, the vertical motion does not exceed $\frac{a^2}{D_0} \simeq 10\,\mu\text{m}$ and can be neglected.

In addition, the centrifugational acceleration $\frac{4\pi^2 f^2 a^2}{D_0} \simeq 1\,\text{m}\,\text{s}^{-2} \simeq 0.1\,g_0$ is much smaller than the vibrational acceleration $4\pi^2 f^2 a$. The cavity containing the fluid is cylindrical (3 mm in diameter), and its axis is horizontal, perpendicular to the coil axis. It is made of sapphire and the cylinder is closed by two parallel sapphire windows 3 mm apart that are sealed by indium rings (see Fig. 23.8b). It is necessary to use sapphire because it prevents heating by eddy currents and Lorentz forces when the sample is vibrated under such a high magnetic field.

23.3.4 Zero-Gravity State

A levitation point is a point where the magnetic force exactly compensates for the Earth's gravitational force. The levitated state occurs when a levitation point coincides with the center of the cell. In our apparatus, the center of the cylindrical sample cell is located on the vertical coil axis (z-axis) at point O, which is at a height of z_0 (see Fig. 23.9a), and this is where the magnetic acceleration a_{mz} reaches its maximum amplitude with $a_{mz} = g_0$. In this situation, the gas phase (bubble) is pushed to the top or bottom cell wall. Therefore, in the levitated state, the magnetic acceleration exactly compensates for Earth's gravity at point O only. Since the vertical force in the vicinity of point O is directed downwards, the center of the cell is an

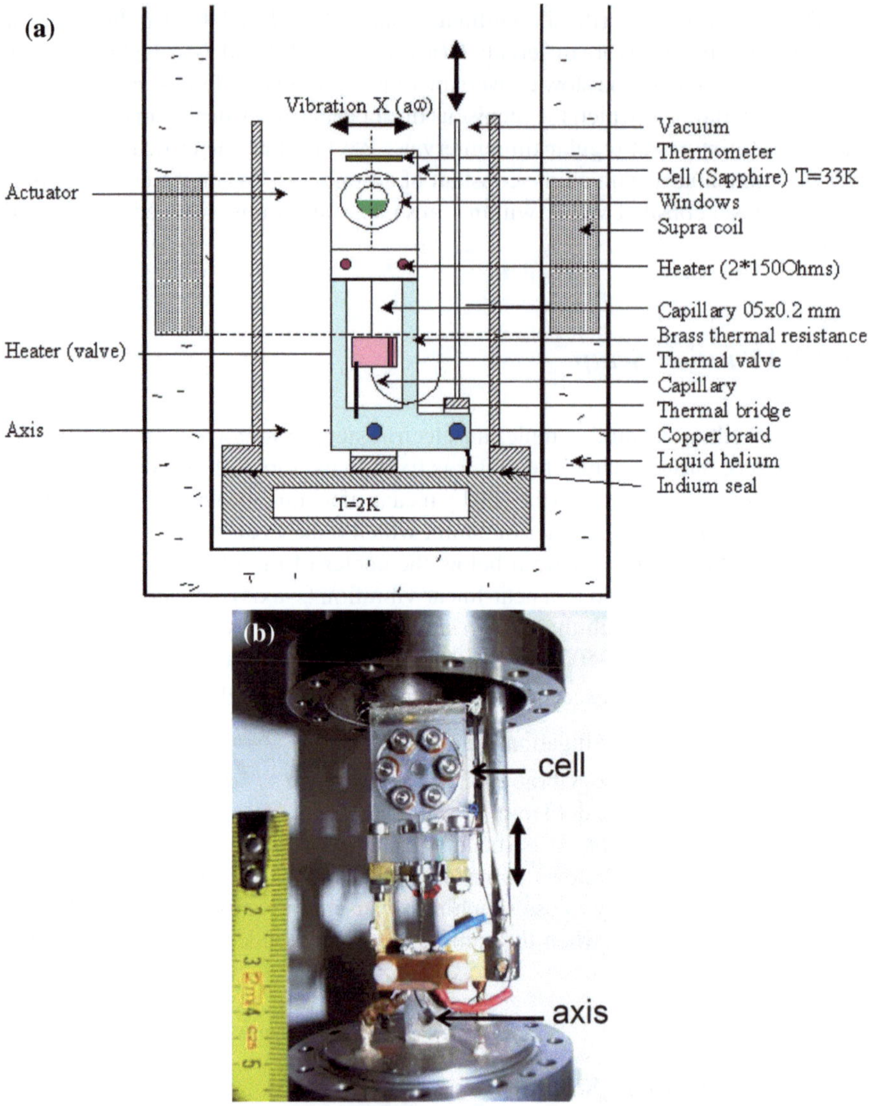

Fig. 23.8 **a** Schema of the vibrating setup. **b** The vibration cell

unstable equilibrium point on the vertical z-axis, and the modulus of the acceleration is a quadratic function of z along the vertical axis, which means that it varies quite slowly with z.

The authors of [32] used a simple method to make a stable levitation equilibrium. The magnetic field was set at a higher value than for compensation at z_0. There are then two compensation points (O_1 and O_2 at z_1 and z_2). The sample was also placed

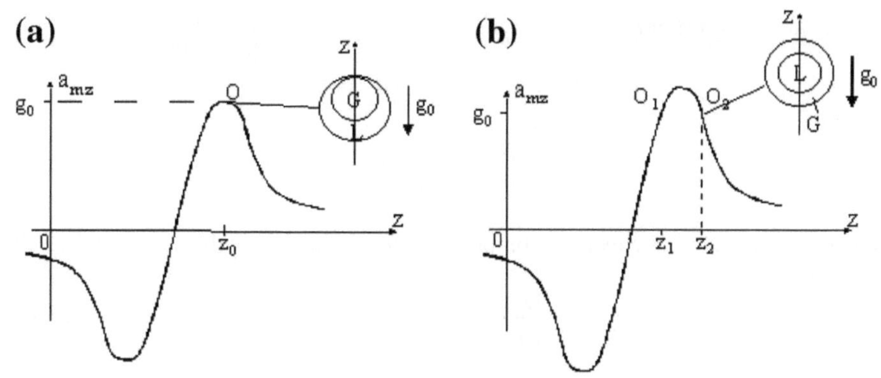

Fig. 23.9 Compensating for the Earth's gravity: two compensation configurations are shown for the magnetic acceleration a_{mz} as a function of the cell altitude z along the z-axis (see text). **a** For minimum magnetic field, the exact compensation occurs at z_0, the height at which the magnetic acceleration value is maximum, $a_{mz} = g_0$. The equilibrium at z_0 is unstable and the gas bubble (G) lifts up. **b** For a higher magnetic field, the exact compensation occurs at two cell altitudes z_1 or z_2. The gas–liquid equilibrium at z_2 is stable, with the liquid drop (L) surrounded by the gas phase (G)

at a height of z_2 (see Fig. 23.9b). In this situation, the levitation state is attained at a slightly larger magnetic field intensity, but the liquid–gas equilibrium is stable with the liquid drop inside the gas phase. (Note that a thin liquid film still wets the container walls.)

However, in this case, the acceleration along the z-axis in the vicinity of the center of the cell is linear in z. Although the levitation is unstable in our configuration $z = z_0$, the inhomogeneities of the acceleration field along the z-axis are reduced as compared with the configuration $z = z_2$ in the case of [32]. In our cell, the walls of the sample cell limit the motion of the fluid around the unstable equilibrium point O.

23.3.5 *Residual Acceleration Field in the Levitation State*

When there is exact compensation for the Earth's gravity, the residual acceleration field can be determined by simple considerations for the magnetic field. In the following, natural cylindrical coordinates (r, z) centered along the vertical coil axis (z-axis) are used.

We assume that the levitation state occurs at point O at a height z_0 (see Fig. 23.9a), which coincides with the center of the sample cell. In the absence of electric currents, the static magnetic field satisfies

$$\begin{aligned} div\mathbf{B} &= 0 \\ curl\mathbf{B} &= 0, \end{aligned} \tag{23.12}$$

which can be expressed as

$$\frac{\partial B_r(r,z)}{\partial r} = -\frac{1}{2}\frac{\partial B_z(0,z)}{\partial z}$$
$$\frac{\partial B_r}{\partial z} = \frac{\partial B_z}{\partial r}. \tag{23.13}$$

Since the magnetic force exactly compensates for the weight at the center of the cell, we write

$$\frac{|\chi_B|}{\rho\mu_0} B_z \frac{\partial B_z}{\partial z} = g_0. \tag{23.14}$$

Noting $B_z(0, z_0) = B_0$, the magnetic field at the center of the cell, the characteristic length scale of the magnetic field gradient is

$$L_B = \frac{|\chi_B|\, B_0^2}{\rho\mu_0 g_0}, \tag{23.15}$$

where $B_0 = 6.65\,\mathrm{T}$ at full weight compensation, $g_0 = 9.81\,\mathrm{m\,s^{-2}}$ and $L_B = 89\,\mathrm{mm}$ for the coil length. Taking into account the fact that in the experiment the magnetic force is maximum at the center of the cell, we obtain

$$\frac{\partial}{\partial z}\left(B_z \frac{\partial}{\partial z} B_z(0, z_0)\right) = 0. \tag{23.16}$$

The expansion of the magnetic field around the center of the cell restricted to the first nonzero order in $\varepsilon = \left[\frac{r}{L_B}, \frac{z-z_0}{L_B}\right]$, and this leads to the following expressions for the components of the effective acceleration field $\mathbf{g}(r, z)$:

$$g_r(r,z) = -g_0 \frac{3r}{4L_B} + O(\varepsilon^3) \tag{23.17}$$

$$g_z(r,z) = g_0 \left(\frac{r^2}{4L_B^2} - \frac{(z-z_0)^2}{4L_B^2}\right) + O(\varepsilon^3). \tag{23.18}$$

The radial acceleration is directed towards the z-axis, and thus the levitation equilibrium is stable to radial perturbations of small amplitude. Therefore, the denser (liquid) phase should go to the middle of the cell, and the lighter (gas) phase should be rejected on the side walls of the cell. The variations of the radial component of the acceleration in the vicinity of the center of the cell are much stronger than those of the axial component, since g_r is linear in r.

In the levitation state, the residual gravity is roughly equal to the value of g_r for r equal to the radius of the cylindrical cell. For a cell diameter of 3 mm, we obtain $g_r \simeq 0.13\,\mathrm{m\,s^{-2}} \simeq 10^{-2}\, g_0$ at $r = 1.5$ mm. Although this value is of the same order

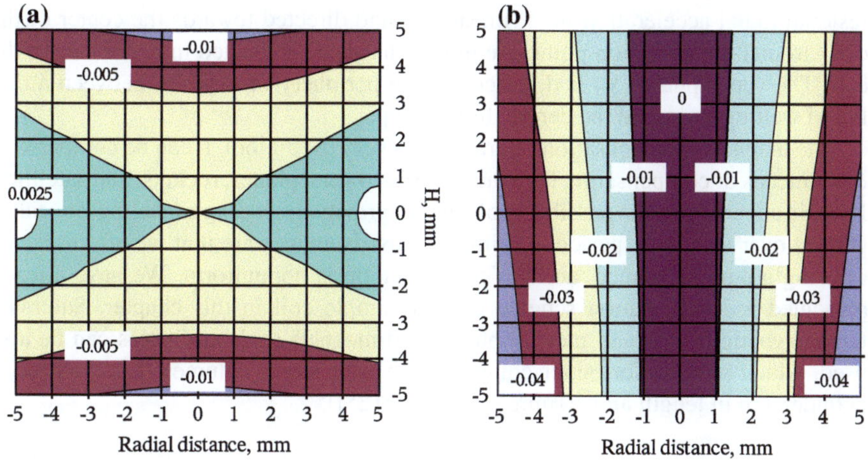

Fig. 23.10 Iso-acceleration curves for the vertical (**a**) and radial (**b**) components of the total acceleration in H_2 gravity compensation. The acceleration values are expressed in units of g_0. (*abscissa* r; *ordinate* $H = z - z_0$)

Fig. 23.11 H_2 at 10 mK below T_c in a state of levitation inside a cell 10 mm in diameter and 2 mm in thickness. The liquid phase is L and the vapor phase is V. The liquid–vapor interface is indicated by a *dashed line*

of magnitude as the residual gravity achieved during aircraft parabolic flights, it is important to note that the residual acceleration achieved by magnetic compensation is permanent, not random. When Earth's gravity is only partially compensated ($g^* \neq 0$, $I_E < 63.1$ A), the acceleration inhomogeneity, which is proportional to I_E^2, has a smaller amplitude than at $g^* = 0$. As a consequence, for $0 < g^* < 1$, the maximal acceleration inhomogeneity is less than $10^{-2}\,g_0$.

The calculated vertical and radial components of g^* are shown in Fig. 23.10. These inhomogeneities can be seen in a two-phase hydrogen cell near its liquid–gas critical point, where the surface tension vanishes (see Sect. 1.7). The shape of the gas–liquid interface then becomes very sensitive to even very weak forces.

Figure 23.11 shows a large sample (10 mm diameter) of H_2 in a state of levitation at 10 mK below T_c. The levitation point is located in the center of the cell. Due to the residual radial acceleration field (linear in r and directed towards the center of the cell) around the levitation point, the liquid phase (L) is located in the center of the cell. The vapor phase (V) is displaced to the periphery of the cell, between a thin liquid wetting layer and the central liquid drop.

The magnetic compensation of gravity is a very useful tool. It can be compared to other means of compensating for gravity: drop towers, planes, rockets, and satellites. Its main advantage is the possibility of varying gravity (including canceling it out altogether) over very long periods of time. Its disadvantages are that we are restricted to using a pure substance, and the compensation is nonuniform. We have already presented a compensation scheme using a simple coil in this chapter. Saturated ferromagnetic inserts and more sophisticated magnets such as dipoles can be also used, which provide for better compensation over larger volumes (for example, a cylinder 1 m in length and 10 mm in diameter [25]).

References

1. AirLiquide (1976) Encyclopedie des Gaz. Elsevier, Paris
2. Barmatz M, Hahn I, Lipa JA, Duncan RV (2007) Critical phenomena in microgravity: past, present, and future. Rev Mod Phys 79(1):1
3. Beaugnon E, Tournier R (1991) Levitation of water and organic substances in high static magnetic fields. J Phys III(1):1423–1428
4. Beaugnon E, Bourgault D, Braithwaite D, de Rango P, Sulpice A, Tournier R, de la Bathie RP (1993) Material processing in high static magnetic field. A review of an experimental study on levitation, phase separation, convection and texturation. J Phys I 3:399
5. Beysens D, Garrabos Y (2001) Critical and supercritical fluids and related phenomena, Chap. 8. Taylor and Francis, New York
6. Beysens D, Chatain D, Evesque P, Garrabos Y (2005) High-frequency driven capillary flows speed up the gas–liquid phase transition in zero-gravity conditions. Phys Rev Lett 95(3):034502
7. Beysens D, Chatain D, Evesque P, Garrabos Y (2008) Nucleation and growth of a bubble pattern under vibrations in weightlessness. Europhys Lett 82:36003
8. Beysens D, Chatain D, Garrabos Y, Lecoutre C, Palencia F, Evesque P, Nikolayev VS (2007) The effect of vibrations on heterogeneous fluids: some studies in weightlessness. Acta Astronaut 61:1002–1009
9. Braunbeck W (1939) Need research. Z Phys 112:735
10. Cork M, Dewandre T, Hueser D (1993) The critical point facility and the IML-2 mission. ESA Bull 74:36–44
11. de Bruijn R (1999) Heat transfer in a critical fluid under microgravity conditions: a spacelab experiment. PhD thesis, University of Amsterdam, Amsterdam, p 197
12. de Bruijn R, van Diest RJJ, Karapantsios TD, Michels AC, Wakeham WA, Trusler JPM (1997) Heat transfer in pure critical fluids surrounded by finitely conducting boundaries in microgravity. Phys A 242(1–2):119–140
13. Frohlich T, Guenoun P, Bonetti M, Perrot F, Beysens D, Garrabos Y, Le Neindre B, Bravais P (1996) Adiabatic versus conductive heat transfer in off-critical SF_6 in the absence of convection. Phys Rev E 54(2):1544–1549
14. Garrabos Y, Lecoutre-Chabot C, Hegseth J, Nikolayev VS, Beysens D, Delville J-P (2001) Gas spreading on a heated wall wetted by liquid. Phys Rev E 64(5):051602
15. Guenoun P, Khalil B, Beysens D, Garrabos Y, Kammoun F, Le Neindre B, Zappoli B (1993) Thermal cycle around the critical point of carbon dioxide under reduced gravity. Phys Rev E 47(3):1531–1540

16. Gurfein V, Beysens D, Garrabos Y, Le Neindre B (1991) Simple grid technique to measure refractive index gradients. Opt Commun 85:147–152
17. Huijser PH, Michels AC, Trappeniers NJ (1983) Optical investigation of critical behavior in simple fluids at reduced gravity. ESA Mater Sci Microgravity SEE N83-31637(20–12):377–384
18. Israelson UE, Larson M (1995) Aiaa 95-0272. In: The 33rd aerospace sciences meeting and exhibit, Reno, NV, USA, 9–12 January 1995
19. Laherrere JM, Koutsikides P (1993) Alice, an instrument for the analysis of fluids close to their critical point in microgravity. Acta Astronaut 29:861–870
20. Lecoutre C, Garrabos Y, Georgin E, Palencia F, Beysens D (2009) Turbidity data of weightless SF_6 near its liquid–gas critical point. Int J Thermophys 30:810–832
21. Lorin C, Mailfert A, Chatain D, Félice H, Beysens D (2009) Magnetogravitational potential revealed near a liquid–vapor critical point. J Appl Phys 106:033905-1–033905-4
22. Marcout R, Zwilling JF, Laherrere JM, Garrabos Y, Beysens D (1995) Alice 2, an advanced facility for the analysis of fluids close to their critical point in microgravity. Microgravity Q 5:162–170
23. Michels AC, de Bruijn R, Karapantsios TD, van Diest RJJ, van den Berg HR, van Deenen B, Sakonidou EP, Wakeham WA, Trusler JPM, Louis A, Papadaki M, Straub J (1995) Adiabatic compressive heating of critical fluids under microgravity conditions. In: The 30th 1995 national heat transfer conference, vol 3. Portland, OR, USA, 6–8 August 1995
24. Morteau C, Salzman M, Garrabos Y, Beysens D (1997) In: Viviani A (ed) The 2nd European symposium on fluids in space, Napoli, Italy, 22–26 April 1997. Roma, Congressi srl, pp 327–333
25. Quettier L, Felice H, Mailfert A, Chatain D, Beysens D (2005) Magnetic compensation of gravity forces in liquid–gas mixtures: surpassing intrinsic limitations of a superconducting magnet by using ferromagnetic inserts. Eur Phys J Appl Phys 32:167–175
26. Valles JM Jr, Lin K, Denegre JM, Mowry KL (1997) Magnetic field gradient levitation of *xenopus laevis*: towards low gravity simulation. Biophys J 73:1130–1133
27. van den Assem D, Huijser RH, Vreeburg JPB (1987) On the development of an optical diagnostic instrument for fluid-physics research in microgravity. J Phys E: Sci Instrum 20:3775
28. van Lieshout Th SH, Michels AC (1997) Microkelvin gradient-reduction thermostats for fluid experiment on microgravity platforms. In: Viviani A (ed) The 2nd European symposium on fluids in space. Congressi srl, Rome, pp 459–462
29. Wakeham WA, Trusler JPM, de Bruijn R, van Diest RJJ, Karapantsios TD, Michels AC (1997) Heat transport near the critical region of a pure fluid in microgravity. In: Viviani A (ed) The 2nd European symposium on fluids in space. Congressi srl, Rome
30. Weilert MA, Whitaker DL, Maris HJ, Seidel GM (1996) Magnetic levitation and noncoalescence of liquid helium. Phys Rev Lett 77(23):4840–4843
31. Weilert MA, Whitaker DL, Maris HJ, Seidel GM (1997) Magnetic levitation of liquid helium. J Low Temp Phys 106:101–131
32. Whitaker DL, Weilert MA, Vicente CL, Maris HJ, Seidel GM (1998) Oscillations of charged helium ii drops. J Low Temp Phys 110:173–178
33. Wilkinson RA (1998) Density relaxation of liquid–vapor critical fluids in earth's gravity. Int J Thermophys 19:1175–1183
34. Wilkinson RA, Zimmerli GA, Hao H, Moldover MR, Berg RF, Johnson WL, Ferrell RA, Gammon RW (1998) Equilibration near the liquid–vapor critical point in microgravity. Phys Rev E 57(1):436–448
35. Zimmerli GA, Wilkinson RA, Ferrell RA, Moldover MR (1999) Electrostriction of a near-critical fluid in microgravity. Phys Rev E 59(5):5862–5869
36. Zimmerli GA, Wilkinson RA, Ferrell RA, Moldover MR (1999) Electrostriction of near-critical SF_6 in microgravity. Phys Rev Lett 82(26):5253–5256

Index

B. Zappoli et al., *Heat Transfers and Related Effects in Supercritical Fluids*,
Fluid Mechanics and Its Applications 108, DOI: 10.1007/978-94-017-9187-8,

MIX
Papier aus verantwortungsvollen Quellen
Paper from responsible sources
FSC® C105338

If you have any concerns about our products,
you can contact us on
ProductSafety@springernature.com

In case Publisher is established outside the EU,
the EU authorized representative is:
Springer Nature Customer Service Center GmbH
Europaplatz 3, 69115 Heidelberg, Germany

Printed by Libri Plureos GmbH
in Hamburg, Germany